Reactivity and Structure Concept in Organic Chemistry

Volume 32

Editors

Klaus Hafner   Jean-Marie Lehn
Charles W. Rees   P. von Ragué Schleyer
Barry M. Trost   Rudolf Zahradník

Springer

Berlin
Heidelberg
New York
Barcelona
Budapest
Hong Kong
London
Milan
Paris
Tokyo

Donald S. Matteson

Stereodirected Synthesis with Organoboranes

With 14 Tables

 Springer

Prof. Dr. Donald S. Matteson
Washington State University
Department of Chemistry
Pullman, WA 99164-4630
USA

ISBN 3-540-59182-6 Springer-Verlag Berlin Heidelberg New York

Cip data applied for

© Springer-Verlag Berlin Heidelberg 1995
Printed in Germany

Typesetting: Data-conversion by M. Schillinger-Dietrich
SPIN:10494277   51/3020-5 4 3 2 1 0 - Printed on acid-free paper

Contents

1 Introduction to Borane Chemistry

1.1 Beginnings

The beautiful green flame of burning triethylborane (1) was first reported by Frankland and Duppa in 1860 [1-2]. Their "boric ethide" was prepared from diethylzinc and triethyl borate. After observing the spontaneous ignition of (1) in air, they tried pure oxygen and watched it explode.

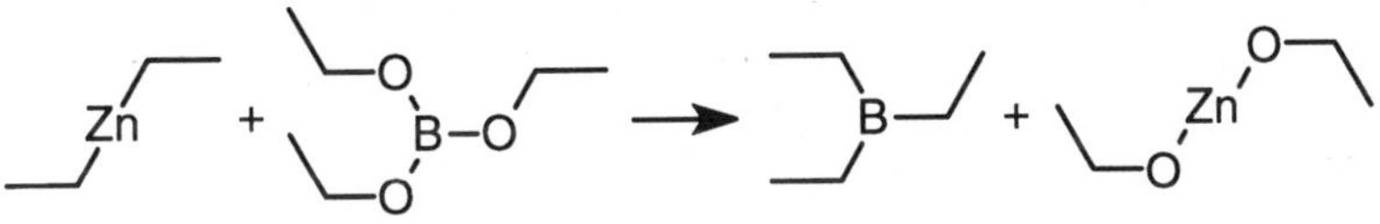

1

Slow air oxidation of **1** yielded the much less reactive diethoxyethylborane, $C_2H_5B(OC_2H_5)_2$, which hydrolyzed readily to sweet—tasting ethylboronic acid, $C_2H_5B(OH)_2$ [1]. Frankland also prepared trimethylborane, a gas that could not be condensed with an ice–salt bath until pressure was applied. Besides its spontaneous ignition, it was a lachrymator with an intolerable odor. The trialkylboranes were found to be stable toward water, to cleave to alkane and dialkylboron chloride with HCl, and to form stable complexes with ammonia and with metal hydroxides [3-4].

These first experiments were well chosen to provide a broad general picture of organoborane properties. Even the method of synthesis, modernized in 1909 by the use of Grignard reagents in place of organozincs [5], remains one of the two generally useful approaches to the synthesis of organoboranes. The other, hydroboration, was not recognized as a practical route until the work of Brown and Subba Rao in 1956 [6]. The first demonstration of the great utility of organoboranes in stereo-controlled synthesis was the work of Brown and Zweifel on diastereospecific [7] and highly enantioselective [8] hydroborations.

1.2 Structure and Bonding in Organoboranes

1.2.1 General Characteristics of Boranes

Boron has a particularly useful combination of properties for stereocontrolled organic synthesis. The boron—carbon bond length is not much greater than that of the carbon—carbon bond and the bond strength is not much less. Boron is more electropositive than carbon, and organoboranes react as relatively stable main—group organometallic compounds (though elemental boron is not metallic). The boron is easily replaceable by electrophilic or oxidizing reagents, often with complete stereochemical control. Organoboron compounds are easily synthesized, easily handled, and ordinarily not serious hazards with regard to reactivity, toxicity, or environmental contamination.

Boron may be regarded as a proton deficient carbon atom, one proton being exactly the difference between the ^{11}B and ^{12}C nuclei. Thus, neutral tricoordinate boron compounds such as trimethylborane (2) are isoelectronic with carbocations, and negatively charged tetracoordinate borates such as borohydride ion (3) are isoelectronic with neutral carbon compounds:

This close relationship to carbon underlies much of the utility of boron, both as a tool for probing theoretical questions and as a heteroatom for making carbon connections in organic synthesis. The ability of boron to form both three-coordinate and four-coordinate species underlies the mechanisms of its most useful synthetic reactions. In some respects this behavior mimics the relationship of 16-electron and 18-electron complexes of transition metals, but boron has no other accessible oxidation states and its behavior is consequently simpler.

1.2.2 Size

Boron is slightly larger than carbon, but small enough that attached groups must interact strongly with groups on an adjacent carbon atom, and as a result, chiral directors linked through boron to the chiral site under construction can interact effectively. For the types of boron compounds of most interest to synthetic chemists, X-ray data are sparse, because most such compounds are liquids at room temperature and difficult to crystallize under any conditions. However, data are available for a few representative examples of boronic esters and related structures. Some typical bond distances involving boron are summarized in Table 1-1.

Table 1-1 Bond Distances in Organoboron Compounds

Atoms	Distance (pm)	Hybridization	Context	Reference
B—C	153–156	$B(sp^2)$—$C(sp^3)$	Structures **4, 5**	[9, 10]
	155–156	$B(sp^2)$—$C(sp^3)$	Structures **6a, b**	[11]
	158	$B(sp^2)$—$C(sp^3)$	Structure **7**	[11]
	161	$B(sp^2)$—$C(sp^3)$	Structure **8**	[12]
	161	$B(sp^3)$—$C(sp^2)$	diethanolamine phenylboronate	[13]
	159	$B(sp^3)$—$C(sp^2)$	chelated B—Ar	[14, 15]
	156–158	$B(sp^2)$—$C(sp^2)$	RB(OR)$_2$	[14, 16, 17]
B—O	137	$B(sp^2)$—O	PhB(OH)$_2$	[18]
	137	$B(sp^2)$—O	PhB(OR)$_2$	[14, 16, 17],
	131–139	$B(sp^2)$—O	Structures **6,7**	[11]
	146–147	$B(sp^3)$—O	diethanolamine phenylboronate	[13]
	146–147	$B(sp^3)$—O	chelated BAr	[14-15]
	143–144	$B(sp^2)$—OR	Structure **8**	[12]
	164	$B(sp^2)$—O=	Structure **8**	[12]

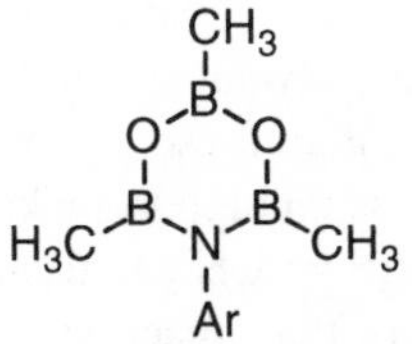

4

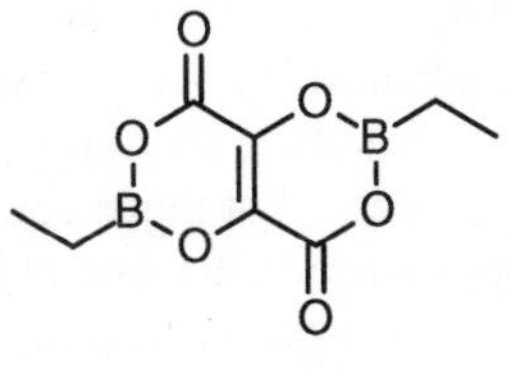

5

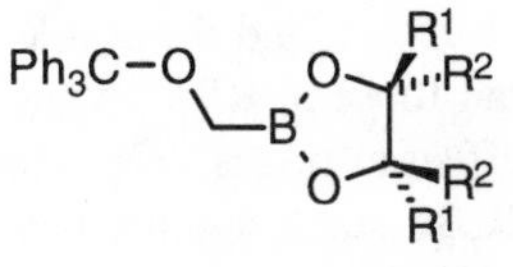

6 a, R^1 = R^2 = CH$_3$
b, R^1 =H, R^2 = cyclohexyl

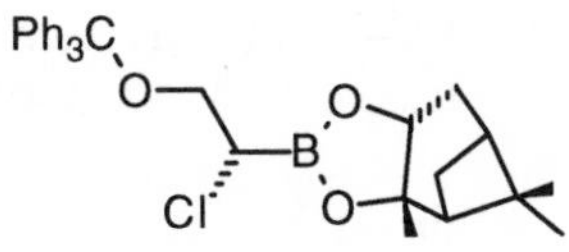

7

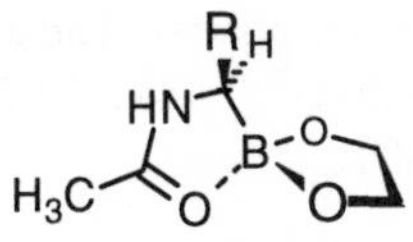

8

Calculations on methylboronic acid with the 6-31G* basis set have yielded the bond lengths B—C, 158.8 pm; B—O with OH *trans*, 135.4 pm; B—O with OH *cis*, 136.6 pm [19].

1.2.3 Bond Strengths

A dominant feature of organoborane chemistry is the ease of oxidative cleavage of boron—carbon bonds. However, B—C bonds are not weak. The average B—C bond energy $E(B—C)$ is estimated to be 322.9 kJ mol^{-1} [20]. The carbon—carbon bond energy $E(C—C)$ is 357.5 kJ mol^{-1}. A much larger difference is found between boron—oxygen bonds, $E(B—O) = 519.2$ kJ mol^{-1}, and carbon—oxygen bonds, $E(C—O) = 383.5$ kJ mol^{-1} [20]. Thus, the thermodynamic reason for the ease of oxidation of organoboranes is readily apparent, though more details are needed in order to apply these numbers to chemical problems.

Bond energies for various BY$_3$ are shown in Table 1-2, and for mixed ligand compounds RBY$_2$ and R$_2$BY in Table 1-3. The "QM" values are based on advanced quantum mechanical calculations [MP4/6-311++G(3df,2p) and HF/6-31G*] and correlation with experimental data [20-21]. The standard gas phase heats of formation $(\Delta H_f^{\circ}{}_{298})$ are believed accurate to within ±4 kJ mol^{-1} for the smaller molecules, ±8 kJ mol^{-1} for the larger ones. The "exp" values are from the critically evaluated experimental data in the JANAF tables where possible [22], otherwise from critical reviews of data for organic compounds [23] or boron compounds [24-25].

The discrepancy between the "QM" and "exp" $E(B—C)$ values reflects a different manner of accounting for the $E(C—H)$ bond energy, multiplied by the number of C—H bonds. This discrepancy arises because bond energies are only approximately additive. Data treatment in the "exp" series assumes all pure σ-bonds of a given type have the same energy in all contexts. In the "QM" treatment, an intricate relationship of $E(C—H)$ and other $E(Y—H)$ to structural details, while keeping other σ-bond energies constant, allows better correlation with ΔH_f values, which are the numbers that have physical reality.

Where Y is a π-bonding ligand and R is not, it is assumed that $E(B—R)$ does not vary with structure, and the B—Y bonds in R$_2$BY are then generally substantially stronger than those in RBY$_2$, which are in turn a little stronger than those in BY$_3$ [20]. The effect averages only a few kJ mol^{-1}, but in general it is enough to make RBY$_2$ and R$_2$BY stable against disproportionation. Reaction of R$_3$B with HY to form RBY$_2$ and HR is generally more favorable thermodynamically than reaction of RBY$_2$ with HY to form BY$_3$ and HR, an effect which is discussed in more detail in Section 1.3.1.

It is of particular interest to know how much π-bond strength is lost when trigonal boranes are converted to tetrahedral borates. The estimated $E(B—O)$ in NH$_3$ complexes of R$_2$B—OR (Table 1-3) averages 497.3 kJ mol^{-1}, a loss of 46-47 kJ mol^{-1} from the value typical for R$_2$B–OR.

Table 1-2. B—Y Bond Energies in BY_3, kJ mol^{-1}

Compound	ΔH_f QM [a]	ΔH_f exp [b]	B–Y	E(B–Y) QM [a]	E(B–Y) exp [b]
BH_3	+91.8	+100.4	B—H	375.0	371.2
$B(CH_3)_3$	−122.6	−122.8	B—C	323.1	350.9
$B(C_2H_5)_3$	−153.1	−148.8	B—C	322.9	344.3
$B(C_6H_5)_3$		+129.9	B—C		434.2
$B[N(CH_3)_2]_3$		−245.6	B—N		422.9
$B_3N_3H_6$		−513.4	B—N		438.9
$B(OH)_3$	−991.6	−992.3	B—O	518.5	521.1
$B(OCH_3)_3$	−897.9	−900.1	B—O	519.8	519.2
$B(OC_2H_5)_3$		−1001.7	B—O		519.3
BF_3	−1135.6	−1135.6	B—F	645.5	644.6
BCl_3		−403.0	B—Cl		442.3
BBr_3		−204.2	B—Br		366.6
BI_3		+71(±50)	B—I		269.7
$B(SCH_3)_3$		−156.1 [c]	B—S		357.4 [d]
$B(SC_2H_5)_3$		−284.9 [c]	B—S		377.3 [d]

[a] Based on advanced quantum mechanical calculations and correlations with experimental data [20]. Essential parameters [22], [21] (kJ mol^{-1}) include ΔH_f (H, g) = 217.999; ΔH_f (C, g) = 716.669; ΔH_f (N, g) = 472.679; ΔH_f (O, g) = 249.170; ΔH_f (B, g) = 562.706 [21]; ΔH_f (F, g) = 79.39 [20]; E(C–C) = 357.48; E(C–O) = 383.51; E(C–H) = 425.35 in BCH_3, 424.05 in BCH_2C, 411.16 in CCH_3, 401.12 in OCH_3; E(O–H) = 466.73 kJ mol^{-1} in BOH [20]; 1 kcal = 4.184 kJ. [b] Derived from experimental data [22], [24] except as noted. Calcd. with ΔH_f° (B, g) = +560 (±12); ΔH_f° (C, g) = 716.7; ΔH_f (F, g) = 79.39; ΔH_f° (Cl, g) = 121.3; ΔH_f° (Br, g) = 111.86 kJ mol^{-1}; ΔH_f° (I, g) = 106.76 kJ mol^{-1} [22]; others same as in footnote a [24]. [c] Ref. [25]. [d] Calcd. using $\Delta H_f^\circ[(CH_3)_2S] = -37.5$ kJ mol^{-1} or $\Delta H_f^\circ [(C_2H_5)_2S] = -83.6$ kJ mol^{-1} [23] and ΔH_f° (S, g) = 276.98 kJ mol^{-1} [22].

Table 1-3. Heats of Formation and Bond Energies E(B—Y) in RBY_2 and R_2BY, kJ mol^{-1}

Compound	ΔH_f QM [a]	ΔH_f exp [b]	Bond	E(B–Y) QM [a]	E(B–Y) exp [b]
HBF$_2$	−738.9	−733.9	B—F	654.3	649.4
H$_2$BF	−320.5		B—F	659.1	
CH$_3$BF$_2$	−832.6	−832.6	B—F	662.9	661.6
C$_2$H$_5$BF$_2$	−846.3	−874.5	B—F	664.6	678.2
(CH$_3$)$_2$BF	−473.5		B—F	659.1	
HBCl$_2$		−248.1	B—Cl		448.7
C$_6$H$_5$BCl$_2$		−266.0	B—Cl		462.6
(C$_6$H$_5$)$_2$BCl		−95.3	B—Cl		489.7
C$_6$H$_5$BBr$_2$		−129.5	B—Br		384.9
(C$_6$H$_5$)$_2$BBr		−9.4	B—Br		394.6
F$_2$B—BF$_2$	−1439.3	−1436.8 [c]	B—B	265.0 [d]	
H$_2$BNH$_2$	−96.3		B—N	475.7	
HB(OH)$_2$	−640.6		B—O	524.2	
HB(OCH$_3$)$_2$	−579.1		B—O	525.9	
H$_2$BOH	−290.4		B—O	541.8	
H$_2$BOCH$_3$	−259.3		B—O	543.2	
HOB(CH$_3$)$_2$	−436.7		B—O	543.3	
HOB(CH$_3$)$_2$←NH$_3$	−499.4 [e]		B—O	496.6 [e]	
CH$_3$OB(CH$_3$)$_2$←NH$_3$	−468.4 [e]		B—O	498.1 [e]	

[a] Based on quantum mechanical calculations and experimental data [20] except as noted. Essential parameters (kJ mol^{-1}) not listed in Table 1-2: E(B—C) = 322.88; E(B—H) = 375.47 in CBH$_2$, 376.23 in C$_2$BH, 386.43 in NBH$_2$, 373.88 in OBH$_2$, assumed same in O$_2$BH, 377.56 in (O)(C)BH, 369.74 in FBH$_2$, assumed same in F$_2$BH, 372.21 in (F)(C)BH; E(N—H) = 377.56 in BNH$_2$. [b] Data from ref. [24] except as noted. [c] Ref. [25]. [d] Assumes E(B—F) same as in HBF$_2$. If average E(B—F) = 659.06 is used, then E(B—B) = 246.0. If average E(B—B) = 285.7 is used, then E(B—F) = 649.1. [e] Ref. [21].

Table 1-4. Theoretical Bond Dissociation Energies $D(B-Y)$, kJ mol^{-1}

Compound	ΔH_f [a]	Bond	$D(B-Y)$ [a]
$H_2B(H_2)BH_2$	+21.3	BH_2B bridge	162.3
$(CH_3)_2B(H_2)B(CH_3)_2$	−211.7	BH_2B bridge	107.5
H_2B-BH_2	+189.7	B—B	445.9
H_2BCH_3	+20.9	B—C	443.8
H_2BNH_2	−96.3	B—N	604.5
H_2BOH	−287.6	B—O	644.4
$HB=CH_2$	+211.6	B=C	617.7
$HB=NH$	+29.5	B=N	777.8
$HB=O$	−263.2	B=O	955.0
$H_3B\leftarrow NH_3$	−73.3	B←N	119.2
$(CH_3)_3B\leftarrow NH_3$	−230.8 [b]	B←N	62.3 [b]
$HOBH_2\leftarrow NH_3$	−383.3	B←N	47.0
$HOB(CH_3)_2\leftarrow NH_3$	−499.4	B←N	19.6
$CH_3OB(CH_3)_2\leftarrow NH_3$	−468.4	B←N	24.2
$H_3B\leftarrow OH_2$	−198.4	B←O	48.4
$H_3B\leftarrow O(CH_3)_2$	−148.4	B←O	56.2
$(CH_3)_3B\leftarrow OH_2$	−374.8	B←O	10.4
$(CH_3)_3B\leftarrow O(CH_3)_2$	−324.3	B←O	17.7
$HOBH_2\leftarrow OH_2$	−518.4 [c]	B←O	−13.8 [c]

[a] Based on quantum mechanical calculations [21]. [b] Experimental values are $\Delta H_f = -226.4$, $D(B\leftarrow N) = 57.5$ kJ mol^{-1}. [c] Calcd. for tetrahedral boron. Molecular complex has $\Delta H_f = -557.6$, $D(B\leftarrow O) = +25.4$ kJ mol^{-1}.

1.3 General Chemical Properties of Organoboranes

1.3.1 Oxidation States of Boron

A useful feature of boron chemistry for synthetic purposes is the accessibility of exothermic, irreversible reactions that occur cleanly with negligible side reactions. The oxidation state of the boron atom is an important factor in determining the course of such reactions.

The ΔH_f of methylboronic acid was not included in the tabulated calculations, but can be estimated by interpolation from the available data by using the calculated value for $HB(OH)_2$ and assuming that the difference in ΔH_f between a B—CH_3 and a B—H group is constant. Comparison with BH_3 and $(CH_3)_3B$ leads to equation (1-1), and comparison with H_2BOH and $(CH_3)_2BOH$ to equation (1-2). Inasmuch as the results are self-consistent to within 2 kJ mol^{-1}, well within the error of the quantum mechanical calculations [20], the assumption of constant $\Delta\Delta H_f$ between B—CH_3 and B—H appears to be justified.

$$\Delta H_f [CH_3B(OH)_2] = (1/3)[(\Delta H_f (B(CH_3)_3) - \Delta H_f (BH_3)] + \Delta H_f (HB(OH)_2) = -712.07$$

$$\Delta H_f [CH_3B(OH)_2] = (1/2)[(\Delta H_f (HOB(CH_3)_2) - \Delta H_f (HOBH_2)] + \Delta H_f (HB(OH)_2) = -713.75 \tag{1-2}$$

Table 1-5. Heats of Reaction, $R_nBY_{(3-n)} + X\text{-}Y \rightarrow RX + R_{(n-1)}BY_{(4-n)}$ (298.15K)^a

$R_nBY_{(3-n)}$	X—Y	RX	$R_{(n-1)}BY_{(4-n)}$	$\Delta H°$, kJ mol^{-1}
$(CH_3)_3B$	H_2O	CH_4	$(CH_3)_2BOH$	−147.1
$(CH_3)_2BOH$	H_2O	CH_4	$CH_3B(OH)_2$	−109.2
$CH_3B(OH)_2$	H_2O	CH_4	$B(OH)_3$	−112.4
$(CH_3)_3B$	$(1/2)O_2 + H_2O$	CH_3OH	$(CH_3)_2BOH$	−273.8
$(CH_3)_2BOH$	$(1/2)O_2 + H_2O$	CH_3OH	$CH_3B(OH)_2$	−235.9
$CH_3B(OH)_2$	$(1/2)O_2 + H_2O$	CH_3OH	$B(OH)_3$	−239.1
$(CH_3)_3B$	H_2O_2	CH_3OH	$(CH_3)_2BOH$	−379.5
$(CH_3)_2BOH$	H_2O_2	CH_3OH	$CH_3B(OH)_2$	−341.6
$CH_3B(OH)_2$	H_2O_2	CH_3OH	$B(OH)_3$	−344.8
$(CH_3)_3B$	HF	CH_4	$(CH_3)_2BF$	−153.2
$(CH_3)_2BF$	HF	CH_4	CH_3BF_2	−161.4
CH_3BF_2	HF	CH_4	BF_3	−105.3
$(C_6H_5)_3B$	HCl	C_6H_6	$(C_6H_5)_2BCl$	−50.3
$(C_6H_5)_2BCl$	HCl	C_6H_6	$C_6H_5BCl_2$	+4.2
$C_6H_5BCl_2$	HCl	C_6H_6	BCl_3	+37.9

^{a} Based on the following ΔH_f values (kJ mol^{-1}): $B(CH_3)_3$ −122.6 [20], $HOB(CH_3)_2$ −436.7, $CH_3B(OH)_2$ −712.9, $B(OH)_3$ −992.28 ($\pm$2.5) [22], H_2O −241.826 ($\pm$0.042) [22], CH_4 −74.873 ($\pm$0.34) [22] {alternative not used, −74.4 ($\pm$0.4) [23]}, O_2 0 by definition, CH_3OH −201.5 ($\pm$0.3) [23], H_2O_2 −136.106 [22], $(CH_3)_2BF$ −473.5 [20], CH_3BF_2 −832.6 [20], BF_3 −1135.6 [22], HF −272.546 [22], $B(C_6H_5)_3$ +129.9 [24], $(C_6H_5)_2BCl$ −95.3 [24], $C_6H_5BCl_2$ −266.0 [24], BCl_3 −403.0 [22], HCl −92.312 [22], C_6H_6 +82.6 [23].

The average ΔH_f [$CH_3B(OH)_2$] = -712.9 kJ mol^{-1} will be used in further calculations. Table 1-5 lists the heats of reaction for several oxidative cleavages of a single R group from R_3B, R_2BY, and RBY_2. There is some redundancy in the tabulation, as replacement of one CH_3 by OH is always 37.9 kJ mol^{-1} more exothermic for $(CH_3)_3B$ than for $(CH_3)_2BOH$, 34.7 kJ mol^{-1} more exothermic for $(CH_3)_3B$ than for $CH_3B(OH)_2$, regardless of the oxidizing agent It is readily apparent from the data in Table 1-5 that the slow rates of reaction of R_3B, R_2BOH, or $RB(OH)_2$ with water are solely a kinetic effect, as are the slow rates of reaction of most $RB(OH)_2$ with oxygen.

Data tabulated for the relative ease of cleavage of successive methyl groups from $(CH_3)_3B$ by HF and of phenyl groups from Ph_3B by HCl also show that cleavage of the first organic group is always more favorable thermodynamically than cleavage of the third. The variable relationship of the second cleavage to the first and third is not easily explainable. The endothermic reaction of $PhBCl_2$ with HCl suggests that it might be possible to prepare $PhBCl_2$ from benzene and BCl_3, which is correct (see Section 2.5).

The thermodynamic data in Table 1-6 indicate that the intermediate oxidation states of boron are stable with respect to most possible disproportionations, but there are some marginal exceptions.

Table 1-6. Heats of Disproportionation of RBY_2 and R_2BY (298.15 K) [a]

Reactant	Disproportionation Products		$\Delta H°$, kJ mol^{-1}
$CH_3B(OH)_2$	(1/3) $(CH_3)_3B$	(2/3) $B(OH)_3$	+10.5
$CH_3B(OH)_2$	(1/2) $(CH_3)_2BOH$	(1/2) $B(OH)_3$	-1.6
$(CH_3)_2BOH$	(2/3) $(CH_3)_3B$	(1/3) $B(OH)_3$	+24.2
$(CH_3)_2BOH$	(1/2) $(CH_3)_3B$	(1/2) $CH_3B(OH)_2$	+19.0
CH_3BF_2	(1/3) $(CH_3)_3B$	(2/3) BF_3	+34.7
CH_3BF_2	(1/2) $(CH_3)_2BF$	(1/2) BF_3	+28.1
$(CH_3)_2BF$	(2/3) $(CH_3)_3B$	(1/3) BF_3	+13.2
$(CH_3)_2BF$	(1/2) $(CH_3)_3B$	(1/2) CH_3BF_2	-4.1
$C_6H_5BCl_2$	(1/3) $B(C_6H_5)_3$	(2/3) BCl_3	+40.6
$C_6H_5BCl_2$	(1/2) $(C_6H_5)_2BCl$	(1/2) BCl_3	+16.9
$(C_6H_5)_2BCl$	(2/3) $B(C_6H_5)_3$	(1/3) BCl_3	+47.6
$(C_6H_5)_2BCl$	(1/2) $B(C_6H_5)_3$	(1/2) $C_6H_5BCl_2$	+27.3

[a] $\Delta H°_f$ data from Table 1-5.

It should be emphasized that the tabulated $\Delta H°$ values are ideal gas phase calculations and do not take into account solvation, phase separations, or entropy. They are probably good models for reactions of higher molecular weight analogues in nonpolar solvents.

Interpretation of the tabulated data may be aided by noting that at 25 °C (298.15 K), $\Delta\Delta H° = -5.7077$ kJ mol^{-1} is the number required to increase an equilibrium constant K_{eq} by 10-fold. (The $\Delta\Delta H°$ that corresponds to a power of 10 in K_{eq} is directly proportional to absolute temperature.) A $\Delta\Delta S° = +19.144$ J mol^{-1} deg^{-1} (a temperature independent figure) produces the same 10-fold increase in K_{eq}. If the total number of molecules is the same in starting materials and products and there are no phase changes, $\Delta S°$ is usually small. A cyclization reaction consumes one less molecule of starting material than does its open chain bimolecular counterpart and thereby gains a favorable entropy effect, which often amounts to 3 or 4 powers of 10 in K_{eq}. Most other entropy effects in a single-phase system are smaller than this, and most of the data in Tables 1-5 and 1-6 clearly indicate which way the equilibria lie.

1.3.2 Ligand Exchange on Boron

Ligands found useful in asymmetric synthesis are mainly confined to those connected by a carbon—boron or oxygen—boron bond. Nitrogen–bound ligands have found some use and may have further potential. Halogen ligands are not chiral directors, but are useful in certain synthetic sequences because of the Lewis acidity and reactivity of boron halides. Sulfur chemistry remains an area that might have some uses, but that most boron chemists would just as soon not explore.

Ligand exchanges involving C—B bonds are usually slow. In accord with the data in Table 1-6, trialkylboranes, R_3B, react with borate esters, $B(OR')_3$, to produce boronic esters, $RB(OR')_2$, at 100 °C if the $B(OR')_2$ group is cyclic [26], preferably with the aid of $BH_3 \cdot THF$ complex as a catalyst [27]. The formation of a cyclic $B(OR')_2$ group is, of course, largely a favorable entropy effect.

In contrast to disproportionations that would yield monomeric BR_3, disproportionations to BH_3 do occur with various HBY_2 and H_2BY because the reaction $2BH_3 \rightarrow B_2H_6$ is strongly exothermic (Table 1-4) and requires little or no activation energy [28]. Many H_2BY and some HBY_2 also dimerize, but not as exothermically as BH_3. Some HBY_2 and H_2BY or their dimers are kinetically stable enough to isolate.

Ligands which have unshared electron pairs generally exchange rapidly between bonding to a boron atom or a proton [29], as shown in the generalized equation (1-3) for RBX_2 and HY.

$$RBX_2 + 2HY \rightleftharpoons RBXY + HX + HY \rightleftharpoons RBY_2 + 2HX \qquad (1\text{-}3)$$

The qualitative term "rapidly" generally means faster than the chemist can manipulate reagent transfers or isolation procedures, and thermodynamic product is to be expected from such exchanges. Carbon–bound ligands that lack unshared electron pairs generally exchange much more slowly, and are kinetically stable in aqueous solutions unless there is a special structural reason, such as stability of the dissociated carbanion or conjugation with a nucleophilic site, that provides a mechanism for ligand exchange. Slow exchange of oxygen–bound ligands is sometimes encountered with cyclic, sterically hindered structures.

Thermodynamic data are especially useful for predicting and understanding ligand exchanges. The gas phase heats of hydrolysis of BX_3 listed in Table 1-7 are calculated for the reaction of equation (1-4). These can be used as a guide to which B—X will react with H—Y to produce B—Y and H—X.

$$(1/3)\ BX_3\ (g) + H_2O\ (g) \rightarrow (1/3)\ B(OH)_3\ (g) + HX\ (g) \tag{1-4}$$

As always with bond energy comparisons, it should be kept in mind that solvation and entropy effects will grossly modify these numbers in the liquid phase, especially for the low molecular weight compounds tabulated. The estimates are probably more reliable as guides to what may happen in nonpolar media. Real world data are not available for many of the species listed, but to give some idea of the magnitude of the deviations from reality, the hydrolysis of (1/3) $B(CH_3)_3$ (ideal gas) in liquid water (ΔH_f –285.830 kJ mol^{-1} [22]) to form methane (ideal gas) and dilute aqueous boric acid (ΔH_f –1071.2 kJ mol^{-1} [22]) yields –105.2 kJ mol^{-1} per C–B bond (instead of the –122.9 tabulated). Also, hydrolysis of boron halides in water will be much more exothermic than indicated because of solvation and ionization of the hydrogen halides produced.

In general, it may be expected that the B—X bonds that have the more negative heats of hydrolysis will react with H—Y to form B—Y bonds having a less negative heats of hydrolysis, but there are several complications. For example, the slight favoring of B—OCH_3 over the B—OH linkage in equilibrium with H_2O and CH_3OH will no doubt vanish in sterically hindered boron compounds. On the other hand,

Table 1-7. Heats of Hydrolysis per B—X bond (298.15 K) [a]

BX_3	ΔH°_f, kJmol^{-1}	HX	ΔH°_f, kJ mol^{-1}	ΔH°, kJ mol^{-1}
$B(OH)_3$	–992.28	H_2O	–241.846	0 by definition
$B(OCH_3)_3$	–900.1	CH_3OH	–201.5	+9.6
$B[N(CH_3)_2]_3$	–245.6	$HN(CH_3)_2$	–18.6	–25.6
BF_3	–1135.6	HF	–272.546	+17.1
BCl_3	–403.0	HCl	–92.312	–46.9
BBr_3	–204.2	HBr	–36.44	–57.3
BI_3	71 (±50)	HI	+26.359	–86 (±17)
$B(SCH_3)_3$	–156.1	$HSCH_3$	–22.9	–59.8
$B(SC_2H_5)_3$	–284.9	HSC_2H_5	–46.3	–40.2
BH_3	+100.4	H_2	0	–122.4
$B(CH_3)_3$	–122.6	CH_4	–74.873	–122.9
$B(C_6H_5)_3$	+129.9	C_6H_6	82.6	–49.6

[a] Based on experimental values [22], [23], [24], [25]. See also Tables 1-2, 1-3, and 1-5 for data sources.

1,2- or 1,3-diols form stable cyclic boronic esters, some of them highly resistant to hydrolysis. Some of the practical aspects of achieving desired alkoxy ligand exchanges are discussed in Section 1.4.2.

It may be generally expected that water or alcohols will displace any chloride, bromide, iodide, amino, or mercaptide ligands from boron. Amino ligands are the closest to being competitive with alkoxy groups, and o-phenylenediamine does form stable, often crystalline cyclic derivatives of boronic acids (9) that have an additional small degree of stabilization by their aromatic character [30-31].

9

Although amines will generally displace halides from boron, it should be kept in mind that the amine hydrohalide will be the byproduct, so that excess amine is required. The thermodynamic data on boron—sulfur compounds are obviously not very accurate, but do imply that boron—sulfur bonds should be comparable to boron—halogen bonds in their qualitative behavior. In general, thioborates can be made from mercaptans and boron chlorides, but the reactions are complicated by acid—base complexing and require heating for completion, or prior metalation of the mercaptan [32].

Although the BF_4^- anion is stable in water, the possibility that B—F bonds may exist in significant concentrations in equilibrium with aqueous or alcoholic solutions of boronic or borinic acid derivatives has found limited application, with the exception of the isolation of the unusually stable fluoride derivatives of α-amido boronic acids (see Section 5.4.2) [33]. Advantage has been taken of the stability of BF_4^- in a method for recovering pinanediol from its ester with boric acid [34].

Thermodynamic data are not readily available, but there are many known instances of mixed ligand derivatives of boron that are thermodynamically stable with respect to disproportionation. For example, dimethoxyboron chloride, $(MeO)_2BCl$, can be distilled, though some decomposition occurs, and it is better prepared for synthetic purposes by mixing appropriate proportions of boron trichloride and trimethyl borate at 0 °C or below [35].

For some synthetic purposes, it is useful to convert boronic esters to alkyldichloroboranes, which are much more acidic and reactive. One method for doing this is to reduce the boronic ester to a borohydride (RBH_3^-), which reacts with hydrogen chloride via alkylborane (RBH_2) to form the alkyldichloroborane ($RBCl_2$) [36]. Note that this replacement is exothermic (Table 1-6), though alkylborane dimerization would diminish the exothermicity to some extent. An older, more direct route utilized exchange with boron trichloride assisted by ferric chloride, which catalyzes the irreversible decomposition of $ROBCl_2$ to RCl and insoluble B_2O_3 [37]. This process has been reexamined, improved, and established as a general method recently [38]. In a sterically hindered system having somewhat labile carbon—boron bonds, exchange has been catalyzed by a small amount of borohydride, but in this case ferric chloride was not tested and a large excess of boron trichloride was required [39].

Haloboranes BX_3, RBX_2, or R_2BX (X = Cl, Br, I) are readily reduced by trialkylsilanes, and hydride/halide exchange between boron atoms of monomeric species appears to be very rapid [40].

It may be noted that the average heat of hydrolysis per C—B bond of trimethylborane (Table 1-6) is more negative than that of methylboronic acid (Section 1.3.1) by an estimated 9.72 kJ mol⁻¹. This value is consistent with the unfavorable thermodynamics of disproportionation of methylboronic acid as well as the generally lower reactivity of boronic esters compared with trialkylboranes.

1.3.3 Acidities

Boron generally functions as a Lewis acid. The stable complexes of triethylborane with ammonia, Et_3BNH_3, and with metal hydroxides, Et_3BOH^-, were among the very first organoboron compounds discovered [3].

Boronic acids (**10**) also react as Lewis acids, not as protic acids. *o*-Nitrophenylboronic acid is anomalously weak because the *o*-nitro group sterically hinders formation of the tetrahedral borate anion [41], and other evidence also indicates that boronic acids $RB(OH)_2$ accept OH⁻ to form $RB(OH)_3^-$ [42]. Further proof of the tetracoordinate character of boronate anions is provided by the contrasting behavior of borazaro compounds (**11**), which are protic acids as indicated by ultraviolet spectra [43] and the unusually broad ¹¹B NMR lines of the anions [44]. In borazaro compounds the boron forms part of an aromatic ring, and the aromaticity would be lost on coordination.

10

11

The pK_a of boric acid is about 9 [45, 49], and boronic acids are usually in the range 9-11 (Table 1-8) [45, 46, 47, 48, 50]. Boronic esters are also weak Lewis acids (apparent pK_a's 9-11 in methanol) [51]. The labile character of borinic acids complicates pK measurements, but a single study found that diphenylborinic acid is 2.6 pK units more acidic than phenylboronic acid [52]. This finding is consistent with recent enzyme inhibition studies [53], as well as the general reactivity of borinic acids and esters. Some examples are illustrated in Table 1-9.

Table 1-8. pK_a's of Boric, Boronic, and Borinic Acids in Water and Aqueous Ethanol

Compound	pK_a in H_2O	pK_a in 25% EtOH	Reference
$B(OH)_3$	9.19; 8.98	9.87	[45, 49]
$CH_3B(OH)_2$	10.60	—	[48]
$CH_3CH_2CH_2CH_2B(OH)_2$	10.74	11.46	[46]
$CH_3CH_2CH(CH_3)B(OH)_2$	10.60	—	[48]
$(CH_3)_3CB(OH)_2$	10.36	—	[48]
$C_6H_5CH_2B(OH)_2$	9.12, 9.14	9.83	[46, 48]
$C_6H_5CH_2CH_2B(OH)_2$	10.00	10.74	[46]
$CH_2=CHB(OH)_2$	9.49	—	[48]
$C_6H_5B(OH)_2$	8.86; 8.8	9.71	[45, 50, 48]
p-$ClC_6H_4CH_2B(OH)_2$	—	9.21	[45]
o-$ClC_6H_4CH_2B(OH)_2$	—	8.85	[45]
p-$O_2NC_6H_4CH_2B(OH)_2$	—	8.01	[47]
o-$O_2NC_6H_4CH_2B(OH)_2$	—	9.25	[47]
$(C_6H_5)_2BOH$	6.2	—	[52, 53]

Table 1-9. Apparent pK_a's of Borate and Boronate Esters in Methanol [a]

Compound	Apparent pK_a	pK_a with $(CH_2OH)_2$ [b]
$B(OCH_3)_3$	8.98	8.54
$CH_3B(OCH_3)_2$	10.62	10.24
$C_6H_5CH_2B(OCH_3)_2$	9.72	9.02
$CH_2[B(OCH_3)_2]_2$	10.83	10.11

[a] Titration was with $LiOCH_3$ in CH_3OH and apparent pH values were uncorrected pH meter readings. Boron compounds were 0.5 mM in 0.048 M NaCl in CH_3OH [51]. [b] Solvent was CH_3OH with 4% $HOCH_2CH_2OH$. Methyl esters are converted to ethylene glycol esters (1,3,2-dioxaborolanes) under these conditions.

Trialkylboranes are well known to be Lewis acids, and the complexing of trimethylborane to various amines was extensively studied in Brown's classical studies of steric strains [54]. The ΔH of dissociation of pyridine—trimethylborane is 71.13 kJ mol^{-1}, and ΔS is 180.75 J deg^{-1} mol^{-1}. The pK_a of pyridinium ion is 5.17, and the stability of the complex suggests that the reaction of $(CH_3)_3B$ with H_2O to form $(CH_3)_3BOH^-$ and H^+ might show a pK_a substantially less than 5.

The calculated dissociation energies of $(CH_3)_3B$—NH_3 differ from those of $HOB(CH_3)_2$—NH_3 or $CH_3OB(CH_3)_2$—NH_3 by ~40 kJ mol^{-1} (Table 1-4). To the extent that the behavior of ammonia parallels that of hydroxide ion, this corresponds to $(CH_3)_3B$ being 7 pK_a units more acidic than $(CH_3)_2BOH$, the pK_a of which is unknown, but is probably not less than 8 in view of the relationship between $PhB(OH)_2$ and Ph_2BOH, and probably not much greater than 10.6, the pK_a of $CH_3B(OH)_2$ (Table 1-8). The doubly extrapolated pK_a of trimethylborane might be in the range 1 to 4, which seems at least qualitatively consistent with the formation of the pyridine complex, as well as with the observation that trialkylboranes will generally dissolve sodium or potassium cyanide in ether (pK_a of HCN ~9) [55].

A recent attempt to measure the pK_a of tributylborane directly has revealed mainly why such data are not available. Tributylborane is insoluble in any solvent that contains much water, and in 1:1 THF—methanol under argon titrated with aqueous sodium hydroxide in the presence of cresol red appeared about as acidic as p-bromophenol, suggesting a pK_a near 9, unexpectedly high [56]. It is unclear whether accidental oxygen oxidized the trialkylborane to borinic ester (and no further), or whether the figure is real and reflects the hardness of hydroxide ion as a base. The hard base pyridine reacts 50 kJ mol^{-1} more exothermically with boron trifluoride than with trimethylborane, but the soft base trimethylphosphine only 10 kJ mol^{-1} more exothermically [54]. Also, steric effects in borane acidities are large [54], and tributylborane might be a much weaker acid than trimethylborane. These questions can only be decided if experimental data are obtained under rigorously controlled conditions, a major project.

In summary, the conventional view is that the general relative order of reactivity of alkylboranes, $R_3B > R_2BOR' > RB(OR')_2$, is a function of both the relative electron donating abilities of the C—B bonds and the relative Lewis acidities of the boron atoms. The thermodynamic analysis presented here is in accord with that view.

1.4 Safety Considerations

1.4.1 General Hazards

The spontaneous flammability and noxious, lachrymatory properties of the most volatile trialkylboranes have already been noted [1, 3]. Tributylborane is not usually spontaneously flammable [57], and its acute toxicity, LD$_{50}$ 80 mg kg^{-1} by intravenous injection in rats [58], is significant but not extreme. In view of the rapidity with which the compound oxidizes in air and the consequent need to handle it in closed systems, the main hazard to laboratory workers is the fire danger in case of a spill. Diborane has sufficient lifetime in air to be a toxic hazard, and an industrial exposure limit of 0.1 parts per million has been established [59].

In contrast to trialkylboranes, boronic acids and esters [$RB(OH)_2$ and $RB(OR')_2$] are generally no more hazardous to handle than typical organic compounds. Studies carried out in connection with the search for suitable agents for the proposed ^{10}B neutron capture therapy of cancer have indicated that water soluble boronic acids tend to have low toxicity, and fat soluble boronic acids are moderately toxic [60, 61, 62]. Water soluble boronic acids are generally excreted unchanged by the kidney [61].

One cautionary note is that some boronic acids and related compounds are specific enzyme inhibitors. For example, (1R)-(1-acetamido-2-phenylethyl)boronic acid, $PhCH_2CH(NHAc)B(OH)_2$, is a good inhibitor of chymotrypsin, K_i = 2.1 x 10^{-6} M at pH 7.5 [63]. The borinic acid analogue of acetylcholine, $Me_3N^+CH_2CH_2CH_2B(OH)CH_3$, is a powerful cholinesterase inhibitor, K_i = 3 x 10^{-8} M at pH 7.5 [63], and may safely be presumed to be highly toxic.

Boric acid was well tolerated at 350 parts per million of boron in the diet of rats and dogs, but toxic effects showed up at higher doses [64]. It has been estimated that 15-20 g of boric acid might be a lethal dose for an adult human, and that 0.5 g per day for six months is likely to cause toxic effects [59]. Boric acid has been tested for carcinogenicity in mice and found inactive at up to 5000 ppm in the diet [65]. Boric acid and sodium borate are not mutagenic in *Salmonella typhimurium* strains TA98 and TA100, with or without added S-9 rat liver enzymes [66].

Long term environmental problems with organoboranes are probably negligible. The rapid air oxidation of trialkylboranes makes their persistence or accumulation unlikely.

Boronic acids do not oxidize nearly so rapidly, especially in the presence of water [57]. The boron—carbon bond is not biodegraded by mammals [61], and probably not by any other type of organism. There is no known natural selection pressure to evolve an enzyme for this purpose, and the only known biogenic boron compounds are oxygen bound chelates such as the antibiotics boromycin [67] and aplasmomycin [68], which in view of the energy cost of reducing boric acid to organoboranes is hardly surprising. Even so, the favorable thermodynamics of oxidation leave the only question one of rate. Small samples of boronic acids or esters stored in closed but unsealed containers on laboratory shelves deteriorate in a few months or years, and inorganic borate is undoubtedly the ultimate form of any boron which reaches the environment.

Boric acid is an essential micronutrient for plants [69-71]. Recent evidence suggests that traces of boron may also be required by animals [72]. In view of this evidence, small amounts of typical organoboron compounds should not be considered a serious environmental hazard.

Additional details regarding the hazards of boron compounds have been reviewed in more detail elsewhere [73].

1.4.2 Laboratory Handling

Many boronic acids and esters can be handled as ordinary organic compounds in the laboratory, though it is well to keep in mind that their stability in air is kinetic rather than thermodynamic. Boronic esters are generally preferred over boronic acids as synthetic intermediates. Simple boronic esters (**12**) are generally stable toward oxygen but are hydrolyzed rapidly by atmospheric moisture. In general, compounds of this class can be made from boronic acids and alcohols by azeotropic distillation of water with a hydrocarbon solvent [29]. Methyl esters can be made from boronic acids and 2,2-dimethoxypropane [74] or by taking advantage of the insolubility of methanol, but solubility of its boronic esters, in pentane [75]. Efficient conversion of **12** back to boronic acids can be accomplished by treatment with water and, if necessary, removal of the alcohol by azeotropic distillation under reduced pressure.

$$R^1\!-\!B\!\!\begin{array}{c}OH\\\\OH\end{array} \;+\; 2\,HOR^2 \;\rightleftharpoons\; R^1\!-\!B\!\!\begin{array}{c}OR^2\\\\OR^2\end{array} \;+\; 2\,H_2O$$

12 R^1 = alkyl or aryl; R^2 = alkyl

Cyclic ester types **13-18** are arranged in approximate increasing order of stability toward hydrolysis. Ethylene glycol esters (**13**) are stable to gas chromatography [76] but generally decompose on silica. Unhindered chiral 2,3-butanediol esters (**14**) are easily hydrolyzed [77], but if R is secondary alkyl they are very resistant to hydrolysis [78]. Most pinacol esters (**15**) are not hydrolyzed by water and are stable to chromatography on silica. Chiral diisopropylethanediol esters (**16**) appear to have similar hydrolytic stability, and are routinely purified by chromatography [79]. Crystalline chelated diethanolamine esters (**17**) have been made by treating unhindered **14** or **16** with diethanolamine [80]. Pinanediol esters (**18**) have been obtained by treatment of ethereal solutions of **14** or **16** with pinanediol [77, 81]. Purification of **18** by chromatography is routine [82], and hydrolysis or transesterification under neutral conditions for preparative purposes is not possible [82-83].

13   **14**   **15**

16   **17**   **18**

Boronic acids are usually crystalline compounds, but their tendency to form anhydrides on drying, even at room temperature under vacuum, often makes it difficult to obtain analytical samples. On occasion, freshly isolated crystalline samples of low molecular weight boronic acids have been observed to darken, get hot, and decompose after a few minutes of exposure to air [57, 83]. To some degree, the problem can be avoided by keeping the sample moist with water, and it has been suggested that borinic anhydrides, R_2B—O—BR_2, present as impurities, are the autoxidation initiators [57]. This instability appears to be associated with the synthesis from Grignard reagents, which also produces small amounts of trialkylboranes as byproducts, and these may also serve as autoxidation initiators.

Whatever the initiators of autoxidation may be, purified samples of boronic acids are generally stable enough in air to permit normal weighing and transfer. Functionally substituted boronic acids, which are not made directly from organometallic reagents, have generally appeared stable in air. However, samples stored unprotected from oxygen will eventually deteriorate, and it is recommended that reactions of boronic acids and esters should be routinely carried out under an inert atmosphere.

Borinic esters, R_2BOR', are much less commonly isolated than boronic esters. Their reactivity is between that of boronic esters and trialkylboranes. Some borinic esters, especially those containing at least one aryl group [84] or a thioether substituent [85], can be handled in air at least briefly. On the other hand, butyl divinylborinate, $BuOB(CH=CH_2)_2$, polymerizes rapidly after brief exposure to air [85]. The dimethylaminoethanol ester, $Me_2NCH_2CH_2OB(CH=CH_2)_2$, which is stabilized by chelation, proved stable in air. Ethylene glycol bis[dibutylborinate], $Bu_2OCH_2CH_2OBBu_2$, became warm when shaken in air, and ignited spontaneously on cotton [86]. Because of the method of synthesis, it is not possible to know whether this material was entirely free from tributylborane, which would tend to initiate radical reactions with oxygen, but from a practical point of view, the synthetic chemist is forewarned.

Trialkylboranes react rapidly with air, as has already been noted, and have to be handled entirely in closed systems under an inert atmosphere, with reagent transfers done by syringe or cannula. The extreme hydrolytic instability of haloboranes requires similar precautions. Since solids are not often involved, Schlenk tubes are only needed in unusual circumstances. Organic chemists who have worked with organolithium reagents or other reactive organometallic compounds are already familiar with the necessary laboratory procedures. The techniques used by Brown and his coworkers have been described in detail in a book [87].

1.5 References

1.   Frankland E, Duppa B (1860) Proc. Royal Soc. (London), 10:568
2.   Frankland E, Duppa BF (1860) Justus Liebigs Ann. Chem. 115:319
3.   Frankland E (1862) J. Chem. Soc. 15:363
4.   Frankland E (1862) Justus Liebigs Ann. Chem. 124:129
5.   Khotinsky E, Melamed M (1909) Chem. Ber. 42:3090

6.   Brown HC, Subba Rao BC (1956) J. Am. Chem. Soc. 78:5694
7.   Brown HC, Zweifel G (1959) J. Am. Chem. Soc. 81:247
8.   Brown HC, Zweifel G (1961) J. Am. Chem. Soc. 83:486
9.   Meller A, Habben C, Noltemeyer M, Sheldrick GM (1982) Z. Naturforsch., Teil B 37: 1504
10. Yalpani M, Boese R, Blaser D (1983) Chem. Ber. 116: 3338
11. Ho O, Soundararajan R, Lu J, Matteson DS, Wang Z, Wei M, Chen X, Willett RD (1994) Manuscript in preparation
12. Matteson DS, Michnick TJ, Willett RD, Patterson CD (1989) Organometallics 8:726
13. Rettig SJ, Trotter J (1975) Can. J. Chem. 53: 1393
14. Kliegel W, Preu L, Rettig SJ, Trotter J (1985) Can. J. Chem. 63: 509
15. Farfan N, Joseph-Nathan P, Chiquete LM, Contreras R (1988) J. Organomet. Chem. 348: 149
16. Gupta A, Kirfel A, Will G, Wulff G (1977) Acta Crystallogr., Section B 33: 637
17. Kliegel W, Preu L, Rettig SJ, Trotter J (1986) Can. J. Chem. 64: 1855
18. Rettig SJ, Trotter J (1977) Can. J. Chem. 55: 3071
19. Chen X, Bartolotti L, Ishaq K, Tropsha A (1994) J. Computational Chem. 15:333
20. Sana M, Leroy G, Wilante C (1991) Organometallics 10:264
21. Sana M, Leroy G, Wilante C (1992) Organometallics 11:781
22. Chase MW Jr, Davies CA, Downey JR Jr., Frurip DJ, McDonald RA, Syverud AN (1985) JANAF Thermochemical Tables, J. Phys. Chem. Ref. Data, Vol. 14, Supplement 1
23. Pedley JB, Naylor RD, Kirby SP (1986) Thermochemical Data of Organic Compounds, 2nd ed., Chapman and Hall
24. Holbrook JB, Smith BC, Housecroft CE, Wade K (1982) Polyhedron 1:701
25. Finch A, Gardner PJ (1970) in: Brotherton RJ, Steinberg H (ed) Progress in Boron Chemistry, Pergamon Press, vol 3, pp 177-210
26. Brown HC, Gupta SK (1971) J. Am. Chem. Soc. 93:1816
27. Brown HC, Gupta SK (1971) J. Am. Chem. Soc. 93:2802
28. Mappes GW, Fridmann SA, Fehlner T, (1970) J. Phys Chem. 74:3307.
29. Lappert MF (1956) Chem. Rev. 56:959
30. Letsinger RL, Hamilton SB (1958) J. Am. Chem. Soc. 80:5411
31. Dewar MJS, Kubba VP, Pettit R (1958) J. Chem. Soc. 3076
32. Mikhailov BM (1970) in: Brotherton RJ, Steinberg H (ed) Progress in Boron Chemistry, Pergamon Press, vol 3, pp 313-370
33. Kinder DH, Katzenellenbogen JA (1985) J. Med. Chem. 28:1917
34. Brown HC, Rangaishenvi MV (1988) J. Organomet. Chem. 358:15
35. Castle RB, Matteson DS (1969) J. Organomet. Chem. 20:19
36. Brown HC, Salunkhe AM, Singaram B (1991) J. Org. Chem. 56:1170
37. Brindley PB, Gerard W, Lappert MF (1956) J. Chem. Soc. 824
38. Brown HC, Salunkhe AM, Argade AB (1992) Organometallics 11:3094
39. Matteson DS, Mattschei PK (1973) Inorg. Chem. 12:2472
40. Soundararajan R, Matteson DS (1990) J. Org. Chem. 55:2274
41. McDaniel DH, Brown HC (1955) J. Am. Chem. Soc. 77:3757
42. Lorand JP, Edwards JO (1959) J. Org. Chem. 24:769
43. Dewar MJS, Dietz R (1961) Tetrahedron 15:26
44. Dewar MJS, Jones R (1967) J. Am. Chem. Soc. 89:2408
45. Branch GEK, Yabroff DL, Bettman B (1934) J. Am. Chem. Soc. 56:937
46. Yabroff DL, Branch GEK, Bettman B (1934) J. Am. Chem. Soc. 56:1850
47. Bettman B, Branch GEK, Yabroff DL (1934) J. Am. Chem. Soc. 56:1865

48. Minato H, Ware JC, Traylor TG (1963) J. Am. Chem. Soc. 85:3024
49. Babcock L, Pizer R (1980) Inorg. Chem. 19:56
50. Juillard J, Geugue N (1967) C. R. Acad. Paris C 264:259
51. Matteson DS, Allies PG (1973) J. Organomet. Chem. 54: 35
52. Chremos GN, Zimmerman HK (1963) Chim. Chronika 28:103
53. Rao G, Philipp M (1991) J. Org. Chem. 56:1505
54. Brown HC (1972) Boranes in Organic Chemistry, Cornell University Press, Ithaca, New York, pp 53-128
55. Pelter A, Smith K, Hutchings MG, Rowe K (1975) J. Chem. Soc., Perkin Trans. I 129
56. Matteson DS, Hagen CB (1993) unpublished results
57. Snyder HR, Kuck JA, Johnson JR (1938) J. Am. Chem. Soc. 60:105
58. Tagaki M, Hisata K, Baba M, Masuhara E (1973) Nippon Kookuuka Gakkai Zasshi 22:533
59. Hughes RL, Smith IC, Lawless EW (1967) Production of the Boranes and Related Research, Holtzman RT, Ed, Academic Press, New York, pp 291-294
60. Soloway AH (1958) Science 128:1572
61. Soloway AH, Whitman B, Messer JR (1962) J. Med. Pharm. Chem. 7:640
62. Matteson DS, Soloway AH, Tomlinson DW, Campbell JD, Nixon GA (1964) J. Med. Chem. 7:640
63. Matteson DS, Sadhu KM, Lienhard GE (1981) J. Am. Chem. Soc. 103:5241
64. Weir RJ Jr, Fisher RS (1972) Toxicol. Appl. Pharmacol. 23:351
65. National Cancer Institute (1989) „Survey of Compounds which Have Been Tested for Carcinogenic Activity," NIH Publication No. 49-468, p. 16
66. Benson WH, Birge WJ, Dorough HW (1984) Environ. Toxicol. Chem. 3:209; Chem. Abstr. 101:124626g
67. Dunitz JD, Hawley DM, Miklos D, White DNJ, Berlin Yu, Marusic R, Prelog V (1971) Helv. Chim. Acta 54:1709
68. Nakamura H, Iitaka Y, Kitahara T, Okazaki T, Okami Y (1977) J. Antibiot. 30:714
69. Warington K (1923) Ann. Bot. 37:629
70. Warington K (1933) Ann. Bot. 47:429
71. Lovatt CJ, Dugger WM (1984) Biochemistry of the Essential Ultratrace Elements, Vol. 3., Chap. 17, Plenum, New York, NY.
72. Hunt CD (1989) Biological Trace Element Research 22:201
73. Matteson DS (1987) in: Hartley F, Patai S (eds) The Chemistry of the Metal-Carbon Bond, John Wiley and Sons, vol 4, p 307
74. Matteson DS, Krämer E. (1968) J. Am. Chem. Soc. 90:7261
75. Brown HC, Bhat NG, Somayaji V (1983) Organometallics 2:1311
76. Matteson DS, Thomas JR (1970) J. Organomet. Chem. 24:263
77. Sadhu KM, Matteson DS, Hurst GD, Kurosky JM (1984) Organometallics 3:804
78. Matteson DS, Campbell JD (1990) Heteroatom Chemistry 1:109
79. Matteson DS, Tripathy PB, Sarkar A, Sadhu KM (1989) J. Am. Chem. Soc. 111:4399
80. Tripathy PB, Matteson DS (1990) Synthesis 200
81. Matteson DS, Kandil AA (1986) Tetrahedron Lett. 27:3831
82. Brown HC, Rangaishenvi MV.(1988) J. Organomet. Chem. 358:15
83. Matteson DS (1960) J. Am. Chem. Soc. 82:4228
84. Matteson DS, Mah RWH (1963) J. Org. Chem. 28:2171
85. Matteson DS (1962) J. Org. Chem. 27:275
86. Letsinger RL, Skoog, I (1954) J. Am. Chem. Soc. 76:4174
87. Brown HC, Kramer GW, Levy AB, Midland MM (1975) Organic Synthesis via Boranes, Wiley–Interscience, New York

2 Sources of Compounds Containing Boron—Carbon Bonds

2.1 Industrial Sources of Boron Compounds

The first step in the synthesis of organoboranes is the conversion of borax (sodium borate, $Na_2B_4O_7 \cdot 10H_2O$), the most common ore of boron [1], into a more tractable derivative for reduction. Acidification of borax with carbon dioxide yields boric acid [$B(OH)_3$] [2], which reacts with various alcohols under dehydrating conditions such as azeotropic codistillation of water to form trialkyl borates [trialkoxyboranes, $B(OR)_3$] [3]. Trimethyl borate [$B(OCH_3)_3$], from boric acid and methanol, forms an azeotrope with methanol. Pure trimethyl borate can be separated from the methanol in the azeotrope by extraction with mineral oil and distillation [2].

The synthesis of boron trichloride cannot be accomplished economically by any direct metathesis, but boric oxide (B_2O_3, from heating H_3BO_3) treated with carbon and chlorine at 600-800 °C will produce BCl_3. Alternatively, boron carbide (B_4C) can be made from carbon and B_2O_3 and treated with Cl_2 or with HCl to produce BCl_3 [2]. Boron trifluoride can be obtained via pyrolysis of metal fluoborates, including $LiBF_4$ or, in the presence of a non-alkali metal halide, sodium or potassium tetrafluoroborates.

Diborane and sodium borohydride can be prepared by the reduction of trimethyl borate with sodium hydride [2]. Dimethoxyborane is an intermediate. Diborane can also be obtained from boron trichloride and hydrogen at 700 °C in the presence of a silver catalyst or at higher temperatures without a catalyst [2]. For laboratory purposes, diborane can be generated easily from sodium borohydride and a Lewis acid such as boron trifluoride or aluminum chloride [4].

2.2 The Organometallic Route

2.2.1 Boronic Esters

2.2.1.1 Grignard and Lithium Reagents. Though the reaction of organozinc reagents with borate esters was the first route to organoboranes to be discovered [5], the corresponding reaction with Grignard reagents is more generally useful [6-7]. An optimized version of the Grignard process for the preparation of phenylboronic acid has been described in *Organic Syntheses* [8]. More recently, a convenient and efficient process for obtaining boronic esters from organolithium reagents and triisopropyl borate has been described [9]. Thus, any Grignard or alkyllithium reagent can be converted to the corresponding boronic acid or ester.

In the Organic Syntheses procedure, trimethyl borate and phenylmagnesium bromide are added simultaneously from separate dropping funnels to diethyl ether that is very vigorously stirred at −78 °C. Phenyl(trimethoxy)borate anion precipitates as a magnesium salt (1), and as a result, further phenylation of the boron does not occur as more than a minor side reaction. In the hydrolysis of 1, hydrogen ion removes methoxide from boron much faster than it attacks phenyl, in spite of the contrary thermodynamic balance of the products, and work up with aqueous acid yields phenylboronic acid (2) [8].

2

This reaction follows a very general pattern for organoborane reactions, inasmuch as trigonal boron in B(OMe)$_3$ reacts with a base to form a tetrahedral borate salt (1), which in a subsequent step loses a ligand to acid to restore trigonal boron in the product (2).

The key to getting good yields in reactions of this class is to have only ligand addition and no ligand dissociation until all of the base has been consumed. A monoalkylborate complex (3) probably cannot react with a second mole of organometallic to form a dialkylborate (5) unless it first dissociates to a boronic ester (4) and an alkoxide. Brown and Cole carried out a systematic study of the reaction of simple alkyllithium and Grignard reagents with various borate esters and found that alkyllithiums and triisopropyl borate provide very clean 3 as shown by NMR evidence, and simple boronic esters 4 are obtained in high yields by treatment of 3 with anhydrous hydrogen chloride [9].

| 3 | 4 | 5 | (M = Li or MgX) |

Brown and Cole found no such selectivity with simple Grignard reagents [9], but even so, Grignard reagents are often more easily accessible than organolithium reagents and often give good yields of boronic esters. Precipitation of the magnesium salt 1 appears to be a factor in the high yield of the *Organic Syntheses* procedure, but low temperatures (below ~20 °C [8]) during the reaction may be more important generally for preventing disproportionation or dialkylation of borates **3**. It is also possible that trialkyl borate could assist the dissociation of **3** via the reaction $R^1B(OR^2)_3^- + B(OR^2)_3 = R^1B(OR^2)_2 + B(OR^2)_4^-$, which is suppressed in the *Organic Syntheses* procedure by keeping the concentration of trimethyl borate low.

The classical approach to preventing dialkylborinate and trialkylborane formation was to add the organometallic reagent to the borate ester at low temperature [7]. However, there is no evidence that the order of mixing is relevant. In one well documented example, (dichloromethyl)boronic acid or its esters, $Cl_2CHB(OR)_2$, can be made in about equally good yields either by adding a borate ester, $B(OR)_3$, to (dichloromethyl)lithium, $LiCHCl_2$ [10], or by generating the (dichloromethyl)-lithium in the presence of triisopropyl borate [11].

Although the classical procedure is successful for a wide variety of boronic acids, the lowest molecular weight boronic acids are too soluble in water and become difficult to isolate. An expedient which works for most of them is to isolate the dibutyl ester, $RB(OBu)_2$. This is accomplished merely by extraction with butanol from an acidic aqueous solution saturated with salt, then distilling the butanol—water azeotrope, butanol, and finally the dibutyl ester. For example, dibutyl vinylboronate [12] and dibutyl ethynylboronate [13] have been made efficiently this way.

Methylboronic acid and its trimeric anhydride are too volatile and water soluble even for the foregoing approach. A Grignard based route to the methylboronic anhydride—pyridine complex [14] generates a hazardous spontaneously flammable byproduct, probably trimethylborane [15]. The reaction of methylmagnesium chloride with 2-methoxy-1,3,2-dioxaborinane yields the methylboronic ester [16].

The method of choice for making methylboronic acid derivatives is the reaction of methyllithium with triisopropyl borate in THF at -78 °C reported by Brown and Cole [9]. The triisopropyl borate intermediate **6** does not disproportionate, and is efficiently converted to diisopropyl methylboronate (**7**) by anhydrous hydrogen chloride. As a result of this process, several methylboronic acid derivatives are now commercially available [17], but in case the preparation is repeated to save money, two precautions should be noted. First, distillation through a fractionating column is required in order to achieve complete purification of diisopropyl methylboronate, and second, attempted use of higher concentrations of reactants than those specified [9] has resulted in stirring difficulties and generation of a volatile spontaneously flammable byproduct, presumably trimethylborane [18].

The Brown-Cole procedure [9] will usually be the method of choice for making boronic esters when the required organolithium reagent is readily available. The lithium reagents tested in addition to methyllithium included *n*-butyl-, *sec*-butyl-, *tert*-butyl-, and phenyllithium. In order to obtain optimum results with *tert*-

butyllithium, it was found necessary to lower the reaction temperature to –98 °C. Methyllithium was tested with a variety of borate esters, and in addition to triisopropyl borate, triisobutyl borate and tri-*sec*-butyl borate yielded highly satisfactory results. Unsatisfactory reagents included BCl_3, $(MeO)_2BF$, $(MeO)_2BCl$, $(MeO)_3B$, and $(tert\text{-}BuO)_3B$ [9].

The Brown–Cole work up procedure is also applicable to the synthesis of (haloalkyl)boronic esters where the requisite (haloalkyl)lithium is unstable and has to be generated in situ. Diisopropyl (chloromethyl)boronate (**8**) [19], diisopropyl (bromomethyl)boronate (**9**) [20] and diisopropyl (dichloromethyl)boronate (**10**) [11, 21] are particularly useful starting materials that can be made in this way, and **10** has become commercially available [17].

$$ICH_2Cl + B(OR)_3 \xrightarrow[\text{[LiCH}_2\text{Cl]}]{\text{add BuLi}} [ClCH_2B(OR)_3]^- \xrightarrow{\text{HCl}} ClCH_2B(OR)_2 + ROH$$

$$[R = (CH_3)_2CH] \quad \textbf{8}$$

$$CH_2Br_2 + B(OR)_3 \xrightarrow[\text{[LiCH}_2\text{Br]}]{\text{add BuLi}} [BrCH_2B(OR)_3]^- \xrightarrow{CH_3SO_3H} BrCH_2B(OR)_2 + ROH$$

$$[R = (CH_3)_2CH] \quad \textbf{9}$$

$$CH_2Cl_2 + B(OR)_3 \xrightarrow[\text{[LiCHCl}_2\text{]}]{\text{add LiNR}_2} [Cl_2CHB(OR)_3]^- \xrightarrow{\text{HCl}} Cl_2CHB(OR)_2 + ROH$$

$$[R = (CH_3)_2CH] \quad \textbf{10}$$

Acetylenic boronic esters, which are very sensitive to water above pH ~7, are easily made via the Brown-Cole procedure [22]. Allylic boronic esters are also easily cleaved or rearranged. For some of the special techniques used in their preparation, see Chapter 7.2.2.

2.2.1.2 Other Organometallics. Hexenylzirconium compounds react with a variety of haloboranes, for example, catechol chloroborane (**11**), to provide hexenylboranes [23].

11

Organotin reagents, R_4Sn, react readily with boron trihalides, BX_3, to form RBX_2 or R_2BX [24-25]. Organomercury compounds can be converted to boron compounds by reaction with boron halides, as in the reaction of ICH_2HgI with BBr_3 to form ICH_2BBr_2 [26]. A recent preparation of vinylboron dichloride, $H_2C=CHBCl_2$, involves reaction of vinyltripropylsilane with boron trichloride at $-78\,°C$ [27]. Because the silyl precursor is very easily prepared from the Grignard reagent, this may be the most convenient laboratory route to vinylboron compounds in general.

2.2.2 Di-, Tri-, and Tetraborylmethanes

Addition of a mixture of di-, tri-, or tetrachloromethane and dimethoxyboron chloride to a vigorously stirred suspension of lithium dispersion in THF at -30 to $-40\,°C$ results in formation of the corresponding bis(dimethoxyboryl)methane **12**, tris-(dimethoxyboryl)methane **13**, or tetrakis(dimethoxyboryl)methane **14** [28-30].

$$CH_2Cl_2 + 4\ Li + 2\ ClB(OMe)_2 \longrightarrow (MeO)_2BCH_2B(OMe)_2$$

12

$$CHCl_3 + 6\ Li + 3\ ClB(OMe)_2 \longrightarrow \underset{\overset{|}{B(OMe)_2}}{\overset{\overset{B(OMe)_2}{|}}{HC}}\!-\!B(OMe)_2$$

13

$$CCl_4 + 8\ Li + 4\ ClB(OMe)_2 \longrightarrow (MeO)_2B\!-\!\underset{\overset{|}{B(OMe)_2}}{\overset{\overset{B(OMe)_2}{|}}{C}}\!-\!B(OMe)_2$$

14

The cyclic haloborane 2-chloro-1,3,2-dioxaborinane fails to yield any of the cyclic analogues of **12**, **13**, or **14**. The reason is not understood. However, trans-esterification of **12**, **13**, or **14** with diols proceeds readily. The resulting cyclic boronic esters are much more efficient than the methyl esters as synthetic intermediates. The conversion of these compounds to other boronic esters via boryl carbanion chemistry is described in Chapter 3.3.

2.2.3 Borinic Esters (Dialkylalkoxyboranes)

The traditional route to dialkylalkoxyboranes, RR'B(OR''), usually referred to as borinic esters, is to treat a boronic ester with a Grignard reagent [3]. This route has been applied successfully to the preparation of several vinylic and other functionalized borinic esters [31-32]. Dibutyl divinylboronate (**15**) was so unstable toward vinyl polymerization that it had to be isolated as the β-dimethylamino-ethanol ester **16**, which is stabilized by internal coordination of the amine nitrogen to boron [32].

$$H_2C=CHB(OBu)_2 \xrightarrow[\text{2. } H_3O^+]{\text{1. } H_2C=CHMgBr} (H_2C=CH)_2BOBu \longrightarrow$$

not isolable

15   **16**

Symmetrical dialkyl or diaryl borinic esters can be obtained in one step from Grignard reagents and triisopropyl borate [33]. The reaction of boronic esters with organolithiums provides a useful general route to borinic esters [34]. The lithium borate complexes can be decomposed by hydrogen chloride in ether [34], by acetyl or benzoyl chloride to form the borinic ester, carboxylic ester, and lithium chloride [34], [35], or by heating to distill volatile borinic esters and leave lithium alkoxide as a residue [35]. Diisopropyl esters of alkenylboronic acids are the best substrates for reaction with alkyllithium or magnesium reagents, THF is the best solvent, and acidification with Me_3SiCl (Li) or ethereal HCl (Mg) should be done at −78 °C [36].

Spontaneous air oxidation of a dibutylborinic ester [37] was reported long ago, and alkylalkenylborinic esters have recently been described as too sensitive to air and moisture to permit mass spectral or elemental analyses [36]. However, after aqueous work up in air, elemental analyses were easily obtained on all but two of a series of vinylarylborinic and other functionalized borinic esters back in the days before NMR [31], [32], and the stable derivative **16** from unstable **15** was analyzed [32].

2.2.4 Trialkylboranes

Organometallic reagents are not often used to make trialkylboranes, as hydroboration is usually more convenient. The conversion of alkenylalanes to alkenyl-9-borabicyclononanes (**17**) has potential utility because the aluminum compounds can provide controlled trisubstituted olefin geometry [38]. The Suzuki coupling reaction (see Chapter 4.3) provides an alternative trisubstituted olefin synthesis.

$$R{=\!\!=}H \xrightarrow[\text{Cp}_2\text{ZrCl}_2]{\text{Al}_2\text{Me}_6} \underset{\text{CH}_3}{R\!\!-\!\!C(H)\!\!=\!\!C\text{-AlMe}_2} \xrightarrow{\text{MeOB}{<}} \underset{\text{CH}_3}{R\!\!-\!\!C(H)\!\!=\!\!C\text{-B}{<}}$$

17

Replacement of tin by boron is a crucial step in the synthesis of the elusive aromatic heterocycle *B*-methylborepin (**18**) [39-40] and related compounds [41].

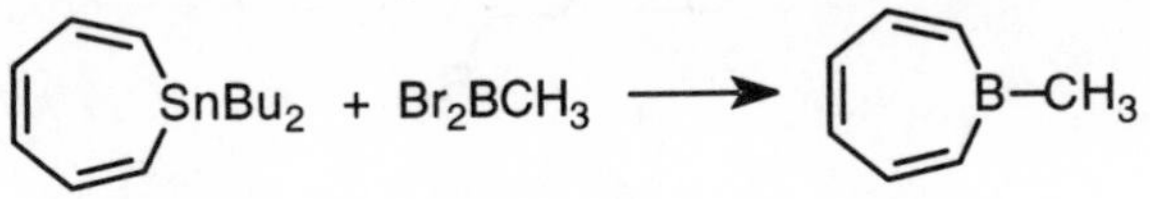

18

2.3 Hydroboration

2.3.1 General Considerations

The discovery of a practical method for adding diborane to double bonds by Brown and Subba Rao provided a revolutionary new route to organoboranes in 1956 [42]. This chemistry has been reviewed a number of times, and only a few of the historical highlights will be touched upon here.

The classical hydroboration is the reaction of B_2H_6 in diethyl ether or BH_3·THF with an alkene to form a trialkylborane (**19**), the most familiar reaction of which is with hydrogen peroxide to form the alcohol (**20**) [4, 43]. The reaction generally shows a very high preference for placing the boron atom on the less hindered carbon of the alkene. With 1-alkenes the reaction generally proceeds rapidly to form the trialkylborane (**19**), but internal alkenes usually form dialkylboranes, as for example in the conversion of 2-methyl-2-butene to "disiamylborane" (**21**). These are too hindered to react further with internal alkenes but will readily hydroborate 1-alkenes to form mixed trialkylboranes (**22**) in a controlled manner. Tetramethylethylene stops at the monoalkylborane "thexylborane" (**23**), which will react with two moles of 1-alkene to form a trialkylborane (**24**), or which with two judiciously chosen alkenes can be used to make a dialkylborane and finally a trialkylborane with three different alkyl groups.

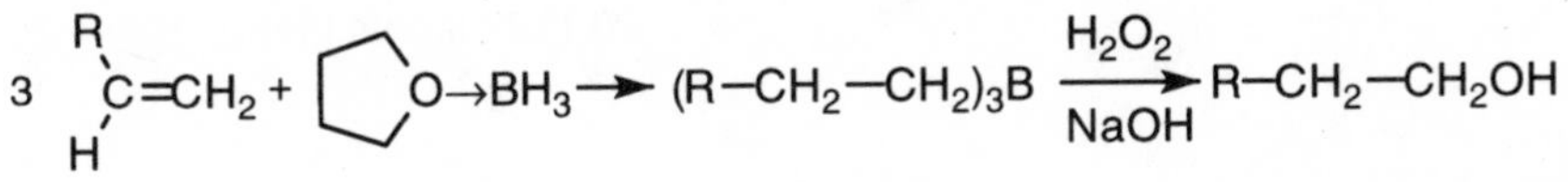

19   **20**

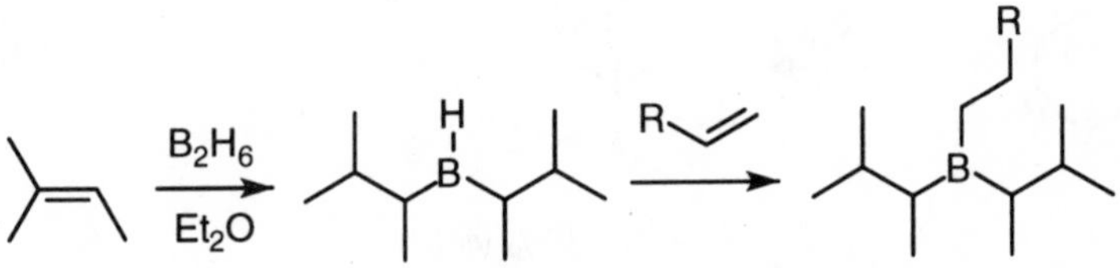

21   **22**

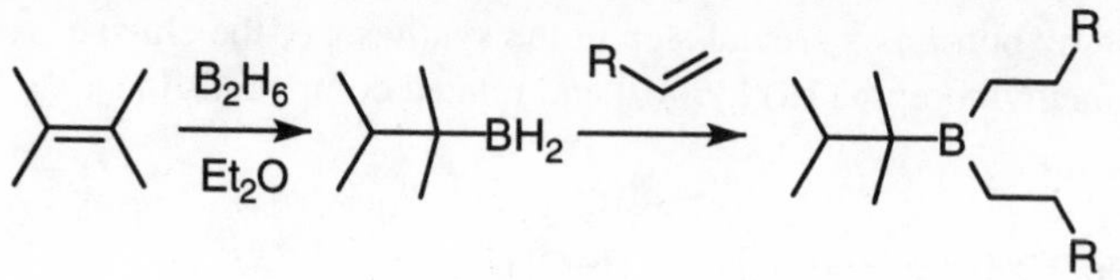

23   **24**

In general, the regiochemistry of hydroboration is determined primarily by steric factors. This rule is not infallible, and hydroboration/oxidation of styrene produces substantial proportions of 1-phenyl-1-borylethanes in competition with the major 1,2-isomer [43]. Enamines present a particularly interesting situation [44]. Unhindered enamines that coordinate well with BH_3 yield mainly α-boryl product such as **25**. More hindered enamines or boranes leave some free enamine at equilibrium in the solution, and the β-boryl products such as **26** and **27** result. However, if BF_3 is added to the hydroboration mixture, the α-boryl products analogous to **25** are formed from all enamines [45]. Rearrangement of α-boryl amines is facile and is discussed in Chapter 3.2.3.3.

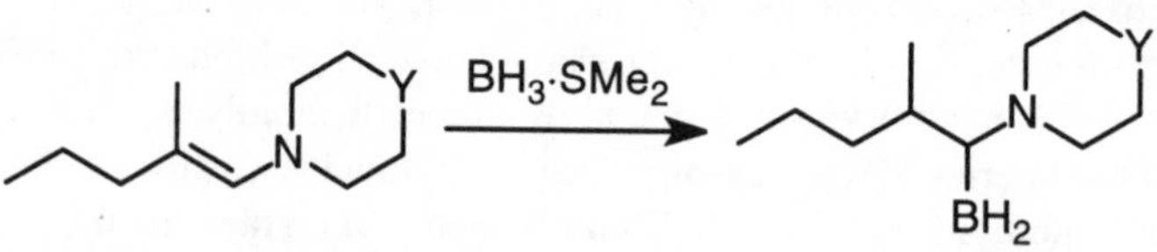

(Y = CH₂ or O)   **25** (80% this isomer)

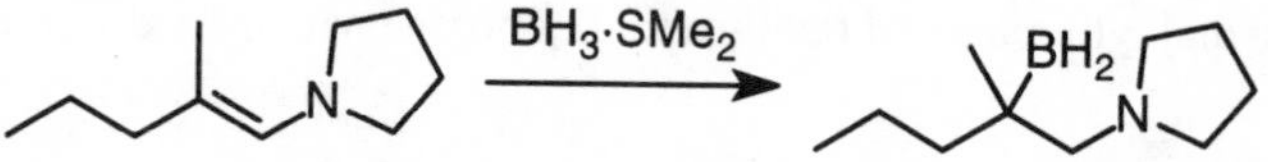

26 (100% this isomer)

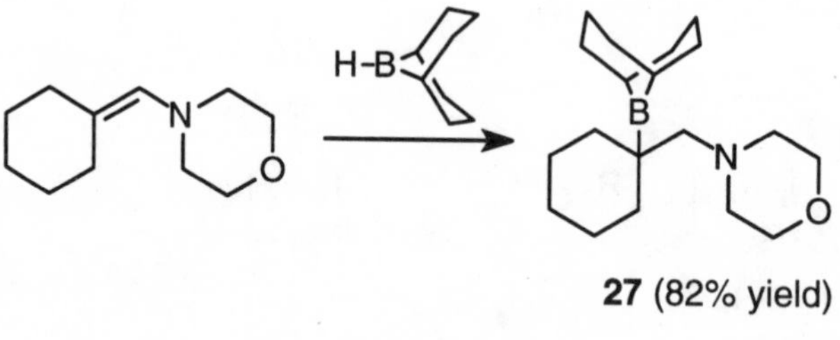

27 (82% yield)

2.3.2 Mechanism of Hydroboration

The essential step in hydroboration is the addition of a monomeric borane (28) to a carbon—carbon double bond via a four–center transition state (30) to form an alkylborane (31).

28   **29**   **30**   **31**

If the borane reagent is dimeric or complexed with a Lewis base, the first step is dissociation to free monomeric borane. As a consequence of the cyclic transition state, hydroborations are diastereospecific. Regioselectivity is usually dominated by steric repulsions. Electronic selectivity corresponds to addition as (imaginary) "$R_2B^+ H^-$", and often parallels but occasionally opposes steric effects.

Free BH_3, generated by thermolysis of BH_3—PF_3, hydroborates ethylene very rapidly in the gas phase, $\Delta H^* = \sim 8$ kJ mol^{-1}, $\Delta S^* = -55$ J deg^{-1} mol^{-1} [46]. The ΔS^* appears to be not as negative as required for the constrained transition state 30 and suggests borane—alkene π-complex 29.

The dissociation of borane dimers as the first step in hydroboration was discovered in kinetic studies with 9-borabicyclo[3.3.1]nonane ("9-BBN") dimer (32) [47], [48]. If the reaction of monomeric 9-BBN (33) with the alkene is very rapid, that is, $k_2 \gg k_{-1}$, then the reaction rate is first-order in 9-BBN dimer and zero-order in the alkene. If dimerization of the monomer 33 to 32 is relatively rapid, that is, $k_{-1} \gg k_2$, as it is in the case of the cyclohexene example illustrated, then the rate of formation of 34 becomes half-order in 9-BBN dimer and first-order in alkene.

32   **33**   **34**

Rate constants first-order in 32 with 1-hexene, 2-methyl-1-pentene, 3,3-dimethyl-1-butene, and cyclopentene at 25 °C are all within the range 1.45-1.54 x 10^{-4} s^{-1}, and cyclohexene yielded a 3/2-order $k = 3.1$ x 10^{-5} M$^{-0.5}$ s^{-1}, all in carbon tetrachloride [49]. In THF the rates for the first three substrates were an order of magnitude faster, 1.37-1.40 x 10^{-3} s^{-1}, and cyclopentene was outside the range where $k_2 \gg k_{-1}$ and showed a composite rate law.

Borinane dimer (35) is a faster hydroborating agent than 32. With all but the least hindered alkenes in heptane, the rate law is half-order in 35 and first-order in alkene, again indicating dissociation to monomeric borinane (36) as the first step in hydroboration [50].

35   **36**

This mechanistic interpretation appears to be consistent with the thermodynamic data tabulated in Chapter 1.2.3. If a free borane **33** or **36** is an intermediate, ΔH^* for hydroboration cannot be less than $\Delta H°$ of the reaction that produces **33** or **36**. The observed $k = 7.5 \times 10^{-5}$ s^{-1} ($0.5k$ for hydroboration of alkenes) for dissociation of **32** at 25 °C corresponds to $\Delta G^* = 96.6$ kJ mol^{-1}. The data listed require that for dissociation of **35**, $k > 2 \times 10^{-3}$ s^{-1} at 0 °C, $\Delta G^* < 81$ kJ mol^{-1}. The true k was described as too fast to measure accurately.

Thermodynamic calculations are available for the dissociation of $(CH_3)_2B(H_2)B(CH_3)_2$ to 2 $(CH_3)_2BH$, $\Delta H° = +107.5$ kJ mol^{-1} (Chapter 1.2.3). In order to compare this with the observed ΔG^* values, a reasonable estimate of the ΔS^* is required. A reasonable model is $Al_2(CH_3)_6 \rightarrow 2 Al(CH_3)_3$, experimental $\Delta S^* = +63$ (±17) J deg^{-1} mol^{-1} in solution [51]. Combined with the $\Delta H°$ value, this suggests $\Delta G^* = 88.7$ kJ mol^{-1} for dissociation of dimethylborane dimer, assuming a small ΔH^* for the reverse reaction, dimerization of Me_2BH. If the heats of dissociation are in the order $32 > (Me_2BH)_2 > 35$, apparently consistent with probable steric strains in the various species, then the entire scheme appears to be self-consistent.

Acceleration of hydroboration by ethers is also consistent with the intermediacy of monomeric borane. For example, the reaction $(CH_3)_2B(H_2)B(CH_3)_2 + (CH_3)_2O \rightarrow (CH_3)_2O—HB(CH_3)_2 + (CH_3)_2BH$ has an estimated $\Delta H° = +77.0$ kJ mol^{-1} based on linear interpolation between values for $(CH_3)_2O—BH_3$ and $(CH_3)_2O—B(CH_3)_3$ (Chapter 1.2.3). In contrast to dimer dissociation, ΔH^* for displacement of borane by ether must be significantly above $\Delta H°$ and ΔS^* should be somewhat negative, but these considerations are in line with the observed $\Delta G^* = 91.1$ kJ mol^{-1} for the THF assisted cleavage of $(9\text{-BBN})_2$ (**32**).

Hydroborations of common alkenes by B_2H_6 in ether at 0 °C are too fast to measure manometrically [43], implying $k \geq 10^{-2}$ s^{-1}, $\Delta G^* \leq 77$ kJ mol^{-1}. Such a low ΔG^* requires an indirect pathway for generating one BH_3 at a time from 1/2 B_2H_6, $\Delta H° = +81.15$ kJ mol^{-1}, $\Delta S°$ approximately +31 kJ deg^{-1} mol^{-1}, $\Delta G^* \geq 73$ kJ mol^{-1}, just within the limit set by experimental observation. Model reaction 1/2 B_2H_6 + Me_2O Æ $Me_2O—BH_3$ has $\Delta H° = +25$ kJ mol^{-1}. This reaction coupled with $Me_2O—BH_3 \rightarrow Me_2O + BH_3$ sums to 1/2 $B_2H_6 \rightarrow BH_3$. The unknown route to $R_2O—BH_3$ might involve some sort of reactive borane species, possibly in a chain reaction. Energy considerations rule out pathways such as $B_2H_6 \rightarrow 2 BH_3$, $\Delta H° = +162.3$ kJ mol^{-1}, or $B_2H_6 + Me_2O \rightarrow Me_2O—BH_3 + BH_3$, $\Delta H° = +106.1$ kJ mol^{-1}.

Diborane in THF is converted largely to BH_3·THF [43]. Dissociation of this complex must be more endothermic than dissociation of 1/2 B_2H_6, but evidently the difference is not large, as BH_3·THF is a rapid hydroborating agent. In contrast, amine boranes are sluggish hydroborating agents, consistent with the estimated $\Delta H° = +119.2$ kJ mol^{-1} for dissociation of $BH_3—NH_3$ (Chapter 1.2.3). Hydroborations by

BH_3—NR_3 or BH_3—SMe_2 are inhibited by excess amine or dimethyl sulfide, consistent with free BH_3 being the reactive intermediate [52]. Dissociation of $HBBr_2 \cdot SMe_2$ is aided by a few per cent of BBr_3, which preferentially binds dimethyl sulfide and liberates the active hydroborating species, $HBBr_2$ [53].

Halogens direct the boron β to the halogen in alkenes [54], but not in alkynes [55]. In either case, the rate is slowed by two orders of magnitude and follows the 3/2-order kinetics, half-order in $(9\text{-BBN})_2$ and first-order in alkene.

2.3.3 Hydroborating Agents

2.3.3.1 Borane Sources. Where BH_3 is desired as a hydroborating agent, the most convenient reagents are $BH_3 \cdot THF$ and $BH_3 \cdot SMe_2$, both of which are commercially available, or the somewhat less obnoxious smelling borane 1,4-thioxane [56]. Stereo- and regioselectivity for addition of BH_3 to double bonds will generally be somewhat less than for additions of various RBH_2 or R_2BH. For stereoselective synthesis, the main utility of the BH_3 reagents is in the synthesis of more selective borane derivatives. Since the chemistry of such hydroborations has been exhaustively reviewed elsewhere [43], it will not be treated further here.

2.3.3.2 Alkylborane reagents. Boranes used for hydroborations will be divided into two general categories, alkyl substituted and heteroatom substituted. The classic approach to making monoalkyl- or dialkylboranes required a choice of alkyl group such that steric factors would allow controlled formation of RBH_2 or R_2BH from an olefin and a BH_3 source. Often used examples include "thexylborane" (**37**) from hydroboration of 2,3-dimethyl-2-butene, "disiamylborane" (**38**) from 2-methyl-2-butene, and dicyclohexylborane (**39**) from cyclohexene [43]. These compounds generally exist as dimers in ethereal solutions.

37   **38**   **39**

Perhaps the most generally useful reagent of this class is "9-BBN" (9-borabicyclo[3.3.1]nonane) (**33**), the dimer (**32**) of which is a stable commercially available reagent derived from hydroboration of 1,5-cyclooctadiene (**40**). Those who want to make the reagent themselves may use the recent optimized *Organic Syntheses* preparation [57]. For most purposes, 9-BBN provides nearly optimum stereoselectivity for the least hindered olefinic site, and with only a few exceptions, subsequent B—C bond reactions proceed fastest at the exocyclic B—C bond, leaving the 9-BBN unit intact.

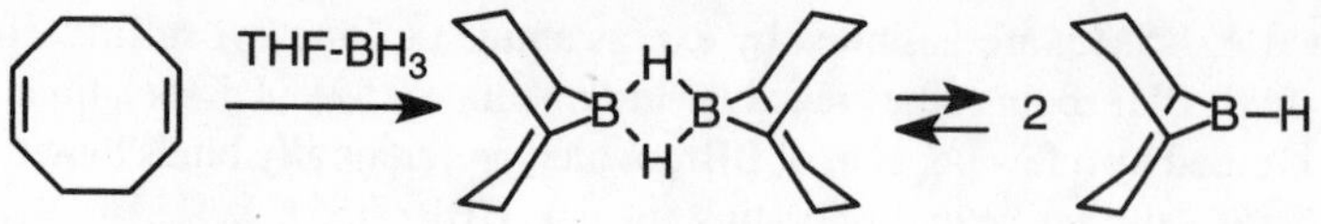

Dimesitylborane (**41**) provides sufficiently long range steric interactions to hydroborate 2-hexyne with 98% regioselectivity at C-2 [58]. Other hydroborating agents generally discriminate poorly between the methyl group and longer alkyl chains.

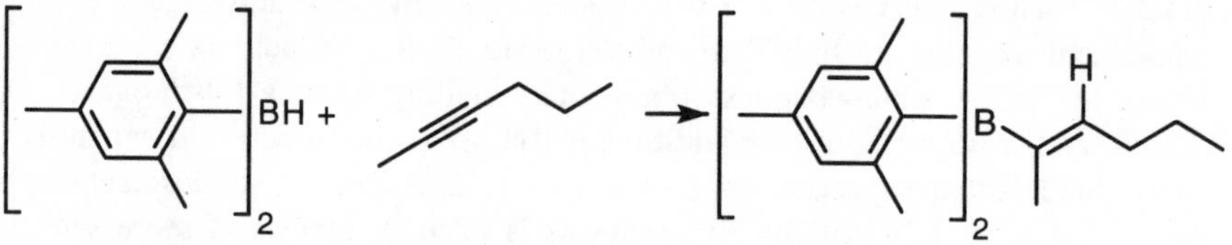

Other alkylboranes have been used for special purposes. For example, bis(2-bicyclo[2.2.2]octyl)borane with internal alkenes yields trialkylboranes which isomerize particularly easily to primary alkylboranes [59-61]. Recent developments in techniques have allowed the straightforward preparation of methylborane, which has been studied as a hydroborating agent [62]. Many other mono- and dialkyl-borane reagents could be catalogued from the literature, but these are generally peripheral to the pursuit of stereocontrolled synthesis.

2.3.3.3 Oxygen substituted hydroborating agents. All oxygen substituted boranes are relatively sluggish hydroborating agents, presumably because π-bonding to oxygen partially fills the vacant p-orbital on boron and thus diminishes its Lewis acidity. Dialkoxyboranes, $(RO)_2BH$, are unstable toward disproportionation, but several cyclic dialkoxyboranes have been used successfully. The most generally used reagent in this class is catecholborane (**42**), which is especially effective for hydroborating terminal acetylenes or terminal alkenes, usually at temperatures in the 65-100 °C range [63, 64, 65, 66, 9].

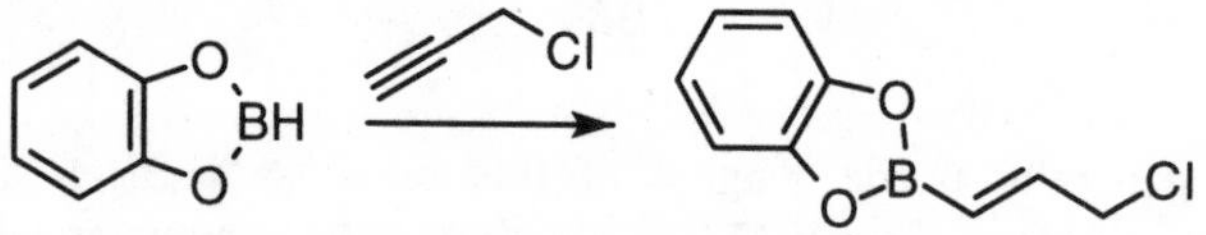

Pinacolborane (**43**) has recently been found to be a faster hydroborating agent than catecholborane [67]. The reagent is prepared from pinacol and borane—dimethyl sulfide in dichloromethane at 0-25 °C. Two equivalents of the reagent used in situ are required for complete hydroboration, though the 63% yield obtained when the reagent was isolated leaves the meaning of this finding ambiguous. Typical examples are illustrated. Product **44** was found to contain 2% (Z)-isomer and 3% regioisomer, and other 1-alkynes generally yielded (E)-1-alkenylboronic esters in even higher purity. With phenylacetylene, there was approximately 4% (Z)-isomer (contrasting with 9% when catecholborane was used). The 93:7 ratio of **45** to **46** was much higher than the 60:40 ratio of analogous products found with catecholborane. The preparation of norbornylboronic ester **47**, like other hydroborations of norbornene, yields the *exo* product with 99% stereoselectivity. The use of alkenes in this procedure was examined only briefly, but it appears that hydroboration generally proceeds smoothly.

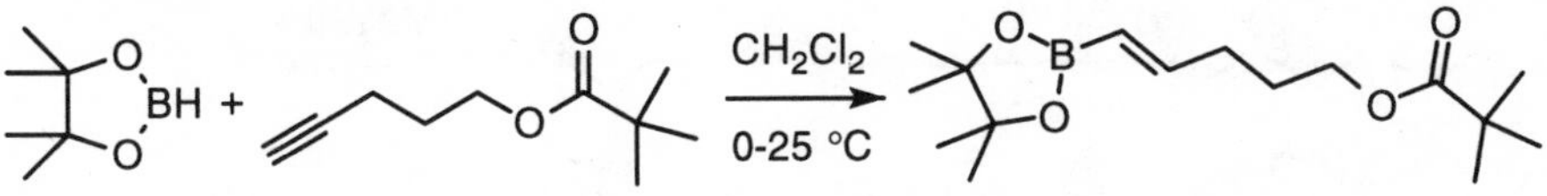

43   **44**

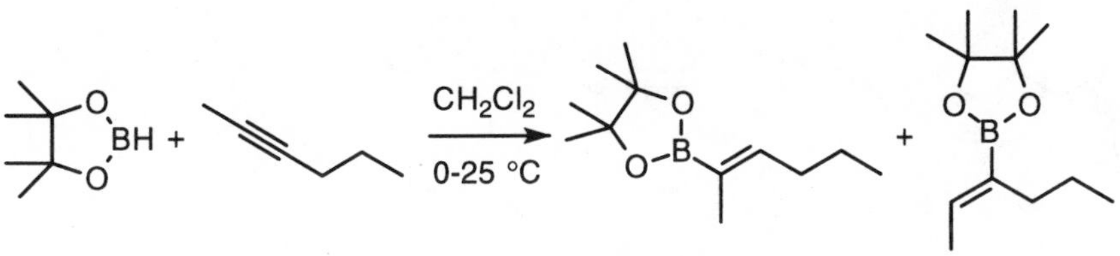

43   **45** (93%)   **46** (7%)

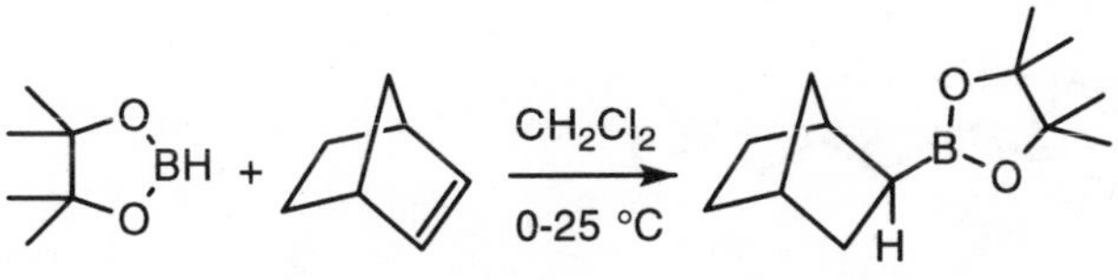

43   **47**

More hindered boranes such as pinanediolborane (**48**) [68] and 4,4,6-trimethyl-1,3,2-dioxaborinane (**49**) [69] require higher temperatures and do not generally give satisfactory yields. An alternative to the oxyboranes which hydroborates alkynes at 50 °C is 1,3,2-dithiaborolane (**50**) [70].

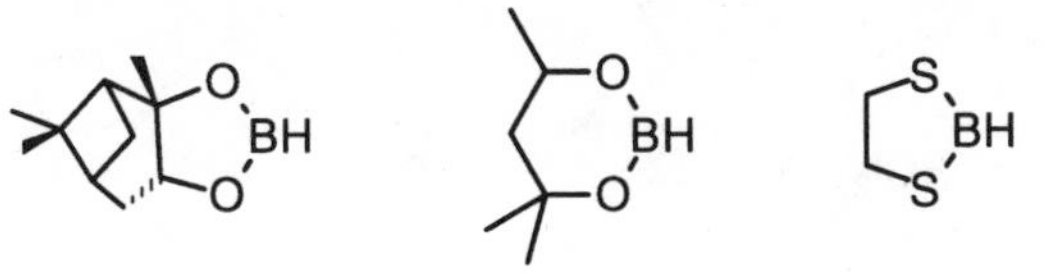

48   **49**   **50**

2.3.3.4 Haloboranes. A particularly useful heteroatom substituted hydroborating agent for alkenes and alkynes not bearing oxygen substituents is solvent free dichloroborane, which is easily generated in situ by adding a mixture of trialkylsilane and alkene or alkyne to boron trichloride [71]. In accord with the mechanism of hydroboration discussed in Section 2.3.2, the solvent free borane is an extremely rapid hydroborating agent. The reaction of 1-hexene with $BHCl_2$ is rapid even below −78 °C, though in general the temperature does not need to be lowered more than enough to keep the boron trichloride liquid (12 °C). The first product of the reaction is an alkyldichloroborane, which can be treated with a different alkene and a second mole of trialkylsilane to form a mixed chlorodialkylborane. An example is the sequence illustrated, in which 1-hexene is converted first to 1-hexyldichloroborane (**51**) and then to (1-hexyl)(cyclohexyl)chloroborane (**52**).

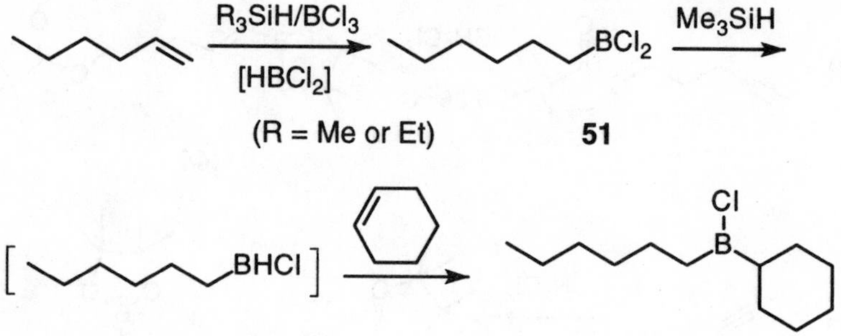

Another simple example is hydroboration of allyl chloride to (3-chloro-propyl)dichloroborane (**53**).

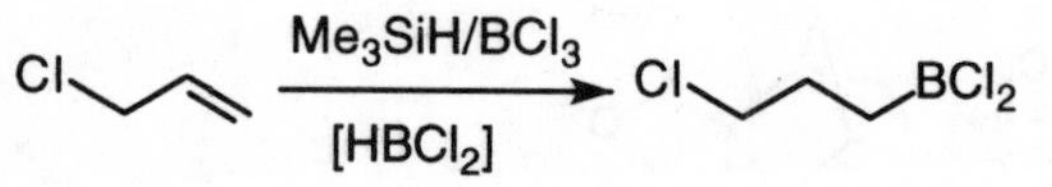

Depending on the ratio of reactants, hydroboration of (+)-α-pinene yields either (isopinocampheyl)dichloroborane (**54**) or (diisopinocampheyl)chloroborane (**55**). The reaction can be controlled cleanly at either stage [71].

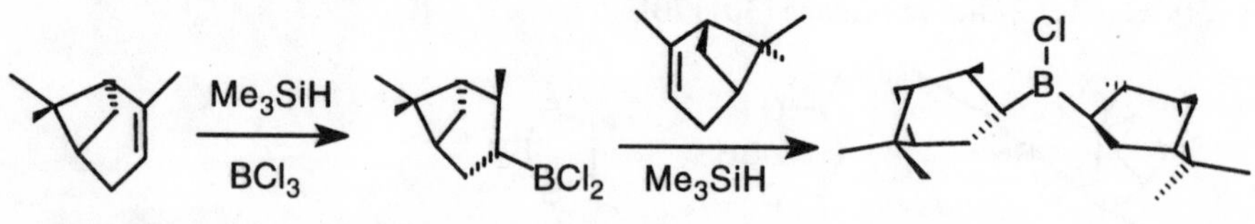

The hydroboration of 1-hexyne proceeds stepwise and can easily be controlled to yield either 1-hexenyldichloroborane (**56**) or 1,1-bis(dichloroboryl)hexane (**57**). The latter was converted to the known 1,1-bis[2-(1,3,2-dioxaborinyl)]hexane (**58**)

for positive identification, and it was found that the product contained 2-3% of the isomer, 1,2-bis[2-(1,3,2-dioxaborinyl)]hexane.

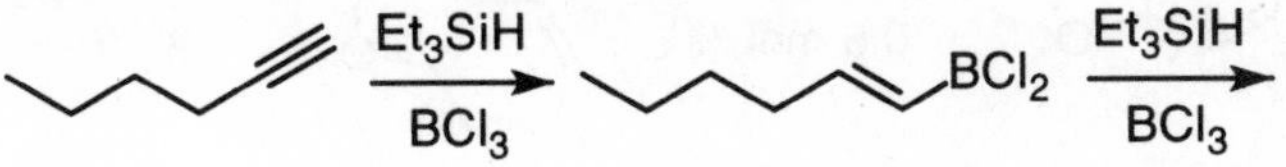

56

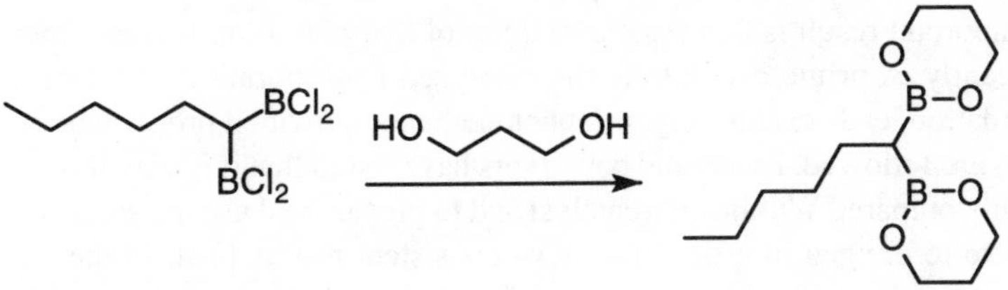

57   **58**

Several of the foregoing processes were duplicated with tribromoborane in place of trichloroborane [71]. It is hard to imagine circumstances under which the storable dimethyl sulfide complex of dibromoborane would work better than solvent free dibromoborane freshly generated from R_3SiH and BBr_3, but if such a need were to arise, it has been noted that a small amount of boron tribromide effectively catalyzes hydroborations by $Me_2S \cdot BHBr_2$ [72]. The catalysis works because BBr_3 complexes Me_2S liberated from dissociation of $Me_2S \cdot BHBr_2$, thus increasing the concentration of the active hydroborating agent, free $HBBr_2$. The initial hydroboration product is not free alkylboron dibromide but its dimethyl sulfide complex [72].

2.3.4 Catalyzed Hydroborations

Although catalysis of a reaction as rapid as hydroboration might seem superfluous if not impossible, Männig and Nöth have inaugurated a new and useful branch of hydroboration chemistry with the observation that the slow reactions of catecholborane are greatly accelerated in the presence of Wilkinson's catalyst, $RhCl(PPh_3)_3$ [73]. One important result of this acceleration is that hydroboration of carbon—carbon double bonds is promoted over competing reduction of carbonyl groups. For example, 5-hexen-2-one is reduced by catecholborane to 5-hexen-2-ol catecholborate (**59**) in the absence of catalyst, but in the presence of 0.5% $RhCl(PPh_3)_3$ yields 83% hydroboration product (**60**) and 17% (**59**) [73].

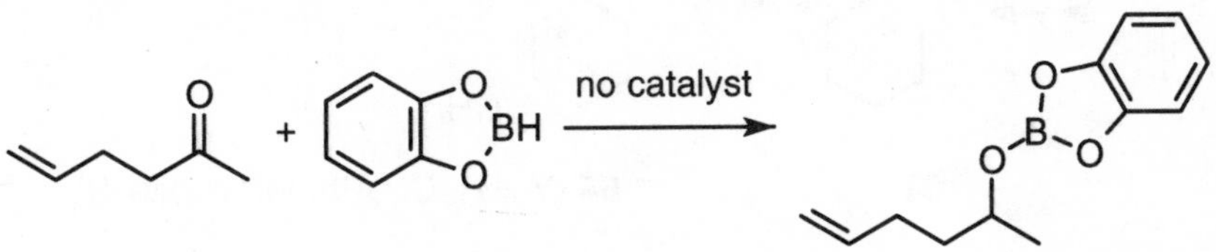

59

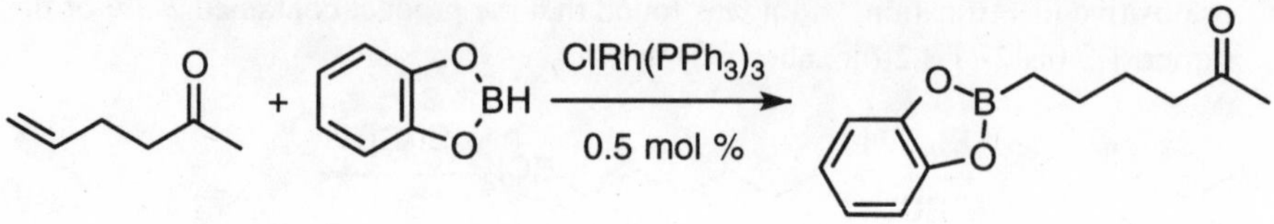

60

A second important result is that regioselectivity of hydroboration is sometimes altered significantly. A prime example is the catalyzed hydroboration of styrene, which after oxidation leads exclusively to 1-phenylethanol, provided proper laboratory procedures are followed. Evans and coworkers have found that it is very important to use freshly prepared Wilkinson's catalyst and to prepare and use the solutions without exposure to oxygen in order to achieve consistent results [74]. In the absence of oxygen exposure, the hydroboration of styrene with catecholborane-*d* in the presence of RhCl(PPh$_3$)$_3$ yields only the (1-phenyl-2-deuteroethyl)boronic ester **61**, which was not isolated but oxidized in situ to 1-phenylethanol-2*d* (**62**). If the catalyst has been stored for some time and consequently exposed to oxygen, or if oxygen is deliberately introduced, then a mixture of 1- and 2-phenylethylboronic esters with considerable deuterium scrambling results [74]. Excess triphenylphosphine can reverse the effect of oxygen at least to some extent [75-76].

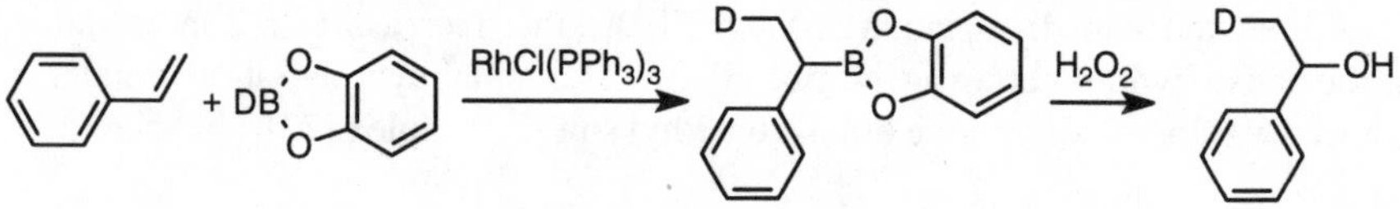

61   **62**

Another useful change of regioselectivity occurs with cyclohexenol derivatives. Uncatalyzed hydroboration with 9-BBN and oxidation with hydrogen peroxide yields mainly the *trans*-1,2-cyclohexanediol derivatives **63** (68-83% of product) but catecholborane with 3 mol % of Wilkinson's catalyst leads to the *trans*-1,3-cyclohexanediol derivatives **64** (72-86% of product) [77-78].

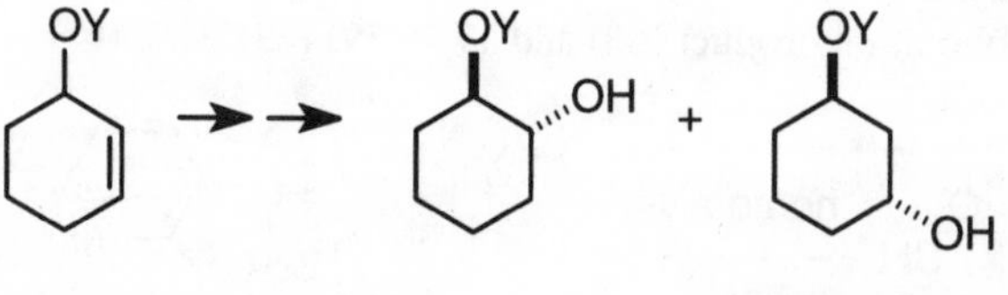

63   **64** (Y = H, CH$_2$Ph, *tert*-BuMe$_2$Si)

Wilkinson's catalyst has not proved successful in attempts to catalyze the hydroboration of alkenes or alkynes by pinacolborane [67].

This is an active research field, and the rate of obsolescence is very rapid. Work prior to 1991 has been reviewed elsewhere [79]. Applications relevant to asymmetric synthesis are discussed in Chapter 6.5.

Männig and Nöth cited the previous isolation of rhodium—boron compound **65** [80] and suggested a reasonable mechanism, which lacks quite a bit of detail but accounts for the gross features of the catalysis [73]. The X-ray structure of a compound closely related to **65** has been determined [81]. An updated version of the Männig-Nöth scheme involves the formation of a rhodium—boron bonded intermediate as **65**, which adds alkene to produce **66**. Rearrangement of **66** to an alkylrhodium compound **67** is followed by migration of the alkyl group from rhodium to boron accompanied by rhodium—boron bond cleavage to regenerate the catalytic intermediate, $ClRh(PPh_3)_2$, and produce the final product **68** [82].

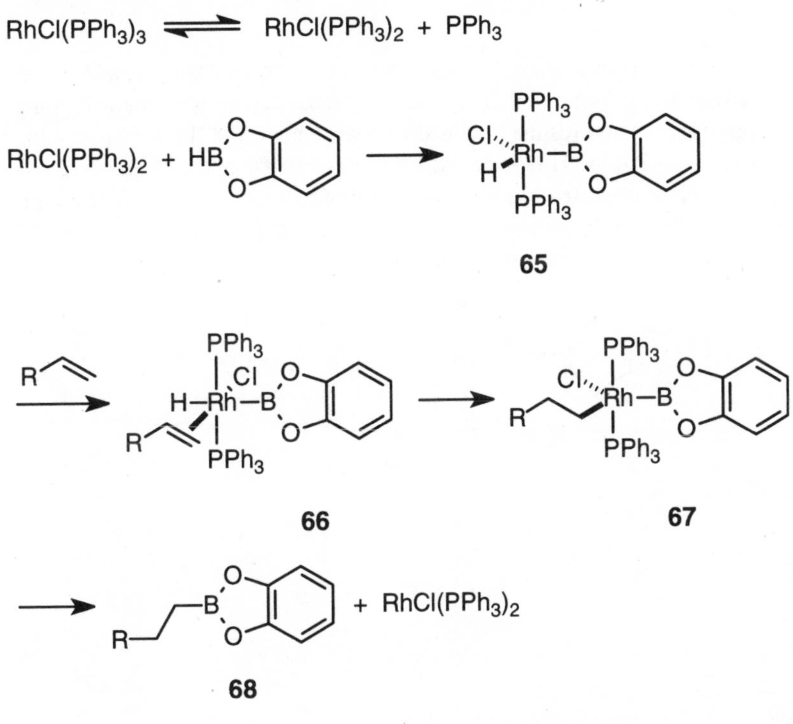

This mechanistic scheme is a useful mnemonic but leaves out a number of details that are important to anyone attempting to improve reaction conditions or design new catalytic systems. Although the deuteroboration of styrene to form **61** appears to be consistent with this simple mechanistic scheme, 1-decene with catecholborane-*d* in the presence of $RhCl(PPh_3)_3$ followed by hydrogen peroxide yields 1-decanol (**69**) having 85% of the deuterium label at C(2) but 15% at C(1) [74]. The Evans group suggested that there might be reversible interconversion of intermediates **66** and **67** of the Männig-Nöth scheme. If the catalyst is oxidized, more extensive scrambling results [74, 76].

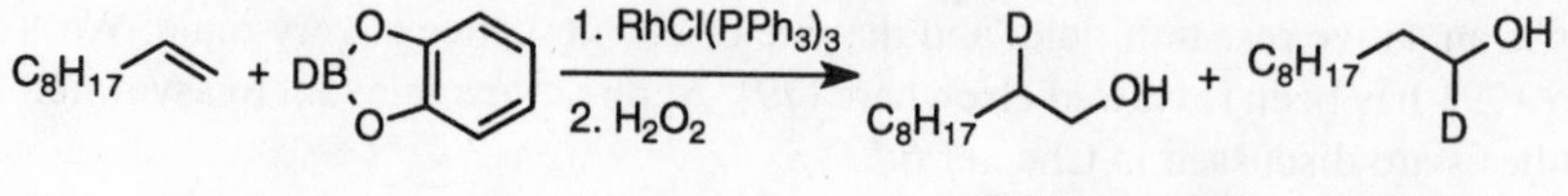

(85%)   **69**   (15%)

If deuterium scrambling occurs by interconversion of intermediates **66** and **67**, then it would be expected that reaction of 4-octene in a similar scheme should yield a mixture of octanols, but with Wilkinson's catalyst the only product is 4-octanol (**70**) [74]. The explanation suggested, that steric hindrance slows the isomerization, is contrary to usual experience with eliminations.

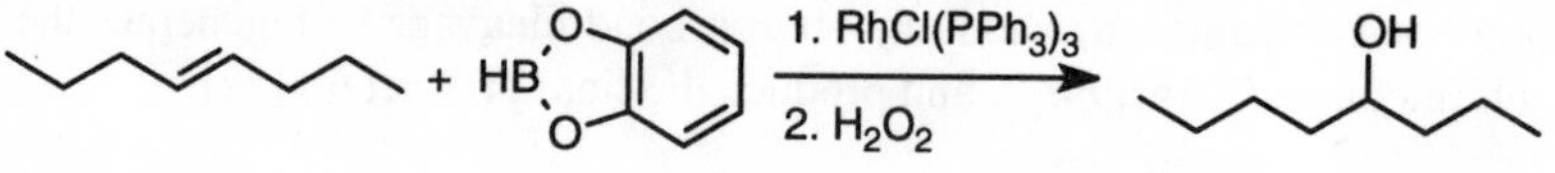

70

A collaborative effort from three laboratories led by Burgess, Marder, and Baker has uncovered several intricacies that are only in partial accord with the foregoing [82]. They agree that earlier results obtained by Evans and Fu [83] are reproducible under anaerobic conditions. However, Wilkinson's catalyst reacts with catecholborane to form a gross mixture of products in addition to **65**, including tris(catechol) diborate (**71**) and the two hydrogenation catalysts **72** and **73**. The structures of **72** and **73** were confirmed by X-ray crystallography.

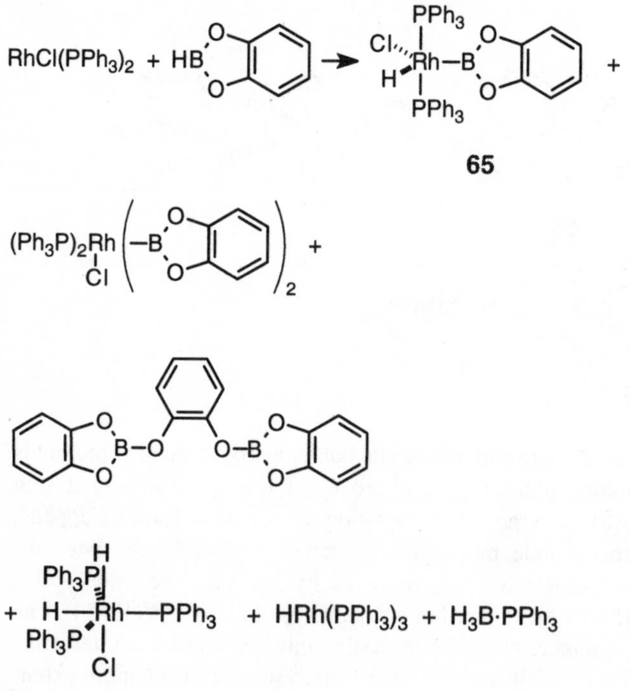

65

71   **72**   **73**

It was noted that some hydrogenation often accompanies catalyzed hydroboration [82]. A more surprising side reaction is the formation of vinylboronate esters, which was attributed to initial alkene insertion into the Rh—B bond of **74** to form **75** followed by Rh—H elimination to yield product **76**, which might undergo either reduction or hydroboration. Products isolated after treatment with pinacol included **78** and **79** in addition to the major product **77** [82].

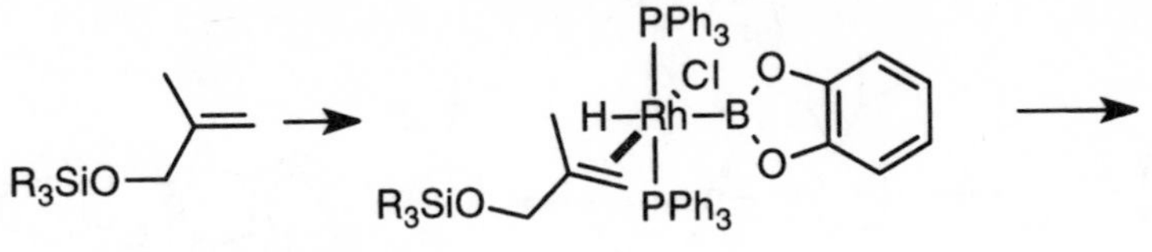

$(R_3Si = \textit{tert-}BuMe_2Si)$ **74**

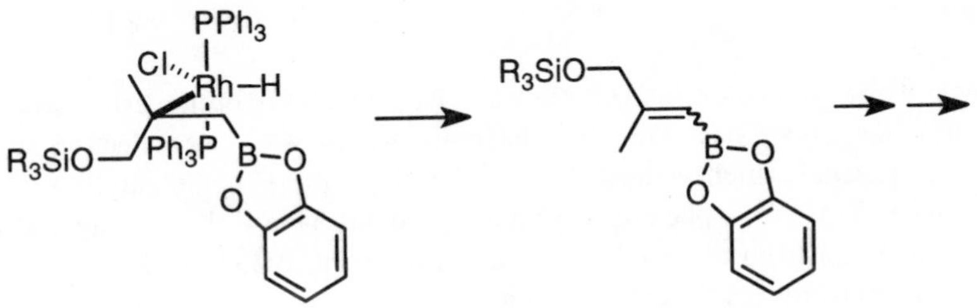

75   **76**

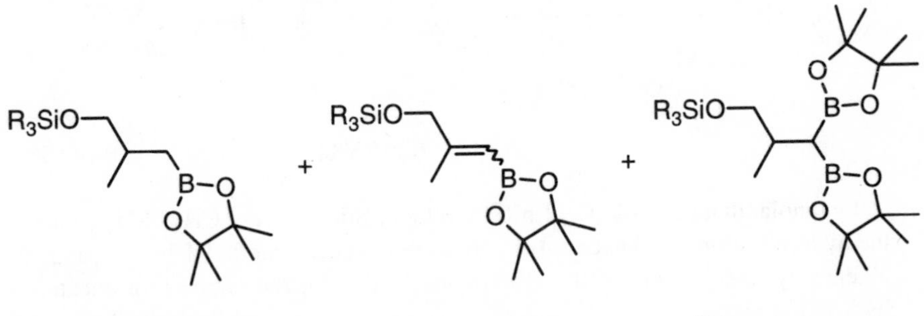

77 (from other pathways)   **78**   **79**

These results provide an obvious source of deuterium label scrambling, since deuteroboration with catecholborane-d would lead to some hydrogenation/deuteration of **76**, with deuterium labels going to either the α- or β-position. These results also account for the formation of some aldehyde among the hydroboration–oxidation products [82]. It was concluded that cleaner catalyst systems are needed in order to optimize catalyzed hydroborations, and that there is good reason to investigate alternative catalyst systems and more stable boron hydrides.

In a test of a series of rhodium complexes for catalytic activity, the π-allyl complex **81** showed particularly promising properties [84]. The COD—rhodium complex **80** is not a useful catalyst, but reacts readily with DiPPE [(i-Pr$_2$)PCH$_2$CH$_2$P(i-Pr)$_2$] to form **81**, which on treatment with catecholborane formed the stable zwitterionic derivative **82**. The catalytic activity of **81** and **82** proved to be the same, and it appears that **82** is the active catalyst. The rhodium in **82** forms an η^6-complex with the benzene ring and the valence shell formally has 18 electrons, but slippage of the benzene ring toward an η^4- or η^2-complex provides the vacant sites necessary for coordination of the metal to substrates [84].

80   **81**   **82**

Catalyst **81/82** yielded exclusively (>99%) (1-phenylalkyl)boronic esters with a variety of alkenylbenzenes. The only example which was less selective was isopropenylbenzene, which yielded 85% of the (1-phenyl-1-methylethyl)boronic ester **83** and 15% of the (2-phenylpropyl)boryl products **84** [84]. It was shown that the latter arose not from the catalyzed reaction but from BH$_3$·THF derived from catalytic decomposition of the catecholborane.

83 (85%)   **84** (15%)

Organolanthanides (Me$_5$C$_5$)$_2$LnR, Ln = La or Sm, R = H or CH(SiMe$_3$)$_2$, catalyze the hydroboration of 1-hexene, 1-methylcyclohexene, 2-methyl-1-butene, and other alkenes by catecholborane [85]. The regioselectivity is the same as for uncatalyzed hydroboration.

Asymmetric catalyzed hydroborations, including diastereoselective reactions, are discussed in Chapter 6.5, and a catalytic hydroboration of dienes that yields allylboronic esters is described in Chapter 7.2.2.4.

2.4 Haloborations

It has been known for half a century that acetylene reacts with boron trichloride in the gas phase over a mercurous chloride on charcoal catalyst to provide chlorovinylboron dichloride [ClCH=CHBCl$_2$] and bis(chlorovinylboron)chloride

[(ClCH=CH)₂BCl] [86-87]. Because of the ease of β-elimination of boron and chlorine, this reaction was not perceived as a useful source of organoboron compounds, though it has been reported that chlorovinylboranes can be hydrogenated to ethylboranes [88]. The reaction of boron tribromide with cyclohexene to form a mixture of 1-cyclohexenylboron dibromide and cyclohexylboron dibromide is another curiosity that has been known for some time [89, 90], as has the reaction of boron tribromide with acetylenes [91].

The utility of the haloboration of acetylenes for producing organoboranes has been demonstrated by Suzuki and coworkers [92]. In general, 1-alkynes with *B*-bromo-9-BBN (**85**) yield (*Z*)-2-bromoalkenylboranes (**86**) [93-96]. Boron tribromide adds similarly, but acetylene itself yields (*E*)-2-bromoethenylboron dibromide (**87**) [97].

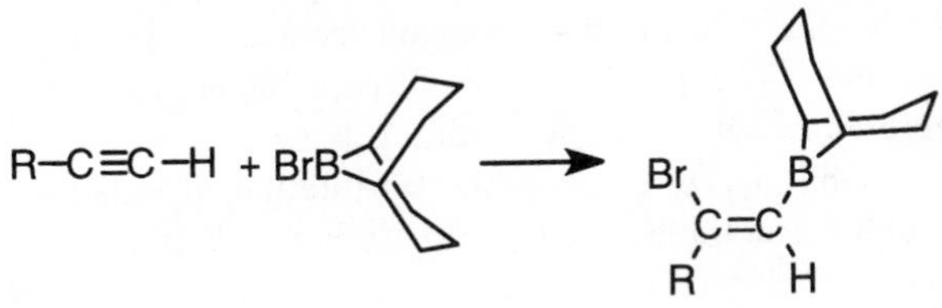

R—C≡C—H + BrB⟨⟩ ⟶ structure **86**

85   **86**

H—C≡C—H —BBr₃→ structure **87**

87

2.5 Other Routes to Carbon—Boron Bonds

Borane carbonyl, made from diborane and carbon monoxide [98], rearranges in water to hydroxymethylboronic acid [99] or in alcohols to esters of hydroxymethylboronic acid, usually isolated as the cyclic dimers **88** [100]. The yield of **88** based on diborane, which was the limiting reagent, was as high as 75%.

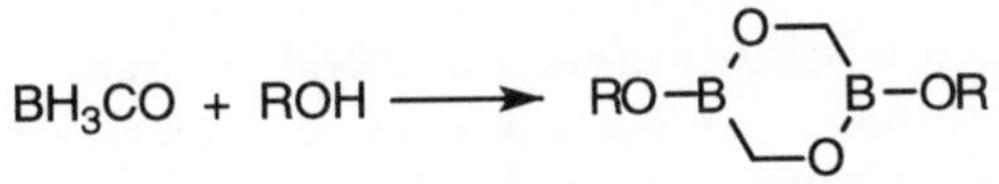

BH₃CO + ROH ⟶ structure **88**

88

Borane carbonyl can also be generated conveniently from borane dimethyl sulfide and carbon monoxide in the presence of a catalytic amount of lithium borohydride. Under these conditions, the carbonyl is rapidly reduced and the product isolated is trimethylboroxin (methylboronic anhydride) (89) [16].

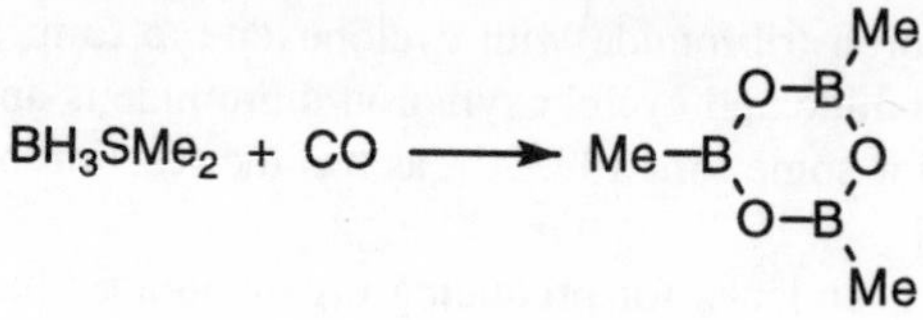

$$BH_3SMe_2 + CO \longrightarrow$$

89

Treatment of borane carbonyl with base yields salts of boranecarboxylate, $[BH_3CO_2]^{2-}$ [98]. Another borane carbonyl derivative is boraglycine, $^+H_3NBH_2CO_2^-$, which forms the basis for a whole branch of borane chemistry that has provided a variety of novel enzyme inhibitors [101], but it has no foreseeable applications in organic synthesis and therefore lies outside the scope of this book.

Aryl compounds offer possibilities for electrophilic substitution. Borylation of benzene and other arenes with boron trichloride to form arylboron dichlorides can be carried out in the presence of aluminum metal as a reducing agent [102-104]. Aryltrimethylsilanes can also be converted to arylboron dichlorides [105-107].

Diboron tetrachloride adds *syn* to multiple bonds to form the corresponding 1,2-diboryl derivatives For example, acetylene yields (Z)-1,2-bis(dichloroboryl)ethene (90) [108-110]. Cyclopentene forms *cis*-1,2-bis(dichloroboryl)cyclopentane (91) [111].

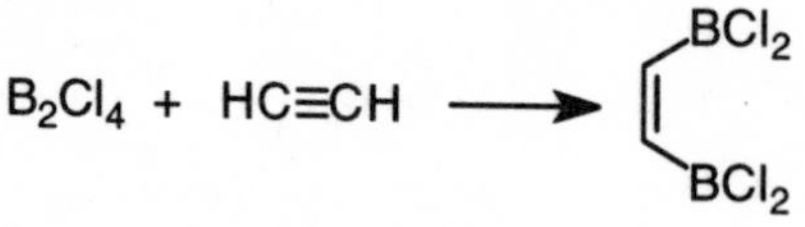

$$B_2Cl_4 + HC{\equiv}CH \longrightarrow$$

90

$$B_2Cl_4 +$$

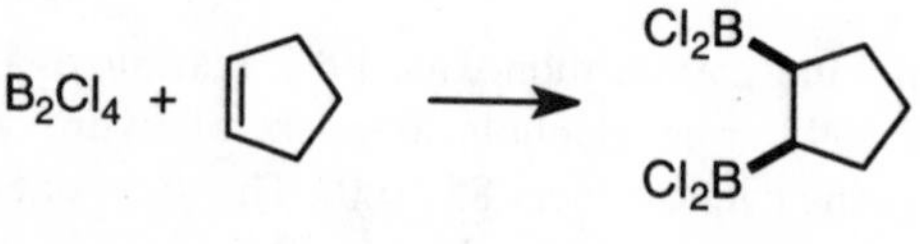

91

Diboron tetrachloride is made from degradation of boron trichloride in an electric discharge, and is not a practical source of boron compounds. Tetrakis(dimethylamino)diboron (93) is easily prepared on a large scale by the reaction of bis(dimethylamino)boron chloride (92) with sodium metal [112], but the boron—boron bond is inert. Treatment of 93 with methanol and hydrogen chloride yields tetramethoxydiboron (94), and other tetraalkoxydiborons can be made similarly [113]. These reagents are relatively inert and do not add to double or triple bonds.

$$(Me_2N)_2BCl + Na \longrightarrow (Me_2N)_2B{-}B(NMe_2)_2 \xrightarrow[HCl]{MeOH} (MeO)_2B{-}B(OMe)_2$$

92   **93**   **94**

In an important recent development, it has been found possible to catalyze the addition of bis(pinacol)diboron (**95**) to alkynes with tetrakis(triphenylphosphine)-platinum(0). A typical example is the addition of **95** to 4-octyne (**96**) to form exclusively (Z)-4,5-bis(4,4,5,5-tetramethyl-1,3,2-dioxaborol-2-yl)-4-octene (**97**) (86%) [114].

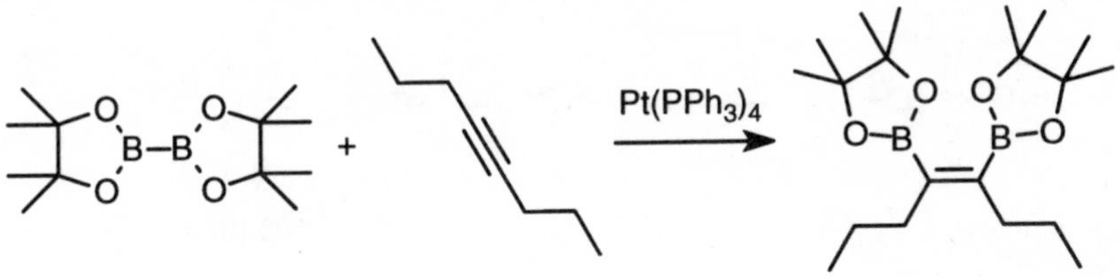

95   **96**   **97**

In another important recent development of fundamental significance, Imamoto and Hikosaka have found that a boron anion can be generated in the form of the boron analogue of a Wittig reagent [115]. Attempts to generate the desired anion **100** by deprotonation of tricyclohexylphosphineborane with butyllithium and potassium *tert*-butoxide resulted in deprotonation of a carbon α to phosphorus instead, which is in accord with the relative electronegativities of carbon and boron. Success was achieved when tricyclohexylphosphine—iodoborane (**98**) was treated with lithium 4,4'-di-*tert*-butylbiphenylide (**99**) in the presence of TMEDA. A threefold excess of benzaldehyde was required in order to obtain optimum results in the synthesis of **101**. Acetaldehyde reacted analogously, yields 47-53%. Benzophenone with **100** yielded the *p*-boryl substituted ketone **102**, the structure of which was confirmed by X-ray crystallography. The analogous *p*-boryl compound was formed as a side product (8%) in the reaction of **100** with benzaldehyde.

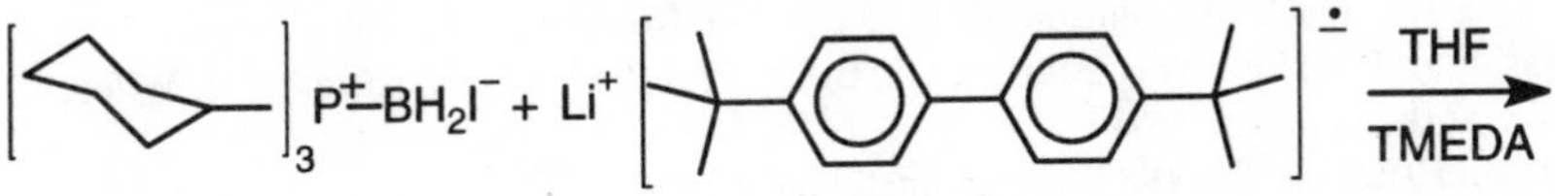

98   **99**

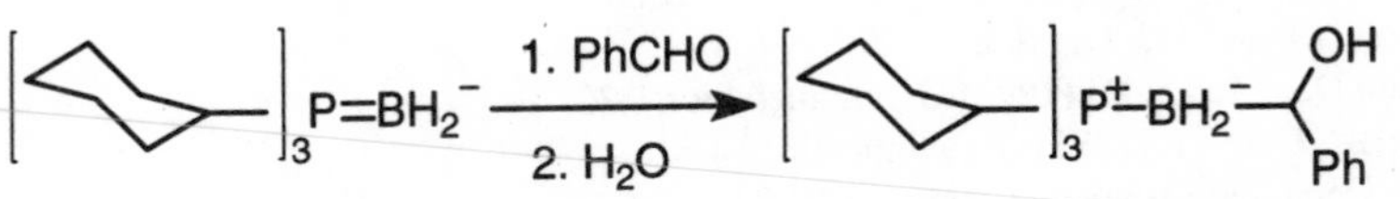

100   **101** (79%)

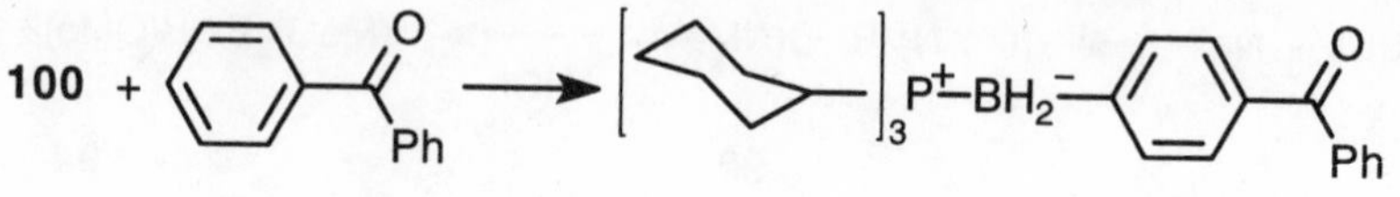

102

Alkylations of **100** required very reactive alkylating agents such as methyl triflate, which yielded **103**, for best results. Acylation by esters was also accomplished, as in the preparation of **104**.

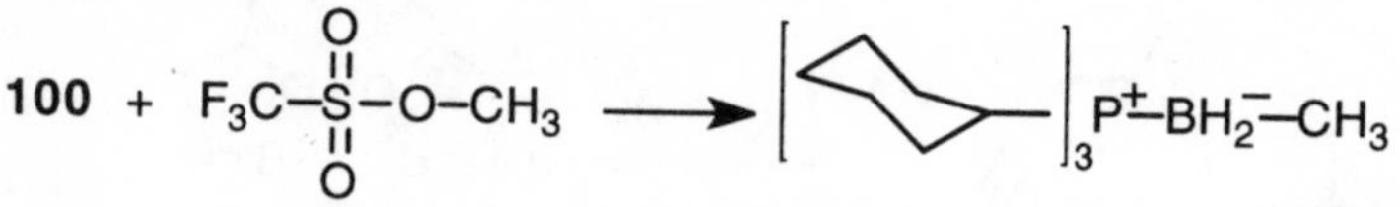

103 (86%)

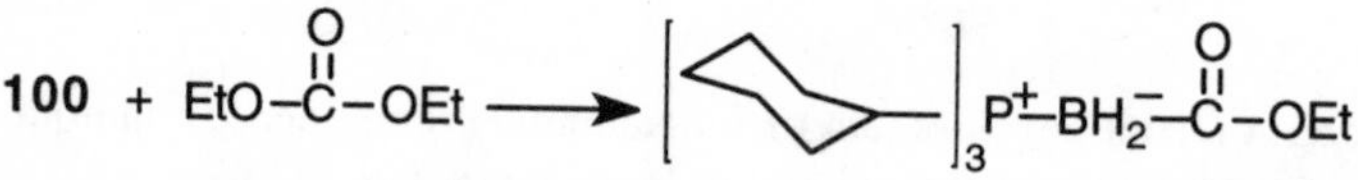

104 (98%)

2.6 References

1.   Doxey, K (1990) Borax Review (U. S. Borax Corporation) 8:16; see also "Glossary", contents page, Borax Review, Gordon, V, Ed., all issues
2.   Hughes RL, Smith IC, Lawless EW (1967) Production of the Boranes and Related Research, Holtzman RT, Ed, Academic Press, New York, pp 2-8, pp 10-52
3.   Lappert MF (1956) Chem. Rev. 56:959
4.   Brown HC (1962) Hydroboration, W. A. Benjamin, New York
5.   Frankland E, Duppa B (1860) Proc. Royal Soc. (London) 10:568
6.   Khotinsky E, Melamed M (1909) Chem. Ber. 42:3090
7.   Snyder HR, Kuck JA, Johnson, JR (1938) J. Am. Chem. Soc. 60:105
8.   Washburn RM, Levens E, Albright CF, Billig, FA (1963) Organic Syntheses, Wiley, New York, Coll. Vol. 4:68
9.   Brown HC, Cole TE (1983) Organometallics 2:1316
10. Rathke MW, Chao E, Wu G (1976) J. Organomet. Chem. 122:145
11. Matteson DS, Hurst GD (1986) Organometallics 5:1465
12. Matteson DS (1960) J. Am. Chem. Soc. 82:4228
13. Matteson DS, Peacock K (1963) J. Org. Chem. 28:369
14. Matteson DS (1964) J. Org. Chem. 29:3399
15. Matteson DS, Moody RJ (1982) Organometallics 1:20
16. Brown HC, Cole TE (1985) Organometallics 4:816
17. Aldrich Chemical Company Catalogue (1993)
18. Sadhu KM unpublished observation

19. Sadhu KM, Matteson DS (1985) Organometallics 4:1687
20. Michnick TJ, Matteson DS (1991) Synlett 631
21. Matteson DS, Hurst GD (1987) U.S. Patent 4,701,545, 20 Oct 1987, 6 pp.; (1988) Chem. Abstr. 109:93315p
22. Brown HC, Bhat NG, Srebnik M (1988) Tetrahedron Lett. 29:2631
23. Cole TE, Quintanilla R, Rodewald S (1991) Organometallics 10:3777
24. Niedenzu K, Dawson JW (1967) Inorg. Synth. 10:126
25. Nöth H, Vahrenkamp H (1968) J. Organomet. Chem. 11:399
26. Matteson DS, Cheng TC (1968) J. Org. Chem. 33:3055
27. Mikhail I, Kaufmann D (1990) J. Organomet. Chem. 398:53
28. Castle RB, Matteson DS (1968) J. Am. Chem. Soc. 90:2194
29. Castle RB, Matteson DS (1969) J. Organomet. Chem. 20:19
30. Matteson DS (1977) Gmelins Handbuch der Anorganischen Chemie, Supplement to 8th Ed., Vol. 48, Borverbindungen, part 16:37
31. Matteson DS, Mah RWH (1963) J. Org. Chem. 28:2171
32. Matteson DS (1962) J. Org. Chem. 27:275
33. Cole TE, Haly BD (1992) Organometallics 11:652
34. Brown HC, Cole TE, Srebnik M (1985) Organometallics 4:1788
35. Brown HC, Srebnik M, Cole TE (1986) Organometallics 5:2300
36. Brown HC, Vasumathi N, Joshi NN (1993) Organometallics 12:1058
37. Letsinger RL, Skoog, I (1954) J. Am. Chem. Soc. 76:4174
38. Negishi E, Boardman LD (1962) Tetrahedron Lett.. 23:3327
39. Nakadaira N, Sato R, Sakurai H (1987) Chem. Lett. (Jpn.) 1451
40. Ashe AJ III, Drone FJ, Kausch CM, Kroker J, Al-Taweel SM (1990) Pure Appl. Chem. 62:513
41. Ashe AJ III, Drone FJ (1987) J. Am. Chem. Soc. 109:1879; (correction) (1988) J. Am. Chem. Soc. 110:6599
42. Brown HC, Subba Rao BC (1956) J. Am. Chem. Soc. 78:5694
43. Brown HC (1972) Boranes in Organic Chemistry, Cornell University Press, Ithaca, New York, pp 227-446
44. Singaram B, Goralski CT, Fisher G (1991) J. Org. Chem. 56:5691
45. Fisher GB, Juarez-Brambila JJ, Goralski CT, Wipke WT, Singaram B (1993) J. Am. Chem. Soc. 115:440
46. Fehlner TP (1971) J. Am. Chem. Soc. 93:6366
47. Brown HC, Scouten CG, Wang KK (1979) J. Org. Chem. 44:2589
48. Brown HC, Wang KK, Scouten CG (1980) Proc. Natl. Acad. Sci. USA 77:698
49. Wang KK, Brown HC (1980) J. Org. Chem. 45:5303
50. Brown HC, Chandrasekharan J, Nelson DJ (1984) J. Am. Chem. Soc. 106:3768
51. Matteson DS (1974) Organometallic Reaction Mechanisms, Wiley, New York, p 39
52. Brown HC, Chandrasekharan J (1984) J. Am. Chem. Soc. 106:1863
53. Brown HC, Chandrasekharan J (1983) Organometallics 2:1261
54. Nelson DJ, Brown HC (1982) J. Am. Chem. Soc. 104:4907
55. Nelson DJ, Blue CD, Brown HC (1982) J. Am. Chem. Soc. 104:4913
56. Brown HC, Mandal AK (1992) J. Org. Chem. 57:4970
57. Soderquist JA, Negron A (1992) Org. Synth. 70:169
58. Pelter A, Singaram S, Brown HC (1983) Tetrahedron Lett. 24:1433
59. Brown HC, Racherla US, Taniguchi H (1981) J. Org. Chem. 46:4313
60. Brown HC, Racherla US (1982) Organometallics 1:765
61. Brown HC, Racherla US (1983) J. Organomet. Chem. 241:C37
62. Srebnik M, Cole TE, Brown HC (1990) J. Org. Chem., 55:5051

63. Brown HC, Gupta SK (1971) J. Am. Chem. Soc. 93:1816
64. Brown HC, Gupta SK (1972) J. Am. Chem. Soc. 94:4370
65. Brown HC, Gupta SK (1975) J. Am. Chem. Soc. 97:5249
66. Brown HC, Chandrasekharan J (1983) J. Org. Chem. 48:5080
67. Tucker CE, Davidson J, Knochel P (1992) J. Org. Chem. 57:3482
68. Matteson DS, Jesthi PK, Sadhu KM (1984) Organometallics 3:1284
69. Woods WG, Strong PL (1966) J. Am. Chem. Soc. 88:4667
70. Thaisrivongs S, Wuest JD (1977) J. Org. Chem. 42:3243
71. Soundararajan R, Matteson DS (1990) J. Org. Chem. 55:2274
72. Brown HC, Chandrasekharan J (1988) J. Org. Chem. 53:4811
73. Männig D, Nöth H (1985) Angew. Chem., Int. Ed. Engl. 24:878
74. Evans DA, Fu GC, Anderson BA (1992) J. Am. Chem. Soc. 114:6679
75. Zhang J, Lou B, Guo G, Dai L (1991) J. Org. Chem. 56:1670
76. Burgess K, van der Donk WA, Kook AM (1991) J. Org. Chem. 56:2949
77. Evans DA, Fu GC, Hoveyda AH (1988) J. Am. Chem. Soc. 110:6917
78. Evans DA, Fu GC, Hoveyda AH (1992) J. Am. Chem. Soc. 114:6671
79. Burgess K, Ohlmeyer MJ (1991) Chem. Rev. 91:1179
80. Kono H, Ito K (1975) Chem. Lett. 1095
81. Taylor NJ, Marder TB, Baker RT, Jones NJ (1991) J. Chem. Soc., Chem. Commun. 304
82. Burgess K, van der Donk WA, Westcott SA, Marder TB, Baker RT, Calabrese JC (1992) J. Am. Chem. Soc. 114:9350
83. Evans DA, Fu GC (1990) J. Org. Chem. 55:2280
84. Westcott SA, Blom HP, Marder TB, Baker RT (1992) J. Am. Chem. Soc. 114:8863
85. Harrison KN, Marks TJ (1992) J. Am. Chem. Soc. 114:9220
86. Arnold HR (1946) U. S. Patents 2,402,589, 2,402,590
87. Lazier WA, Salzberg PL (1946) U. S. Patent 2,402,591; Chem. Abstr. 40:5769
88. Hughes RL, Smith IC, Lawless EW (1967) Production of the Boranes and Related Research, Holtzmann, RT, Ed., Academic Press, New York
89. Willcockson GW (1962) U. S. Patent 3,060,218
90. Joy F, Lappert MF, Prokai B (1966) J. Organomet. Chem. 5:506
91. Lappert MF, Prokai B (1966) J. Organomet. Chem. 1:384.
92. Suzuki A (1986) Pure Appl. Chem. 58:629.
93. Hara S, Dojo H, Takinami S, Suzuki A (1983) Tetrahedron Lett. 731
94. Satoh Y, Serizawa H, Hara S, Suzuki A (1984) Synthetic Commun. 14:313
95. Hara S, Kato T, Shimizu H, Suzuki A (1985) Tetrahedron Lett. 26:1065
96. Satoh Y, Tayano T, Hara S, Suzuki A (1985) Synthesis 406
97. Hyuga S, Chiba Y, Yamashina N, Hara S, Suzuki A (1987) Chemistry Lett. 1757
98. Malone LJ, Parry RW (1967) Inorg. Chem. 6:817
99. Malone LJ, Manley MR (1967) Inorg. Chem. 6:2260
100. Malone, L. J. (1968) Inorg. Chem. 7:1039
101. Sood A, Sood CK, Spielvogel BF, Hall IH, Wong OT, Mittakanti M, Morse K (1991) Arch. Pharm. (Weinheim, Ger.) 324:423
102. Muetterties EL, Tebbe FN (1968) Inorg. Chem. 7:2663
103. Muetterties EL (1959) J. Am. Chem. Soc. 81:2597
104. Muetterties EL (1960) J. Am. Chem. Soc. 82:4163
105. Fleming I (1981) Chem. Soc. Rev. 10:83
106. Ishibashi H, Sakashita H, Ikeda M (1992) J. Chem. Soc., Perkin Trans. 1, 1953.
107. Bennetau B, Dunogues J (1993) Synlett 171
108. Rudolph RW (1967) J. Am. Chem. Soc. 89:4216
109. Zeldin M, Gatti AR, Wartik T (1967) J. Am. Chem. Soc. 89:4217

110. Coyle TD, Ritter JJ (1967) J. Am. Chem. Soc. 89:5739
111. Rosen A, Zeldin M (1971) J. Organomet. Chem. 31:319
112. Brotherton RJ, McCloskey AL, Petterson LL, Steinberg H (1960) J. Am. Chem. Soc. 82:6242
113. Brotherton RJ, McCloskey AL, Boone JL, Manasevit HM (1960) J. Am. Chem. Soc. 82:6245
114. Ishiyama T, Matsuda N, Miyaura N, Suzuki A (1993) J. Am. Chem. Soc. 115:11018
115. Imamoto T, Hikosaka T (1994) J. Org. Chem. 59:6753

3 General Reactions of Organoboranes

3.1 Introduction

This chapter provides a brief overview of organoborane reaction types. Reaction mechanisms are discussed, particularly with reference to whether they may or may not allow stereocontrol, but stereocontrolled applications are deferred to later chapters. Reactions that replace boron by atoms other than carbon are discussed first (Section 3.2), followed by boron substituted carbanions (Section 3.3) and reactions that have the net effect of replacing boron—carbon bonds by carbon—carbon bonds (Section 3.4). However, β-eliminations, which also create a carbon—carbon bond at the expense of a boron—carbon bond, are deferred to Section 4.2. Finally, a variety of reactions of organoboranes that leave the boron—carbon bond intact and affect other functionality are described in Section 3.5.

3.2 Oxidative Replacement of Boron

3.2.1 Introduction

Several oxidative boron—carbon bond cleavage reactions replace the boron atom stereospecifically with retention of configuration of the carbon atom, and therefore are particularly useful in asymmetric synthesis. These share the following general mechanistic features: (1) Coordination of an oxidizing agent having a nucleophilic site, $(:Y\!-\!Z)^{-n}$ ($n = 0$ or 1), to the boron to form a borate complex (**1**). (2) Intramolecular migration of carbon from the electron-rich boron atom to the relatively electron-deficient atom Y with concurrent displacement of nucleofuge Z^{-n}, as illustrated by transition state **2** and product **3**. (3) If the boron-linked atom of Y is oxygen or nitrogen, the product **3** will usually react with a weak proton source to form the deboronated product, R—Y—H, and boric acid or a derivative. The mechanistically analogous reactions where Y is carbon centered (Section 3.3) stop at **3**.

In addition to stereospecific replacements of boron with retention, there are some known halogenations with good stereoselectivity for inversion and a number of radical processes which are not inherently stereoselective.

$$X\underset{R}{\overset{X}{B}} + (:Y-Z)^{-n} \rightarrow X-\underset{R}{\overset{X}{B}}=\overset{\underset{Z}{|}}{Y}^{+1-n} \rightarrow \left[X-\overset{X}{\underset{R}{B}}\overset{Y}{\underset{\diagdown\diagup}{\cdots}}Z\right]^{-n} \rightarrow X-\overset{X}{\underset{R}{B}}-Y + Z^{-n} \rightarrow \underset{R}{YH}$$

1   **2**   **3**

(X = alkoxy, alkyl, halogen, etc.; R = alkyl, aryl; :Y—Z⁻, see text)

The order of presentation in this section will be (1) oxygen, (2) nitrogen, (3) halogen, (4) sulfur, (5) proton, (6) metal cations. Where reagents related to the foregoing can transform other functionality without disturbing the boron—carbon linkage, these will be covered in Section 3.5.

Electrophilic carbon can also replace boron oxidatively via transition states resembling **2**. These are among the most important stereocontrolled reactions of boron compounds, and are discussed separately in Section 3.4.

3.2.2 Oxygen Electrophiles

3.2.2.1 Hydrogen peroxide. Alkaline hydrogen peroxide is the most commonly used reagent for replacement of boron by oxygen with stereospecific retention of configuration. A wide variety of boranes react rapidly and quantitatively.

Oxidations of phenylboronic acid (**4**) [1] and butylboronic acid [2] by hydrogen peroxide were reported in the 1930s. The reaction of **4** was studied in detail by Kuivila in the 1950s. In acetate buffered aqueous solution, the rate is first-order each in **4**, OH⁻, and H_2O_2, which was interpreted as first-order each in **4** and OOH⁻, indicating that the reaction proceeds via intermediate **5** and transition state **6** to borate ester **7**, which hydrolyzes to the final products, phenol and inorganic borate [3]. The transition state **6** represents an intramolecular migration of carbon from electron-rich boron to electron-deficient oxygen.

4   **5**   **6**   **7**

A Hammett correlation of ten arylboronic acids showed essentially no substituent effect ($\rho = 0.07$), with all of the k_2 values near 3×10^{-3} $1\,mol^{-1}\,s^{-1}$ at pH 5.42 in aqueous 25% ethanol at 25 °C, the increased concentration of $ArB(OH)_2OOH^-$ with increasing acidity being just offset by the rate retardation by electron withdrawal for rearrangement of this intermediate [4].

In addition to the mechanism via **5**, there is a term second-order in **4** and first-order in OOH⁻, which corresponds to reaction via [PhB(OH)$_2$OOB(Ph)OH]⁻ in place of **5**, with the leaving group PhB(OH)O⁻ in place of OH⁻ [3]. If the concentration of **4** is greater than ~0.1 M and the pH is above 5, conditions likely to be used for synthetic purposes, extrapolation of the kinetic data indicates that this pathway will predominate. Weaker catalysis by boric acid was also observed [3]. In strong acid, pH <1, there is an acid catalyzed term, and there a neutral intermediate in the range between pH 1 and 3, which yields the minimal rate, k_2 <10⁻³ l mol⁻¹ s⁻¹ [5].

Interpreting the kinetic data from the synthetic chemist's point of view, the minimum rate extrapolates to a half-life of ~15 minutes for **4** in aqueous 1 M hydrogen peroxide at 25 °C at pH 1-3. The alkaline conditions generally used for synthetic purposes will generally result in complete deboronation in a fraction of a second. Trialkylboranes and dialkylborinates do not lend themselves to detailed kinetic study, but it may be presumed that their reactions with hydrogen peroxide are even faster. The exothermicity of the process has been noted in Section 1.3.1, Table 1-5. Reactions of sterically hindered boronic esters in less aqueous media do sometimes require substantial time for completion, and monitoring by thin layer chromatography is prudent practice for such substrates.

This mechanism clearly requires total retention of configuration at carbon, but in the absence of any sort of stereocontrolled boronic acid preparation, the potential synthetic utility was not appreciated. The significance of peroxidic deboronation was subsequently recognized when near stereospecificity was demonstrated by Brown and Zweifel in their landmark asymmetric hydroboration of *cis*-2-butene with the dialkylborane derived from (+)-α-pinene, which was followed by its oxidation to (*R*)-2-butanol [6]. Recent experimental evidence has shown that the degree of retention of configuration in the peroxidic deboronation of a secondary alkylboronic ester is >99.8% [7], which implies that there are no competing pathways in the system examined.

The rule that hydrogen peroxide replaces boron by a hydroxyl group regiospecifically and stereospecifically has become so well established that chemists can be caught unaware by the occasional exceptions. Two caveats follow. First, it is important to note that trialkylboranes require alkaline conditions for this process. With neutral hydrogen peroxide, alkyl radical dimers predominate in a gross mixture of products [8]. Second, even though boronic acids and esters usually give clean results at any pH, this expectation has sometimes failed with unsaturated boronic esters, even in strongly basic media. Sodium perborate has been found less prone to cause side reactions (Section 3.2.2.3), but it, too, can lead to anomalous rearrangements with certain structures.

In view of the anomalies just mentioned, there are bound to be a few errors in interpretation in the large body of literature on hydroboration in which it is taken for granted that the borane structures are accurately indicated by the structures of the alcohols obtained after peroxidic oxidation. The problem becomes particularly acute in situations where functionalized trialkylboranes have yielded mixtures of alcohols, sometimes including carbonyl compounds, and these may not always accurately reflect the structures of the borane precursors. Fortunately, the literature of this type may be largely ignored for purposes of understanding and designing stereocontrolled syntheses.

3.2.2.2 Migratory aptitudes. In order to estimate relative migratory aptitudes in deboronations, it is necessary to correct for the relative concentrations of the reactive intermediates $RB(OH)_2OOH^-$. After this correction for the acidity of the boronic acid, relative migratory aptitudes of organic groups from boron to oxygen in oxidations by hydrogen peroxide are *tert*-alkyl > *sec*-alkyl > *n*-alkyl ~ benzyl > vinyl ~ phenyl > CH_3 (Table 3-1) [9]. This rather unusual order of migratory aptitudes evidently reflects the considerable electron demand of oxygen as the destination atom. Analogous migrations where carbon is the destination atom have not been studied quantitatively, but qualitatively show relatively much more rapid migrations by phenyl, vinyl, and benzyl, with methyl apparently remaining slow (see Section 3.4 and Chapter 5).

Table 3-1. Migratory Aptitudes in Deboronation with Aqueous Hydrogen Peroxide at $H_0 = 5.23$, 25 °C [9]

Boronic Acid	$10^3 k_2$ $1\ mol^{-1}\ s^{-1}$	pK_a	Rel. k_2/K_a $MeB(OH)_2 = 1$
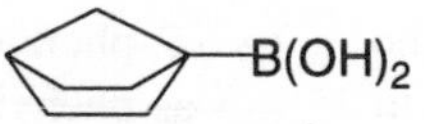 ⬠—$B(OH)_2$	110	10.52	715
$(CH_3)_3CB(OH)_2$	71.8	10.36	330
$C_2H_5CH(CH_3)B(OH)_2$	23.3	10.60	185
$CH_3(CH_2)_3B(OH)_2$	4.8	10.74	52
$PhCH_2B(OH)_2$	87.5	9.14	24
$H_2C=CHB(OH)_2$	6.8	9.49	4.2
$PhB(OH)_2$	16	8.86	2.3
$CH_3B(OH)_2$	0.127	10.60	1

3.2.2.3 Sodium perborate and other peroxides. Although the foregoing mechanism of peroxidic oxidation appears to be very general for all types of boranes, there are a few known instances in which unselective competing mechanisms do occur.

One example is the hydrogen peroxide oxidation of 2-phenylpropeneboronic acid (8), which yielded a mixture of B—C bond cleavage product 9 and C=C bond cleavage product 10, the latter reaching as much as 40% in weakly basic solutions buffered with bicarbonate [10]. Alkenylboronic acids that were unsubstituted at the 2-position yielded <5% C=C cleavage under similar conditions.

8   9   10

Recalling Kuivila's observation of the term second-order in PhB(OH)$_2$ (4) in the deboronation of 4 [3] suggested the use of sodium perborate in place of hydrogen peroxide. Sodium perborate caused less carbon—carbon bond cleavage under all conditions tested, and in strongly basic solution the amount of cleavage product was reduced to ~1-2% [10]. Furthermore, sodium perborate is a stable, inexpensive compound with a long shelf life and can be used in stoichiometric quantities without need for titration to determine active peroxide content, and is safer to handle than hydrogen peroxide. Also, it appears that sodium perborate can be used under less alkaline conditions than hydrogen peroxide, which results in improved yields if the products are sensitive to base. A disadvantage is that it is not very soluble in mixed aqueous solvents and react relatively slowly, sometimes inconveniently so.

Even sodium perborate has given an anomalous result in the oxidation of a borinic ester intermediate formed from 1,1-bis(1,3,2-dioxaborin-2-yl)-2-phenyl-ethene 11 and believed to be 12 [11]. The major product (up to 75%) was phenyl-acetone, and it was shown by NMR that the carbon skeleton did not rearrange prior to the oxidation step. The compound having a cyclohexyl group in place of the phenyl of 11 behaved similarly. These bizarre but well documented results should give pause to anyone who attempts to interpret a complex sequence of borane chemistry solely on the basis of the peroxidic oxidation products.

11   12 (not isolated)

The use of sodium perborate to oxidize trialkylboranes has been reported recently [12], and sodium percarbonate appears to be equally useful [13]. Sodium percarbonate has essentially the same advantages and limitations as sodium perborate.

Peracids also cleave B—C bonds [14]. With certain β-chlorocycloalkylboranes, *m*-chloroperbenzoic acid yields oxidative cleavage to β-chloro alcohols and base-catalyzed elimination is avoided [15, 16].

The molybdenum peroxide reagent MoO_5-py-HMPA oxidizes carbon-boron bonds efficiently [17] with complete retention of configuration at carbon [18].

3.2.2.4 Trimethylamine *N*-oxide. Anhydrous trimethylamine *N*-oxide is a relatively slow and selective oxidizing agent for trialkylboranes, and leads to alkoxyboranes without hydrolysis [19]. It is particularly useful for selectively cleaving an alkyl group in the presence of alkenyl groups [20, 21]. Hydroboration of silylacetylenes to silylated alkenylboranes followed by deboronation with trimethylamine oxide and hydrolysis yields acylsilanes [22]. A particularly interesting example is the conversion of $Me_3SiC \equiv CSiMe_3$ to $Me_3SiCH_2COSiMe_3$, a useful synthetic reagent [23]. Trimethylamine *N*-oxide with resolved α-methylbenzylboronic acid yields 1-phenylethanol with complete retention of configuration [24]. An improved preparation of pure anhydrous trimethylamine *N*-oxide uses dimethylformamide as the solvent [25].

Trimethylamine oxide dihydrate oxidizes trialkylboranes at relatively high temperatures. In refluxing THF, two of three alkyl groups react, and all three groups react in diglyme [26]. A single *n*-hexyl group can be cleaved from trihexylborane at 25 °C, but similar selectivity was not achieved with tricyclohexylborane [27]. In view of the simplicity of the improved dehydration process [25], there seems little reason to use the hydrate.

3.2.2.5 Molecular oxygen. The first controlled boron—carbon bond cleavage was the admission of oxygen to triethylborane sufficiently slowly to avoid ignition, resulting in formation of diethyl ethylboronate [28]. The mechanism involves radical intermediates, and resolved α-methylbenzylboronic acid yields racemic 1-phenylethanol [29]. Accordingly, the role of molecular oxygen in stereocontrolled synthesis with organoboranes is mainly a problem to be avoided.

Iodine is a highly effective inhibitor of the autoxidation of trialkylboranes [30]. Oxygen uptake from air by 0.5 M tributylborane in THF was delayed for a few minutes by 5 mol % of iodine, and more hindered boranes such as tri-*sec*-butylborane were protected from air for somewhat longer periods by 1 mol % of iodine. With 0.2 M iodine, initiation of oxygen uptake by 0.5 M trialkylborane solutions under air could be delayed several days, though the iodine was consumed slowly with formation of alkyl iodides.

The extreme rapidity of the reaction of trialkylboranes with oxygen has proved useful for the preparation of butanol-¹⁵O, isotope half-life 2.04 min, as a medical diagnostic agent [31, 32]. The reaction can also be used to synthesize organic hydroperoxides, efficiently yielding one mole of peroxide per mole of trialkylborane [33].

3.2.2.6 Other reagents. Chromate ion at pH 3-7 oxidizes *tert*-butylboronic acid four orders of magnitude faster than ethylboronic acid to the corresponding alcohol, which is stable to the reagent under these conditions [34]. In strong acid, dichromate oxidizes secondary alkylboranes directly to ketones [35]. Pyridinium chloro-chromate oxidizes primary trialkylboranes to aldehydes [36, 37].

Sodium chlorite, a reagent known to oxidize aldehydes to carboxylic acids, has been used to oxidize α–chloro boronic esters directly to carboxylic acids [38].

N-Chlorosuccinimide in methanol with a suitable base specifically oxidizes (α-phenylthio)boronic esters to monothioacetals, and a second mole of oxidizing agent converts these to methyl acetals [39, 40].

3.2.3 Nitrogen Electrophiles

3.2.3.1 Chloramine and hydroxylaminesulfonic acid. Replacement of boron by nitrogen with retention of configuration occurs under restricted circumstances, and has not found much application in synthesis to date. The major problem is that none of the known nitrogen electrophiles is reactive toward boronic esters, and only one or at most two alkyl groups of a trialkylborane can be utilized.

The prototype nitrogen electrophile is chloramine, $ClNH_2$. Preformed reagent is difficult to prepare and gives unsatisfactory yields [41], but generation from ammonia and aqueous sodium hypochlorite in the presence of the trialkylborane substrate gives good results [42] and is particularly useful for making ^{13}N-labeled amines [43, 44]. Dimethylalkylboranes (**13**) provide a way around the problem of utilizing alkyl groups, since the B-methyl groups in the postulated intermediate complex **14** migrate relatively slowly, though yields of the primary amines **15** are only 51-78%. Norbornyldimethylborane (**16**) yields exo-norbornylamine (**17**) [45]. This and similar earlier results [41] indicate that the reaction is stereospecific.

13   **14**   **15**

16   **17**

Hydroxylaminesulfonic acid, H_3N^+—OSO_3^-, is a stable alternative to chloramine and provides efficient utilization of one alkyl group of a trialkylborane [46]. Partial reaction of a second alkyl group occurs, but best results have been obtained by the use of dimethylalkylboranes such as **18** [47].

18   (87%)

The hydroxylaminesulfonic acid reaction has been used for conversion of boronic acid **19** (see Section 3.4.3.9 and 3.4.5.2) to the antiviral compound rimantidine (**20**) [48].

19   **20** (37% from **19**)

a, MeLi; AcCl. b, H_3N^+—OSO_3^-; OH^-, H_2O; HCl

The use of methylamine or other primary alkyl primary amines with hypochlorite allows the synthesis of secondary amines from symmetrical trialkylboranes [49] or alkyldimethylboranes [50]. Yields vary, but are typically 60-80%. This reaction can be carried out in the presence of preexisting silylated amino group to provide *N*-substituted unsymmetrical diamines such as **21** [51].

21

(a) $(CH_3)_2BH$. (b) $CH_3CH_2CH_2NH_2$, NaOCl.

The use of *N*-chlorosulfonamides with trialkylboranes has yielded *N*-alkyl-sulfonamides [52]. Reaction of tributylborane with $ClNMe_2$ results in a mixture of radical chlorination and amination product. Galvinoxyl inhibits the radical chlorination in favor of amination [53].

Nitroso alkane dimers, (R—NO)$_2$, with trialkylboranes, R'$_3$B, have yielded *N,N*-dialkylhydroxylamines, RR'NOH [54, 55].

3.2.3.2 Alkyl azides. Azides provide another route to nitrogen substitution. Again, only one alkyl group of a trialkylborane can be utilized [56], but alkyldichloroboranes can be used [57] and work much better than the trialkylboranes [58]. Retention of configuration is shown in the reaction of **22** with cyclohexyl azide, presumably via intermediate **23**, to form **24** [57]. Retention in the migration to boron is similarly exhibited in the synthesis of aziridine **25**, which of course inverts the carbon from which iodide is displaced in the ring closure [58, 59]. Trialkylboranes with hydrazoic acid [60], which may be generated conveniently in situ from trimethylsilyl azide and methanol [61], yield primary amines.

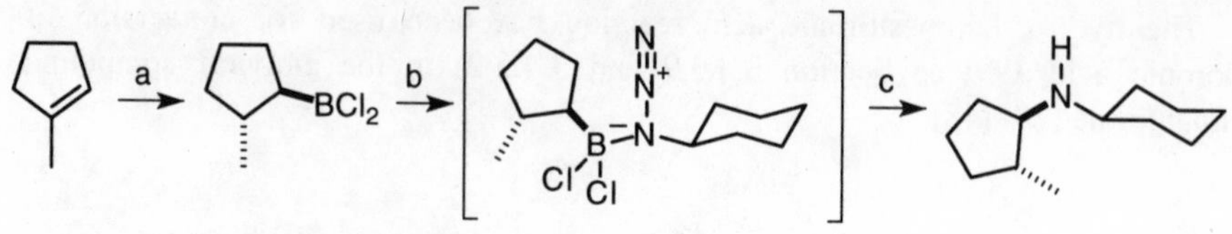

22 (racemic)   **23**   **24**

(a) HBCl₂. (b) C₆H₁₁N₃. (c) N₂ evolution; aqueous work up.

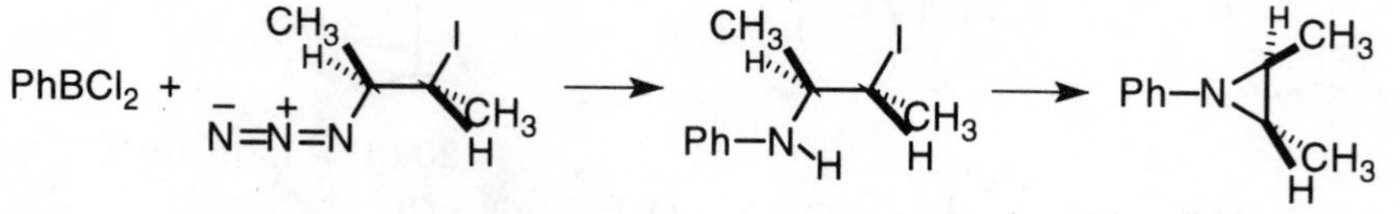

(racemic)   **25**

The reaction of alkyl azides with boranes provides the key step in the conversion of boraadamantane–THF (**26**) (see Section 3.4.5.2) to azaadamantane (**27**) reported by Bubnov and coworkers [62, 63].

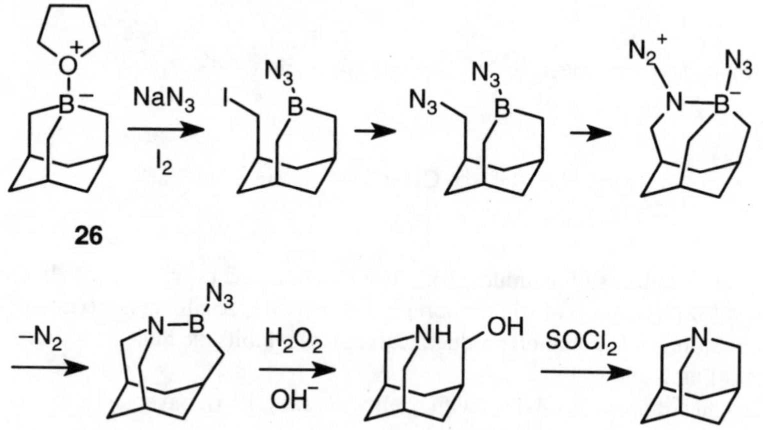

26

27

A number of applications of reactions of alkyl azides with boranes have been reported by Vaultier and coworkers. For example, 4-azidoalkylboronic esters such as **28** treated with boron trichloride cyclize efficiently to pyrrolidines, and the 5-azido analogues yield piperidines [64]. The azido boronic esters were obtained from the corresponding bromo compounds, which were prepared by standard routes. The cyclization is believed to proceed via a boron halide intermediate, either RBCl(OEt) or RBCl₂. Hydroboration of alkenyl azides with dicyclohexylborane or thexyl-chloroborane [65] provides an alternative route to pyrrolidines.

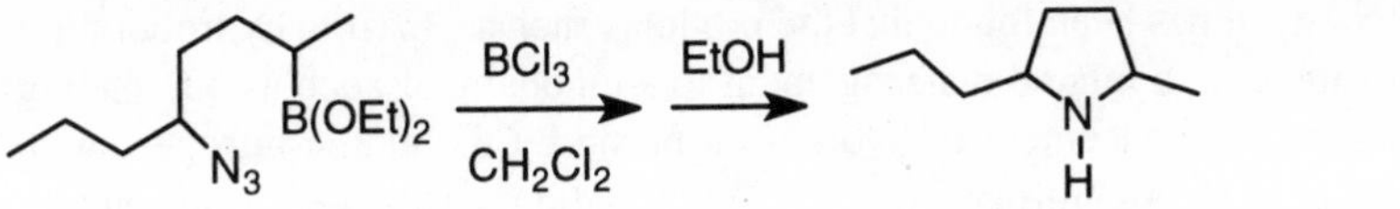

28   86%, 2 diastereomers

Although boronic esters do not react with azides, it has been found that PhBCl(OMe) reacts with PhCH₂N₃ to form PhNHCH₂Ph, though not as rapidly or efficiently as PhBCl₂ does [66]. However, PhBBr₂ yields a gross mixture of products, including the product of bromide migration and attack by a second azide, a tetraazaboroline (**29**). It was also found that tertiary alkyl groups on boron will alkylate primary but not secondary azides.

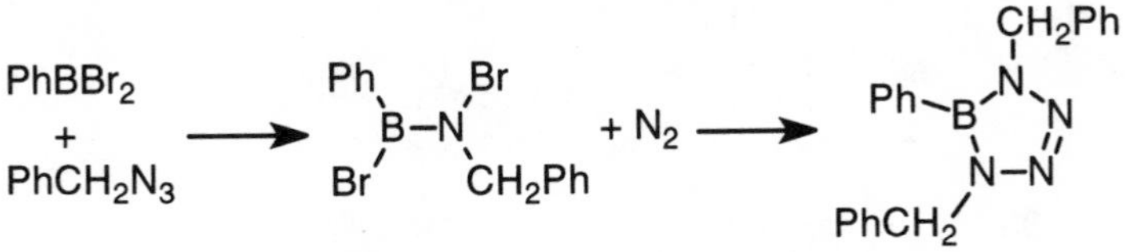

PhBBr₂ + PhCH₂N₃

29

Organobis(diisopropylamino)boranes, (iPr₂N)₂BR, react with azides in the presence of anhydrous HCl [67]. It is also possible to utilize both alkyl groups of R₂BCl to alkylate R'N₃ if HCl is used to cleave the intermediate RBClNRR', which is otherwise inert. Reaction of H₂N(CH₂)₃N₃ with Br(CH₂)₃BCl₂ in the presence of HCl yields H₂N(CH₂)₃NH(CH₂)₃Br·2HCl, which can in turn be converted to the azide for repetition of the process [68].

The use of azido boronic esters in the synthesis of amino acids or amino alcohols is discussed in Section 5.2.3.2 and 5.4.5, and the use of the borane—azide reaction in asymmetric synthesis in Section 6.3.3 and 6.4.1.2.

3.2.3.3 Rearrangement of α-aminoboranes. α-Amino boronic acids or esters are unstable toward migration of boron from carbon to nitrogen over periods of minutes or hours, and yield the corresponding amines in protic solvents [69, 70, 71]. An example is the rearrangement of 1-amino-2-phenylethylboronic ester **30** to the 2-phenylethylamine derivative **31** [69]. Boronic esters bearing tertiary amino groups in the α-position do not undergo this type of deboronation [72, 73]. Stable α-amido boronic acids and a route to them are described in Section 5.4.2.

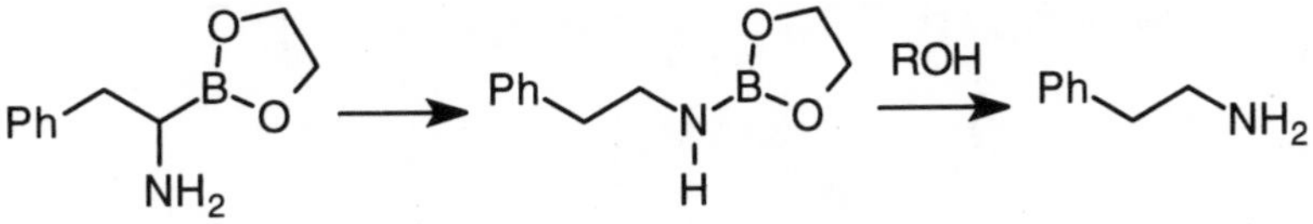

30   **31**

More recently, it has been found that the products such as **32** from hydroboration of enamines undergo a related rearrangement to aminoboranes such as **33**, leading after hydrolysis to replacement of hydride on boron by the amine nitrogen, with breaking of the carbon—nitrogen bond [74]. Oxidation yields the corresponding alcohol (**34**).

32

33 **34**

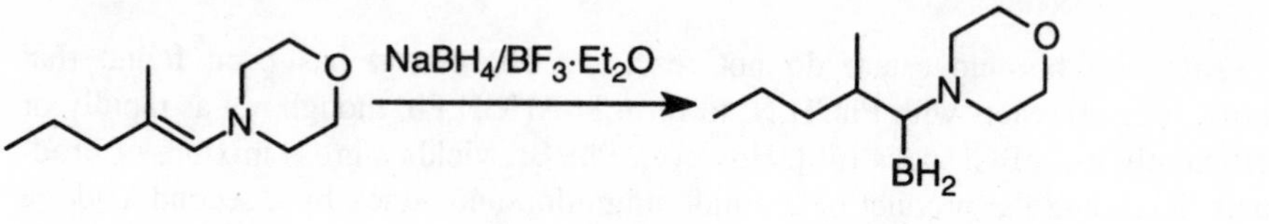

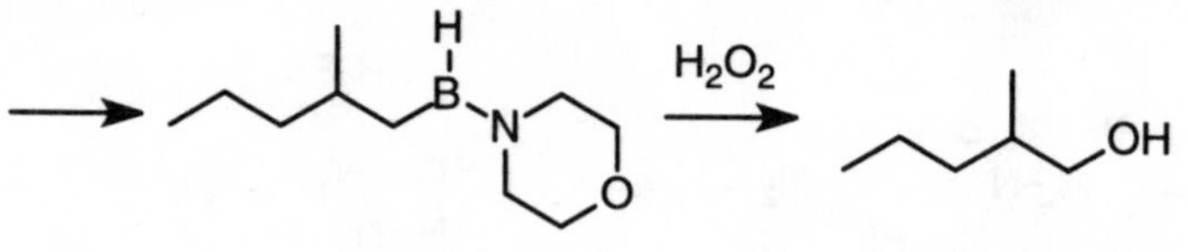

It should be noted that only those boronic esters containing a basic amino group undergo this deboronation. α-Amido boronic esters, bis(trimethylsilyl)amino boronic esters, and salts of amino boronic esters all appear to be stable indefinitely. It therefore appears likely that the rearrangement occurs via intramolecular nucleophilic attack of the amino group of **30** on boron. Ring opening of **35** to **36** might then occur, or proton migration to form **31** might be concerted with ring opening. The analogous closure of α-aminoborane **32** would lead to intermediate **37**, which might undergo ring opening to zwitterion **38**, in this case concerted with or followed by hydride migration to form **33**.

35 **36** (or concerted H^+ migration)

37 **38** (or concerted H^- migration)

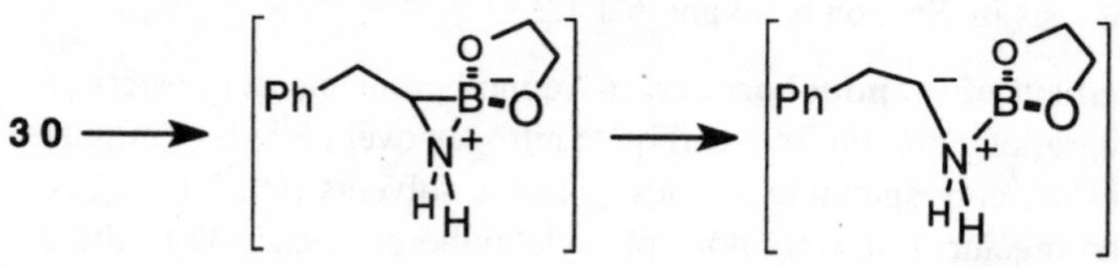

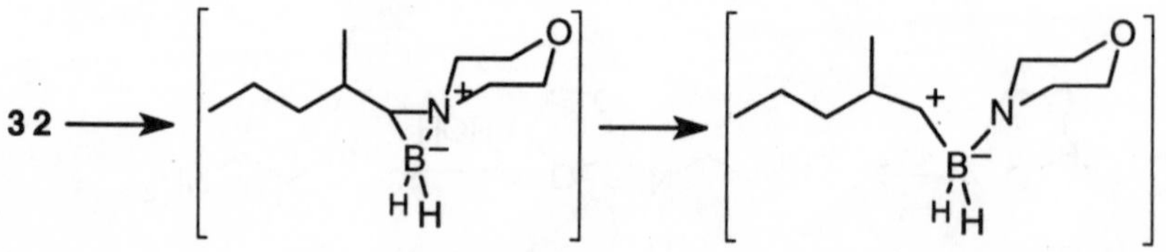

3.2.4 Halogenation

3.2.4.1 Stereoselective replacements. Because halogens are poor nucleophiles, the reaction mechanisms that allow replacement of boron with retention of configuration at chiral carbon are inaccessible with halogenating agents. Bromine with sodium methoxide cleaves all three alkyl groups from trialkylboranes [75] with predominant but incomplete inversion [76]. Iodine with sodium methoxide cleaves only two alkyl groups [77] but can give almost complete inversion [78]. For example, bis[(−)-pinanyl]-[(2S)-2-butyl]borane (**39**) with iodine and sodium methoxide yielded (R)-2-iodobutane (**40**) with 97-98% net inversion [78].

39   **40**

From a practical synthetic point of view, the only obvious reason an enantiomerically pure secondary alkyl halide would be desired would be for purposes of a subsequent S_N2 displacement. Sulfonate esters of ohols made via stereospecific peroxidic deboronation are generally more useful for this purpose.

1-Alkenylboronic acids can be cleaved by iodine and base with retention of configuration [79]. Bromination leads to inversion via an addition-elimination mechanism (see Section 4.2.1.2). These are among the most useful of stereoselective replacements of boron by halogen. Arylboronic acids can also be cleaved by halogens, but since arylboronic acids are usually made via Grignard reagents from haloarenes, this has limited synthetic potential.

3.2.4.2 Halogenation at carbon adjacent to boron. Cleavage of carbon—boron bonds by halogenating agents is greatly accelerated by neighboring sulfur [80]. An α-(phenylthio)alkylboronic ester (**41**) with sulfuryl chloride at 25 °C or N-chlorosuccinimide in refluxing carbon tetrachloride yields an α-chloroalkyl phenyl thioether (**42**). With one equivalent of N-chlorosuccinimide and triethylamine in methanol a monothioacetal (**43**) is obtained. Two equivalents of N-chlorosuccinimide yielded the dimethyl acetal, $PhCH_2CH(OMe)_2$.

41   **42**   **43**

The foregoing cleavage requires the sulfur substituent, as shown by the 95% recovery of a methylenediboronic ester, $(CH_2)_3O_2B-CH_2-BO_2(CH_2)_3$, after refluxing with sulfuryl chloride in dichloromethane for 2 h [80]. Trialkylboranes containing one α-phenylthioalkyl and two ordinary alkyl groups were cleaved preferentially at the sulfur-substituted group by N-chlorosuccinimide. It was concluded that initial halogen attack is probably at sulfur.

Bromination of trialkylboranes in dichloromethane results in cleavage of one alkyl group. The mechanism involves replacement of one α-hydrogen by bromine by the usual radical chain process [81]. An example is the conversion of [(i-Pr)CH$_2$]$_3$B to [(i-Pr)CH$_2$]$_2$BCHBr(i-Pr) followed by cleavage of the halogenated group by the HBr formed in the reaction to yield [(i-Pr)CH$_2$]$_2$BBr and BrCH$_2$(i-Pr). Bromine does not cleave the carbon—boron bond of boronic esters under neutral conditions (see Section 3.5.3).

3.2.5 Sulfur

Sulfides are too weakly nucleophilic toward boron to promote carbon—boron bond cleavage, and a radical chain mechanism prevails in the reaction of R$_3$B with MeS—SMe to form RSMe + B(SMe)$_3$ [82]. Iron(III) thiocyanate or selenocyanate preferentially cleaves the most highly branched alkyl group from unsymmetrical trialkylboranes, yielding RSCN or RSeCN [83].

3.2.6 Protonolysis

Organoboranes show a wide range of susceptibility to protonolysis, ranging from too slow to be generally useful for synthetic operations to so fast that premature cleavage is a serious problem. The relative reactivities of the various types of boron compounds generally correlate with their acidity, or with the acidity of a proton at the position occupied by boron.

Saturated alkylboranes are relatively resistant. An hour of refluxing with 48% hydrobromic acid cleaves only one alkyl group from tributylborane, yielding butane and dibutylborinic acid, Bu$_2$BOH [84]. The very slow hydrolysis of triethylborane by water at 25 °C is unaffected by hydrochloric acid and is retarded by sodium hydroxide, Et$_3$BOH$^-$ evidently being stable [85]. However, carboxylic acids cleave triethylborane efficiently to ethane and Et$_2$BOH at 25 °C in the presence of water, or slowly to EtB(OAc)$_2$ in acetic anhydride [85]. Propionic acid in refluxing diglyme cleaves all three alkyl groups from typical trialkylboranes, R$_3$B, to form RH and $(CH_3CH_2CO_2)_3B$ [86].

A particularly useful acid cleavage is the reaction of trialkylboranes with methanesulfonic acid to form dialkylboron mesylates, R$_2$BOSO$_2$CH$_3$, which are valuable synthetic intermediates [87].

A kinetic study of trialkylborane cleavage by a series of carboxylic acids, varying R of RCO$_2$H, has yielded $\rho^* = -0.94$ [88], implying nucleophilic involvement of the carbonyl oxygen in the transition state. Consistent with this mechanism is the observation of retention of configuration in the cleavage of trinorbornylborane (**44**), presumably via transition states resembling **45**, to norbornane-*exo*-2-*d* (**46**) by deuterated propionic acid [89].

44   **45** (R = norbornyl or EtCO₂)   **46**

Unsaturated groups are cleaved much more readily. Z-Alkenyldialkylboranes (**47**) are converted cleanly to Z-alkenes by acetic acid at 0 °C [90]. Methanol alone or with a small amount of carboxylic acid added is sufficient to cleave a variety of alkenylboranes [91]. Arylboronic acids undergo protonolysis by several different mechanisms, including a process involving one molecule of boronic acid and one of carboxylic acid in the transition state, again presumably cyclic [92].

47 (R groups are alkyl)

Benzylic and allylic boranes also undergo particularly facile cleavage. Attempted preparations of allylboronic acid or its esters via the conventional route from the Grignard reagent have given erratic yields, and hydrolysis during aqueous work up appears to be the source of the problem [93]. Enantiomerically enriched 1-phenylethylboronic acid is converted to ethylbenzene-1-*d*, PhCHDCH₃, by OD⁻ in D₂O with 54% net inversion at carbon [94].

Ethynylboronic acid is cleaved rapidly to acetylene by aqueous sodium bicarbonate, though it survives exposure to slightly acidic water at 25 °C at least for some minutes or hours [95].

Borane functionality in positions where a proton would be more acidic than in acetylene is generally very labile or unobservable. Reactions of enolates of haloesters with trialkylboranes yield rearrangement products that strongly imply that the carbon of the enolate is bonded to the tetrahedral borate complex [96], but it is highly probable that as soon as a tricoordinate boron is formed in the context B—C—C=O, it migrates to the carbonyl oxygen to produce the enolate C=C—O—B. The enolate is clearly favored by thermodynamics, and the vacant *p*-orbital of the boron atom allows the 1,3-migration to occur intramolecularly without violating orbital symmetry rules. However, in the special circumstance of hydroboration of propargylic esters, the α-dialkylboryl ester does not enolize to the allenic enolate (see Section 4.2.2.2) [97].

3.2.7 Metalation

Boron is not highly electropositive, and most metal cations do not have enough oxidation potential to effect replacement. However, metalations or couplings are known that involve mercury(II), lead(II), silver(I), gold(III), platinum(IV), palladium(0), copper(I), and one example of an organomagnesium compound. In all examples that have been studied, the boron has to be converted to a tetracoordinate borate before it can be displaced.

Though the practical synthetic utility of organomercury compounds would be limited even without the high cost and the toxic waste disposal problem, these reactions with their well defined products are of some mechanistic interest as possible models for other metalations. The reaction of mercury(II) chloride with phenylboronic acid to form phenylmercuric chloride was the first metalation of a boron compound found [98], and the similar conversion of benzylboronic acid to benzylmercuric chloride is also long known [100]. The kinetics are first-order each in benzylboronic acid, mercuric chloride, and sodium hydroxide, suggesting attack of $HgCl_2$ on $PhCH_2B(OH)_3^-$ as the rate-determining step [101]. A small degree of net retention of configuration was observed in the analogous reaction of partially resolved 1-phenylethylboronic acid [102]. Mercuration of alkenylboron compounds with mercuric acetate is rapid and proceeds with retention of alkenyl geometry [103].

gem-Diboronic esters react much faster than simple boronic esters with mercury(II) [104]. An example is the methylenediboronic ester $CH_2[B(OMe_2)_2]_2$, which is easily converted to $ClHgCH_2B(OMe_2)_2$ or $CH_2(HgCl)_2$ [105].

Triethylborane reacts with mercuric chloride and sodium hydroxide to form diethylmercury, or with lead(II) oxide to form tetraethyllead [106]. Mercuric acetate and trialkylboranes yield alkylmercuric acetates at 0 °C [107]. In contrast to the results with 1-phenylethylboronic acid, mercuration of deuterated tri-*n*-alkylboranes has been found to proceed with 90% inversion [108, 109]. One or two alkyl groups can be cleaved from tri-*sec*-alkylboranes react with mercury(II) alkoxides via a nonstereoselective radical mechanism [110, 111], but mercury(II) acetate fails to react [107].

α,β-Unsaturated boronic esters undergo the most facile reactions with mercury(II), and olefinic configuration is retained [112, 113]. Both boron atoms of alkenyl-1,1-diboronic esters are easily replaced by mercuric chloride and base [114, 115].

Silver(I) as $Ag(NH_3)_2^+$ couples alkyl groups of boronic acids, $RB(OH)_2$, to hydrocarbons, R—R [116]. Triethylborane with silver(I) oxide and sodium hydroxide yields 72% butane, 9% ethylene, and 9% ethane, total 90% based on the ethyl groups [117]. Ethylsilver was postulated to be an intermediate. Gold(III) oxide yields similar results. With platinum dioxide, triethylborane is converted mainly to ethane [117]. Silver nitrate and sodium hydroxide couple the alkyl groups of tri-*n*-alkylboranes and tricycloalkylboranes [118]. Mixtures of trialkylboranes lead to random alkyl coupling [119], and the process with 1-phenylethylboronic acid is not stereoselective [94]. Instead of coupling, arylboronic acids, $ArB(OH)_2$, are deboronated by $Ag(NH_3)_2^+$ to arenes, ArH [98, 99].

Palladium(II) acetate replaces boron from (*E*)- or (*Z*)-1-hexeneboronic acid with stereospecific retention, and the resulting alkenylpalladium (not isolated) substitutes methyl acrylate to form methyl (*E,E*)- or (2*E*,4*Z*)-nona-2,4-dienoate [120]. Palladium(0) triphenylphosphine complex catalyzes stereospecific coupling of unsaturated organoboranes with alkenyl and aryl halides [121]. Methylcopper participates in some similar couplings [122, 123]. The presumed organometallic intermediates in these reactions have not been isolated. Synthetic applications are discussed in Section 4.3.

Organoboranes have been converted to Grignard reagents by treatment with pentamethylenedimagnesium bromide (**48**) [124]. The equilibrium is shifted in favor of the stable spiroborate (**49**).

$$R_3B \ + \ 2 \ BrMg\diagdown\diagup\diagdown\diagup MgBr \ \longrightarrow \ 3 \ RMgBr \ + \ \left[\text{cyclohexylspiroborate}\right]B^- \quad MgBr^+$$

48   **49**

If a stable carbanion would result from cleavage of boron by reaction with a base, an organoborane will generally undergo such cleavage under mild conditions. Several examples have been described as protonolyses in Section 3.2.6. The most useful examples are reactions with alkyllithiums that generate boron–substituted carbanions by removal of one boryl group from compounds containing two or more boryl groups, reviewed in the following Section 3.3.

Other carbanions obtained by deboronation can usually be obtained by deprotonation of a more easily accessible substrate, and thus are of little importance except as obstacles to synthesis. For example, the pinanediol ester $PhSCH_2B(O_2C_{10}H_{16})$ does not undergo chain extension with $LiCHCl_2$, probably because it generates $PhSCH_2Li$ and $Cl_2CHB(O_2C_{10}H_{16})$ instead, though this was never rigorously proved [125].

3.3 Boron Substituted Carbanions

3.3.1 By Deboronation

Formation of boron-substituted carbanions was first accomplished via reaction of a di-, tri-, or tetraboryl compound with an alkyllithium to form a boron–substituted carbanion, a version of the replacement of boron by metals described in Section

3.2.7. However, the most useful boron substituted carbanions are those that can be made via deprotonation. Once the anion has been made, reaction with an electrophile can follow, and this may connect either carbon or a heteroatom. Thus, this section does not fit into any single theoretical or simply defined category, but bears some relation to Sections 3.2 and 3.4.

The first examples of boron stabilized carbanions were discovered in the trialkylborane series. Dihydroboration products from acetylenes are deboronated by alkyllithiums, and the resulting (1-borylalkyl)lithiums **50** react with electrophiles such as aldehydes [126, 127].

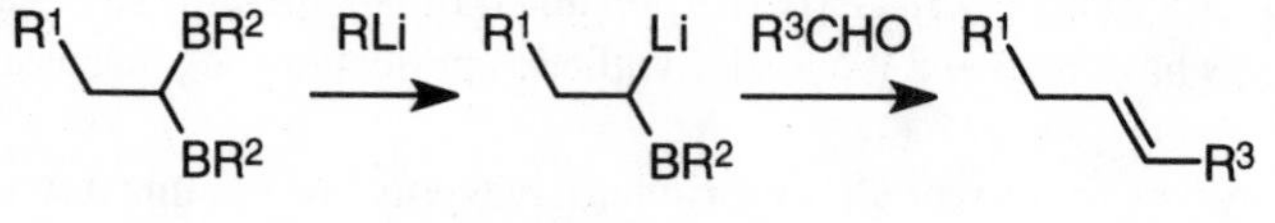

50

Bis(dialkoxyboryl)carbanions do have sufficient stability to serve as useful synthetic intermediates. The first route to these was the reaction of tris- and tetrakis(dialkoxyboryl)methanes with alkyllithiums to produce lithium bis- and tris(dialkoxyboryl)methides [128, 129], which yield derivatives with carbonyl compounds [130, 131], alkyl halides [132], group 14 metal halides [133, 134], and bromine [135]. The first work was done with dimethoxyboryl compounds, but the reactions were found to be much more efficient with cyclic boronic esters. Lithium tris(1,3,2-dioxaborinyl)methide (**51**) can be partially purified by precipitation [136], and reacts efficiently with aldehydes or ketones to form alkylidenediboronic esters **52** [137]. With bromine, **51** yields $BrC[BO_2(CH_2)_3]_3$, or with triphenyltin chloride, $Ph_3SnC[BO_2(CH_2)_3]_3$ [137]. Reaction of several **52** with mercuric chloride has yielded $R^1R^2C=C[BO_2(CH_2)_3]_2$ [115].

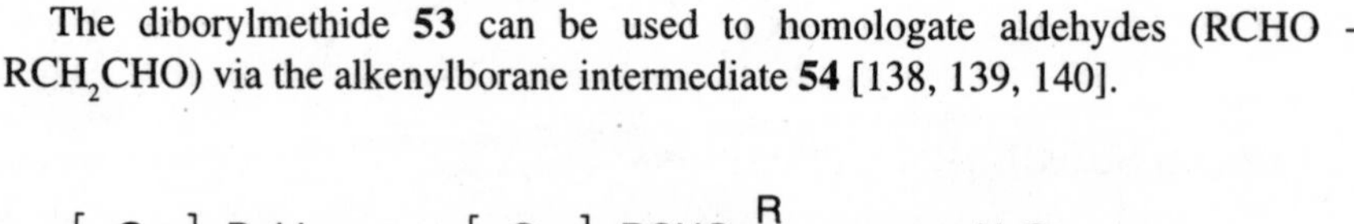

51   **52**

The diborylmethide **53** can be used to homologate aldehydes (RCHO → RCH_2CHO) via the alkenylborane intermediate **54** [138, 139, 140].

53   **54**

3.3.2 By Destannylation

A single boronic ester group is not sufficient to stabilize an adjacent carbanionic center to the point where the α-lithio boronic ester can be a useful synthetic intermediate. Pinacol (1-lithioethyl)boronate (56) has been generated from the trimethyltin precursor 55 and has been coupled with pinacol (1-chloroethyl)boronate (57) to produce the corresponding diboronic ester 58 [141]. The two diastereomers of 58 are produced in approximately equal amounts. Starting from enantiomerically enriched 55 and 57 yielded optically active product but the same diastereomeric ratio, indicating that the carbanionic intermediate 56 does not retain its configuration. Reaction of 56 with acetophenone yielded (Z)- and (E)-2-phenyl-2-butene (59) in a ~1:1 ratio. However, attempts to methylate 56 with methyl iodide were unsuccessful. Although cross coupling of 56 with pinacol (α-chlorobenzyl)boronate was successful [141], a more recent attempt to generate pinacol (lithiomethyl)boronate and couple it with 57 yielded a gross mixture of products, apparently including 58 and 1,2-bis(4,4,5,5-tetramethyl-1,3,2-dioxaborolyl)ethane as well as the expected 1,2-bis(4,4,5,5-tetramethyl-1,3,2-dioxaborolyl)propane, as if halogen-metal exchange was faster than addition of the α-lithio boronic ester to the α-halo boronic ester [142].

55   **56**

58 (2 diastereomers)   **57**   **59** [(Z)/(E)-mixture]

3.3.3 By Deprotonation

Deprotonation of the methyl group of B-CH$_3$-9-BBN can be accomplished with lithium 2,2,6,6-tetramethylpiperidide [143]. The resulting anion was alkylated with butyllithium, though in poor yield.

A better route to diborylmethide anions is provided by deprotonation of bis(1,3,2-dioxaborinyl)methane (60) [144, 145]. In addition to the types of reactions already described, the diborylmethide anion 61 has been treated with alkyl halides to form alkylidenediboronic esters 62. In turn, 62 (R = n-pentyl) has been deprotonated to 63. If acylation of 63 with a carboxylic ester such as methyl benzoate were to yield the acyldiboronic ester 64, this should undergo migration of boron from carbon to oxygen to form an enolate such as 65, and in any event the product isolated was the ketone **66** [145]. Reaction of **63** with benzaldehyde followed by oxidation yielded the expected ketone **67**, isomeric to **66**.

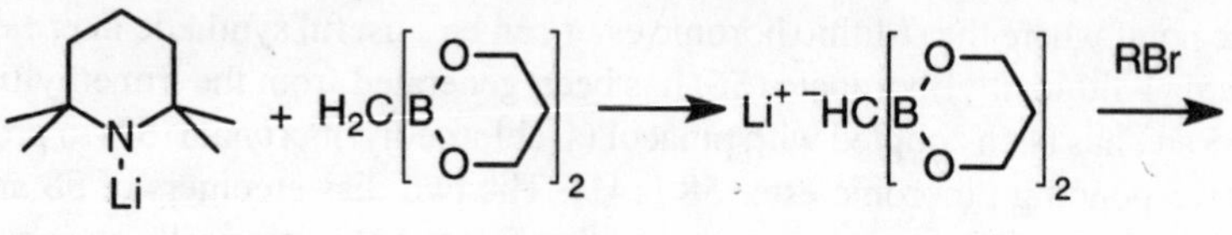

60   **61**

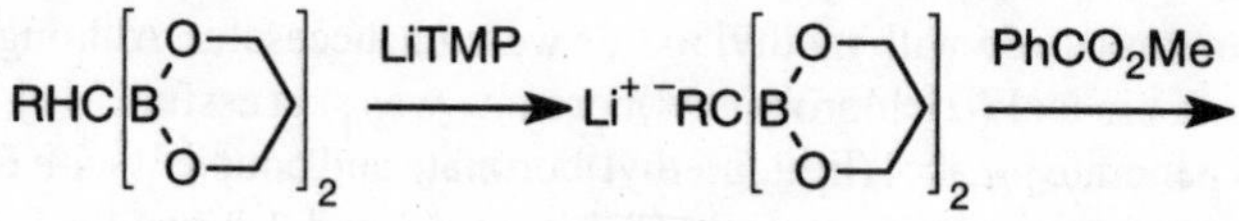

62   **63** (R = pentyl)

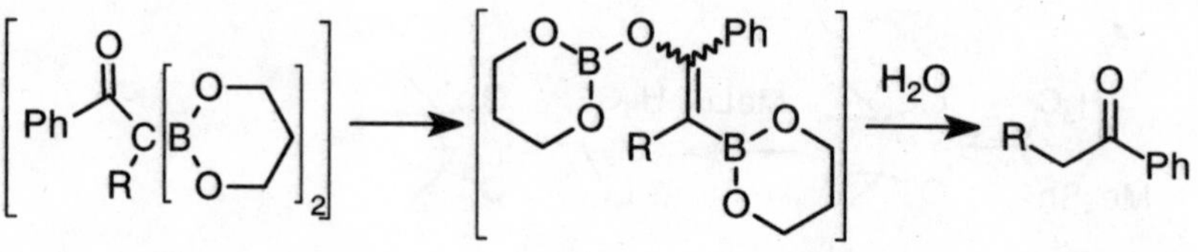

64   **65**   **66**

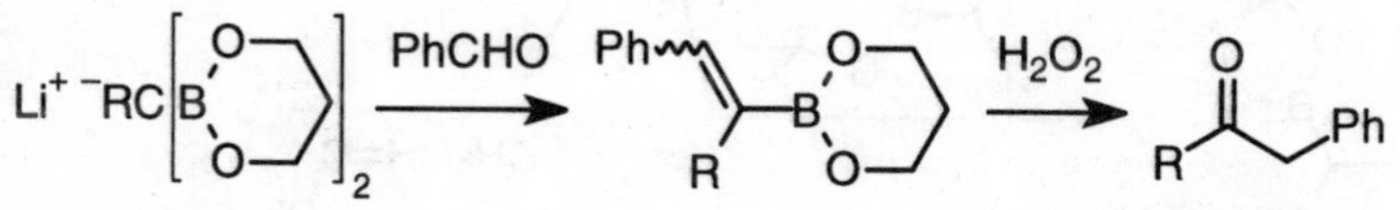

63 (R = pentyl)   **67**

The route to bis(dialkoxyboryl)methanes has been described in Section 2.2.2. It should be noted that 1,1-bis(dialkylboryl)alkanes, $RCH_2CH[B(OR')_2]_2$, are also readily available from the dihydroboration of acetylenes, preferably with dichloroborane (see Section 2.3.3.4).

It is easier to prepare sulfur or silicon substituted borylmethanes than the diborylmethanes, and for some purposes the carbanions derived from these are equally useful. Pinacol phenylthiomethylboronate (**68**) is deprotonated by LDA and TMEDA (= $Me_2NCH_2CH_2NMe_2$), and the resulting anion **69** reacts with all of the usual electrophiles [146, 147]. Examples include alkylation to **70**, acylation/deboronation to **71**, epoxide opening to form **72**, and Wittig-like reaction with a carbonyl compound to form **73**. Alkylated derivatives can also be deprotonated, as in the conversion of **70** to **74**, which also reacted with a variety of electrophiles, for example with butyrolactone, which yielded the deboronated product **75** [147].

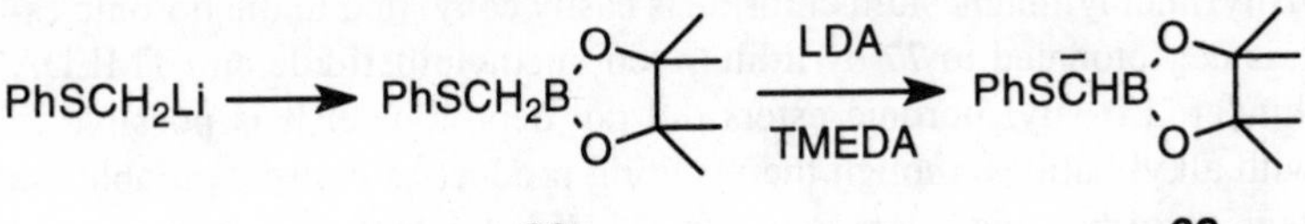

68   **69**

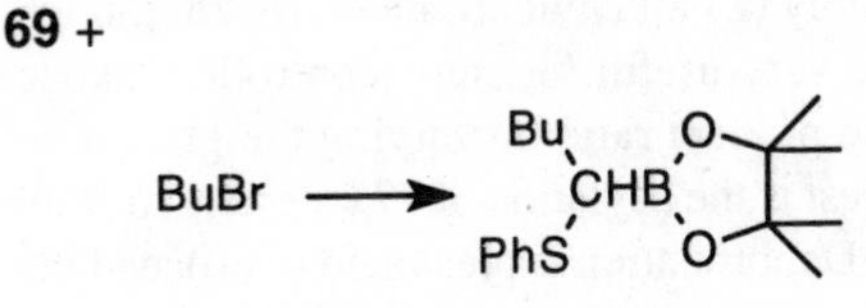

69 +

BuBr →

70

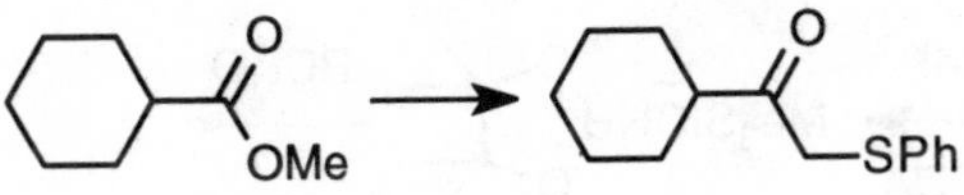

71

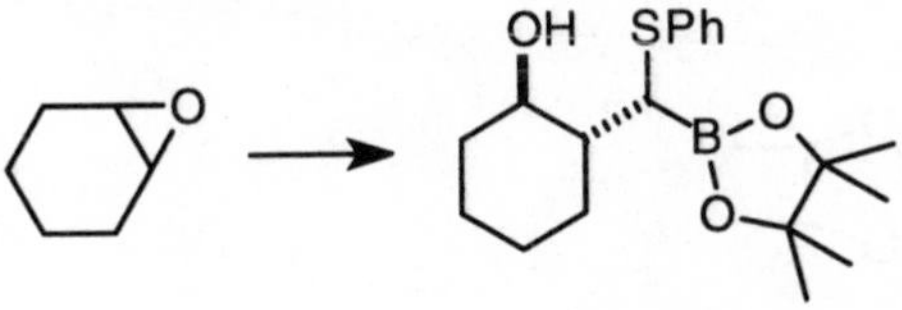

72

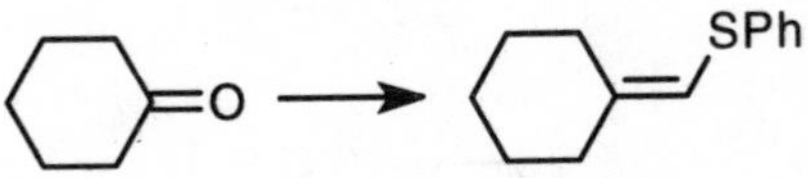

73

70 →

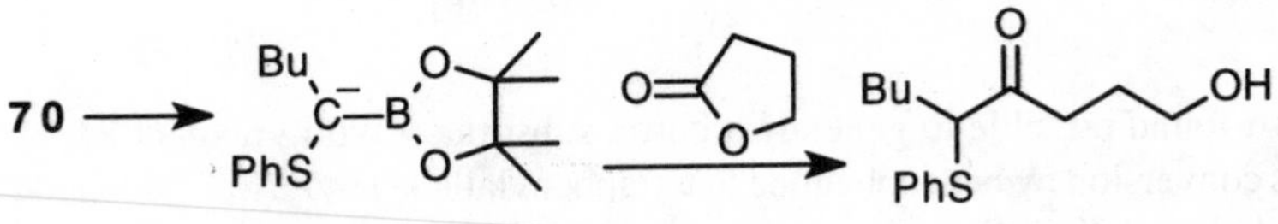

74   **75**

Pinacol bis(phenylthio)methylboronate, $(PhS)_2CHBO_2C_2Me_4$, can be deprotonated similarly, and with aldehydes or ketones RCOR' yields ketene thioacetals, $RR'C=C(SPh)_2$ [148].

(Trimethylsilyl)methylmagnesium chloride is easily converted to the boronic ester **76**, which is deprotonated to **77** by lithium tetramethylpiperidide and TMEDA [149]. Less hindered α-silyl boronic esters did not deprotonate. It is possible to alkylate **77** with alkyl halides, though the resulting products are also available via the chain extension of boronic esters with $Me_3SiCHLiCl$, which is a more general process with respect to possible substituents. It was also disappointing that the alkylation products were inert toward deprotonating agents. More interesting are the reactions of **77** with aldehydes to form mainly (Z)-alkenylboronic esters **78**, though with too much of the (E)-isomer **79** to be very useful for stereocontrolled alkene synthesis. The possibility of modifying the product ratio by varying the groups on silicon has not been explored. Also of interest is the acylation of **77** by esters to form (trimethylsilyl)methyl ketones such as **80**. Deboronation is presumed to proceed via the enolate as described for **64** → **65** above.

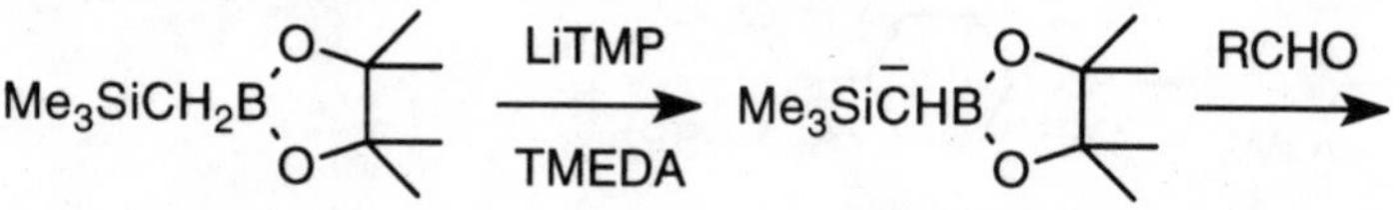

76   **77**

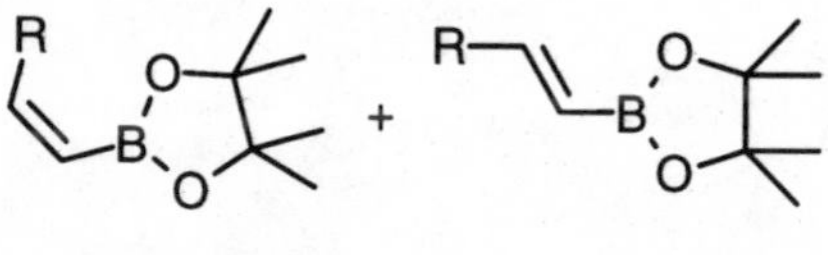

78 (70%)   **79 (30%)**

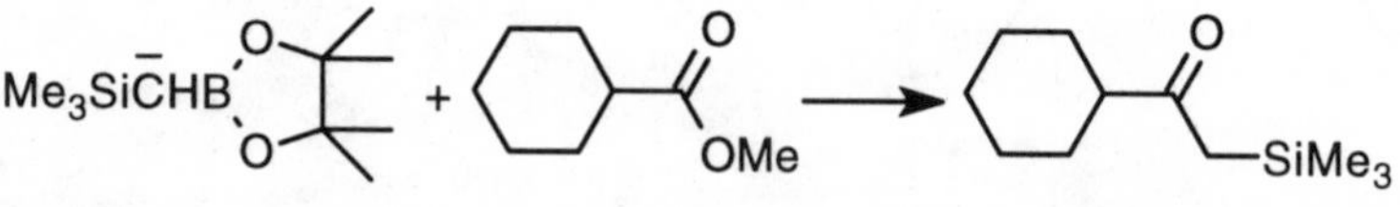

77   **80**

It was also found possible to generate a boron substituted Wittig reagent **81**, as shown by its conversion by benzophenone to tetraphenylallene [149]. Benzaldehyde yielded 1,3-diphenylallene, but it appeared that carbonyl compounds having acidic hydrogens did not form allenes.

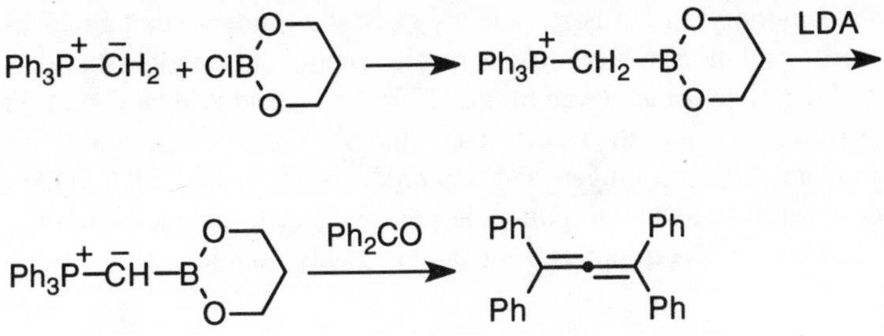

81

Mesitylboranes offer the advantage of a more electrophilic boron *p*-orbital than the boronic esters, and the mesityl group blocks attack on the boron atom by any but very small nucleophiles. As a result, dimesitylmethylborane (**82**) can be deprotonated by lithium dicyclohexylamide, and the resulting anion (**83**) has been alkylated with methyl iodide [150]. Wilson reported that the alkylation product (**WC**) could be deprotonated and alkylated, and the process could be repeated to provide the *tert*-butylborane **85**.

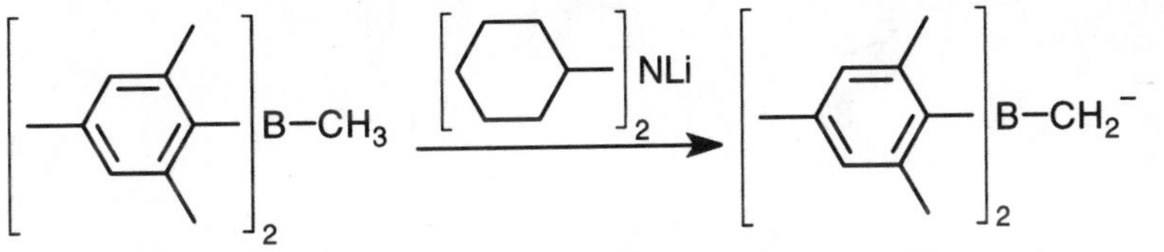

82 **83**

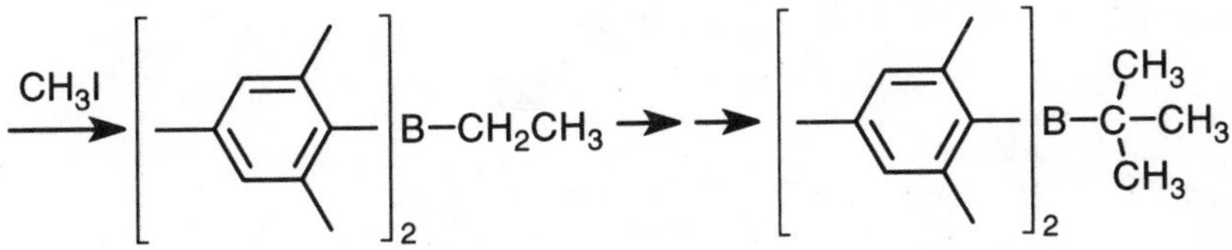

84 **85**

A series of papers by Pelter, Wilson, and coworkers followed in which it was shown that the anion **83** reacted with primary alkyl bromides [151], aldehydes or ketones in a Wittig-like reaction [152], and other electrophiles such as Me$_3$SiCl, Me$_3$SnCl, PhSCl, and (mesityl)$_2$BCl [153].

3.3.4 By Michael Addition

Ordinarily, anions do not add in a conjugate manner to α,β-unsaturated boranes, and an early attempt to observe such behavior with a vinylboronic ester failed [154].

However, the dimesitylboryl group is sufficiently sterically hindered that attack at boron does not occur, and the boron is sufficiently electrophilic to stabilize the adjacent carbanion and facilitate conjugate additions [155]. Cooke and Widener reported that α-trimethylsilylvinylborane **86** readily adds alkyllithiums, lithiodithiane, or *tert*-butyl lithioacetate to form conjugate addition products, for example **87**, in high yields after hydrolysis. 6-Iodo-1-hexenylborane (**88**) cyclizes to the carbanion **89**, which reacts with common electrophiles, for example methyl iodide.

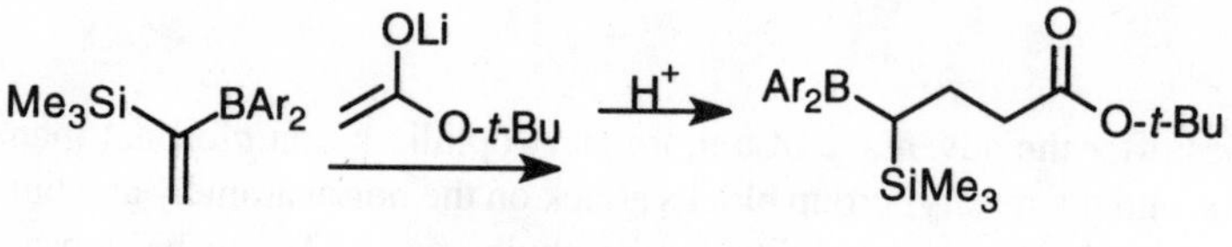

86   **87**

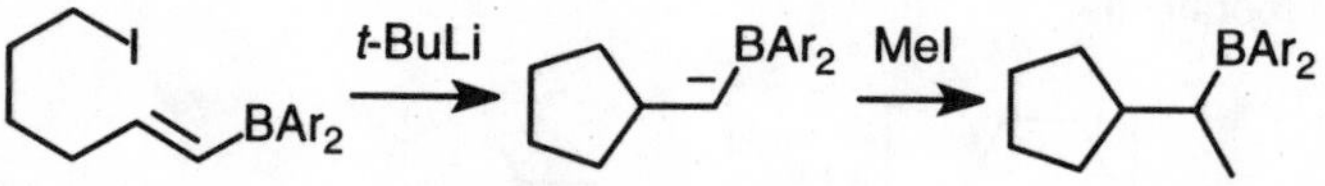

88   **89**   (55%)

(5-Iodo-1-pentynyl)dimesitylborane (**90**) cyclizes under lithium/halogen exchange conditions to form the boron stabilized carbanion **91** exocyclic to a four-membered ring, leading to the borylalkylidenecyclobutane **92** [156]. The five-membered ring compound **93** was similarly prepared.

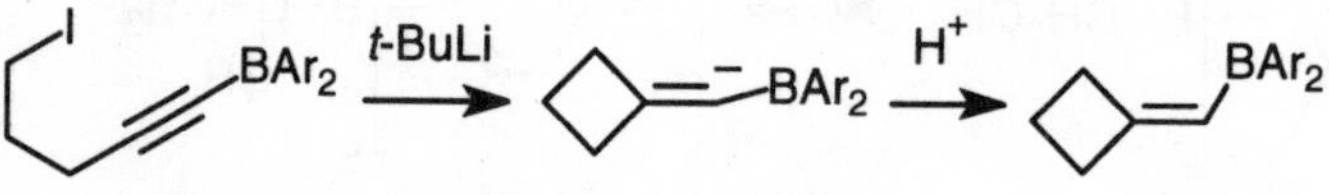

90   **91**   **92**

(Ar = 2,4,6-triphenylmethyl)

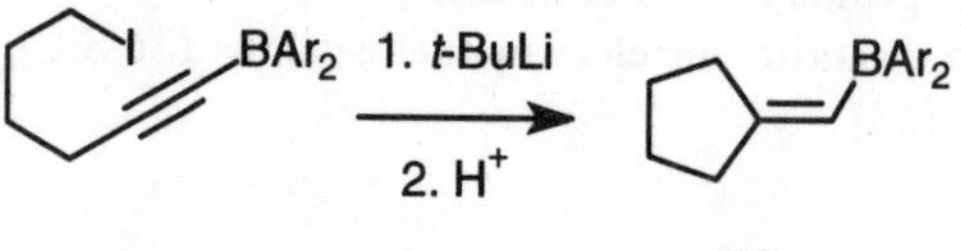

93

3.4 Replacement of Boron by Carbon

3.4.1 Introduction

The most common mechanism for replacement of a boron—carbon bond by a carbon—carbon bond involves formation of a borate complex containing two boron—carbon bonds. The requisite intermediates are most often formed by addition of an anionic carbon to an organoboron compound. In order for rearrangement to occur, one of the carbon atoms must also have electrophilic character, the ability to become a carbocation equivalent of some type. These rearrangements form the fundamental basis for the majority of reactions that are useful in stereocontrolled synthesis. Displacements of boron that appear to involve other mechanisms conclude this section.

A potential obstacle to complex synthetic schemes, β-elimination of boron and an electronegative substituent, is mechanistically somewhat related to reactions that replace a boron—carbon bond by a carbon—carbon bond. Because of the need of the synthetic chemist never to forget this potentially devastating diversion of synthetic intermediates, the nuisance aspects of β-eliminations are discussed very briefly in Section 3.3.2 below. Discussion of the utility of β-eliminations for stereocontrolled olefin synthesis is deferred to Section 4.2.

3.4.2 β-Eliminations as Nuisance Reactions

β-Eliminations in a sense replace a boron—carbon bond by a carbon—carbon bond. These reactions are most often base catalyzed, and can be extremely facile under certain circumstances [157, 158]. β-Eliminations render certain types of synthetic pathways unworkable. However, it can also be said that one of the advantages of boranes as organometallic intermediates is that so many of the thermodynamically favored β-eliminations of elements have turned out to be very slow. In this regard, β-substituted boronic esters have shown much greater stability than the analogous trialkylboranes.

In spite of the facility of elimination reactions of trialkylboranes containing a β-halogen substituent [157], β-chlorocyclohexylboranes where the boron and halogen are both equatorial have sufficient stability to undergo replacement of boron by peracid oxidation [159]. β-Alkoxy substituted trialkylboranes undergo *syn* elimination under neutral conditions, presumably via an intramolecular mechanism, and *anti* elimination in the presence of acid or base catalysis [160]. These reactions are of no known synthetic interest except as traps that have to be avoided by proper choice of synthetic strategy.

In sharp contrast to the behavior of the trialkylborane series, β-alkoxy boronic esters have considerable kinetic stability. Attempted distillation of an early example resulted in partial decomposition at 80-100 °C [161], but by keeping temperatures below that level, a great variety of β-alkoxy boronic esters have been utilized successfully in asymmetric synthesis, for example, in the syntheses of L-ribose [162] and deuterium labeled glycerol [163].

β-Elimination of an oxygen function and one of a geminal pair of boron atoms is very facile [126, 127].

3.4.3 Migrations to Electron–Deficient α-Carbon

3.4.3.1 Introduction. Organoborane chemistry provides a variety of ligand migrations from boron to carbon that are of major utility in asymmetric synthesis. These generally involve the formation of a borate complex (**94**), which rearranges with migration of a ligand R^1 from boron to electron deficient carbon via generalized transition state **95** to form the product **96**. (This mechanism is merely a special case of rearrangement of an intermediate borate **1**, Section 3.2.) Assembly of **94** is generally accomplished by addition of a base to a tricoordinate boron compound. The base may be the unit which contains the electrophilic carbon $(XCR^2R^3)^-$, an organometallic or metal salt R^1M of the ligand R^1 that ultimately migrates, or a ligand L^- added to a borane which already contains all of the groups that will react.

94   **95**   **96**

This very general scheme is applicable to a wide variety of systems. The migrating ligand R^1 may be connected through carbon or any of a wide variety of heteroatoms capable of acting as nucleophiles. The leaving group X^- is usually halide, but displacements of N_2, various sulfur groups, methanesulfonate, and others are also known. The boron ligands L that do not migrate may be alkyl, alkoxy, or others [164].

In a conceptual sense, these rearrangements might be considered to proceed via a virtual "boron ylid", Y_3B^-—$C^+R_2 \rightarrow Y_2B$—CYR_2. This idea was once articulated as a rationalization for the messy array of 4-octene, pentylbenzene, octylborane, and other products from tetrabutylborate ion and benzyl chloride in ether at 120 °C [165]. It might be more appropriate as a description of the oldest examples of such rearrangement, the well known polymerizations of diazomethane by halo or alkoxy boranes [166], where the methylene insertion step requires intermediate **97** and might well involve the "boron ylid" **98**. However, it is unlikely that there is any energy barrier to the rearrangement of such a "boron ylid" [167], and the closest approach to **98** might still have the nitrogen molecule loosely connected as the polymethylene chain migrates to form **99**.

$$Y_2B-(CH_2)_nY \xrightarrow{CH_2N_2} \left[\begin{array}{c} CH_2\text{-}N\overset{+}{\equiv}N \\ Y_2\overset{|}{B}=(CH_2)_nY \end{array} \right] \xrightarrow{-N_2}$$

97

$$\left[\begin{array}{c} CH_2^+ \\ Y_2\overset{|}{B}=(CH_2)_nY \end{array} \right] \longrightarrow \begin{array}{c} CH_2-(CH_2)_nY \\ Y_2\overset{|}{B} \end{array}$$

98   **99**

3.4.3.2 Discovery of process and mechanism. The first well defined example of an α-haloalkylborate rearrangement was discovered serendipitously [167]. The α-halo boronic ester **100** was obtained from a free radical addition of bromotrichloromethane to dibutyl vinylboronate (see Section 3.5.5.1) [168]. Treatment with Grignard reagents (R = Ph, 2,5-Me$_2$C$_6$H$_3$, mesityl, Et) yielded rearrangement products **102** instead of the expected α-bromo borinic ester **103** [167]. It was then found that prompt work up with acid at low temperatures did yield α-bromo borinic esters **103** [154], which rearranged to boronic esters **102** on treatment with aqueous sodium bicarbonate (R = Ar) or hydroxide (R = Et) and butanol at room temperature.

$$\begin{array}{c} BuO \\ BuO' \end{array}\overset{H}{\underset{Br}{B-\overset{|}{C}}}-CH_2CCl_3 \xrightarrow[-78\,°C]{RMgBr} \left[\begin{array}{c} R\ H \\ BuO-B-\overset{|}{C}-CH_2CCl_3 \\ BuO\ Br \end{array} \right]^- \xrightarrow{25\,°C} \begin{array}{c} BuO \\ BuO' \end{array}\overset{R}{\underset{H}{B-\overset{|}{C}}}-CH_2CCl_3$$

100   **101**   **102**

$$H^+ \Big\Updownarrow \begin{array}{c} \text{aq. base} \\ -78\,°C \quad + BuOH \end{array}$$

$$\begin{array}{c} R\ H \\ B-\overset{|}{C}-CH_2CCl_3 \\ BuO'\ Br \end{array} \quad \left(\begin{array}{c} R \\ B-CH=CH_2 \\ BuO' \end{array} \xleftarrow{BrCCl_3} \right)$$

103   **104**

The structures of **103** (R = Ar) were confirmed by synthesis from butyl (aryl)(vinyl)borinates (**104**) and bromotrichloromethane [154]. Thus, the expected borate **101** must be an intermediate which can react by either of two pathways, rearrangement or dissociation.

3.4.3.3 Favored reaction pathways. It is an important rule that carbon ligands migrate in preference to oxygen ligands. This rule has only one known exception, which occurs when unusual chirality factors force oxygen migration, and even this exception is of benefit in synthesis (Section 5.3.4). The underlying reason is the much greater thermodynamic stability of **102** compared to **103**, which was conjectured to be ~120-160 kJ mol^{-1} [167].

Thermodynamic data and calculations now allow a more precise estimate that model boronic ester **105** is more stable than isomeric borinic ester **106** by approximately 138 kJ mol^{-1}. This estimate is based on the sum of typical B—O and C—C bond energies (**105**) versus the sum of B—C and C—O bond energies (**106**), with the typical difference in C—H bond energies between BCH$_3$ and CCH$_3$ as well as OCH and CCH groups taken into account (see Section 1.2.3). Only a small fraction of this energy difference needs to be reflected in the transition states leading to the respective products in order to account for the kinetic preference for formation of **105**.

$$\begin{array}{cc} \text{H} \quad \text{OCH}_3 & \text{H} \quad \text{OCH}_3 \\ \text{H}_3\text{C}-\overset{|}{\text{C}}-\text{B} & \text{H}_3\text{C}-\overset{|}{\text{C}}-\text{B} \\ \overset{|}{\text{CH}_3}\;\text{OCH}_3 & \overset{|}{\text{CH}_3\text{O}}\;\;\text{CH}_3 \end{array}$$

105   **106**

The foregoing chemistry serves as a model for all of the more recent work on halo boronic esters. Nucleophiles other than alkyl or aryl, including iodide, alkoxide, and butylsulfide, also coordinate to the boron atom before displacing bromide from **100** [167], and are discussed further in Section 3.5.2.1. It was found that displacement of the α-bromide was usually the preferred reaction course. Elimination of hydrogen bromide from **100** proved unexpectedly difficult, in spite of the fact that the trichloromethyl group should serve as an activator for elimination. Dehydrobromination was accomplished by the use of *tert*-butylamine as the base [169].

3.4.3.4 Routes to α-haloalkylboron compounds. At first, all of the reactions of α-haloalkylboron compounds remained mere mechanistic curiosities because there was no practical general synthesis of these compounds. It did prove possible to add liquid hydrogen bromide or iodide to some alkenylboronic esters to make α-haloalkylboronic esters [170, 171] (see Section 3.5.5.2), but this approach fell short of general practical synthetic utility. Some transformations did appear to have synthetic potential in a limited way, such as the radical addition of bromomalononitrile to dibutyl vinylboronate to form the α-bromoalkylboronic ester **107** (see Section

3.5.5.1), which on treatment with any base immediately cyclized to the dicyano-cyclopropylboronic ester **108** [171]. An attempt was made to add hydrogen halides to (aryl)(vinyl)boronic esters **109** for rearrangement to the (1-arylethyl)boronic esters **110**, since these compounds were hard to make via 1-arylethyl Grignard reagents, but the sequence was only partially successful because considerable vinyl group cleavage occurred to form arylboronic esters **111** [172].

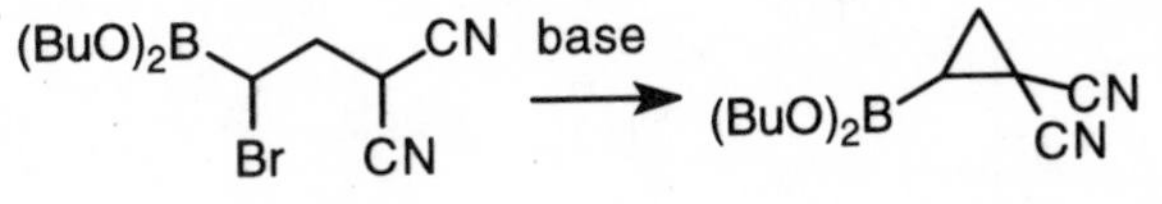

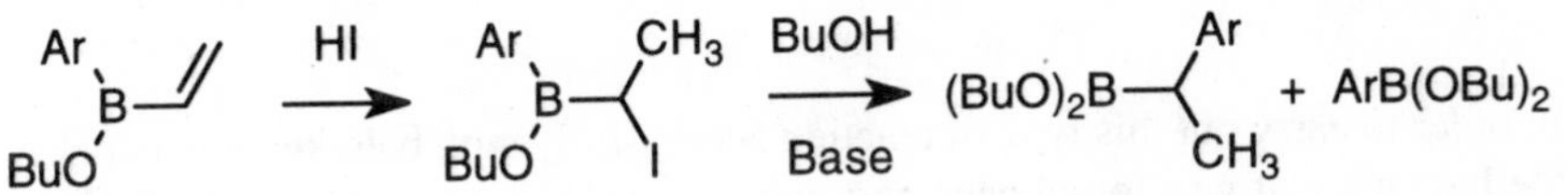

The first successful synthetic application of (α-haloalkyl)borate complexes was the reaction of the enolate of ethyl bromoacetate with trialkylboranes by Brown and coworkers [96]. The rearrangement was shown to proceed with retention of configuration of the migrating group, as would be expected based on the mechanism of the process. The product of hydroboration of 1-methylcyclopentene with 9-BBN (**112**) reacts with the bromoenolate [BrCHCO$_2$Et]$^-$ generated in situ from ethyl acetate and potassium *tert*-butoxide to form a borate complex **113**, which presumably rearranges to the unstable α-boryl ester **114** and then the boron enolate **115**, which reacts with the *tert*-butyl alcohol present in the mixture to form ethyl *trans*-2-methyl-1-cyclopentaneacetate (**116**) [173].

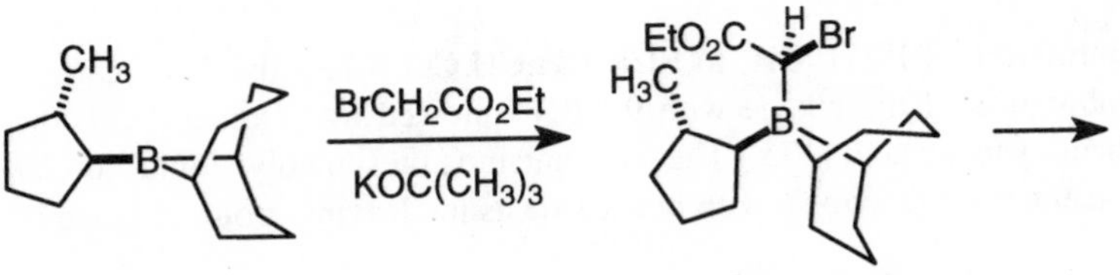

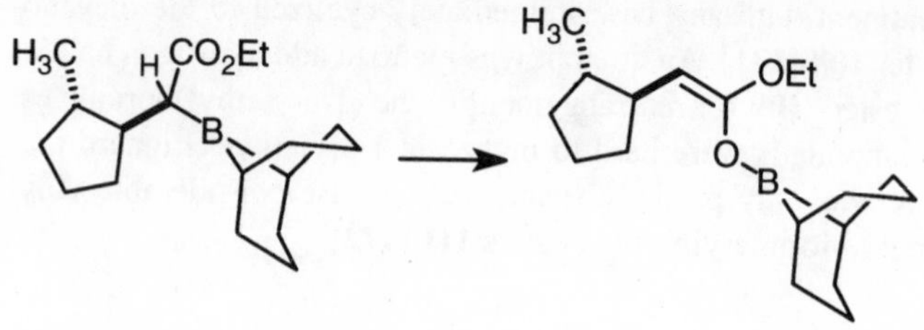

114   **115**

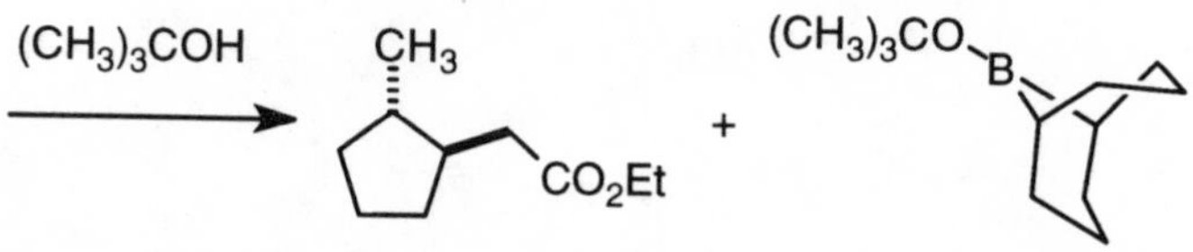

116

In order to carry out this type of reaction successfully with halo ketones in place of the halo ester, it was found necessary to use a weaker but highly sterically hindered base, potassium 2,6-di-*tert*-butylphenoxide (**117**) [174].

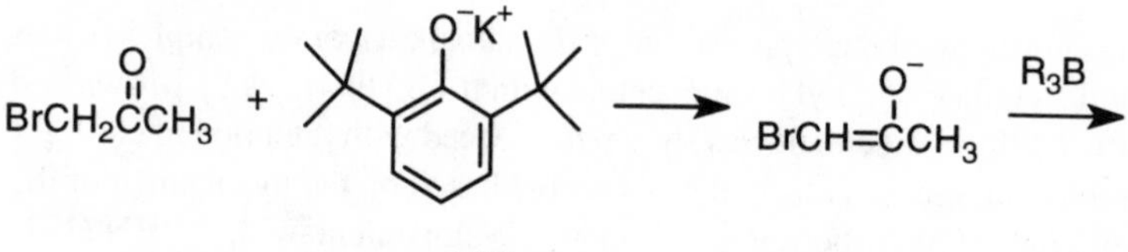

117

The transformation of $RCH=CH_2$ to $RCH_2CH_2CH_2CO_2H$ has also been carried out via hydroboration of the alkene with 9-BBN followed by treatment with the dianion of phenoxyacetic acid [175]. The mechanism is presumably similar to that of the bromoacetic ester reactions, with phenoxide as the leaving group in place of bromide.

Another synthesis in which an α-haloalkylboronate anion is an intermediate is the reaction of trialkylboranes or borinic esters with lithiated dichloromethyl methyl ether. An example is the reaction of (dicyclohexyl)methoxyborane (**118**) with

Cl$_2$CHOMe and LiOCEt$_3$ to form dimethyl [(dicyclohexyl)methoxymethyl]-boronate (**119**) [176]. Peroxidic oxidation of **119** yields dicyclohexyl ketone.

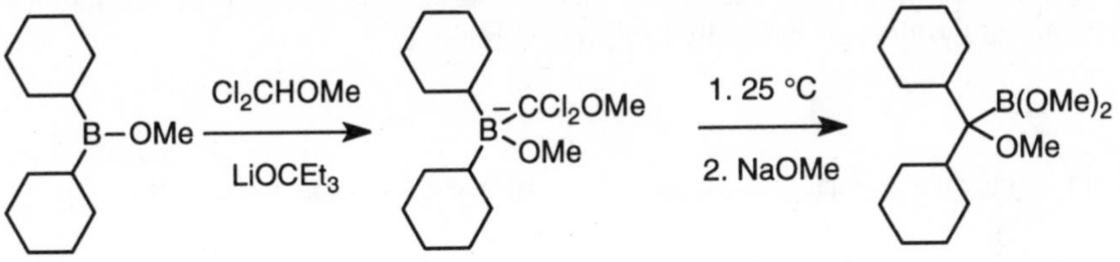

118   **119**

An interesting application of this reaction is its use in the conversion of medium-ring boracyclanes such as *B*-methoxyborecane (**120**) to the corresponding ketones such as cyclodecanone **121** [177].

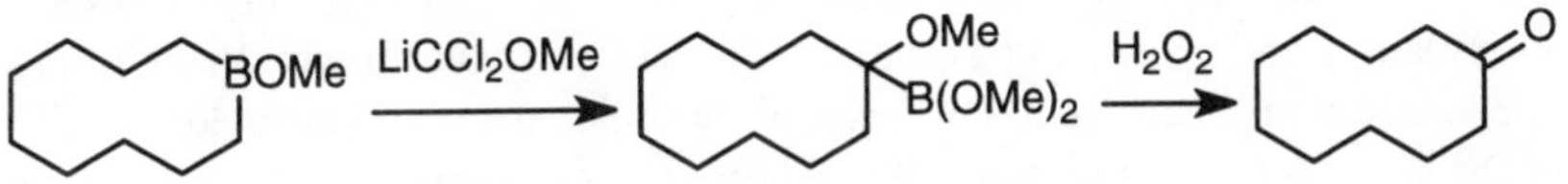

120   **121**

Preparations of α-haloalkylboronic esters via free radical additions of halocarbons or polar additions of hydrogen halides to double bonds are described in Section 3.5.5. Chlorination of a di-*tert*-butyl methylboronate with *tert*-butyl hypochlorite gave poor yields of chloromethylboronic esters [178], but radical bromination of *sec*-alkylboronic esters proved facile [179, 180]. As refined by Brown, Yamamoto, and coworkers, this reaction provides a useful route to achiral or racemic α-bromoalkylboronic esters such as 2-(1-bromocyclohexyl)-1,3,2-dioxaborinane (**122**) [181].

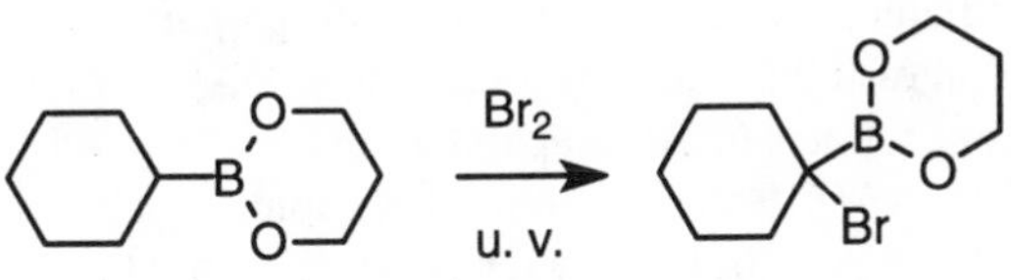

122

3.4.3.5 The stereocontrollable route to halo boronic esters. The fundamental reaction that is of utility in asymmetric synthesis is the reaction of (dichloromethyl)-lithium with boronic esters. (Dichloromethyl)lithium was discovered by Köbrich and coworkers [182], and Köbrich and Merkle found that triphenylborane reacts with α-chloro lithium reagents [183]. The assembly and rearrangement of (dichloromethyl)borates (**123**) from (dichloromethyl)boronic esters and alkyllithiums was reported by Rathke, Chao, and Wu [184], who did not isolate the resulting (α-chloroalkyl)boronic esters (**124**) but oxidized them to aldehydes. Inefficiencies in the oxidation step may well be the factor that obscured how efficient the formation of (α-chloroalkyl)boronic esters by this process really is.

$$R^1Li + Cl_2CH-B(OR^2)_2 \longrightarrow \underset{Cl_2CH}{\overset{R^1}{\diagdown}}\bar{B}(OR^2)_2 \longrightarrow R^1-\underset{Cl}{\overset{H}{\underset{|}{\overset{|}{C}}}}-B(OR^2)_2$$

123   **124**

The breakthrough to efficient synthesis of (α-haloalkyl)boronic esters (**125**) was the reaction of (dichloromethyl)lithium with boronic esters reported by Matteson and Majumdar [161, 185]. Yields were generally excellent with a variety of R groups, including primary and secondary alkyl, *tert*-butyl, aryl, alkenyl, (α-alkoxy)alkyl, and one having a remote carboxylic ester substituent. In addition to the 1,3,2-dioxaborolanes (ethylene glycol boronic esters) illustrated, 4,4,5,5-tetramethyl-1,3,2-dioxaborolanes (pinacol boronic esters) and 1,3,2-dioxaborinanes (1,3-propanediol esters) worked well, and open chain boronic esters were also successfully used.

125

Although –100 °C is indicated for the first step, this is merely the temperature required in order to preform (dichloromethyl)lithium. In situ preparation of (dichloromethyl)lithium by addition of lithium diisopropylamide (LDA) to a mixture of dichloromethane and other substrates in the temperature range –78 to –5 °C had been reported previously [186, 187]. Boronic esters also capture (dichloromethyl)lithium generated in situ at –78 °C [161] or any temperature within the –78 to –5 °C range [188]. The in situ preparation and capture method also provides the best synthesis of diisopropyl (dichloromethyl)boronate, $Cl_2CHB(O\text{-}iPr)_2$, from triisopropyl borate, dichloromethane, and LDA, and can be carried out at temperatures as high as –5 °C [189, 190].

3.4.3.6 Alkylation of (halomethyl)boronic esters. The practical synthesis of (halomethyl)boronic esters (Section 2.2.1.1) allows alkylation of these compounds to form boronic esters that might be difficult to prepare otherwise. An example is the synthesis of *tert*-butyl 4,4,5,5-tetramethyl-1,3,2-dioxaborolane-2-propanoate (**126**)

[38, 191]. In this case, the (chloromethyl)boronic ester is not reactive enough to give a good yield of **126** before the ester undergoes Claisen condensation. Whiting has extended this type of synthesis to other ester thiol ester, ketone, and amide enolates, and obtained fair to good yields [191]. One of the more favorable examples is the preparation of **127** from cyclopentanone enolate (83%).

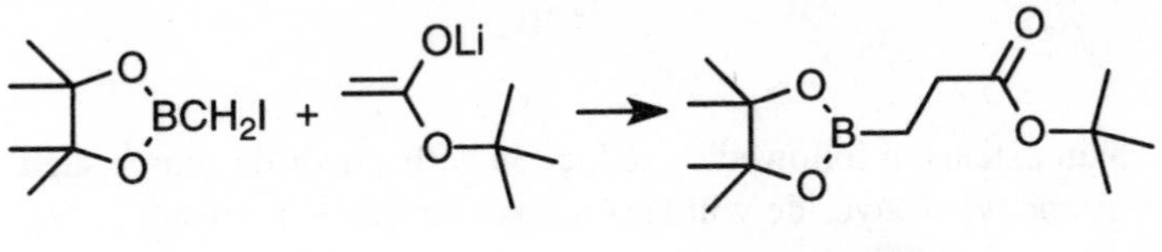

126

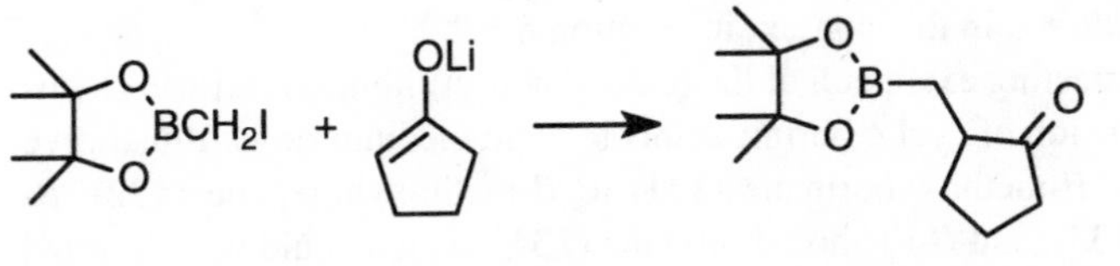

127

Precedents for the foregoing reactions are found in the reactions of methyl cyanoacetate anion with dibutyl iodomethylboronate [72] and malonic ester anion with pinacol (chloromethyl)boronate [192]. Also related is the use of pinacol (chloromethyl)boronate with alkenyl Grignard or lithium reagents to provide allylic boronic esters [193].

3.4.3.7 Simple homologation of boronic and borinic esters. The conversion of a boronic ester to its next higher homologue can be accomplished by reaction with (chloromethyl)lithium [194]. It was shown that the 1-phenyl-1-methylethyl group of **128** migrates with retention of configuration. (Bromomethyl)lithium [195] and (iodomethyl)lithium [196] have also been used. Inexplicably, BrCH$_2$Li gave poor yields with some functionalized boronic esters where ClCH$_2$Li worked well [195]. On the other hand, with the isoxazoline **129**, ClCH$_2$Li failed, BrCH$_2$Li yielded 40%, and ICH$_2$Li yielded 69% of **130** [196].

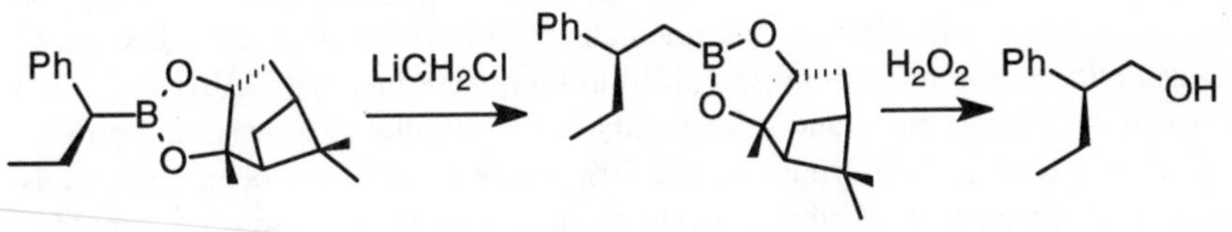

128

129   **130**

Alternatively, chain extension followed by reduction with a hydride source such as potassium triisopropoxyborohydride will accomplish the same net result [197]. This alternative route is valuable for preparing asymmetrically deuterated compounds (see Section 5.4.4.2 and 5.4.5). Chain extension by either route is valuable for replacing boron by carbon at chiral centers, and application and comparison of the two routes is discussed in this context in Section 6.3.2.2.

A particularly interesting extension of the (chloromethyl)lithium insertion chemistry is the ring expansion of cyclic borinic esters to form medium rings. Illustrative is the conversion of B-methoxyborinane (**131**) to B-methoxyborepane (**132**), B-methoxyborocane (**133**), and B-methoxyboronane (**134**), which could be converted by (dichloromethyl)lithium to the cyclononylboronic ester (**135**) as illustrated, or by treatment with LiCCl$_2$OMe (see Section 3.4.3.2) to cyclononanone [177, 198]. The ring expansion has been continued all the way up to the 12-membered ring.

131   **132**

133   **134**   **135**

3.4.3.8 Rearrangement of α,β-unsaturated borates. The Zweifel olefin synthesis converts alkenyldialkylboranes (**136**) to alkenes (**139**) via formation and rearrangement of an iodonium ion (**137**) which undergoes electrophilically induced alkyl migration to form a β-iodoborane intermediate (**138**), which then undergoes base catalyzed elimination to form alkene (**139**) in high stereopurity [199]. There are a number of useful variations and refinements of the original synthesis, though side reactions that tend to limit yields to the 70% range have never been completely overcome. Alternative synthetic pathways to similar objectives are often provided by Suzuki coupling of suitable borane intermediates. Further discussion of both of these alternatives will be deferred to Chapter 4.

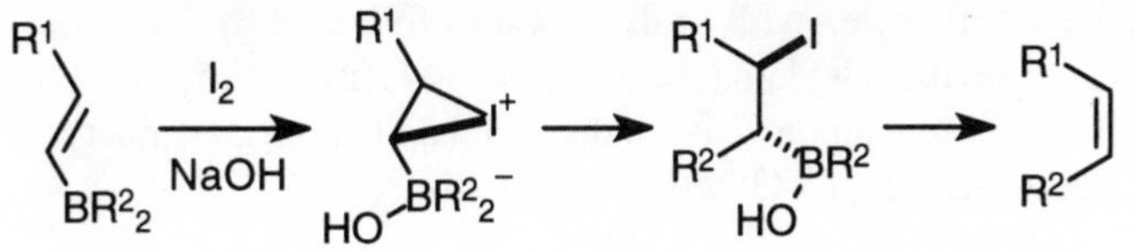

136   **137**   **138**   **139**

A vinylic carbon is made electrophilic enough to induce alkyl migration by conjugation with an ester, as in the base initiated rearrangement of **140** [200]. Rearrangement of (α-methoxyvinyl)trialkylborate anions to **141** (as its lithium methoxide salt) can occur spontaneously, or the Zweifel reaction can be induced by treatment with iodine [201], and a more recent route to **141** utilizes trimethylsilyl chloride to remove methoxide and promote rearrangement [202].

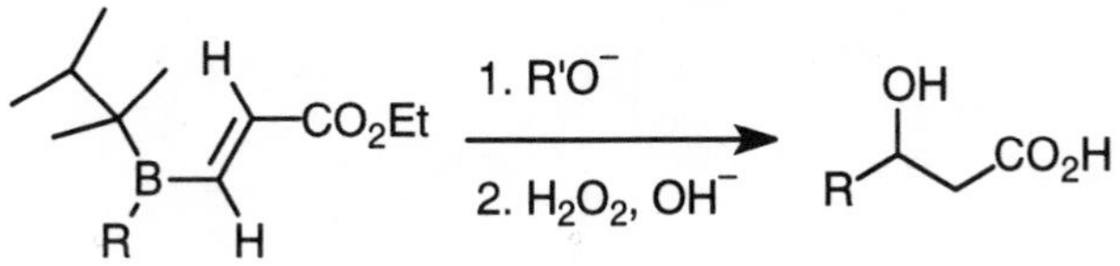

140

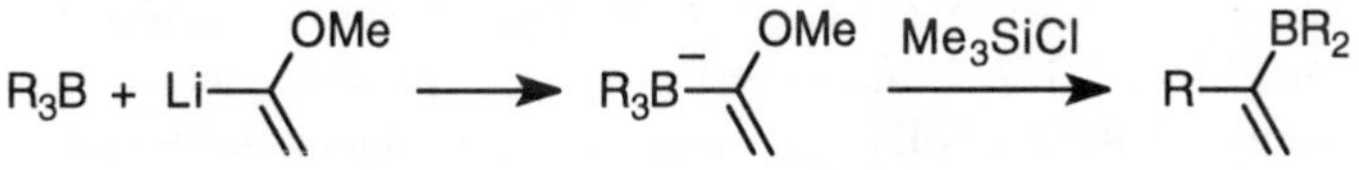

141

Alkynylborate salts undergo rearrangement to alkenylboranes on treatment with a variety of electrophilic reagents [203, 204]. An early stereoselective example was the reaction of the alkynylborate **142** with triethyloxonium fluoborate to produce the (E)-alkenylborane **143** in 97% isomeric purity [205]. Other electrophiles that have given good (E)-selectivity in analogous rearrangements include propionic acid [206], Me$_3$SiCl [207], and Bu$_3$SnCl [208]. Additional examples, not necessarily stereoselective, include hydrogen chloride [209, 210], dimethyl sulfate [211], epoxides [212], and Me$_2$N$^+$=CH$_2$ [213]. This is by no means an exhaustive list. An additional example is cited in Section 4.2.5.

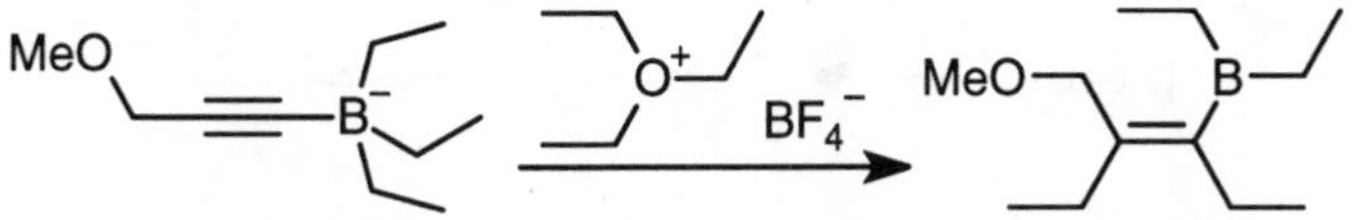

142   **143**

Reaction of lithiated heterocycles with trialkylboranes followed by iodination results in rearrangement to the alkylated heterocycle **144** [214, 215]. Carbon electrophiles such as iodoacetamide can induce related rearrangement in indolylborates, as in the formation of **145** [216].

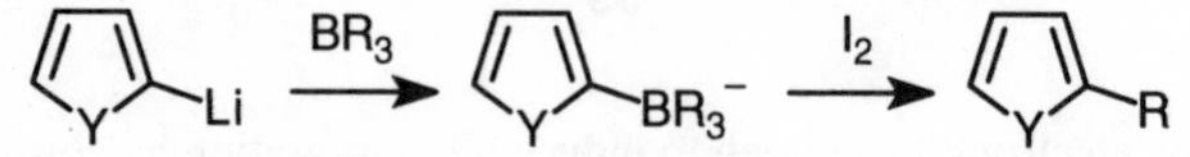

(Y = O, S, NCH₃)

144

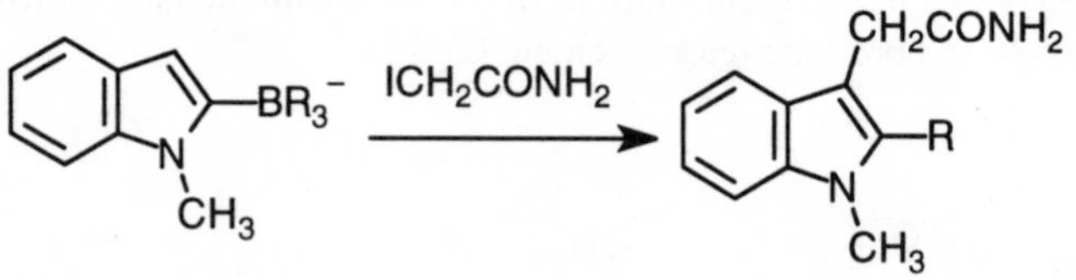

145

3.4.3.9 Carbonylation and cyanidation. Hillman reported that treatment of triethylborane (**146**) with carbon monoxide in aqueous medium under pressure yields carbonylation-rearrangement product **150** at 50 °C, followed by further rearrangement product **151** at 150 °C [217]. Reaction in the presence of diols yielded diol boronic esters, for example, the 1,3,2-dioxaborolane **152** [218], and in the presence of aldehydes, the cyclic borinic ester **150** forms cyclic acetals such as **153** [219]. Intermediates **147** and **148** were postulated by Hillman [217], and **149** is analogous to the first isolable rearrangement product from the reaction of isonitriles with trialkylboranes. Hillman also showed that other primary trialkylboranes behaved similarly, but he never determined whether the very high carbon monoxide pressures were really necessary.

A recent application of carbonylation is the conversion of boraadamantane–THF (**26**) to adamantylboronic ester **154** [48]. Chain extension of **154** with LiCHCl₂ then yielded chloro boronic ester **155**, which was methylated to **19**. The conversion of **19** to the antiviral compound rimantidine (**20**) has been described in Section 3.2.3.2, and the preparation of boraadamantane (**26**) is described briefly in Section 3.4.5.2.

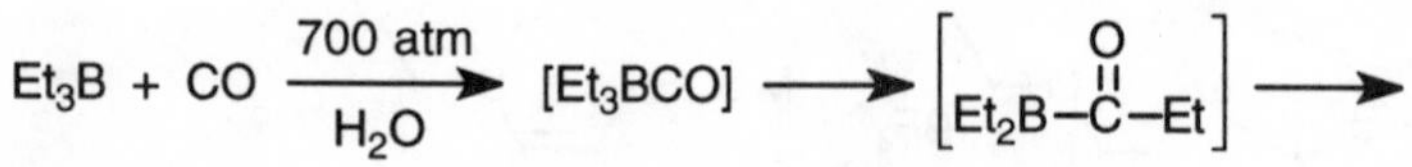

146   **147**   **148**

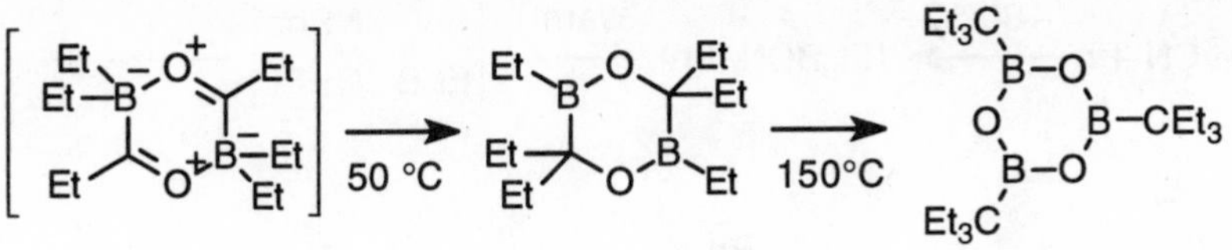

149   **150**   **151**

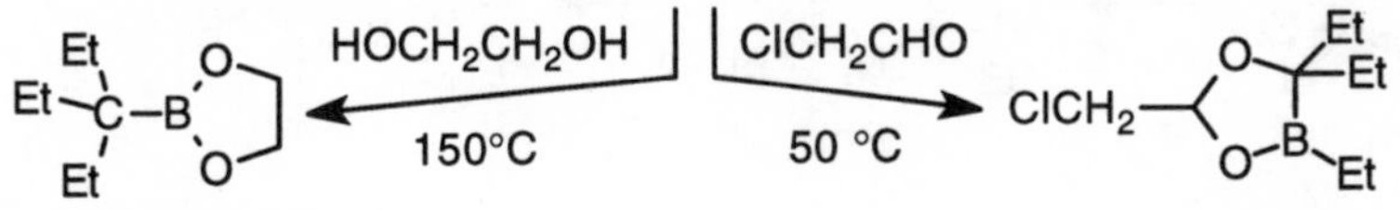

152   **153**

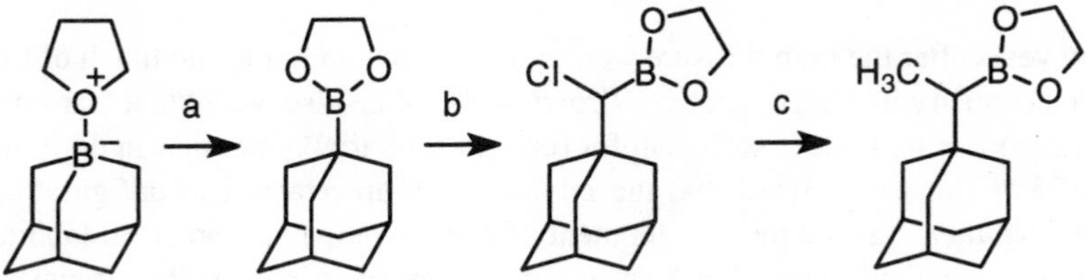

26   **154**   **155**   **19**

a, CO, HOCH₂CH₂OH. b, LiCHCl₂. c, MeLi.

The first isolable product from *tert*-butyl isocyanide and trimethylborane is the simple adduct **156**, which rearranges to stable monomer **157**, confirmed by mass spectroscopic measurement [220]. Triphenylborane and cyclohexyl isocyanide also formed an isolable monomer analogous to **156** [221]. The less hindered phenyl isocyanide with triethylborane in refluxing diethyl ether yielded dimer **158** as the first isolable product [222, 223, 224]. The ethyl isocyanide analogue of **158** was formed together with an enamine tautomer, which were separated and characterized [225]. Heating **158** to 180-200 °C resulted in rearrangement to **159**. Compounds **156**, **157**, and **158** are analogues of postulated intermediates **147**, **148**, and **149**, respectively, in the carbon monoxide sequence. At 300 °C, **159** rearranges to the 1,3,4,2-diazadiborolidine **160** and resists further rearrangement [226]. The rearrangement of an analogue of **159** to an analogue of **160** having butyl groups in place of ethyl groups was also obtained by aluminum chloride catalyzed rearrangement [227].

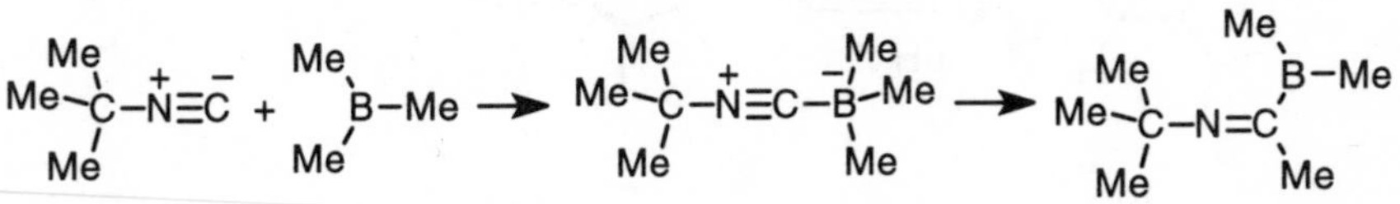

156   **157**

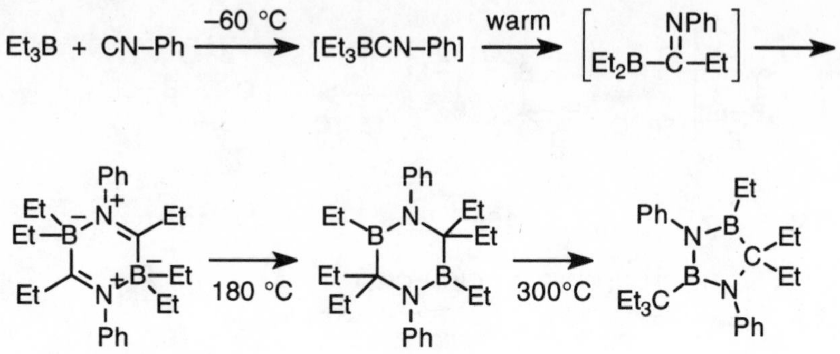

158   **159**   **160**

A few years after the high pressure work with carbon monoxide and much of the isonitrile chemistry had been reported, Brown and Rathke discovered that 1 atmosphere of carbon monoxide is sufficient for reaction with trialkylboranes in diglyme at 100-125 °C, and confirmed that the migrating group retains its configuration [228]. Brown then pursued the development of carbonylation as a practical laboratory synthetic method. This chemistry has been reviewed repeatedly elsewhere [229, 230, 231], and what follows is only a very brief summary of the most important points.

The best system found for carbonylation to form aldehydes or carbinols appears to be boranes prepared from 9-BBN in the presence of potassium triisopropoxyborohydride [232]. For example, the racemic borane **161** can be converted stereospecifically to the borinic ester **162**, which is oxidized to the aldehyde **163** or hydrolyzed with base to the ketone **164**.

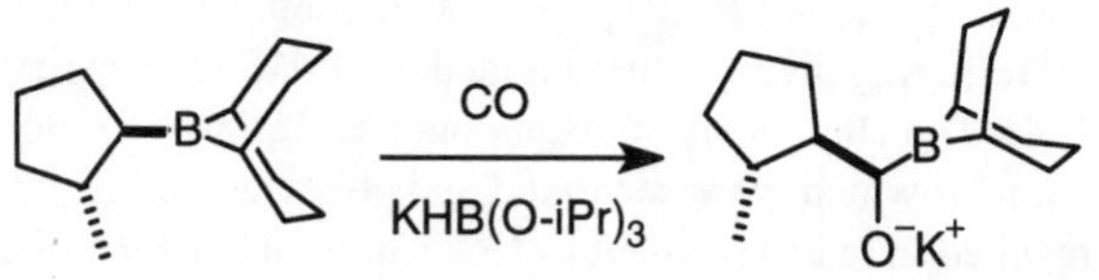

161   **162** (probably dimeric)

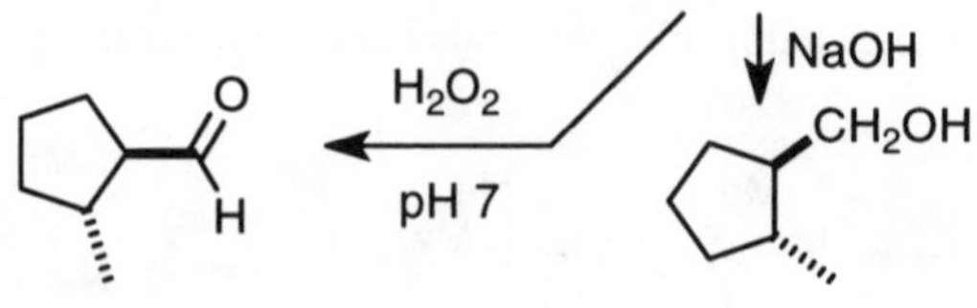

163   **164**

Thexyl groups generally migrate more slowly than other alkyl groups. Hydroboration with a thexylmonoalkylborane, for example **165**, to a thexyl-dialkylborane, **166**, can be followed by the carbon monoxide reaction to make ketones, as in the synthesis of juvabione (**167**) [233].

The carbon monoxide reactions described in this section are obsolete for most synthetic purposes, as carbonylation with LiCCl$_2$OMe (see Section 3.4.3.4) works with borinic esters derived from asymmetric organoboron compounds. Borinic esters are usually easier to make and react more efficiently than trialkylboranes as intermediates for asymmetric synthesis.

Carbon monoxide is useful for the synthesis of ^{11}C, ^{13}C and ^{14}C labeled compounds, since labeled CO is a usual source of such isotopes [234]. The 34-minute half-life of ^{11}C allows its use only in the faster hydride-induced carbonylations.

A useful alternative to carbonylation which is often easier to carry out under laboratory conditions is the cyanidation sequence discovered by Pelter and coworkers [235, 236, 237, 238]. For example, tricyclopentylborane (**168**) in THF dissolves sodium or potassium cyanide to form the cyanoborate **169**, which on acylation rearranges to intermediate heterocycle **170** [235]. Oxidation of **170** with hydrogen peroxide yields dicyclopentyl ketone (**171**). With benzoyl chloride as the acylating agent, intermediate heterocycle **170** (R = Ph) was isolated as the propionate salt and as the hydrate. However, trifluoroacetic anhydride consistently gave better yields of ketones (84-100%) from a series of trialkylboranes (R = *n*-butyl, *n*-octyl, cyclopentyl, cyclohexyl, *exo*-2-norbornyl).

Thexylborane (**172**) was used to prepare mixed trialkylboranes for the cyanidation process, making possible such sequences as the preparation of cyclopentyl *n*-alkyl ketones (**173**) [235]. Thexylchloroborane provides a wider range of possible substrates for this reaction [239].

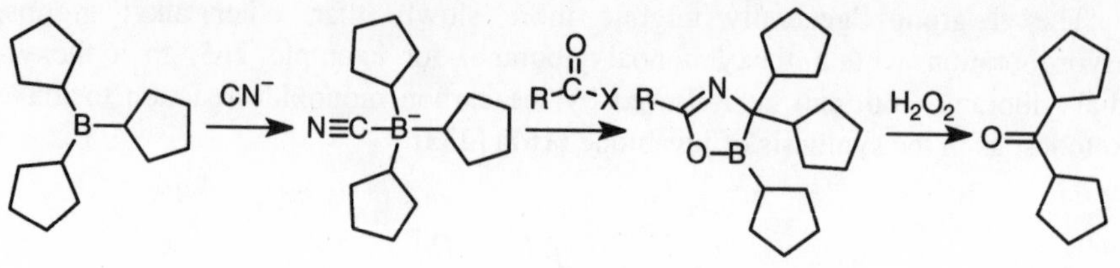

168   **169**   (R = Ph or CF₃)   **170**   **171**

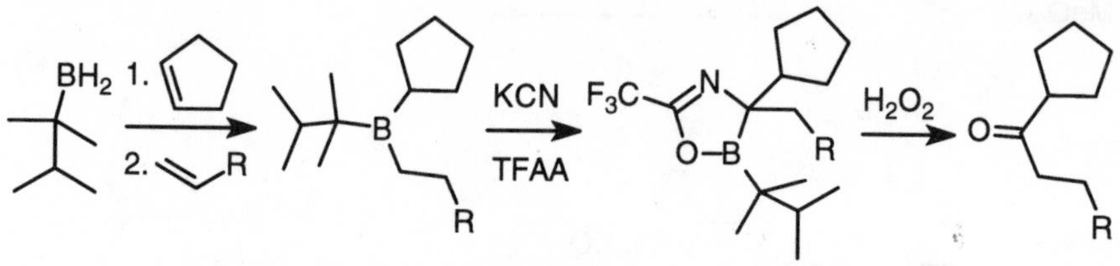

172   [R = n-C₆H₁₃, (CH₂)₆Cl; TFAA = (CF₃CO)₂O]   **173**

The use of three moles of trifluoroacetic anhydride with R_3BCN^- results in (trialkylmethyl)boron intermediates (**174**), which on peroxidic oxidation yield trialkylcarbinols (**175**) [236].

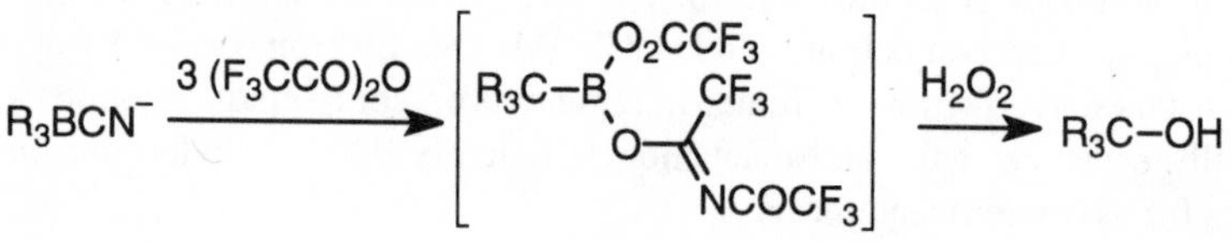

174   **175**

It was shown that protonation of R_3BCN^- causes rearrangement, but an intermediate traps cyanide ion and restricts the yield to 50% maximum [237]. It was also shown that a *trans*-2-methylcyclopentyl group retains configuration during migration, and that the migratory aptitudes are primary > secondary > tertiary [238].

Cyanidation has also proved to be a useful method for introducing isotopic carbon labels [234]. Ketones can be obtained rapidly enough to permit ¹¹C labeling by this route [240].

3.4.3.10 Other chain extensions. Boronic esters, for example **176**, are converted stereoselectively to aldehydes such as **177** by a sequence utilizing [(methoxy)-(phenylthio)methyl]lithium [241]. This route was believed at the time to be more efficient that oxidation of a α-chloro boronic esters, but the rearrangement step requires mercuric chloride catalysis, and recent developments have indicated that the α-chloro boronic ester route can be easier and equally effective (see Section 5.4.3.5).

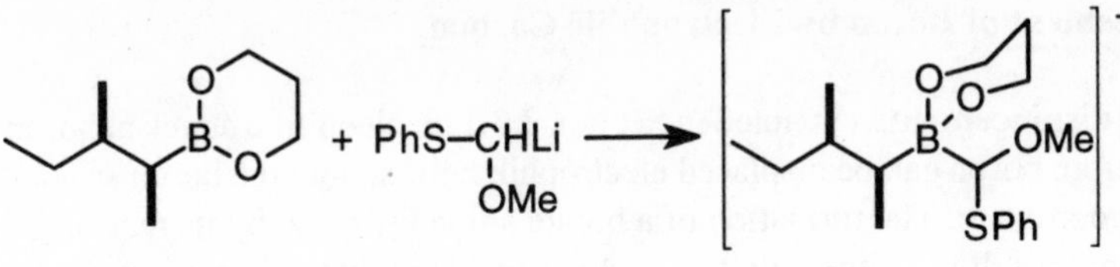

176

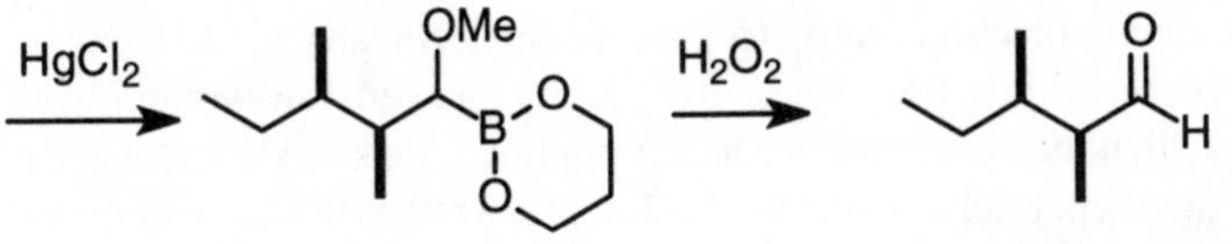

177 (96% *erythro*)

Chain extension of a series of boronic esters (**178**) with [(trimethylsilyl)-(chloro)methyl]lithium to form α-trimethylsilyl boronic esters (**179**) proceeded in 77-86% yields [149].

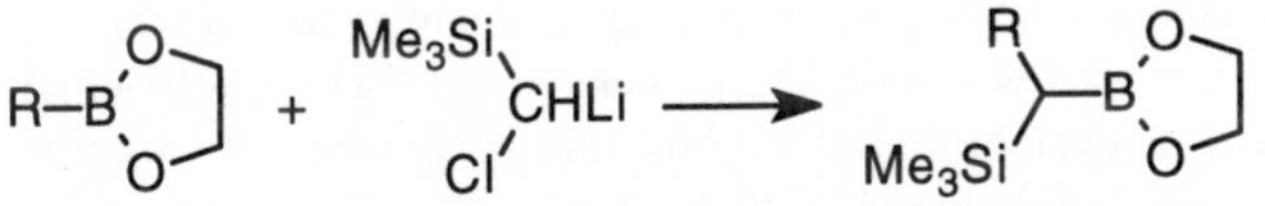

178   **179**

(R = Bu, EtMeCH-, octyl, cyclopentyl, cyclohexyl, Ph, PhCH$_2$, PhSCH$_2$)

The use of a pinanediol boronic ester (**180**) with Me$_3$SiCHClLi led to the known (S)-(–)-α-(trimethylsilyl)benzyl alcohol (**181**) in 46% ee [242]. The chirality is determined by enantioface selection in the attack of Me$_3$SiCHClLi on the asymmetrically substituted boron atom, and contrasts with the much better stereoselection in the intramolecular S$_N$2 reaction of alkyl(dichloromethyl)borates described in Chapter 5. If there were motivation to do so, it would seem likely that the mediocre stereoselection seen in the only example tried to date could be considerably improved upon.

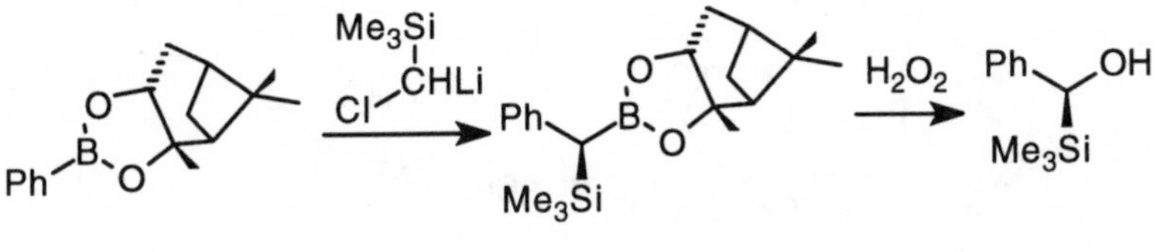

180   **181** (46% ee)

3.4.4 Displacement of Boron by Electrophilic Carbon

3.4.4.1 Polar displacements. As noted in Section 3.3.2, nucleophilic attack at boron is required before boron can be displaced electrophilically, so that the boron species cleaved is tricoordinate. The formation of a borate anion followed by its rearrangement has the net result of connecting two carbons and leaving the boron connected to one of them. Reactions considered here are those that create a carbon—carbon bond at the expense of a boron—carbon bond and transfer the boron to a more electronegative ligand. Because of the unfavorable energetics of forming any divalent BY_2^+ species, these reactions have to provide some means of delivering a nucleophile Y^- to boron so that BY_3 can be the species cleaved, but without using such nucleophilic Y^- that the attacking carbon electrophile is deactivated. Outside of allylic systems, treated separately in Section 3.4.5, this is a rather limited category of reactions.

Unsaturated boranes are the best candidates for direct attack by electrophilic carbon, and can function in intermolecular processes. Brown and Jacob found that alkenyl-9-BBN's react with aldehydes with direct displacement of the boron [243]. The best yield reported (86%) was with benzaldehyde to form **182**. Brown and coworkers have also reported the reaction of *B*-(1-alkynyl)-9-BBN's, which are prepared from *B*-methoxy-9-BBN and the alkynyllithium [244], to aldehydes and ketones to form propargyl alcohols [245]. Representative examples include the reaction of *B*-5-chloro-1-pentynyl-9-BBN (**183**) with propionaldehyde to form the corresponding propargylic alcohol **184** and the similar behavior of *B*-[1-(3,3-dimethyl-butynyl)]-9-BBN (**185**) with cyclohexanone to yield **186**. This process has recently been extended to *B*-(trimethylsilylethynyl)-9-BBN [246].

182

183   **184** (78%)

185   **186** (97%)

A very important recent development is the catalyzed asymmetric alkynylation of aldehydes by boranes reported by Corey and Cimprich (see Section 9.4.1).

Presumed enol ether boronic ester intermediate **187** derived from *B*-iodo-9-BBN and ethoxyacetylene reacts readily with aldehydes to form postulated intermediate **188**, which hydrolyzes to α,β-unsaturated esters (**189**) [247]. The group R can be saturated, unsaturated, or aromatic.

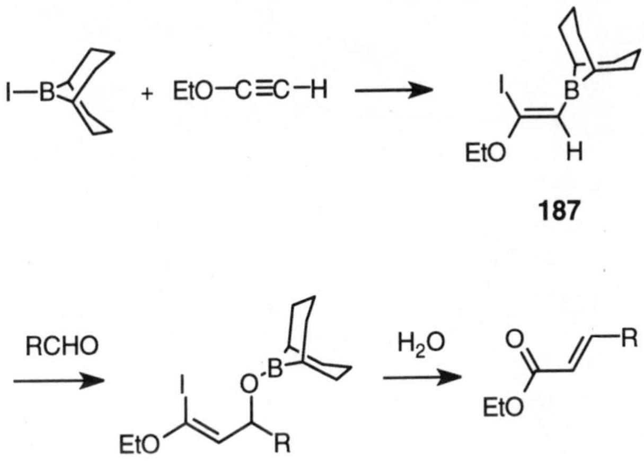

187

188   **189**

α,β-Unsaturated ketones also replace boron from **187** to yield the Michael adducts **190** [248]. A cyclic transition state was suggested, and this is supported by the fact that cyclohexenone fails to react. The reaction is noteworthy because there appears to be only 1,4-addition and no 1,2-addition of **187** to ketones. This is a special case of the more general Michael addition of alkenylboranes to α,β-unsaturated ketones, which leads to controlled alkene geometry and is discussed in Section 4.4.2.

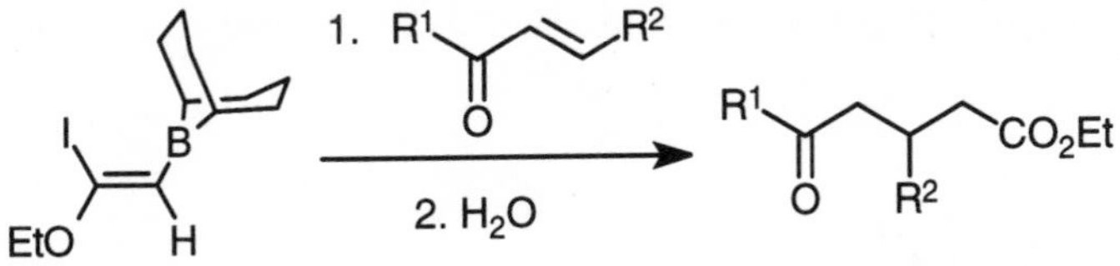

187   **190**

R^1 = alkyl or Ph; R^2 = H, Me, Ph

3.4.4.2 Small ring formation. Cyclopropane is a major product from the hydroboration of allyl chloride followed by treatment with sodium hydroxide [157]. The yield of cyclopropane is much improved if 9-BBN is used as the hydroborating agent [249], and a synthesis of more general interest can be carried out with the dihydroboration product (191) from propargyl chloride, which yields a cyclo-propylborane (192) [250]. Similar reaction of the dihydroboration product (193) of homopropargyl tosylate yields the analogous cyclobutylborane (194).

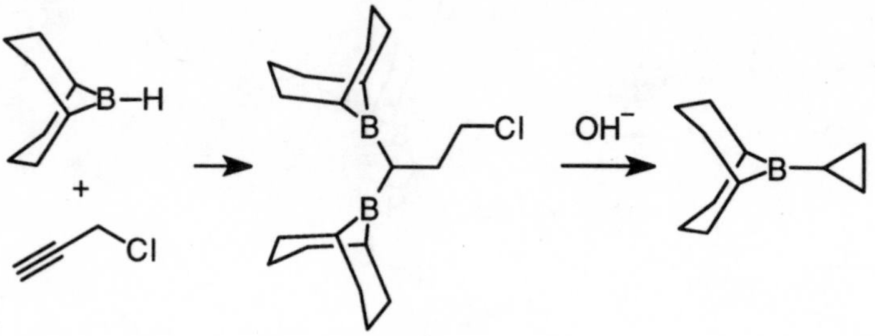

191   192

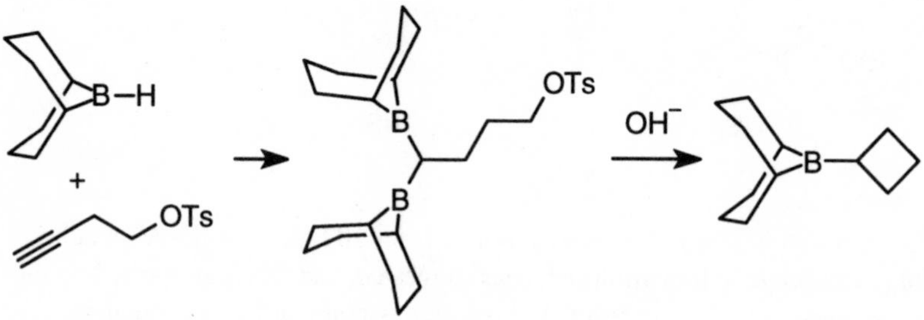

193   194

Of interest for possible asymmetric applications is the observation by Goering and Trenbeath that the boron is displaced with inversion of configuration, as illustrated by the behavior of the diastereomeric β-chloroalkylboranes 195 and 196 [251].

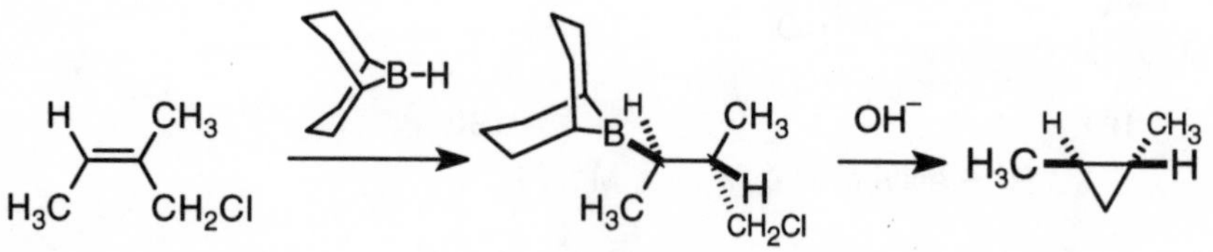

195

196

There is some mechanistic analogy between the foregoing and the earlier observation that electrophilic ring closure of *exo*-norbornenylboronic acid (**197**) in its reaction with mercuric chloride produces nortricyclylmercuric chloride (**198**) [252]. The *endo*-isomer of **197** also yielded **198** but at a 400-fold slower rate at 25 °C in buffered aqueous acetone [253]. The isotope effect indicated B—C bond breaking in the rate determining step [254]. Thus, displacement of boron by carbon from carbon with inversion was faster than similar displacement with retention. Parallel reactions of the analogous bicyclooctyl compounds with mercuric chloride [255] and an inefficient reaction of **197** (but not its *endo*-isomer) with 2,4-dinitrobenzenesulfenyl chloride [252] were found.

197   **198**

3.4.4.3 Radical displacement mechanisms. Reaction of α,β-unsaturated carbonyl compounds with trialkylboranes has the net effect of displacing boron from carbon by carbon, with concomitant transfer of boron to oxygen. A simple example is the reaction of acrolein with triethylborane to form the diethylboron enolate of pentanal (**199**) [256]. Free radical intermediates are involved, as shown by inhibition of the reaction by galvinoxyl. There is inherently no way that the stereochemistry of the carbon—carbon connection can be directed by groups on boron in reactions of this type, and they will therefore not be reviewed further here.

199

Nitrones such as **200** are alkylated at carbon by trialkylboranes at 110 °C in THF in sealed tubes to produce hydroxylamines (**201**) [257]. Two of the three alkyl groups of the borane can react. The stereochemistry was not tested and there is thus no mechanistic evidence, but a radical process is a likely possibility.

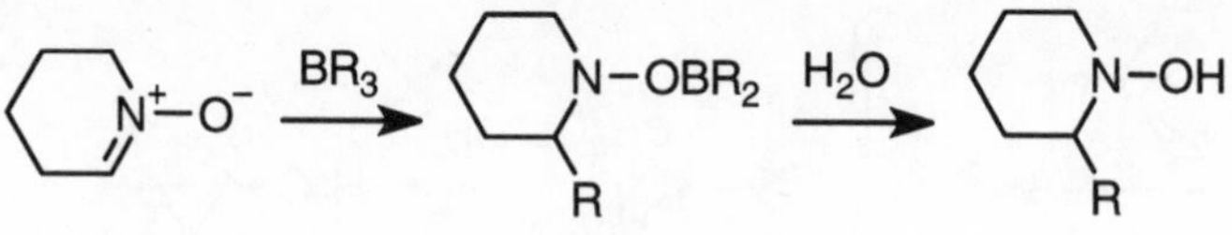

200   (R = Et, Bu, *sec*-Bu)   **201**

3.4.5 Allylborane Chemistry

3.4.5.1 Introduction. Boron transfers allyl groups to electrophilic carbon via the S_E2' mechanism, which results in allylic rearrangement. Usually cyclic transition states are involved, and excellent stereocontrol can result. The use of such stereo-controlled reactions in synthesis is fully reviewed in Chapter 7. Allylborane chemistry also includes a number of interesting reactions that are not known to provide stereocontrol, and these are reviewed in this section.

3.4.5.2 Triallylborane. The reaction of triallylborane with aldehydes and ketones to form boronic esters of homoallyl alcohols was discovered by Mikhailov and Bubnov [258], and allylic inversion in the reaction of tricrotylborane with formaldehyde was noted soon afterward [259]. The general reaction of aldehydes with allylic boronic esters was discovered by a French group [260]. Subsequent Russian work showed that allylic inversion is the usual pathway for electrophilic replacements of boron from allylic carbon [261, 262, 263, 264].

Recently reported reactions of triallylborane include the diastereoselective reductive diallylation of pyridine that occurs when pyridine—triallylborane (**202**) is warmed in water, methanol, or ethanol at 40-60 °C for 2 h [265, 266, 267]. The hydroxylic solvent is necessary for the reaction, and in its absence **202** can be recovered unchanged after 6 h at 160 °C. The initial product, *trans*-2,6-diallyl-1,2,5,6-tetrahydropyridine (**203**), was hydrogenated to **204**.

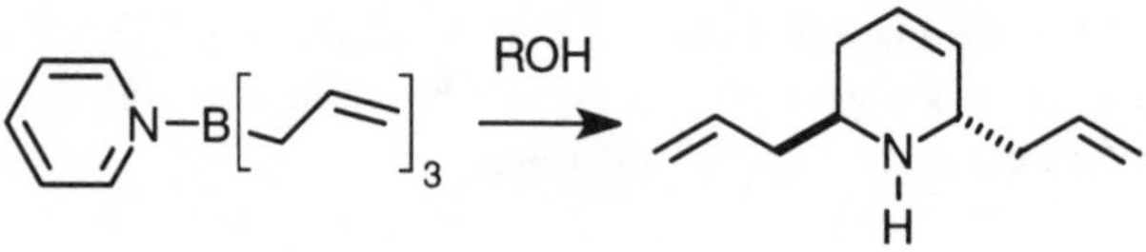

202   **203** (40-80%)

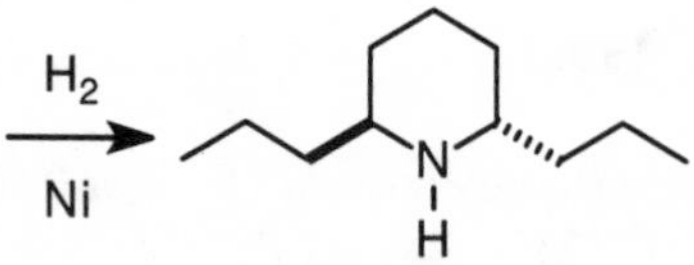

204

Heating **203** or 3-bromo-**203** with triallylborane 6 h at 130 °C followed by hydrolytic work up results in nearly quantitative conversion to the corresponding *cis*-2,6-diallyl-1,2,5,6-tetrahydropyridine [268]. Though the detailed mechanism is obscure, the position of the equilibrium is readily understood when it is considered that the diallylboryl derivatives **205** and **206** are probably the species equilibrated.

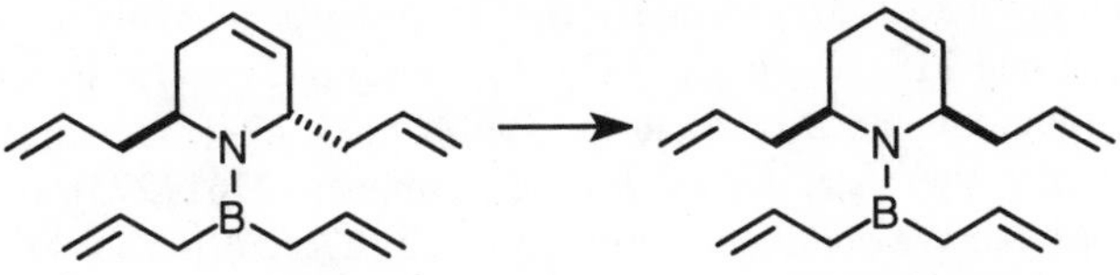

205   **206**

Quinoline and phenanthrene undergo monoallylation by triallylborane [269]. Isoquinoline is monoallylated by either triallylborane or allyldipropylborane at 0-20 °C. Treatment of the allyldipropylborane derivative with ethanol leads to simple replacement of N—BPr$_2$ by N—H, but treatment of the triallylborane derivative **207** with methanol results in diallylation to **208** [270].

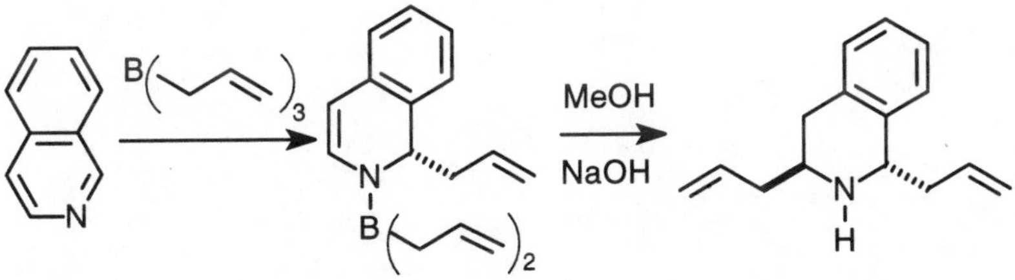

207   **208**

Pyrrole and indole are more reactive than quinoline toward allylboranes. Pyrrole with triallylborane yields mainly *trans*-2,5-diallylpyrrolidine (**209**), but also some 2-allyl-2,5-dihydropyrrole (**210**), the product of double bond migration, and indole yields 2-allylindoline [271]. Pyrrole can be monoallylated to **211** with allyldipropylborane, or converted to 1,1-dimethylallyl derivatives by treatment with triprenylborane and hydrolytic work up, as illustrated by the conversion of **211** to **212** [272].

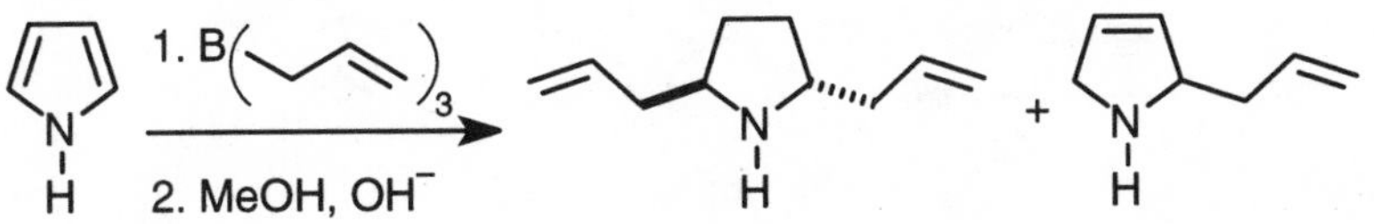

209 (61%)   **210** (15%)

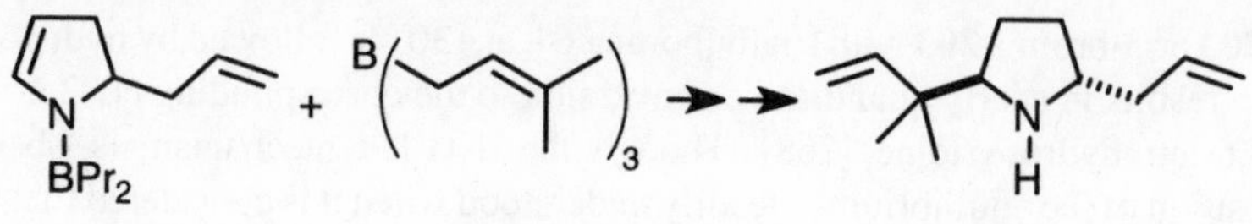

211   **212**

The reaction of triallylborane (**213**) with acetylenes to produce intermediates such as the illustrated **214** and **215**, which are useful for building complex structures, was reported by Mikhailov and Bubnov in the 1960's [273, 274]. The most interesting products from this chemistry are the boraadamantanes (**216**) [275], on which Bubnov and coworkers have continued active research [276-279]. The conversion of boraadamantanes to adamantylboranes has been noted in Section 3.4.3.9, and conversion to azaadamantane in Section 3.2.3.2.

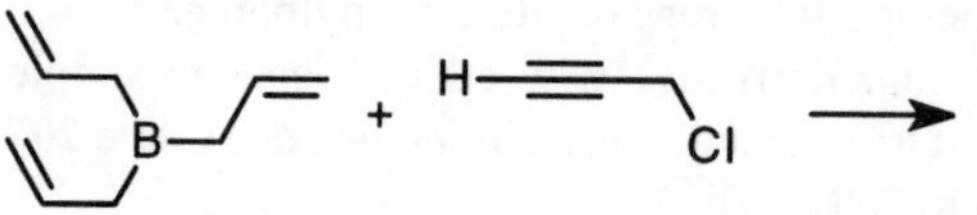

213

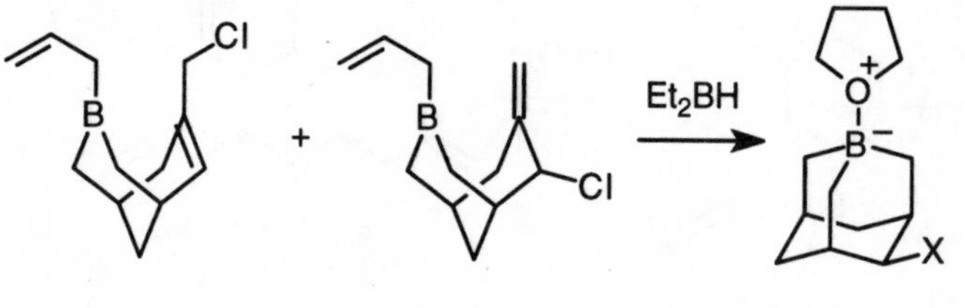

214   **215**   **216** (X = H or Cl)

Ethoxyacetylene reacts with allylic boranes to form 2-ethoxy-1,4-pentadienyl-boranes such as **217** [280], and these can be converted to allylacetylenes by treatment with an aluminum alkyl [281]. A recent improvement is the use of ethoxy-trimethylsilylacetylene. Adducts such as **218** eliminate ethoxyborane spontaneously at 20 °C and provide allyltrimethylsilylacetylenes such as **219** efficiently with a variety of allylic borane substrates [282]. The net result of these reactions is to replace the dialkylboryl group, with allylic rearrangement, by an acetylenic group.

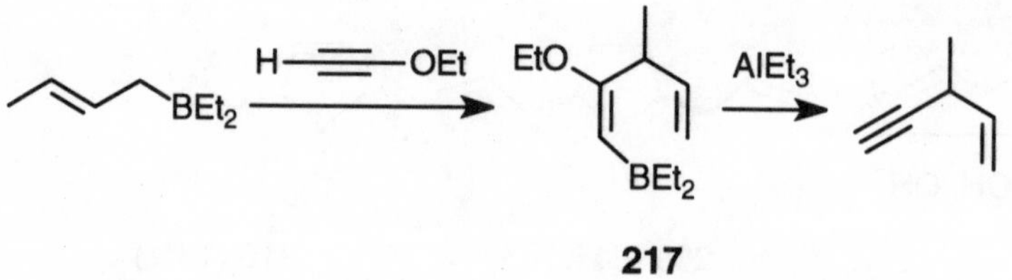

217

218   **219 (73%)**

3.4.5.3 Intramolecular allylic rearrangements. A primary consideration that must be kept in mind when designing systems for the stereocontrolled use of allylic boranes is the ease of allylic rearrangement, which has been known since the early days of allylborane chemistry [283]. Since rearrangement of a (Z)-crotylborane (**220**) to a less stable methylallylborane (**221**) results in equilibration with the (E)-crotylborane (**222**) [284], it is necessary to avoid such scrambling if a stereo-selective reaction is desired.

220   **221**   **222**

The thermodynamic parameters for such rearrangements have recently been measured by Bubnov and coworkers using NMR methods [285, 286, 287, 288]. For example, (E)-5-(dipropylboryl)-1,3-pentadiene (**223**) isomerizes via 3-(dipropyl-boryl)-1,4-pentadiene (**224**) to regenerate itself with the boron at the other end (**225**), or intermediate **224** rearranges via conformer **226** to (Z)-5-(dipropylboryl)-1,3-pentadiene (**227**) or its reflection (**228**) [287]. There was not enough **224** present at equilibrium to measure, but $\Delta G°$ for conversion of **223** to **227** was found to be +3.5 kJ mol^{-1} at 366 K. The measured value of $\Delta G^{\ddagger}$ for interconversion of **223** and **225** was 82.3 kJ mol^{-1} at 366 K. Conversion of **223** to **227** requires a bit higher activation energy, which reflects the higher energy of conformer **226** compared to **224**. The figures given [287] for the **223** to **227** and **227** to **228** interconversions are not entirely self consistent, but apparently lie in the range 85.5-91.5 kJ mol^{-1}.

223   **224**   **225**

227   **226**   **228**

Naive application of the Woodward-Hoffmann rules might suggest that the suprafacial 1,3-shift of boron outlined is forbidden, but the vacant *p*-orbital of boron enters into the process and the Woodward-Hoffmann rules do not apply. It is not possible that the boron undergoes direct 1,5-shift, as the interconversion of **223** and **225**, for which such a mechanism is sterically impossible, is faster than the interconversion of **227** and **228** [287].

The major consideration here from a synthetic point of view is that these reactions measurable on the NMR time scale are practically instantaneous on the synthetic operations time scale at temperatures above 25 °C. It was also noted that the dienyl system rearranges somewhat more slowly than simple crotyl.

Other kinds of rearrangements occur at higher temperatures. Thus, at 135 °C, **223** cyclizes with loss of propene to the borinene **229**, which then rearranges to other products [287]. (3*E*,5*E*)-7-(Dipropylboryl)-1,3,5-heptatriene (**230**) isomerizes via a series of 1,3-boron migrations at rates measurable on the NMR time scale at 87 °C to form (3*Z*,5*Z*)-7-(dipropylboryl)-1,3,5-heptatriene (**231**), which undergoes antarafacial 1,7-hydrogen transfer to form (1*E*,3*Z*,5*Z*)-1-(dipropylboryl)-1,3,5-heptatriene (**232**), together with 5% of its (*Z*,*Z*,*Z*)-isomer, within ~ 3 h [288].

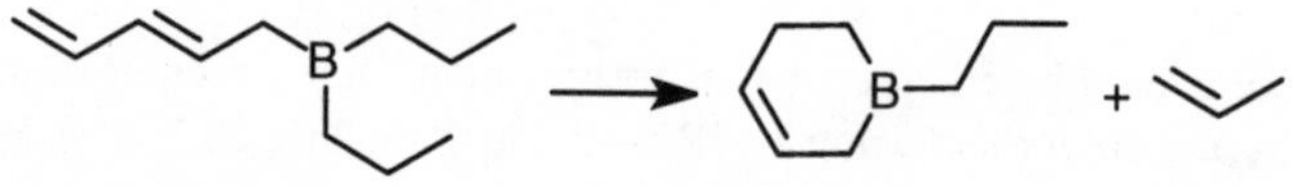

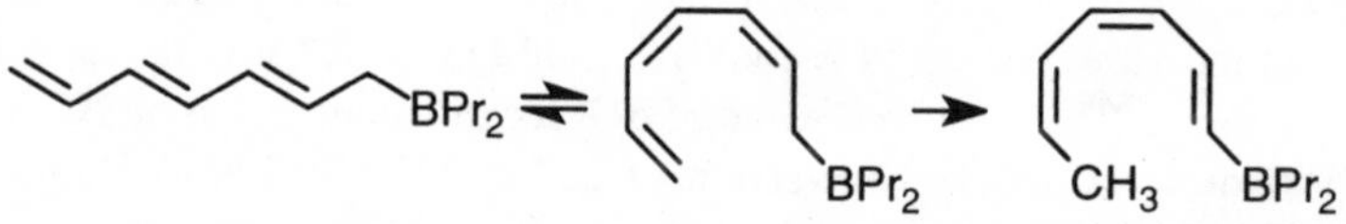

3.4.5.4 Allylic bisboranes. Allylic bisboranes are another invention of Bubnov and coworkers [289]. These compounds can provide routes to rapid construction of complex structures. For example, the bisborane **233** derived from dilithioisobutylene reacts with aldehydes or ketones.

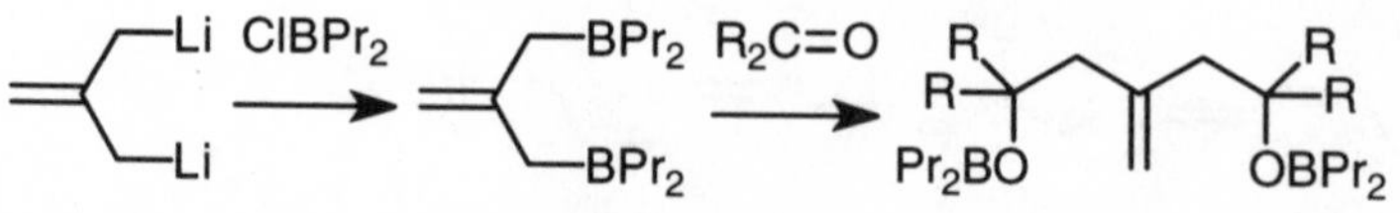

Boronic ester **234** was prepared from the corresponding bis(dipropylboryl) compound and trimethyl borate and found to undergo stereoselective reactions with diketones. The resulting dienes such as **235** are potentially useful in Diels-Alder reactions [289, 290]. The bis(dipropylboryl) analogue of **234** undergoes analogous reactions with 1,2-diimines [291].

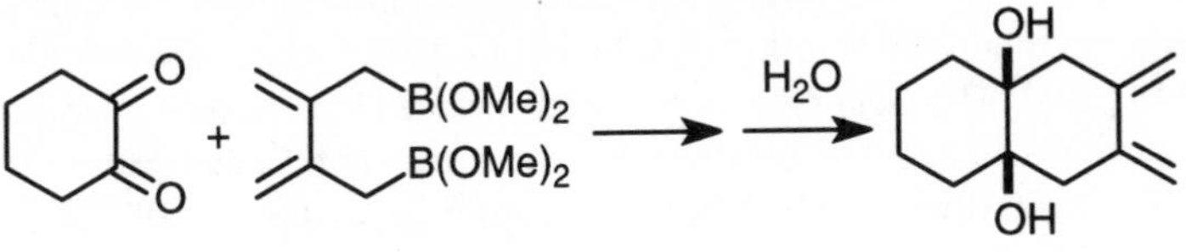

234   **235**

Reaction of 1,6-bis(dipropylboryl)-2,4-hexadiene (**236**), which is a 4:5 mixture of (*E,E*)- and (*E,Z*)-isomers, with acetone yields only the (*R**,*R**)-2,3-divinyl-1,4-diols **237**. Treatment of **237** with iodine and sodium bicarbonate leads to dioxabicyclooctane **238** [292]. Analogous sequences starting from the dipropylboryl analogue of **234** yield mixtures of 5- and 6-membered ring products [293].

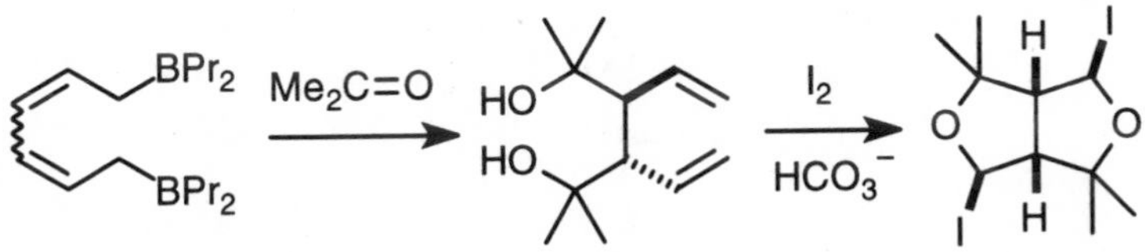

236   **237**   **238**

B-(Cycloalkenylmethyl)-*B,B*-(dipropyl)boranes react as typical allylic boranes with a variety of electrophilic substrates to form methylenecycloalkyl derivatives [294]. For example, *B*-(cyclohexenylmethyl)-*B,B*-(dipropyl)borane with benzonitrile yields the adduct **239**. More reactive substrates such as phenyl isocyanate, phenyl isothiocyanate, and the methylimine derivative of benzaldehyde react in a similar pattern, and *B*-(cyclopentenylmethyl)-*B,B*-(dipropyl)borane is also a useful substrate.

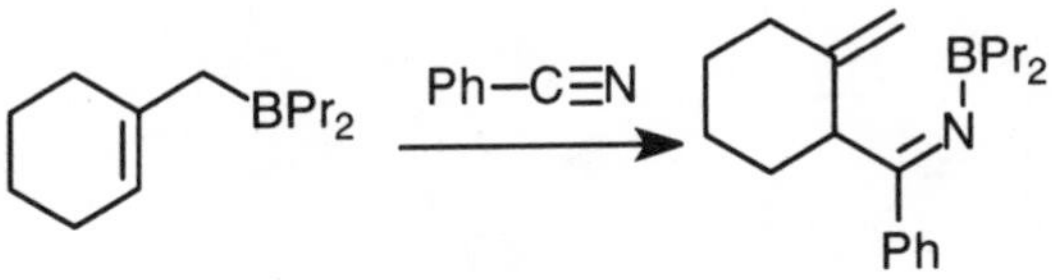

239

3.4.5.5 Three-carbon chain extension of boronic esters. Allylic boronic esters can rearrange, and this property has recently been used by Brown and coworkers to provide a very interesting and potentially useful three-carbon chain extension of boronic esters [295]. The first step is the in situ generation of (α-chloroallyl)lithium from allyl chloride and LDA [296]. A typical example is the reaction of *trans*-2-methylcyclopentylboronic ester **240** with (α-chloroallyl)lithium to form postulated borate intermediate **241**, which rearranges to isolable intermediate boronic ester **242**, which rearranges thermally in refluxing toluene (24-36 h) to the chain extended boronic ester **243** [295]. The (*E*)/(*Z*) ratio of the product **243** is about 9:1, and of course the double bond can be hydrogenated if a simple chain extension by insertion of –(CH$_2$)$_3$– is the desired net result. Yields are improved by the use of lithium dicyclohexylamide in place of LDA [297]. Although demonstrated with racemic substrates, there is no apparent loss of stereochemistry at the chiral center where chain extension takes place.

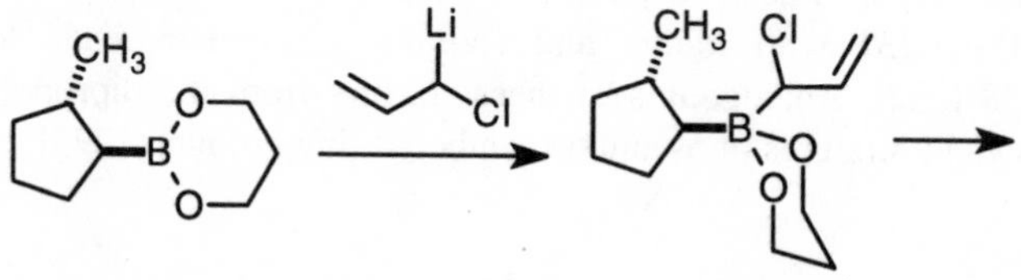

240   **241**

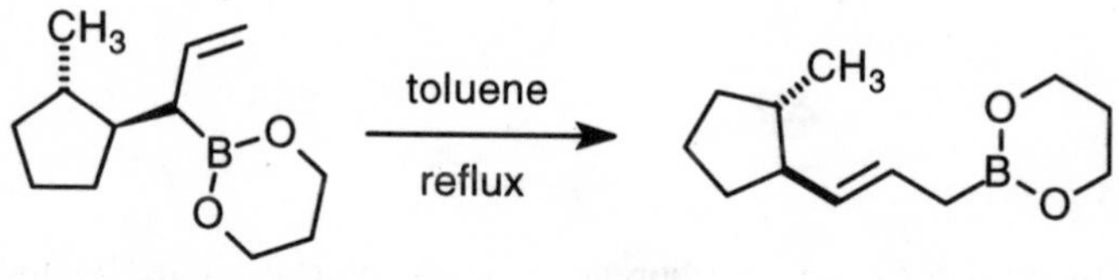

242   **243**

Naturally, it is also possible to approach allylic boronic ester intermediates **245** via (α-chloroallyl)boronic esters such as **244**, and rearrangement to the expected **246** follows [298]. The preparation of **244** was accomplished via reaction of triisopropyl borate with (α-chloroallyl)lithium generated in situ [299].

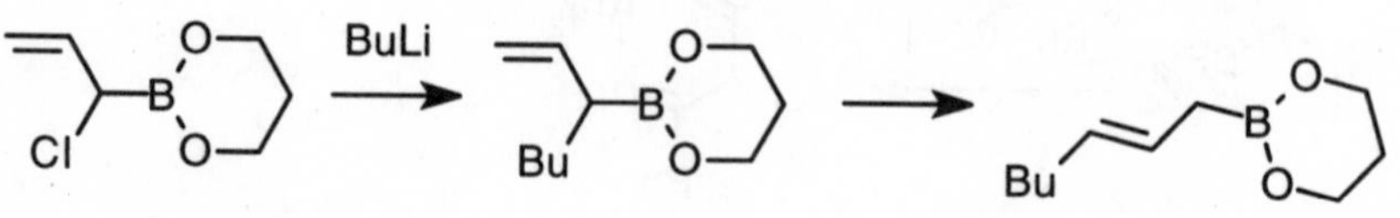

244   **245**   **246**

3.4.5.6 3-Borolenes. 3-Borolenes such as **248** are readily obtained via a reductive rearrangement of 1-iodo-1-borylbutadienes (**247**). Alkylation first with an organometallic and second with a reactive alkyl halide results in ring opening and after deboronation yields the homoallylic alcohol (**249**) diastereoselectively [300]. In the example illustrated the *anti/syn* ratio was 94:6.

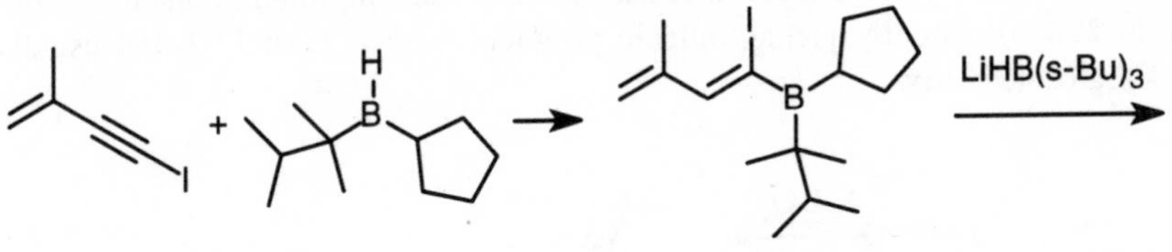

247

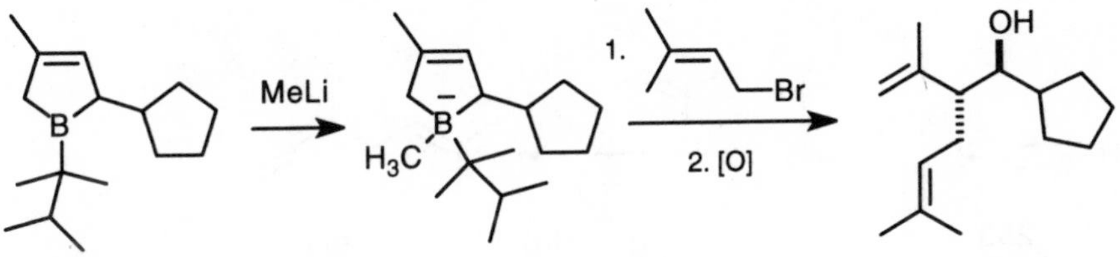

248   **249**

Reaction of 3-borolene **250** with aldehydes leads to a diastereoselective synthesis of 1,3-diols (**251**), which can be converted by conventional means to potentially useful intermediates such as **252** [301].

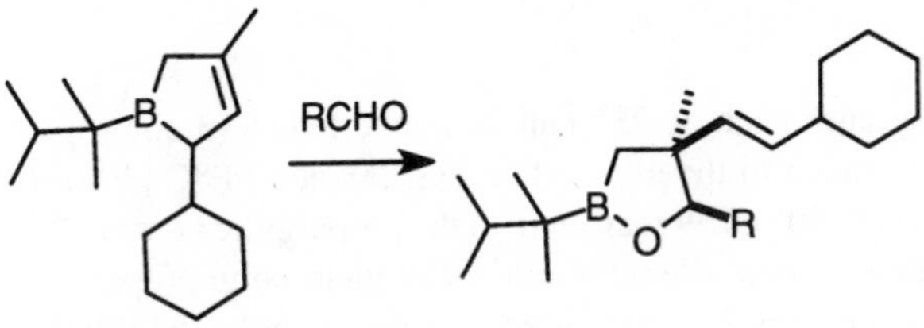

250

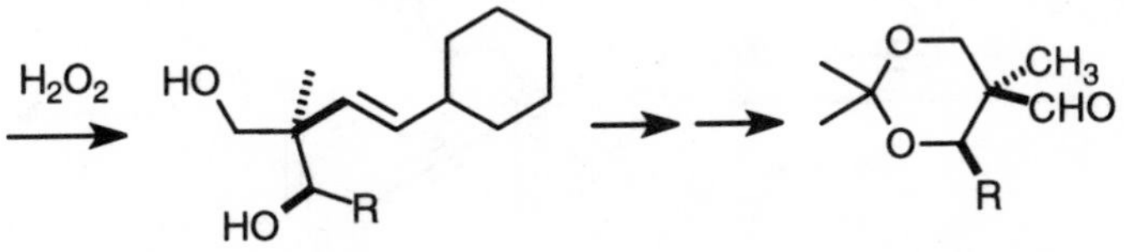

251   **252**

3.4.5.7 Allenyl and propargyl boranes. Propargylic borate **253** (R = butyl or octyl) can be carboxylated to provide the allenic acid **254**, treated with 1-bromo-3-methyl-2-butene to form the allene **255**, or treated with propanal to form the homoallenyl alcohol **256** [302]. In these cases, the allene/propargyl isomer ratios were ≥50:1. However, reaction of **253** (R = Bu) with acetone yielded a gross mixture of allenic and propargylic alcohols, as did **253** (R = Me₃C−) with propanal, and **253** (R = Bu) with allyl bromide yielded a 5:1 allene/acetylene ratio. Aluminum reagents analogous to **253** consistently yielded allenic products, with at least 11:1 and usually ≥20:1 regioselectivity.

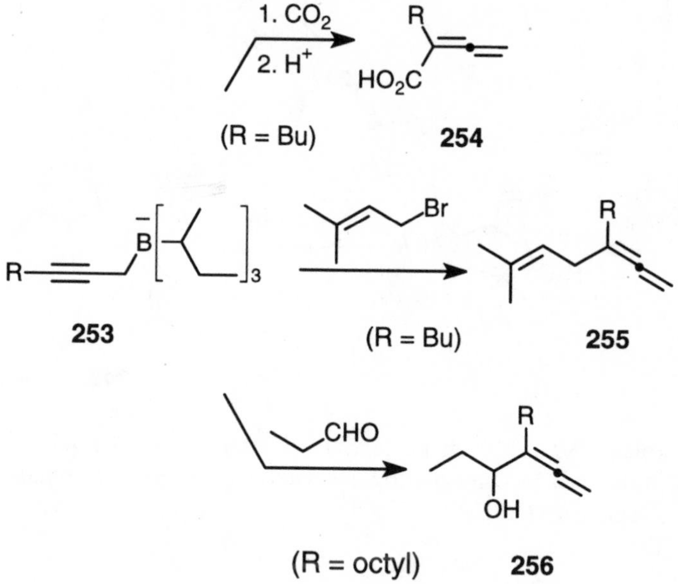

$$1.\ CO_2 \qquad 2.\ H^+$$

(R = Bu)   **254**

253

(R = Bu)   **255**

(R = octyl)   **256**

(Thexyl)(alkenyl)chloroboranes such as **257** can be converted to (3-chloropropynyl)borates **258**, which rearrange to the allenic boranes **259** at −78 °C [303]. In accord with earlier findings [304, 305], **259** rearrange to the propargylic boranes **260** at 25 °C. On treatment with aldehydes, allenic boranes **259** yield homopropargylic alcohols **261**, and propargylic boranes **260** yield homoallenic alcohols **262** [303].

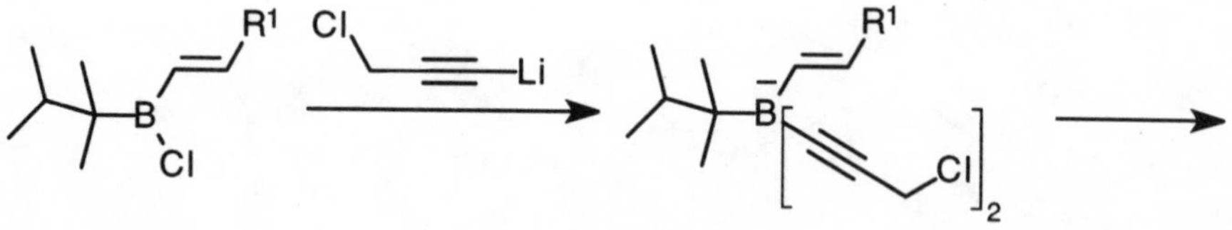

257   **258**

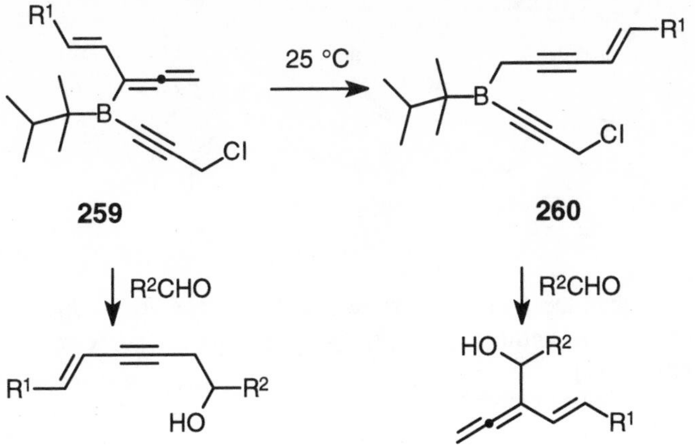

259   25 °C   **260**

R²CHO   R²CHO

261   **262**

3.4.6 Photochemical Rearrangements

Boron compounds are transparent to ultraviolet light longer than 200 nm unless there is another chromophore present. As a consequence, the photochemistry of boron compounds has been limited to aryl and unsaturated derivatives. The most interesting results from a synthetic viewpoint have produced some exotic three-membered rings containing boron. Otherwise, the foreseeable synthetic utility is limited and the treatment is very brief.

Photochemical rearrangement of tetraarylborate ions in aqueous solution in the absence of oxygen yields arylcyclohexadienes, as illustrated with tetraphenylborate (**263**) [306, 307, 308]. In air, the products are essentially all aromatized, but some unusual rearrangements occur, as shown for the photolysis of potassium diphenyldimesitylborate (**264**) in aqueous dimethoxyethane [309]. Rearrangements of this type were shown to be intramolecular [310].

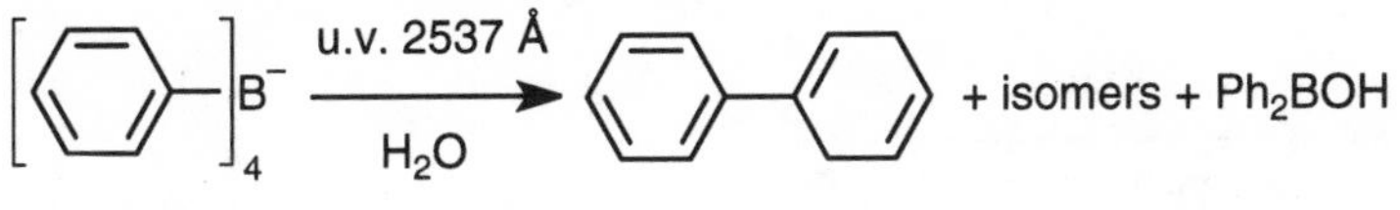

$$\left[\langle \bigcirc \rangle \right]_4 B^- \xrightarrow[\text{H}_2\text{O}]{\text{u.v. 2537 Å}} \langle \bigcirc \rangle - \langle \bigcirc \rangle + \text{isomers} + \text{Ph}_2\text{BOH}$$

263

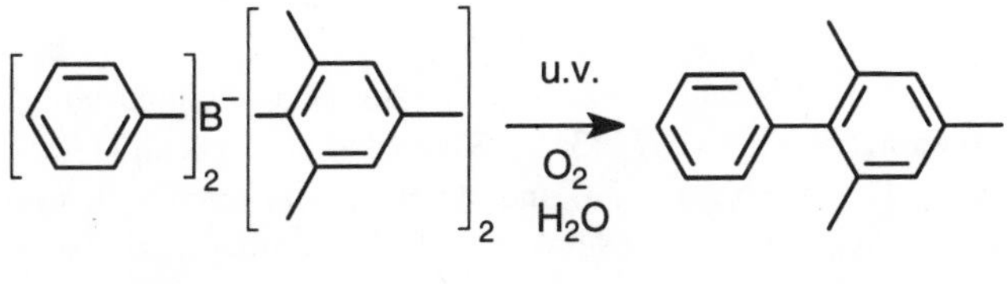

264

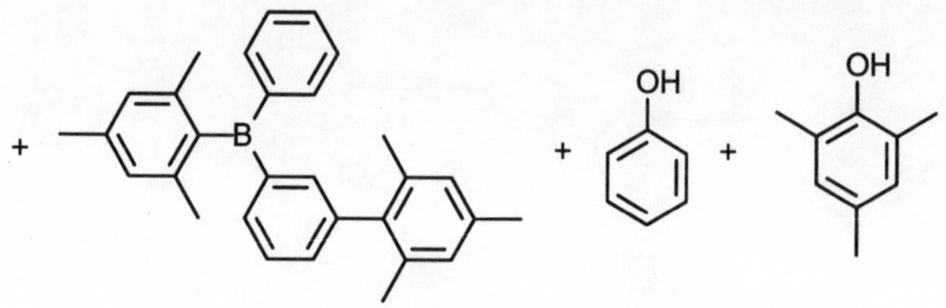

In a simpler reaction than the foregoing, aryl-aryl coupling has been achieved with the product from *o*-iodoaniline and diphenylboron chloride to form the borazaro compound **265** [311].

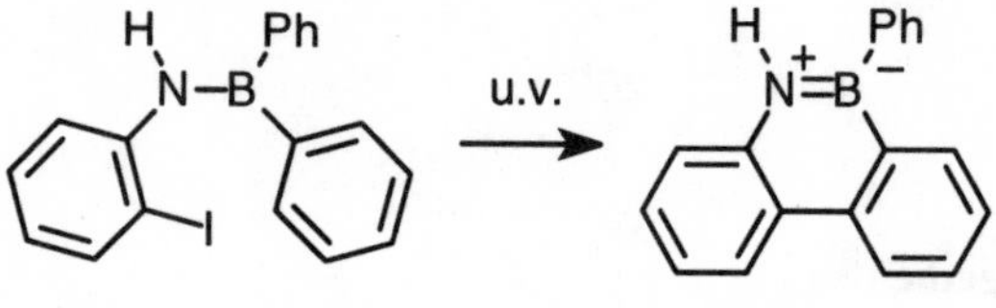

265

Eisch and coworkers have photolyzed dimesityl(mesitylethynyl)borane to trimesitylborirene (**266**), the structure of which has been confirmed by X-ray crystallography [312, 313]. Though extremely strained, this compound appears to have some stability attributable to the aromaticity of the borirene ring. Nucleophiles such as pyridine, which destroy the aromatic character by coordinating with the boron atom, lead to ring opening reactions. In earlier work, it appeared that triphenylborirene had been generated, but the reactivity of the compound had thwarted isolation [314], and it now appears that the 2,6-dimethyl substituents on the benzene rings are essential in shielding the borirene ring from attack.

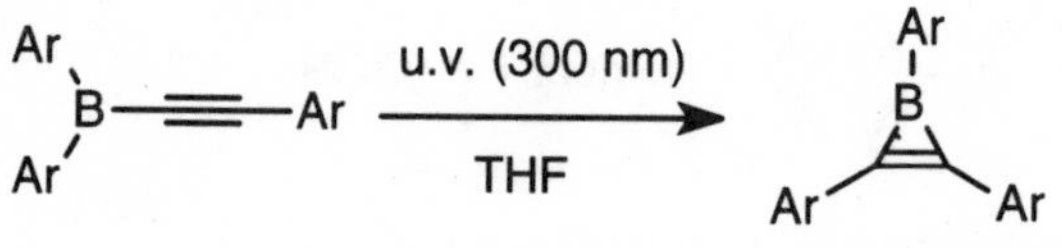

266 (Ar = 2,4,6-Me$_3$C$_6$H$_2$)

Schuster and coworkers have photolyzed suitable borates and obtained tetra-methylammonium boratanorcaradiene **267** [315, 316] and potassium tetra-phenylboratirene **268** [317]. The structures of **267** and **268** were confirmed by X-ray crystallography. The photolysis of **267** and **268** has been carefully investigated and it was concluded that there is no credible evidence that diphenylborene, Ph$_2$B, is produced as a reactive intermediate in such photolyses [318], in contradiction of earlier claims by Eisch's group [319, 320]. Further reactions of **268** have been studied, and ring opening by electrophiles has been observed, including methylation—protonolysis to **269** [321].

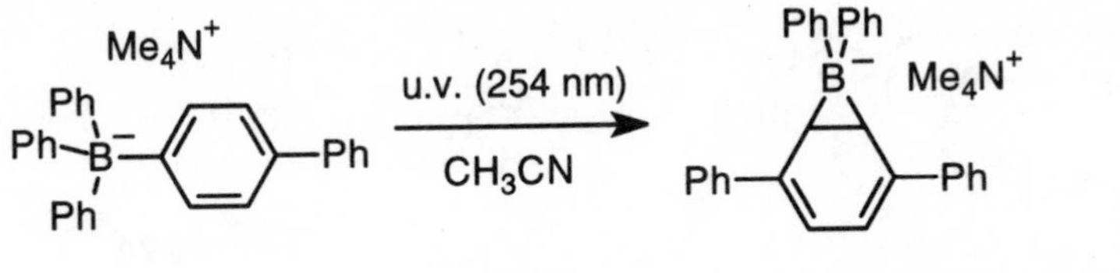

267 (and isomers)

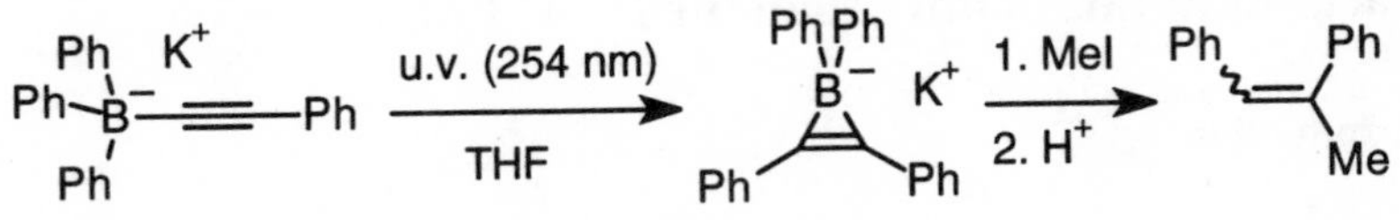

268   **269**

Photochemical rearrangement of the β-styryldiphenylborane nicotine complex **270** has yielded the separable borirane diastereomers **271** and **272** [322]. Heating **271** with pyridine in a sealed tube at 150 °C yielded the pyridine complex **273** without loss of stereochemistry. It was concluded that the free borirane that is probably an intermediate in this exchange is stable toward electrocyclic ring opening, which would have led to racemization.

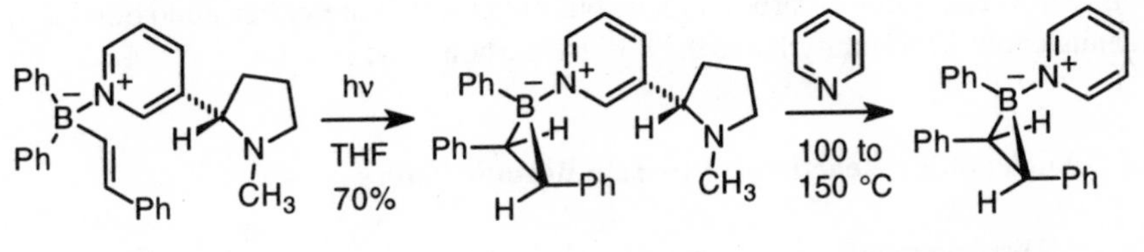

270   **271**   **273**

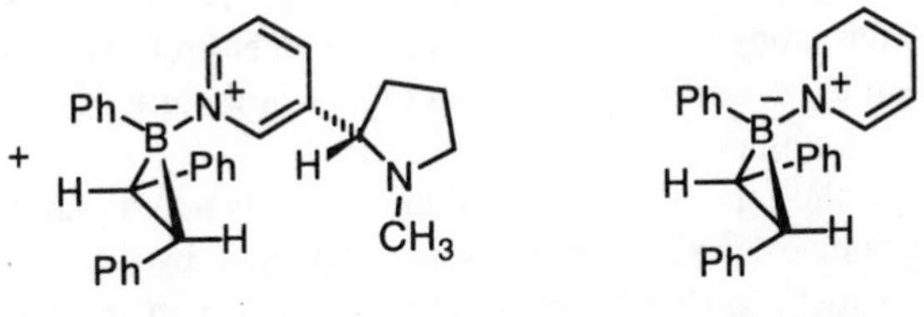

272   not formed from **271**

(1E)-Dienylboranes rearrange photochemically to borolenes (**274**) [323, 324]. Since (1Z)-dienylboranes generated chemically (Section 4.2.3) cyclize similarly, it appears that the photochemical step may merely be isomerization of (E)-diene to (Z)-diene.

274

3.5 Reactions at Sites Other than the B—C Bond

3.5.1 Introduction

The reactions in this miscellaneous category most useful for asymmetric synthesis are displacements of halide from α-halo boronic esters, Section 3.5.2 and Chapter 5. Diels-Alder reactions, Section 3.5.5.3 and Chapter 8, can be diastereoselective but have not yet been made enantioselective. Most of the rest have little or no potential for stereodirected synthesis, though they may serve as useful sources of achiral boronic esters or provide useful secondary transformations of asymmetric boronic esters.

Electrophilic reagents or oxidizing agents that lack the requisite basicity to bind to boron may leave a boron—carbon bond untouched while attacking a functional group elsewhere in the molecule. Some types of oxidizing radicals, most notably those containing oxygen with an available electron pair for nucleophilic coordination, will break boron—carbon bonds, but radicals that are either nonoxidizing or nonnucleophilic normally leave the boron—carbon bond intact.

3.5.2 Nucleophilic Reactions of α-Halo Boronic Esters

3.5.2.1 Displacements. Reactions of α-halo boronic esters with carbon nucleophiles have been classified as carbon—boron substitution reactions and described in Section 3.3.3, and make up the majority of the reactions discussed in Chapter 5. Heteroatom nucleophiles behave in an entirely analogous fashion to carbon nucleophiles, except that there is a higher probability of competition from migration of oxygen ligands on boron.

In the earliest examples studied, it was noted that displacement of bromide from boronic ester **97** by iodide was accelerated by the neighboring boronic ester group. When α-bromo boronic ester **97** reacted with sodium butanethiolate in butanol to form butylthio product **277**, there was a small amount of butoxide migration to form

278, which was considered evidence that there was a borate intermediate **275** leading to **277** and a parallel borate **276** as precursor to **278** [167]. Presumably, **275** has the greater rate constant for rearrangement, but **276** is present in higher concentration. Sodium butoxide in butanol readily converted **97** to **278**. Because of the ease with which simple alkyl $RCHBrCH_2CCl_3$ undergo elimination with alkoxides to form $RCH{=}CHCCl_3$, the formation of **278** is remarkable evidence for the facilitation of S_N2 type displacements by neighboring boronic ester groups, and further support for intermediate **276**.

97 **275** (major path)

276 (minor path) **277** (major) **278** (minor)

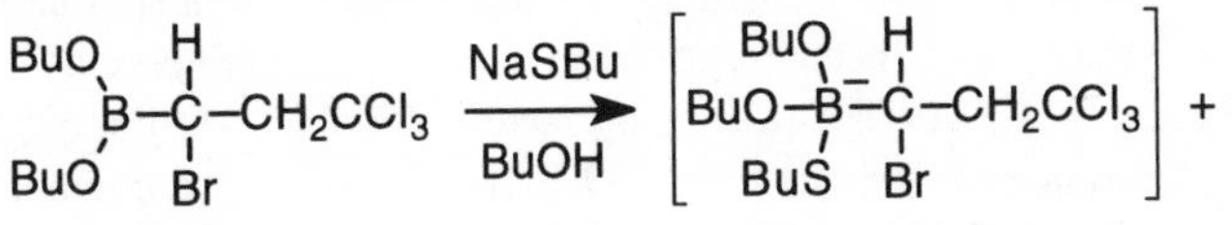

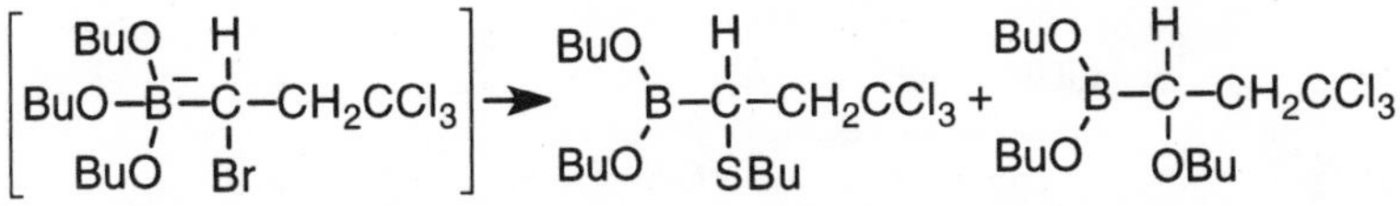

In subsequent work, it was found that the 2-bromoisopropylboronic ester **279** with sodium phenylthiolate in butanol yields only the butoxy substitution product **281**, presumably via the borate intermediate **280** [171]. The relative nucleophilicities of the migrating groups evidently play a smaller role than the relative basicities in this case.

279 **280** **281**

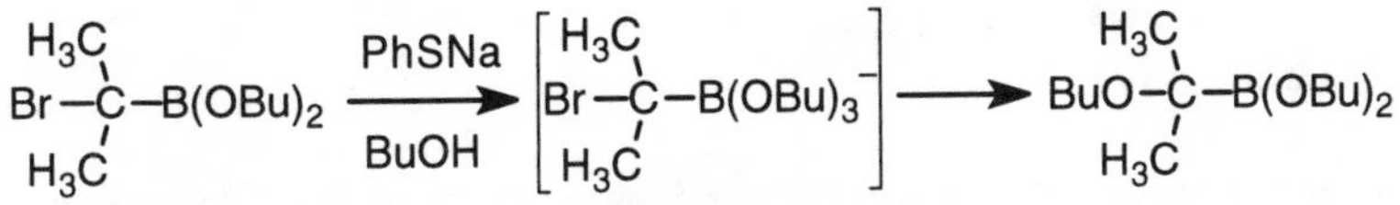

In contrast to the oxygen substitution seen in the alcohol solvent, thiourea in acetonitrile converted **279** to the normal sulfur substitution product **282** [171].

279 **282**

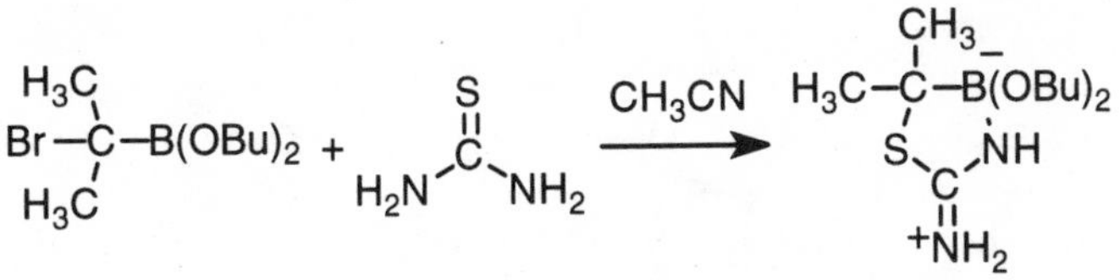

Nitrogen nucleophiles also displace halides from α-halo boronic esters. The first examples, conversion of $ICH_2B(OR)_2$ to $R'_2NCH_2B(OR)_2$ [72, 73] have been noted in Section 3.2.4.3. Further examples are noted in Chapter 5, where application of heteroatom substitutions to asymmetric synthesis is described.

3.5.2.2 Eliminations. If the halide is at the β-position, the only reaction with nucleophiles that is normally observed is boron—halide elimination, which is described in Section 4.2.1. Minor exceptions are noted in Section 3.3.2.

In contrast, halides in the α-position usually undergo only displacement reactions, and conversion to alkenylboranes via hydrogen—halide elimination is usually difficult. Eliminations that have been reported include the conversion of $Cl_3CCH_2CHBrB(OBu)_2$ to $Cl_3CCH=CHB(OBu)_2$ by *tert*-butylamine [169] and the conversion of the 1,1-dichloroethylboronic ester **283** to the chlorovinyl compound **284** by lithium chloride in DMF [325]. Elimination was also encountered in the reaction of **285** with lithium benzyl oxide, which yielded a mixture of unsaturated isomers [326].

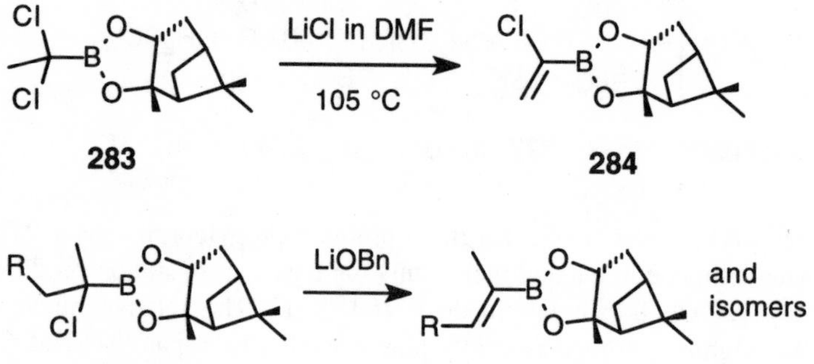

3.5.3 α-Bromination

Bromine does not cleave the carbon—boron bond of boronic esters under neutral conditions. An early attempt to brominate methylboronic esters failed, and even chlorination with *tert*-butyl hypochlorite was too sluggish and indiscriminate to be useful [178]. However, if the alkyl group is secondary, the hydrogen atom adjacent to the boron is easily photobrominated [327, 328], and this provides a useful synthetic route to α-bromo boronic esters such as **286** [122], though it is without any stereocontrol.

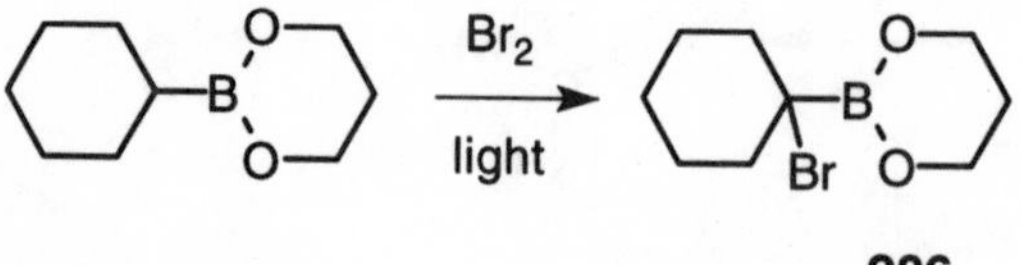

Bromination of trialkylboranes in dichloromethane proceeds via a radical mechanism to form the α-bromoalkylborane, as in the conversion of $(CH_3CH_2)_3B$ to $(CH_3CH_2)_2BCH(Br)CH_3$ [329]. In order to prevent cleavage of the initially formed α-bromoalkylborane to alkyl halide and bromoborane, it is necessary to sweep the HBr formed in the reaction out of the mixture. In the presence of water, the bromoborane rearranges rapidly to a borinic acid, $Et_2BCH(Br)CH_3$ yielding $EtB(OH)CH(Et)CH_3$.

3.5.4 Oxidations of Other Functions in the Presence of C—B Bonds

It is sometimes useful in synthesis to oxidize other functionality without disturbing a carbon—boron bond. Even though boronic esters are very susceptible to oxidation, the nearly universal requirement for nucleophilic attack at boron prior to boron—carbon bond rupture retards or prevents attack by several types of strong oxidants, and several oxidations of other carbon functionality without disturbing the boronic ester group are known.

The oldest example of oxidation of other functionality is the permanganate oxidation of the methyl group of p-tolylboronic acid (287) to the corresponding carboxylic acid (288) [330, 331, 332].

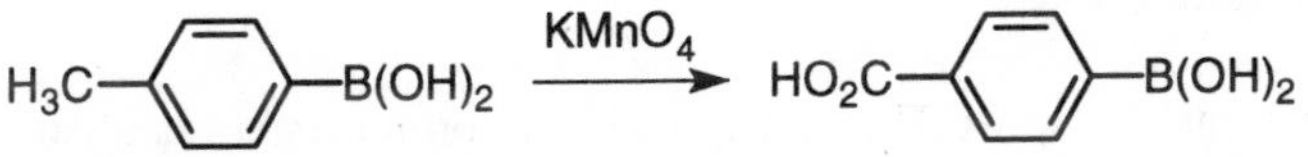

Methoxybenzyl ethers used as hydroxyl protecting groups [333] can be cleaved with dichlorodicyanoquinone (DDQ) without loss of boronic ester functionality [334], and Swern's reagent oxidizes a primary alcohol to an aldehyde in the presence of a boronic ester group, illustrated by the conversion of 289 via 290 to 291 [335].

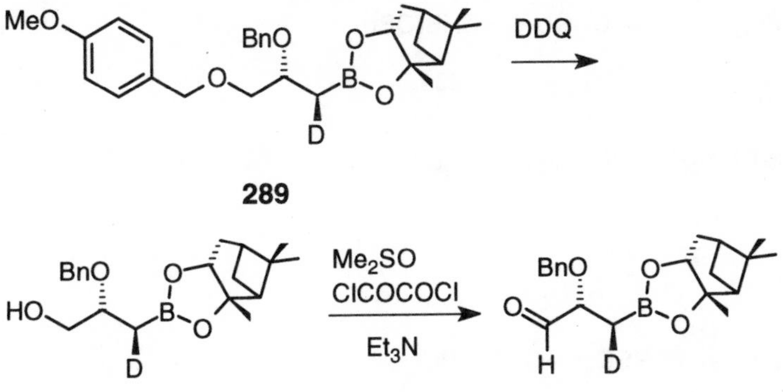

Although aqueous dichromate oxidizes boronic acids and pyridinium chlorochromate oxidizes trialkylboranes (see Section 3.2.2.6), it has proved feasible to oxidize a secondary alcohol to a ketone with pyridinium dichromate without disturbing a pinacol boronic ester [336].

Osmium tetraoxide catalyzed dihydroxylations of carbon—carbon double bonds with *N*-methylmorpholine *N*-oxide can be carried out in the presence of phenylboronic acid to yield 1,3,2-dioxaborinanes, for example **293** from **292**, without affecting the carbon—boron bond [337]. Yields for a series of six olefins, including indene, ethyl cinnamate, 2-octene, cyclooctene, cyclooctadiene, phenylcyclohexene, and **292**, were 81-93%.

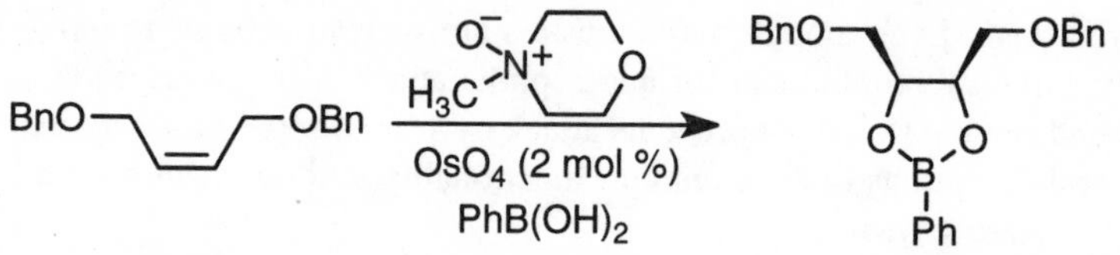

292 **293**

3.5.5 Additions to Double Bonds

3.5.5.1 Radical additions. Early work on functionalized boronic ester chemistry was greatly facilitated by the discovery that a variety of reagents could be added to dibutyl vinylboronate under free radical conditions, with initiation by azobisisobutyronitrile or ultraviolet light [168, 338]. For example, bromotrichloromethane adds to dibutyl vinylboronate to form bromo boronic ester **97**, presumably in a radical chain reaction involving $Cl_3C\cdot$ and intermediate **294** as chain carriers [168]. Various thiols add via the usual $RS\cdot$ as chain carrier to produce **296** via **295** [168, 339], and hydrogen bromide similarly yields $BrCH_2CH_2B(OBu)_2$ [170].

$$CH_2{=}CHB(OBu)_2 \xrightarrow{Cl_3C\cdot} Cl_3CCH_2\text{–}\dot{C}HB(OBu)_2$$

294

$$\xrightarrow{Cl_3CBr} Cl_3CCH_2\text{–}CH(Br)B(OBu)_2 + Cl_3C\cdot$$

97

$$CH_2{=}CHB(OBu)_2 \xrightarrow{RS\cdot} RSCH_2\text{–}\dot{C}HB(OBu)_2$$

295

$$\xrightarrow{\text{RSH}} \text{RSCH}_2\text{–CH}_2\text{B(OBu)}_2 + \text{RS}^\bullet$$

296

Addition of methanethiol to the pinanediol vinylboronic ester has been reported recently, though the adduct did not prove useful, and addition to the allylic boronic ester CH$_2$=CHCH(NHAc)B(O$_2$C$_{10}$H$_{16}$) was used instead in the synthesis of the boronic acid analogue of *N*-acetylmethionine [340].

Bromomalononitrile adds to dibutyl vinylboronate in the presence of azobis-isobutyronitrile to form the adduct **107** (85%) via intermediate **297** [171]. Bromo-malononitrile did not add to the other test compounds 1-octene, vinyl acetate, ethyl acrylate, or acrylonitrile. Addition of sodium bisulfite to vinylboronic acid in methanol–water to make (HO)$_2$BCH$_2$CH$_2$SO$_3^-$ further illustrates the range of structures accessible via radical reactions [339].

$$(\text{BuO})_2\text{BCH}=\text{CH}_2 + \overset{\bullet}{\text{HC}}\underset{\text{CN}}{\overset{\text{CN}}{\diagdown}} \longrightarrow (\text{BuO})_2\overset{\bullet}{\text{BCH}}-\text{CH}_2-\overset{\text{CN}}{\underset{\text{CN}}{\text{CH}}}$$

297

$$\underset{\text{Br}}{\overset{\text{H}}{\diagdown}}\overset{\text{CN}}{\underset{\text{CN}}{\diagup}}\text{C}\xrightarrow{} (\text{BuO})_2\text{BCH}-\text{CH}_2-\underset{\text{CN}}{\overset{\text{CN}}{\text{CH}}} + \overset{\bullet}{\text{HC}}\underset{\text{CN}}{\overset{\text{CN}}{\diagdown}}$$
$$\overset{|}{\text{Br}}$$

107

Dibutyl ethynylboronate adds radical reagents in a manner similar to the vinylboronate [95].

3.5.5.2 Polar and catalytic additions. Halogenation of a carbon—carbon double bond in the presence of a boron—carbon bond occurs readily, as in the facile bromination of H$_2$C=CHB(OBu)$_2$ to BrCH$_2$CHBrB(OBu)$_2$ [341].

Polar hydrogen bromide or hydrogen iodide addition to α,β-unsaturated boronic esters occurs in liquid HBr or HI [170, 171]. The boron has less directing influence than the carbon skeleton. Reaction of dialkyl vinylboronates with liquid HBr or HI yields mainly the α-halo boronic esters, CH$_3$CHXB(OR)$_2$, mixed with some β-isomer, XCH$_2$CH$_2$B(OR)$_2$, which can be removed by treatment with water to convert them to ethylene and boric acid under conditions where the α-isomers are stable [171]. Dibutyl 1-propenylboronate with HBr yields the β-halo boronic ester CH$_3$CHBrCH$_2$B(OBu)$_2$ as the only isolable product, and the isopropenylboronic ester H$_2$C=C(CH$_3$)B(OBu)$_2$ yields only the α-halo product, (CH$_3$)$_2$CHBrB(OBu)$_2$ [170].

Hydroboration of dibutyl vinylboronate with $BH_3 \cdot THF$ yields mainly the 1,1-diborylethane product [342, 343, 104]. Reactions of 1,1-diboryl compounds are described in Sections 3.3 and 3.4.4.2.

Catalytic hydrogenation of $PhC(Me)=C[BO_2(CH_2)_3]_2$ over palladium proceeded normally to furnish $PhCH(Me)CH[BO_2(CH_2)_3]_2$ [145].

3.5.5.3 Cycloadditions. Diels-Alder reactions of dibutyl vinylboronate and ethynylboronate were first reported by Matteson and coworkers [95, 344]. For example, the reaction of dibutyl vinylboronate with cyclopentadiene yielded a mixture of dibutyl *endo*-norborneneboronate (**298**) and its *exo*-isomer (**299**) in roughly equal amounts [344]. By the use of the sterically hindered *tert*-butyl vinylboronate, the proportion of *exo*-isomer was increased somewhat. Cyclohexadiene was found to furnish the bicyclooctyl analogues of (**298**) and (**299**) [255].

298   **299**

In related work elsewhere, 2-ethynyl-1,3,2-dioxaborinane with butadiene heated in a sealed tube yielded 59% of the Diels-Alder adduct [345]. Dimethyl allylboronate reacted poorly with cyclopentadiene and satisfactorily with perchlorocyclopentadiene [346]. *C*-Substituted vinylboronic esters were investigated by Evans and coworkers and found to be relatively sluggish dienophiles, [347, 348]. Also, the appearance of good regioselectivity in the reaction of dibutyl vinylboronate with isoprene [344] was found to be incorrect [347]. Vinyldichloroborane gave low yields with cyclopentadiene [349], and a vinylidene-1,1-diboronic ester reacted with cyclopentadiene or perchlorocyclopentadiene [137].

None of this earlier work provided any incentive for further development Diels-Alder reactions of organoboranes for purposes of stereoselective synthesis. However, in recent times, several approaches to diastereocontrolled Diels-Alder reactions of organoboranes have been devised, and asymmetric borane-catalyzed Diels-Alder reactions have become highly significant. These developments comprise Chapter 8.

The 1,3-dipolar additions of reagents such as ethyl diazoacetate to dibutyl vinylboronate to form **300**, which rearranges spontaneously to **301**, was found soon after the Diels-Alder reactions [350]. Disproportionation of **301** liberates tributyl borate, and the disproportionated mixture is converted by ethanol to the pyrazoline **302**. Dibutyl ethynylboronate yielded the stable pyrazole **303**.

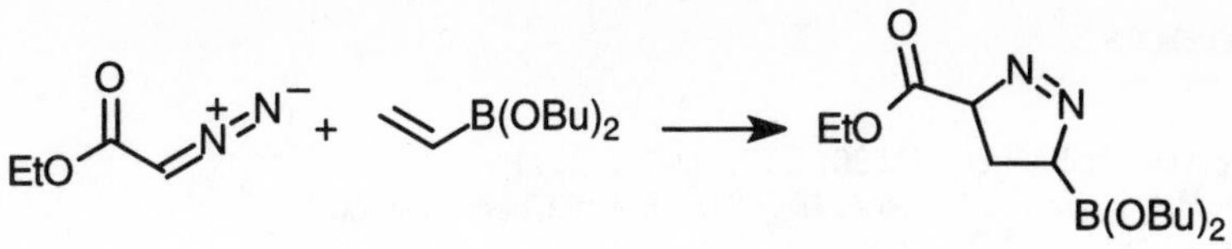

300

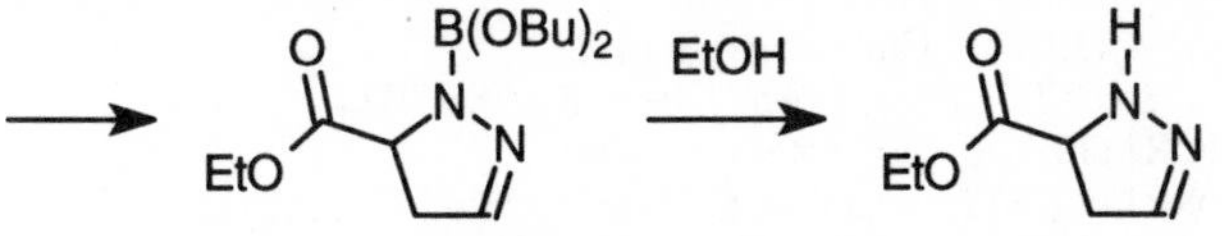

301   EtOH   **302**

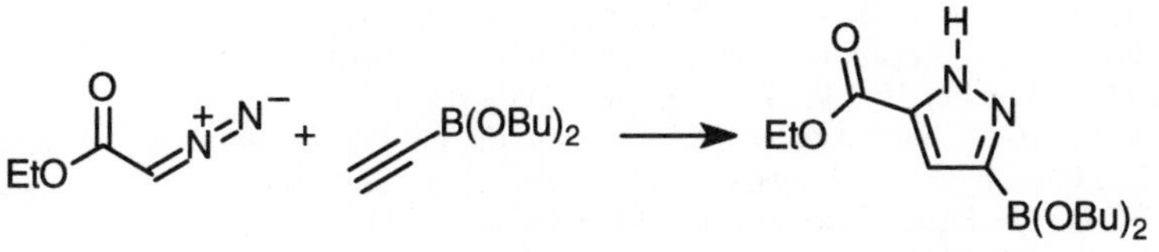

303

In related work elsewhere, nitrile oxides were added to dibutyl vinylboronate and ethynylboronate to make isoxazole derivatives [351, 352]. Improved conditions and the use of pinacol vinylboronate have been reported recently [196].

3.5.5.4 Cyclopropanation. Addition of dichlorocarbene to a vinylboronic ester to form the dichlorocyclopropylboronic ester has been reported [353].

Cyclopropanation of alkenylboronates can preserve the olefinic geometry in the product. Use of either diazomethane and palladium(II) acetate or the Simmons-Smith reagent derived from diiodomethane and diethylzinc cyclopropanates a number of substituted vinylboronic esters to form the cyclopropylboronic esters with full retention of stereochemistry **304** [354, 355]. The palladium(II) catalyzed process works well only with monosubstituted alkenes, but the Simmons-Smith reagent cyclopropanates disubstituted alkenylboronic esters as well.

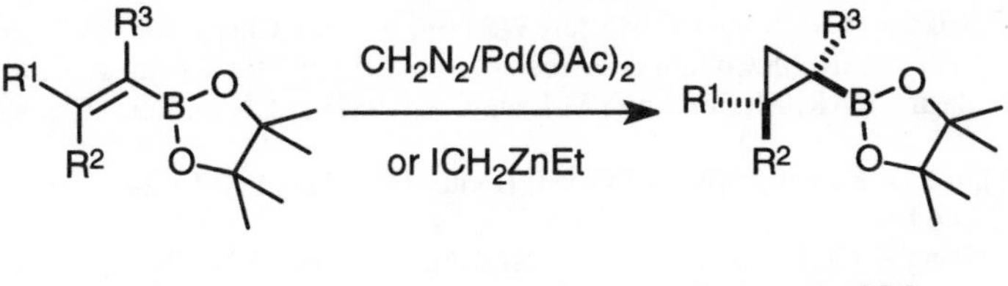

304

R^1 = H, alkyl, MeO_2C, Me_3Si; R^2 = H, Me, Bu; R^3 = H, Me

3.6 References

1. Ainley AD, Challenger F (1930) J. Chem. Soc. 2171
2. Snyder HR, Kuck JA, Johnson, JR (1938) J. Am. Chem. Soc. 60:105
3. Kuivila HG (1954) J. Am. Chem. Soc. 76:870
4. Kuivila HG, Armour AG (1957) J. Am. Chem. Soc. 79:5659
5. Kuivila HG (1955) J. Am. Chem. Soc. 77:4014
6. Brown HC, Zweifel G (1961) J. Am. Chem. Soc. 83:2544
7. Tripathy PB, Matteson DS (1990) Synthesis 200
8. Bigley D B, Payling DW (1970) J. Chem. Soc. B 1811
9. Minato H, Ware JC, Traylor TG (1963) J. Am. Chem. Soc. 85:3024
10. Matteson DS, Moody RJ (1980) J. Org. Chem. 45:1091
11. Matteson DS, Moody RJ (1978) J. Organomet. Chem. 152:265
12. Kabalka GW, Shoup TM, Goudgaon, NM (1989) J. Org. Chem. 54:5930
13. Kabalka GW, Wadgaonkar PP, Shoup TM (1990) Organometallics 9:1316
14. Johnson JR, Van Campen MG Jr (1938) J. Am. Chem. Soc. 60:121
15. Cristol SJ, Durango FP, Plorde DE (1965) J. Am. Chem. Soc. 87:2870
16. Pasto DJ, Hickman J (1968) J. Am. Chem. Soc. 90:4445
17. Evans DA, Vogel E, Nelson JV (1979) J. Am. Chem. Soc. 101:6120
18. Midland MM, Preston SB (1980) J. Org. Chem. 45:4514
19. Köster R, Morita Y (1966) Angew. Chem. 78:589; Int. Ed. Engl. 5:580
20. Zweifel G, Polston NL, Whitney CC (1968) J. Am. Chem. Soc. 90:6243
21. Fisher RP, On HP, Snow JT, Zweifel G (1982) Synthesis 127
22. Miller JA, Zweifel G (1981) Synthesis 288
23. Miller JA, Zweifel G (1981) J. Am. Chem. Soc. 103:6217
24. Davies AG, Roberts BP (1968) J. Chem. Soc. C 1474
25. Soderquist JA, Anderson CL (1986) Tetrahedron Lett. 27:3961
26. Kabalka GW, Hedgecock HC Jr (1975) J. Org. Chem. 40:1776
27. Kabalka GW, Slayden SW (1977) J. Organomet. Chem. 125:273
28. Frankland E, Duppa B (1860) Proc. Royal Soc. (London), 10:568
29. Davies AG, Roberts BP (1967) J. Chem. Soc. B 17
30. Midland MM, Brown HC (1971) J. Am. Chem. Soc. 93:1506
31. Kabalka GW, McCollum GW, Fabrikiewicz AS, Lambrecht RM, Fowler JS, Sajjad M (1984) J. Labelled Compd. Radiopharm. 21:1247
32. Kabalka GW (1988) J. Appl. Radiat Isot. 39:537
33. Brown HC, Midland MM (1987) Tetrahedron 43:4059
34. Ware JC, Traylor TG (1963) J. Am. Chem. Soc. 85:3026
35. Brown HC, Garg CP (1961) J. Am. Chem. Soc. 83:2951,2952
36. Rao CG, Kulkarni SU, Brown HC (1979) J. Organomet. Chem. 172:C20
37. Brown HC, Kulkarni SU, Rao CG (1980) Synthesis 151
38. Matteson DS, Beedle EC (1987) Tetrahedron Lett. 28:4499
39. Mendoza A, Matteson DS (1978) J. Chem. Soc., Chem. Commun. 357
40. Mendoza A, Matteson DS (1978) J. Organomet. Chem. 156:149
41. Brown HC, Heydkamp WR, Breuer E, Murphy WS (1964) J. Am. Chem. Soc. 86:3565
42. Kabalka GW, Sastry KAR, McCollum GW, Yoshioka H (1981) J. Org. Chem. 46:4296
43. Kabalka GW, Sastry KAR, McCollum GW, Lane CA (1982) J. Chem. Soc., Chem. Commun. 62
44. Kothari PJ, Finn RD, Kabalka GW, Vora MM, Boothe TE, Emran AM (1986) Appl. Radiat. Isot. 37:469
45. Kabalka GW, Wang Z, Goudgaon NM (1989) Synthetic Commun. 19:2409
46. Rathke MW, Inoue N, Varma KR, Brown HC (1966) J. Am. Chem. Soc. 88:2870
47. Brown HC, Kim KW, Srebnik M, Singaram B (1987) Tetrahedron 43:4071
48. Gurskii ME, Potapova TV, Bubnov YuN (1993) Mendeleev Commun. 56

49. Kabalka GW, Wang Z (1989) Organometallics 8:1093
50. Kabalka GW, Wang Z (1990) Synthetic Commun. 20:231
51. Kabalka GW, Wang Z (1990) Synthetic Commun. 20:2113
52. Jigajinni VB, Pelter A, Smith K (1978) Tetrahedron Lett. 181
53. Davies AG, Hook SCW, Roberts BP (1970) J. Organomet. Chem. 23:C11
54. Foot KG, Roberts BP (1971) J. Chem. Soc. C 3475
55. Kabalka GW, Gai YZ, Goudgaon NM, Varma RS, Gooch EE (1988) Organometallics 7:493
56. Suzuki A, Sono S, Itoh M, Brown HC, Midland MM (1971) J. Am. Chem. Soc. 93:4329
57. Brown HC, Midland MM, Levy AB (1973) J. Am. Chem. Soc. 95:2394
58. Brown HC, Midland MM, Levy AB, Suzuki A, Sono S, Itoh M (1987) Tetrahedron 43:4079
59. Levy AB, Brown HC (1973) J. Am. Chem. Soc. 95:4067
60. Kabalka GW, Henderson DA, Varma RS (1987) Organometallics 6:1369
61. Kabalka GW, Goudgaon NM, Liang Y (1988) Synthetic Commun. 18:1363
62. Bubnov YuN, Gurskii ME, Pershin DG (1994) Mendeleev Commun. 43
63. Bubnov YuN, Gurskii ME, Pershin DG (1991) J. Organomet. Chem. 412:1
64. Jego JM, Carboni B, Vaultier M, Carrié R (1989) J. Chem. Soc., Chem. Commun. 142
65. Jego JM, Carboni B, Vaultier M (1992) Bull. Soc. Chim. Fr. 129:554
66. Carboni B, Vaultier M, Courgeon T, Carrié R (1989) Bull. Soc. Chim. Fr. 844
67. Chavant PY, Lhermitte F, Vaultier M (1993) Synlett 519
68. Carboni B, Benalil A, Vaultier M (1993) J. Org. Chem. 58:3736
69. Matteson DS, Sadhu KM (1984) Organometallics 3:614
70. Amiri P, Lindquist RN, Matteson DS, Sadhu KM (1984) Arch. Biochem. Biophys. 234:531
71. Duncan K, Faraci SW, Matteson DS, Walsh CT (1989) Biochemistry 28:3541
72. Matteson DS, Cheng TC (1968) J. Org. Chem. 33:3055
73. Matteson, D. S.; Majumdar, D. (1979) J. Organomet. Chem. 170:259
74. Fisher GB, Juarez-Brambila JJ, Goralski CT, Wipke WT, Singaram B (1993) J. Am. Chem. Soc. 115:440
75. Brown HC, Lane CF (1970) J. Am. Chem. Soc. 92:6660
76. Brown HC, Lane CF (1970) J. Chem. Soc. D 521
77. Brown HC, Rathke MW, Rogic MM (1968) J. Am. Chem. Soc. 90:5038
78. Brown HC, De Lue NR, Kabalka GW, Hedgecock HC Jr (1976) J. Am. Chem. Soc. 98:1290
79. Brown HC, Hamaoka T, Ravindran N (1973) J. Am. Chem. Soc. 95:5786
80. Matteson DS, Mendoza A (1978) J. Organomet. Chem. 156:149
81. Lane CF, Brown HC (1970) J. Am. Chem. Soc. 92:7212
82. Brown HC, Midland MM (1971) J. Am. Chem. Soc. 93:3291
83. Arase A, Masuda Y (1976) Chem. Lett. 1115
84. Johnson JR, Snyder HR, Van Campen MG (1938) J. Am. Chem. Soc. 60:115
85. Brown HC, Hébert NC (1983) J. Organomet. Chem. 255:135
86. Brown HC, Murray K (1959) J. Am. Chem. Soc. 81:4108
87. Trofimenko S (1969) J. Am. Chem. Soc. 91:2139
88. Toporcer LH, Dessy RE, Green SIE (1965) J. Am. Chem. Soc. 87:1236
89. Brown HC, Murray K (1961) J. Org. Chem. 26:631
90. Brown HC, Zweifel G (1959) J. Am. Chem. Soc. 81:1512
91. Brown HC, Molander GA (1986) J. Org. Chem. 51:4512
92. Kuivila HG, Nahabedian KV (1961) J. Am. Chem. Soc. 83:2159, 2164, 2167
93. Matteson DS, Campbell JD (1990) Heteroatom Chemistry 1:109
94. Davies AG, Roberts BP (1968) J. Chem. Soc. C 1474
95. Matteson DS, Peacock K (1963) J. Org. Chem. 28:369
96. Brown HC, Rogic MM, Rathke MW, Kabalka GW (1968) J. Am. Chem. Soc. 90:818
97. Plamondon J, Snow JT, Zweifel G (1971) Organomet. Chem. Syn. 1:249

98.   Michaelis A, Becker P (1882) Chem. Ber. 15:180
99.   Seaman W, Johnson JR (1931) J. Am. Chem. Soc. 53:711
100. Khotinsky E, Melamed M (1909) Chem. Ber. 42:3090
101. Matteson DS, Krämer E (1968) J. Am. Chem. Soc. 90:7261
102. Matteson DS, Bowie RA (1965) J. Am. Chem. Soc. 87:2587
103. Brown HC, Larock RC, Gupta SK, Rajagopalan S, Bhat NG (1989) J. Org. Chem. 54:6079
104. Matteson DS, Shdo JG (1964) J. Org. Chem. 29:2742
105. Matteson DS, Allies PG (1973) J. Organomet. Chem. 54:35
106. Honeycutt JB Jr, Riddle JM (1960) J. Am. Chem. Soc. 82:3051
107. Larock RC, Brown HC (1970) J. Am. Chem. Soc. 92:2467
108. Gielen M, Fosty R (1974) Bull. Soc. Chim Belg. 83:333
109. Bergbreiter DE, Rainville DP (1976) J. Org. Chem. 41:3031
110. Larock RC (1974) J. Organomet. Chem. 67:353
111. Larock RC (1974) J. Organomet. Chem. 72:35
112. Larock RC, Gupta SK, Brown HC (1972) J. Am. Chem. Soc. 94:4371
113. Kunda SA, Varma RS, Kabalka GW (1984) Synth. Commun. 14:755
114. Matteson DS, Tripathy PB (1974) J. Organomet. Chem. 69:53
115. Mendoza A, Matteson DS (1978) J. Organomet. Chem. 152:1
116. Snyder HR, Kuck JA, Johnson JR (1938) J. Am. Chem. Soc. 60:105
117. Brown HC, Hébert NC, Snyder CH (1961) J. Am. Chem. Soc. 83:1001
118. Brown HC, Snyder CH (1961) J. Am. Chem. Soc. 83:1001
119. Brown HC, Verbrugge C, Snyder CH (1961) J. Am. Chem. Soc. 83:1002
120. Dieck HA, Heck RF (1975) J. Org. Chem. 40:1083
121. Suzuki A (1982) Acc. Chem. Res. 15:178
122. Yamamoto Y, Yatagai H, Maruyama K, Sonoda A, Murahashi S (1977) J. Am. Chem. Soc. 99:5652
123. Miyaura N, Sasaki N, Itoh M, Suzuki A (1977) Tetrahedron Lett. 3369
124. Kondo K, Murahashi S. (1979) Tetrahedron Lett. 1237
125. Matteson DS, Sadhu KM, Peterson ML (1986) J. Am. Chem. Soc. 108:812
126. Cainelli G, Dal Bello G, Zubiani G (1965) Tetrahedron Lett. 4329; (1966) Tetrahedron Lett. 4315
127. Zweifel G, Arzoumanian H (1966) Tetrahedron Lett. 2535
128. Castle RB, Matteson DS (1968) J. Am. Chem. Soc. 90:2194
129. Castle RB, Matteson DS (1969) J. Organomet. Chem. 20:19
130. Matteson DS, Tripathy PB (1970) J. Organomet. Chem. 21:P6
131. Matteson, D. S.; Tripathy, P. B. (1974) J. Organomet. Chem. 69:53
132. Matteson DS, Thomas JR (1970) J. Organomet. Chem. 24:263
133. Matteson DS, Larson GL (1973) J. Organomet. Chem. 57:225
134. Matteson DS, Wilcsek RJ (1973) J. Organomet. Chem. 57:231
135. Matteson DS, Davis RA, Hagelee LA (1974) J. Organomet. Chem. 69:45
136. Matteson DS, Hagelee LA, Wilcsek RJ. (1973) J. Am. Chem. Soc. 95:5096
137. Matteson DS, Hagelee LA (1975) J. Organomet. Chem. 93:21
138. Matteson DS, Moody RJ, Jesthi PK (1975) J. Am. Chem. Soc. 97:5608
139. Matteson DS, Jesthi PK (1976) J. Organomet. Chem. 110:25-37
140. Matteson DS, Moody RJ (1980) J. Org. Chem. 45:1091
141. Matteson DS, Wilson JW (1985) Organometallics 4:1690
142. Matteson DS, Gatzweiler W (1991) unpublished observations
143. Rathke MW, Kow R (1972) J. Am. Chem. Soc. 94:6854
144. Matteson DS, Moody RJ (1977) J. Am. Chem. Soc. 99:3196
145. Matteson DS, Moody RJ (1982) Organometallics 1:20
146. Matteson DS, Arne K (1978) J. Am. Chem. Soc. 100:1325
147. Matteson DS, Arne KH (1982) Organometallics 1:280
148. Mendoza A, Matteson DS (1979) J. Org. Chem. 44:1352

149. Matteson DS, Majumdar D (1983) Organometallics 2:230
150. Wilson JW (1980) J. Organomet. Chem. 186:297
151. Pelter A, Williams L, Wilson JW (1983) Tetrahedron Lett. 24:627
152. Pelter A, Singaram B, Wilson JW (1983) Tetrahedron Lett. 24:635
153. Garad MK, Pelter A, Singaram B, Wilson JW (1983) Tetrahedron Lett. 24:637
154. Matteson DS, Mah, RWH (1963) J. Org. Chem. 28:2171
155. Cooke MP Jr, Widener RK (1987) J. Am. Chem. Soc. 109:931
156. Cooke MP Jr (1994) J. Org. Chem. 59:2930
157. Hawthorne MF, Dupont JA (1958) J. Am. Chem. Soc. 80:5830.
158. Matteson DS, Liedtke JD (1963) J. Org. Chem. 28:1924.
159. Pasto DJ, Hickman J (1968) J. Am. Chem. Soc. 90:4445
160. Pasto DJ, Snyder R (1966) J. Org. Chem. 31:2777
161. Matteson DS, Majumdar D (1983) Organometallics 2:1529
162. Matteson DS, Peterson ML (1987) J. Org. Chem. 52:5116
163. Matteson DS, Kandil AA, Soundararajan R (1990) J. Am. Chem. Soc. 112:3964
164. Matteson DS (1989) Chem. Rev. 89:1535
165. Jäger H, Hesse G (1962) Chem. Ber. 95:345
166. Bawn CEH, Ledwith A (1964) Progress in Boron Chemistry 1:345
167. Matteson DS, Mah, RWH (1963) J. Am. Chem. Soc. 85:2599
168. Matteson DS (1960) J. Am. Chem. Soc. 82:4228.
169. Matteson DS, Mah, RWH (1963) J. Org. Chem. 28:2174
170. Matteson DS, Liedtke J (1963) Chem. Ind. (London) 1241
171. Matteson DS, Schaumberg GD (1966) J. Org. Chem. 31:726
172. Matteson DS, Bowie RA, Srivastava G (1969) J. Organomet. Chem. 16:33
173. Brown HC, Rogic MM, Rathke MW, Kabalka GW (1969) J. Am. Chem. Soc. 91:2150
174. Brown HC, Nambu H, Rogic MM (1969) J. Am. Chem. Soc. 91:6852
175. Hara S, Kishimura K, Suzuki A, Dhillon RS (1990) J. Org. Chem. 55:6356
176. Katz JJ, Carlson BA, Brown HC (1974) J. Org. Chem. 39:2817
177. Brown HC, Phadke AS, Rangaishenvi MV (1990) Heteroatom Chem. 1:83
178. Matteson DS (1964) J. Org. Chem. 29:3399
179. Pasto DJ, Chow J, Arora SK (1969) Tetrahedron 25:1557
180. Pasto DJ, McReynolds K (1971) Tetrahedron Lett. 801
181. Brown HC, De Lue NR, Yamamoto Y, Maruyama K (1977) J. Org. Chem. 42:3252
182. Köbrich G, Flory K, Drischel W (1964) Angew. Chem. 76:536; Angew. Chem. Int. Ed. Engl. 3:513
183. Köbrich G, Merkle HR (1967) Chem. Ber. 100:3371
184. Rathke MW, Chao E, Wu G (1976) J. Organomet. Chem. 122:145
185. Matteson DS, Majumdar D (1980) J. Am. Chem. Soc. 102:7588
186. Corey EJ, Jautelat M, Oppolzer W (1967) Tetrahedron Lett. 2325
187. Taguchi H, Yamamoto H, Nozaki H (1974) J. Am. Chem. Soc. 96:3010
188. Matteson DS, Sadhu KM (1985) U.S. Patent 4,525,309, 25 Jun 1985.
189. Matteson DS, Hurst GD (1986) Organometallics 5:1465
190. Matteson DS, Hurst GD (1987) U.S. Patent 4,701,545, 20 Oct 1987, 6 pp.; (1988) Chem. Abstr. 109:93315p
191. Whiting, A (1991) Tetrahedron Lett. 32:1503
192. Kinder DH, Ames MM (1987) J. Org. Chem. 52:2452
193. Wuts PGM, Thompson PA, Callen, GR (1983) J. Org. Chem. 48:5398
194. Sadhu KM, Matteson DS (1985) Organometallics 4:1687
195. Michnick TJ, Matteson DS (1991) Synlett 631
196. Wallace RH, Zong KK (1992) Tetrahedron Lett. 33:6941
197. Brown HC, Imai T, Perumal PT, Singaram B (1985) J. Org. Chem. 50:4032
198. Brown HC, Phadke AS, Rangaishenvi MV (1988) J. Am. Chem. Soc. 110:6263
199. Zweifel G, Arzoumanian H, Whitney CC (1967) J. Am. Chem. Soc. 89:3652
200. Negishi E, Yoshida T (1973) J. Am. Chem. Soc. 95:6837

201. Levy AB, Schwartz SJ, Wilson N, Christie B (1978) J. Organomet. Chem. 156:123
202. Soderquist JA, Rivera I (1989) Tetrahedron Lett. 30:3919
203. Binger P (1967) Angew. Chem., Int. Ed. Engl. 6:84
204. Binger P, Benedikt G, Rotermund GW, Köster R (1968) Justus Liebigs Ann. Chem. 717:21
205. Binger P, Köster R (1974) Synthesis 350
206. Miyaura N, Yoshinari T, Itoh M, Suzuki A (1975) Tetrahedron Lett. 2961
207. Binger P, Köster R (1973) Synthesis 309
208. Hooz J, Mortimer R (1976) Tetrahedron Lett. 805
209. Midland M, Brown HC (1975) J. Org. Chem. 40:2848
210. Brown HC, Levy AB, Midland M (1975) J. Am. Chem. Soc. 97:5017
211. Pelter A, Subrahmanyam C, Laub RJ, Gould KJ, Harrison CR (1975) Tetrahedron Lett. 1633
212. Naruse M, Utimoto K, Nozaki H (1975) Tetrahedron 30:3037
213. Binger P, Köster R (1975) Chem. Ber. 108:395
214. Akimoto I, Suzuki A (1979) Synthesis 146
215. Sotoyama T, Hara S, Suzuki A (1979) Bull. Chem. Soc. Jpn. 52:1865
216. Levy AB (1979) Tetrahedron Lett. 4021
217. Hillman MED (1962) J. Am. Chem. Soc. 84:4715
218. Hillman MED (1963) J. Am. Chem. Soc. 85:982
219. Hillman MED (1963) J. Am. Chem. Soc. 85:1626
220. Casanova J Jr, Schuster RE (1964) Tetrahedron Lett. 405
221. Hesse G, Witte H, Bittner G (1965) Justus Liebigs Ann. Chem. 687:9
222. Hesse G, Witte H (1963) Angew. Chem. 75:791; Int. Ed. 2:617
223. Hesse G, Witte H (1965) Justus Liebigs Ann. Chem. 687:1
224. Casanova J Jr, Kiefer H, Kuwada D, Boulton A (1965) Tetrahedron Lett. 703
225. Bresadola S, Carraro G, Pecile C, Turco A (1964) Tetrahedron Lett. 3185
226. Casanova J Jr, Kiefer H (1969) J. Org. Chem. 34:2579
227. Witte H (1965) Tetrahedron Lett. 1127
228. Brown HC, Rathke MW (1967) J. Am. Chem. Soc. 89:2737
229. Brown HC (1972) "Boranes in Organic Chemistry," Cornell University Press, Ithaca, NY pp 301-446
230. Brown HC, Kramer GW, Levy AB, Midland MM (1975) "Organic Synthesis via Boranes," Wiley–Interscience, New York
231. Brown HC (1980) Science 210:485
232. Brown HC, Hubbard JL, Smith K (1979) Synthesis 701
233. Kulkarni SU, Lee HD, Brown HC (1980) J. Org. Chem. 45:4542
234. Kabalka G, Varma RS (1989) Tetrahedron 45:6601
235. Pelter A, Smith K, Hutchings MG, Rowe K (1975) J. Chem. Soc., Perkin Trans. I 129
236. Pelter A, Hutchings MG, Rowe K, Smith K (1975) J. Chem. Soc., Perkin Trans. I 138
237. Pelter A, Hutchings MG, Smith K (1975) J. Chem. Soc., Perkin Trans. I 142
238. Pelter A, Hutchings MG, Smith K, Williams DJ (1975) J. Chem. Soc., Perkin Trans. I 145
239. Zweifel G, Pearson NR (1980) J. Am. Chem. Soc. 102:5919
240. Kothari PJ, Finn RD, Kabalka GW, Vora MM, Boothe TE, Emran AM, Mohammadi M (1986) Appl. Radiat. Isot. 37:471
241. Brown HC, Imai T (1983) J. Am. Chem. Soc. 105:6285
242. Tsai DJS, Matteson DS (1983) Organometallics 2:236
243. Brown HC, Jacob P III (1977) J. Org. Chem. 42:579
244. Sinclair JA, Brown HC (1977) J. Organomet. Chem. 131:163
245. Brown HC, Molander GA, Singh SM, Racherla US (1985) J. Org. Chem. 50:1577
246. Evans JC, Goralski CT, Hasha DL (1992) J. Org. Chem. 57:2941
247. Satoh Y, Tayano T, Hara S, Suzuki A (1989) Tetrahedron Lett. 30:5153
248. Kawamura F, Tayano T, Satoh Y, Hara S, Suzuki A (1989) Chemistry Lett. 1723

249. Brown HC, Rhodes SP (1969) J. Am. Chem. Soc. 91:2149
250. Brown HC, Rhodes SP (1969) J. Am. Chem. Soc. 91:4306
251. Goering HL, Trenbeath SL (1976) J. Am. Chem. Soc. 98:5016
252. Matteson DS, Waldbillig JO (1964) J. Am. Chem. Soc. 86:3778
253. Matteson DS, Talbot ML (1967) J. Am. Chem. Soc. 89:1119
254. Matteson DS, Waldbillig JO, Peterson SW (1964) J. Am. Chem. Soc. 86:3781
255. Matteson DS, Talbot ML (1967) J. Am. Chem. Soc. 89:1123
256. Kabalka GW, Brown HC, Suzuki A, Honma S, Arase A, Itoh M (1970) J. Am. Chem. Soc. 92:710
257. Hollis WG Jr, Smith PL, Hood DK, Cook SM (1994) J. Org. Chem. 59:3485
258. Mikhailov BM, Bubnov YuN (1964) Izv. Akad. Nauk SSSR, Ser. Khim. 1874
259. Mikhailov BM, Pozdnev VF (1967) Izv. Akad. Nauk SSSR, Ser. Khim. 1477
260. Blais J, L'Honoré A, Soulié J, Cadiot P (1974) J. Organomet. Chem. 78:323
261. Mikhailov BM, Bubnov YuN, Tsyban' AV (1978) J. Organomet. Chem. 154:113
262. Mikhailov BM, Bubnov YuN, Tsyban' AV, Grigoryan MSh (1978) J. Organomet. Chem. 154:131
263. Mikhailov BM, Bubnov YuN, Tsyban' AV, Base K (1978) Izv. Akad. Nauk SSSR, Ser. Khim. 1586
264. Mikhailov BM, Bubnov YuN, Tsyban' AV (1978) Izv. Akad. Nauk SSSR, Ser. Khim. 1892
265. Bubnov YuN, Shagova EA, Evchenko SV, Ignatenko AV, Gridnev ID (1991) Izv. Akad. Nauk SSSR, Ser. Khim. 2644; Engl. Transl. 2315
266. Bubnov YuN, Shagova EA, Evchenko SV, Ignatenko AV (1994) Izv. Akad. Nauk, Ser. Khim. 693
267. Bubnov YuN (1994) Pure Appl. Chem. 66:235
268. Bubnov YuN, Shagova EA, Evchenko SV, Ignatenko AV (1993) Izv. Akad. Nauk, Ser. Khim. 1672; Engl. Transl., Russian Chem. Bull. 42:1610
269. Bubnov YuN, Evchenko SV, Ignatenko AV (1992) Izv. Akad. Nauk, Ser. Khim. 2815; Engl. Transl. 2239
270. Bubnov YuN, Evchenko SV, Ignatenko AV (1993) Izv. Akad. Nauk, Ser. Khim. 1325; Engl. Transl., Russian Chem. Bull. 42:1268
271. Bubnov YuN, Lavrinovich LI, Zykov AYu, Klimkina EV, Ignatenko AV (1993) Izv. Akad. Nauk, Ser. Khim. 1327; Engl. Transl., Russian Chem. Bull. 42:1269
272. Bubnov YuN, Zykov AYu, Lavrinovich LI, Ignatenko AV (1993) Izv. Akad. Nauk, Ser. Khim. 1329; Engl. Transl., Russian Chem. Bull. 42:1271
273. Mikhailov BM, Bubnov YuN (1965) Izv. Akad. Nauk SSSR, Otd. Khim. Nauk 1310
274. Mikhailov BM, Bubnov YuN (1967) Izv. Akad. Nauk SSSR, Otd. Khim. Nauk 2290
275. Mikhailov BM (1983) Pure Appl. Chem. 55:1439
276. Gurskii ME, Potapova TV, Bubnov YuN (1993) Mendeleev Commun. 56; (1994) Chem. Abstr. 119:159741q
277. Bubnov YuN, Gurskii ME, Pershin DG (1991) J. Organomet. Chem. 412:1
278. Bubnov YuN, Potapova TV, Gurskii ME (1991) J. Organomet. Chem. 412:311
279. Gurskii ME, Pershin DG, Bubnov YuN (1992) Mendeleev Commun. 153
280. Bubnov YuN, Grigorian MSh, Tsyban' AV, Mikhailov BM (1980) Synthesis 902
281. Bubnov YuN, Tsyban' AV, Mikhailov BM (1980) Synthesis 904
282. Bubnov YuN, Geiderikh AV, Golovin SB, Gursky ME, Etinger MYu, Lavrinovich LI, Meleshkin PYu, Ponomarev SV, Zheludeva VI (1994) Mendeleev Commun. 55
283. Mikhailov BM, Bogdanov VS, Lagodzinskaya GV, Pozdnev VF (1966) Izv. Akad. Nauk SSSR, Ser. Khim. 386
284. Bogdanov VS, Pozdnev VF, Bubnov YuN, Mikhailov BM (1970) Dokl. Akad. Nauk SSSR 193:586
285. Bubnov YuN, Gurskii ME, Gridnev ID, Ignatenko AV, Ustynyuk YuN, Mstislavsky VI (1992) J. Organomet. Chem. 424:127

286. Gurskii ME, Gridnev ID, Il'ichev YuV, Ignatenko AV, Bubnov YuN (1992) Angew. Chem. 104:762; Angew. Chem. Internat. Ed. Engl. 31:781
287. Gurskii ME, Gridnev ID, Geiderikh AV, Ignatenko AV, Bubnov YuN, Mstislavsky VI, Ustynyuk YuN (1992) Organometallics 11:4056
288. Gridnev ID, Gurskii ME, Ignatenko AV, Bubnov YuN, Il'ichev YuV (1993) Organometallics 12:2487
289. Bubnov YuN (1991) Pure Appl. Chem. 63:361
290. Gursky ME, Golovin SB, Ignatenko AV, Bubnov YuN (1993) Izv. Akad. Nauk, Ser. Khim. 155; Russ. Chem. Bull. (Engl. Transl.) 42:139
291. Gursky ME, Golovin SB, Ignatenko AV, Bubnov YuN (1992) Izv. Akad. Nauk, Ser. Khim. 2198; Engl. Transl. 1724
292. Gursky ME, Geiderikh AV, Ignatenko AV, Bubnov YuN (1993) Izv. Akad. Nauk, Ser. Khim. 160; Russ. Chem. Bull. (Engl. Transl.) 42:144
293. Gursky ME, Geiderikh AV, Golovin SB, Ignatenko AV, Bubnov YuN (1993) Izv. Akad. Nauk, Ser. Khim. 236; Russ. Chem. Bull. (Engl. Transl.) 42:215
294. Lavrinovich LI, Ignatenko AV, Bubnov YuN (1992) Izv. Akad. Nauk, Ser. Khim. 2597; Engl. Transl. 2051
295. Brown HC, Rangaishenvi MV, Jayaraman S (1992) Organometallics 11:1948
296. Macdonald TL, Narayanan BA, O'Dell DE (1981) J. Org. Chem. 46:1504
297. Brown HC, Jayaraman S (1993) J. Org. Chem. 58:6791
298. Brown HC, Rangaishenvi MV (1990) Tetrahedron Lett. 31:7115
299. Brown HC, Rangaishenvi MV (1990) Tetrahedron Lett. 31:7113
300. Zweifel G, Hahn G, Shoup TM (1987) J. Org Chem. 52:5484
301. Zweifel G, Shoup TM (1988) J. Am. Chem. Soc. 110:5578
302. Pearson NR, Hahn G, Zweifel G (1982) J. Org Chem. 47:3364
303. Zweifel G, Pearson NR (1981) J. Org. Chem. 46:829
304. Leung T, Zweifel G (1974) J. Am. Chem. Soc. 96:5620
305. Zweifel G, Backlund SJ, Leung T (1978) J. Am. Chem. Soc. 100:5561
306. Williams JLR, Grisdale PJ, Doty JC (1967) J. Am. Chem. Soc. 89:4538
307. Williams JLR, Doty JC, Grisdale PJ, Searle R, Regan TH, Happ GP, Maier DP (1967) J. Am. Chem. Soc. 89:5153
308. Williams JLR, Doty JC, Grisdale PJ, Regan TH, Happ GP, Maier DP (1968) J. Am. Chem. Soc. 90:53
309. Grisdale PJ, Babb BE, Doty JC, Regan TH, Maier DP, Williams JLR (1968) J. Organomet Chem. 14:63
310. Williams JLR, Grisdale PJ, Doty JC, Glogowski ME, Babb BE, Maier DP (1968) J. Organomet Chem. 14:53
311. Grisdale PJ, Williams JLR (1969) J. Org. Chem. 34:1675
312. Eisch JJ, Shafi B, Rheingold AL (1987) J. Am. Chem. Soc. 109:2526
313. Eisch JJ, Shafi B, Odom JD, Rheingold AL (1990) J. Am. Chem. Soc. 112:1847
314. Eisch JJ, Shen F, Tamao K (1982) Heterocycles 18:245
315. Wilkey JD, Schuster GB (1988) J. Am. Chem. Soc. 110:7569
316. Wilkey JD, Schuster GB (1991) J. Am. Chem. Soc. 113:2149
317. Kropp MA, Schuster GB (1989) J. Am. Chem. Soc. 111:2316
318. Boyatzis S, Wilkey JD, Schuster GB (1990) J. Org. Chem. 55:4537
319. Eisch JJ, Tamao K, Wilcsek RJ (1975) J. Am. Chem. Soc. 97:895
320. Eisch JJ, Boleslawski MP, Tamao K (1989) J. Org. Chem. 54:1627
321. Kropp MA, Baillargeon M, Park KM, Bhamidapaty K, Schuster GB (1991) J. Am. Chem. Soc. 113:2155
322. Denmark SE, Nishide K, Faucher AM (1991) J. Am. Chem. Soc. 113:6675
323. Zweifel G, Horng A (1973) Synthesis 672
324. Clark GM, Hancock KG, Zweifel G (1971) J. Am. Chem. Soc. 93:1308
325. Matteson DS, Beedle EC (1990) Heteroatom Chem. 1:135
326. Matteson DS, Hurst GD (1990) Heteroatom Chem. 1:65

327. Pasto DJ, Chow J, Arora SK (1969) Tetrahedron 25:1557
328. Pasto DJ, McReynolds K (1971) Tetrahedron Lett. 801.
329. Brown HC, Yamamoto Y (1971) J. Am. Chem. Soc. 93:2796
330. Michaelis A (1901) Justus Liebigs Ann. Chem. 315:19
331. König W, Scharrnbeck W (1930) J. Prakt. Chem. 128:153
332. Bettman B, Branch GEK, Yabroff DL (1934) J. Am. Chem. Soc. 56:1865
333. Oikawa Y, Yoshioka T, Yonemitsu O (1982) Tetrahedron Lett. 23:885
334. Matteson DS, Kandil AA (1987) J. Org. Chem. 52:5121
335. Matteson DS, Kandil AA, Soundararajan R. (1990) J. Am. Chem. Soc. 112:3964
336. Matteson DS, Man HW, Ho O (1994) Manuscript in preparation
337. Iwasawa N, Kato T, Narasaka K 1988 Chem. Lett. (Jpn.) 1721
338. Matteson DS (1966) Organomet. Chem. Rev. 1:1
339. Matteson DS, Soloway AH, Tomlinson DW, Campbell JD Nixon GA (1964) J. Med. Chem. 7:640
340. Matteson DS, Michnick TJ, Willett RD, Patterson CD (1989) Organometallics 8:726
341. Mikhailov BM, Aronovich PM (1961) Izv. Akad. Nauk SSSR, Otd. Khim. Nauk 927
342. Mikhailov BM, Aronovich PM (1961) Izv. Akad. Nauk SSSR, Otd. Khim. Nauk 1233
343. Matteson DS, Shdo JG (1963) J. Am. Chem. Soc. 85:2684
344. Matteson DS, Waldbillig JO (1963) J. Org. Chem. 28:366
345. Woods WG, Strong PL (1967) J. Organomet. Chem. 7:371
346. Mikhailov BM, Bubnov YuN (1964) Izv. Akad. Nauk SSSR, Ser. Khim. 12:2170
347. Evans DA, Scott WL, Truesdale LK (1972) Tetrahedron Lett. 121
348. Evans DA, Golob AM, Mandel NS, Mandel GS (1978) J. Am. Chem. Soc. 100:8170
349. Coindard G, Braun J (1972) Bull. Soc. Chim. Fr. 817
350. Matteson DS (1962) J. Org. Chem. 27:4293
351. Bianchi G, Cogoli A, Grünanger P (1966) Ric. Sci. 36:132; Chem. Abstr. 64:19650e
352. Bianchi G, Cogoli A, Grünanger P (1966) J. Organomet. Chem. 6:598
353. Woods WG, Bengelsdorf IS (1966) J. Org. Chem. 31:2769
354. Fontani P, Carboni B, Vaultier M, Maas G (1991) Synthesis 605
355. Imai T, Mineta H, Nishida S (1990) J. Org. Chem. 55:4986

4 Alkenylboranes and Control of Olefinic Geometry

4.1 Introduction

There are two fundamental ways in which boron compounds can be used for stereocontrolled olefin synthesis. One is via the β-elimination of boron and a nucleofuge, such as halide, alkoxide, amino, or alkylthio. The other is via direct electrophilic replacement of boron from an alkenylborane. Many known examples in the first category derive the stereospecifically β-substituted borane from an alkenylborane, though there are significant exceptions that will be noted.

The most generally useful source of stereocontrolled alkenylboranes is hydroboration of alkynes (Section 2.3.3). Boron replacement of a more electropositive metal from an alkenylmetal compound is the second most useful route (Section 2.2.1 and 2.2.4). Further examples will be added here.

The most interesting alkene synthesis involving β-elimination is Zweifel's method, which has several variants that can lead to a wide variety of structures, though yields seldom exceed the 70% range. For synthetic purposes, the coupling of alkenylboranes with organic halides in the presence of a soluble palladium catalyst and base discovered by Miyaura, Suzuki, and coworkers tends to be more useful in a broad range of synthetic contexts.

In view of recent enantioselective oxidations of alkenes, especially the Sharpless dihydroxylation [1, 2], Sharpless epoxidation of allylic alcohols [3, 4, 5], and Jacobsen epoxidation of alkenes [6, 7], stereocontrolled alkene synthesis can be a key step in asymmetric synthesis.

4.2 β-Elimination Routes to Unsaturated Compounds

4.2.1 Simple Alkenes

4.2.1.1 Introduction. Thermodynamics strongly favors the elimination of boron and any β-substituent that can function as a nucleofugic group. As usual, the boron has to become tetracoordinate before boron-carbon bond cleavage can take place, though with tris-(β-haloalkyl)boranes elimination can occur under the usual conditions of hydroboration in an ethereal solvent. These reactions usually give essentially 100% stereocontrol.

The first example of β-elimination in a β-haloalkylborane was found in the hydroborations of vinyl chloride and allyl chloride by Hawthorne and Dupont [8]. Elimination from $ClCH_2CHRBR'_2$ formed $CH_2=CHR$ (R = H or Me), which was immediately rehydroborated to $R'_2BCH_2CH_2R$.

The facility of β-elimination is further illustrated by the behavior of dibutyl 2-bromoethylboronate, $BrCH_2CH_2B(OBu)_2$, which is easily prepared by light initiated addition of hydrogen bromide to dibutyl vinylboronate but worthless as a synthetic intermediate [9]. Every nucleophile tested except iodide ion caused deboronation to ethylene, bromide ion, and boric acid or a derivative. Kinetic studies under solvolytic conditions showed that the elimination involved nucleophilic attack at boron [10].

4.2.1.2 Stereochemistry. Of more interest is the observation that the elimination of the boronic ester group and bromide ion is *anti*, as shown by the *anti* bromination of **1** to **2**, which with base undergoes *anti* elimination to **3**, and the complementary sequence with the opposite stereoisomers **4**, **5**, and **6** [10]. The net result is replacement of boron by bromine with inversion of olefinic geometry, and the reactions appeared to be stereospecific within the limits of the analytical tools of the time [10].

This process has been updated and placed in the framework of hydroboration chemistry by Brown's [11, 12, 13].and Kabalka's [14] groups, who used a wide variety of alkenylboranes or boronic esters and obtained the corresponding (Z)-1-haloalkenes (**7**) (X = Cl, Br) in high isomeric purity [11]. It should be noted that X can also be iodine with boronic esters or acids if the reaction is carried out in the presence of pyridine, but under some conditions and with some structures, especially R = aryl or branched alkyl, iodine cleaves (E)-alkenylboranes with retention to yield (E)-1-iodoalkenes, or (Z)- to (Z)- [15], and the details of obtaining **7** with X = I via this route are complex. Analogous sequences have been tested successfully for making (E)-1-haloalkenes [16] and (E)- or (Z)-bromo internal alkenes [17] in high stereopurity.

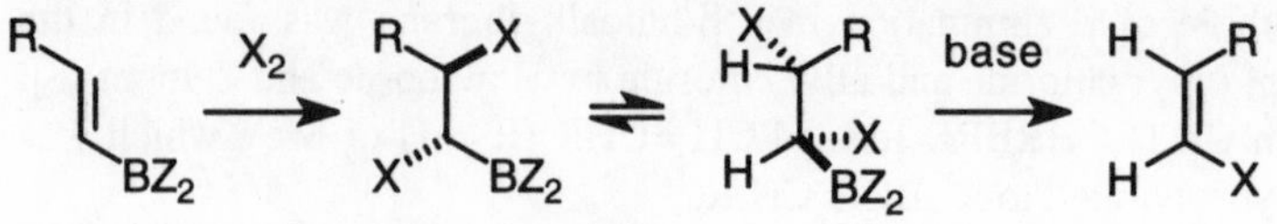

Z = alkyl, alkoxy, OH; X = Cl, Br   **7**
R = *n*, *sec*, or *tert*-alkyl, aryl

4.2.1.3 Fragmentation. An early application of boron β-elimination was a fragmentation reaction that resulted in expansion of fused six-membered rings to a ten-membered ring [18, 19, 20]. An example is the conversion of **8** to **11**. The ratio of β- to α-hydroboration products **9** to **10** is not known, but either one leads to the same elimination product **11** because of the relative orientations of the hydrogen atoms at each site and the requirement for *anti* elimination.

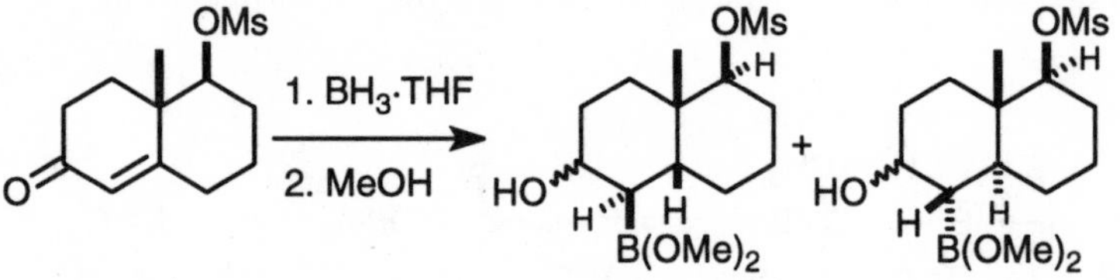

8   **9**   **10**

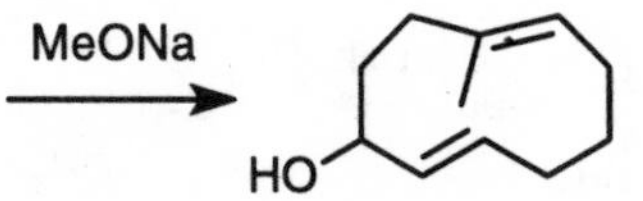

11

4.2.1.4 Alkenes from enamines. Hydroboration of isomerically pure enamines leads to diastereomerically pure β-aminoborane intermediates. Boronic ester groups are stable in the presence of β-amino functionality [21], but dialkylboryl groups are more easily eliminated. By proper choice of hydroborating agent and subsequent steps, either a pure (*E*)-olefin or a pure (*Z*)-olefin can be produced [22, 23]. For example, the morpholine enamine of propiophenone (**12**) can be hydroborated to form a boronic ester (**13**), which is converted to the amine oxide **14** by neutral (unbuffered) hydrogen peroxide without affecting the boronic ester. No direct evidence for **14** was obtained, since it eliminates to (*E*)-1-phenylpropene (**15**) (75%) under the reaction conditions, but the reaction does not proceed with base alone, and unbuffered hydrogen peroxide worked best among the several oxidizing agents tried. If the hydroboration is carried out with 9-BBN, the intermediate **16** undergoes elimination in methanol to form (*Z*)-1-phenylpropene (**17**) (81%).

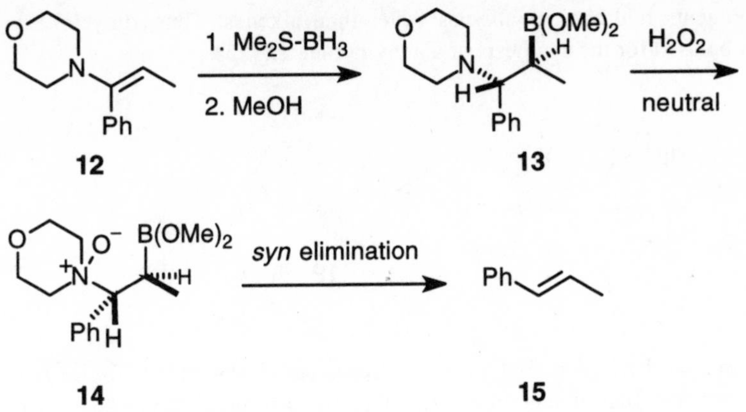

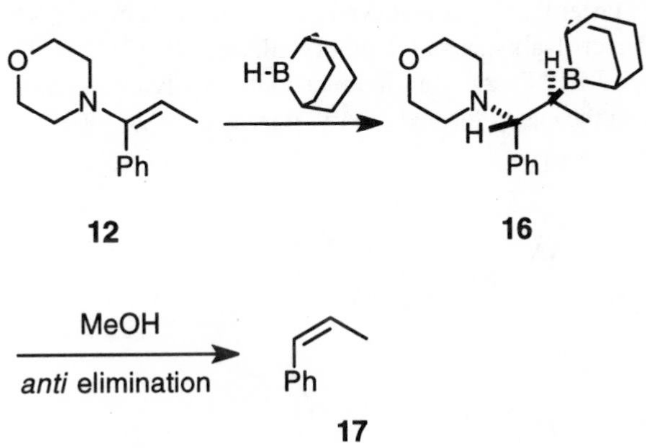

12   **13**

14   **15**

12   **16**

17

4.2.2 Routes to Alkenylboranes

4.2.2.1 (E)-1-Alkenylboranes from 1-alkynes. The *syn* addition of dialkylboranes to acetylenes, first reported by Brown and Zweifel in 1961 [24], is one of the classical examples of the value of hydroboration in stereocontrolled synthesis. 1-Alkynes are particularly valuable substrates, since they yield (E)-1-borylalkenes (**18**) with high regioselectivity in addition to the absolute stereospecificity inherent in the reaction. This hydroboration process and its major variants have been described in part in Section 2.3.3.

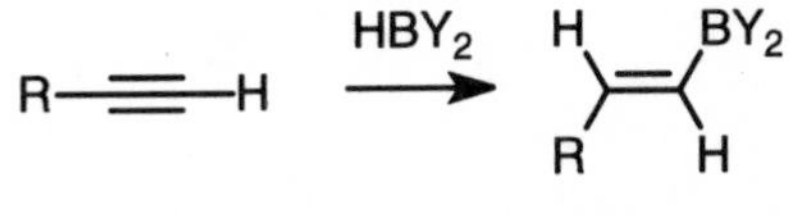

18

(R = alkyl, aryl; Y_2 = catechol, dihalo, dialkyl)

Some reagents hydroborate alkynes faster than alkenes. Thus, dicyclohexyl-borane can be used for the synthesis of alkenylborane **19** [25].

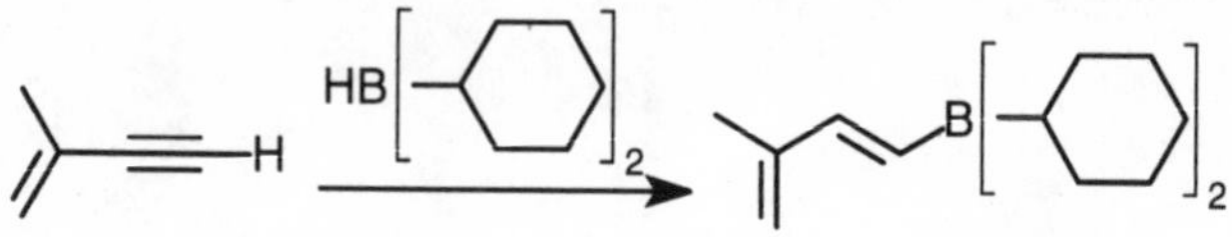

19

Catecholborane hydroborates typical 1-alkynes and 1-alkenes at similar rates, which are only one order of magnitude faster than (E)-3-hexene or cycloalkenes and two orders of magnitude faster than 1-methylcyclohexene [26]. Reaction of 9-BBN with 1-hexene is seven times faster than with 1-hexyne [27], which is in accord with the tendency of 9-BBN to convert 1-alkynes to 1,1-diborylalkanes (**20**) [28, 29]. The best synthesis of (E)-alkenyl-9-BBN's **21** is an indirect route involving oxidative elimination of one boron from the easily formed diboryl intermediate **20** [30].

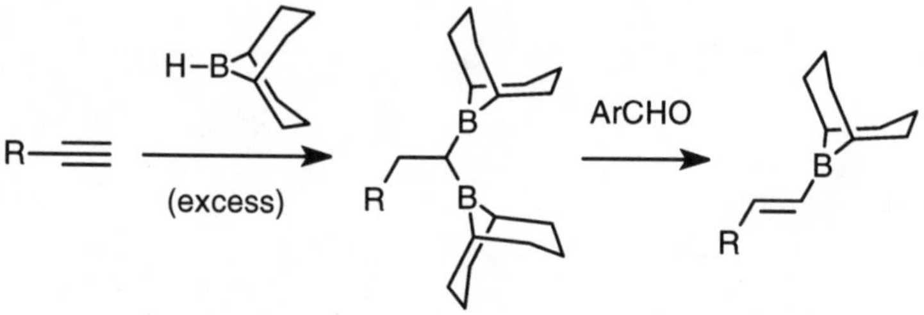

20   **21**

Although the relative rates have not been measured, the available data indicate that hydroboration with unsolvated dichloroborane generated from boron trichloride and trialkylsilanes yields alkenyldichloroboranes **22** very cleanly [31], and may well be the method of choice for access to many types of alkenylboranes. This route is incompatible with substrates containing oxygen functionality.

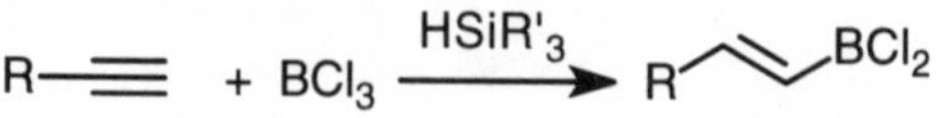

22

4.2.2.2 Alkenylboranes from disubstituted alkynes. Regioselectivity is a problem in the hydroboration of internal alkynes, as noted in Section 2.3.3.2. Zweifel and coworkers found that electronic effects direct boron α to the substituent in 1-haloalkynes [32], alkynylsilanes [33], and propargylic esters [34] as illustrated for the formation of **23**, **24**, and **25**. The α-disiamylboryl heptenoic ester **25** is an exception to the rule that boryl groups α to carbonyl groups rearrange rapidly to the corresponding enolates, which in this case would be allenic. The structure of intermediate **25** was proved by protonolysis to the (Z)-heptenoic ester, deuterium labeling, and peroxidic oxidation to the α-ketoheptanoic ester. However, it should be noted that ethyl propargylate itself with sterically hindered dialkylboranes places the boron β to the ester group, as illustrated by **26** [35].

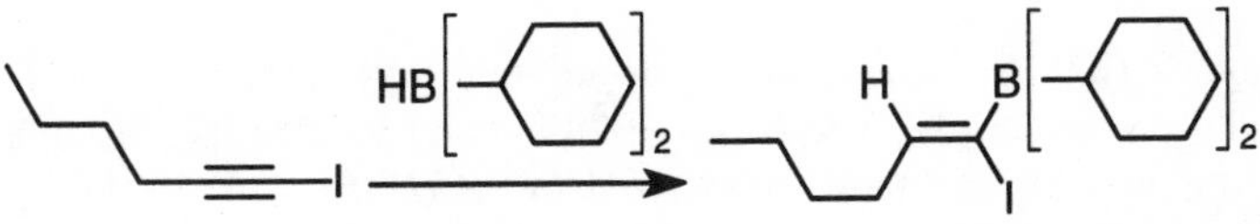

23

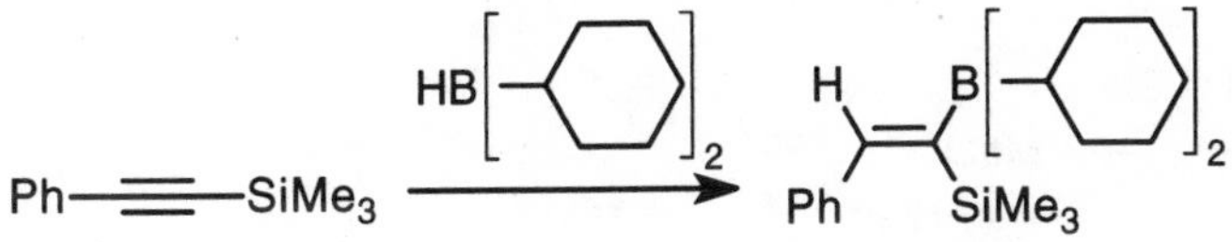

24

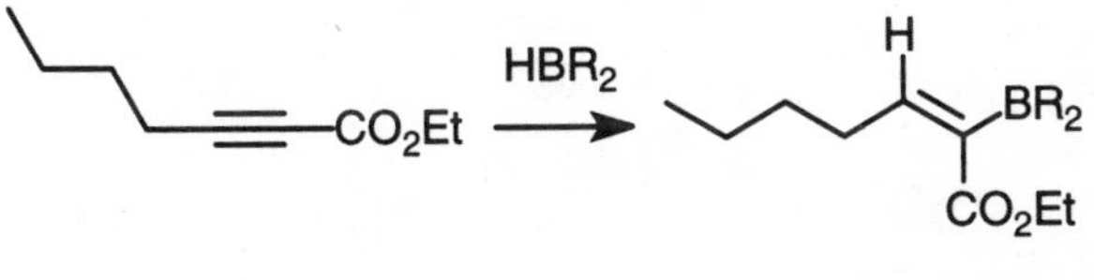

25   R = –CH(CH₃)CH(CH₃)₂

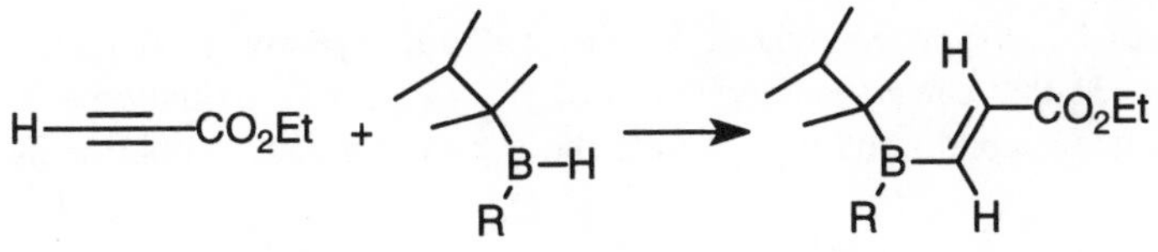

26

(R = cyclohexyl, EtCH(Me)–, 2-Me-cyclopentyl, etc.)

Of the foregoing examples, it is the 1-haloalkenylborane **23** that has had the greatest impact on synthesis. One of the simplest applications is acidic hydrolysis to provide the (Z)-1-haloalkene [32]. This synthesis has been generalized and optimized to provide a variety of (Z)-1-haloalkenes in 98-100% stereopurity [36].

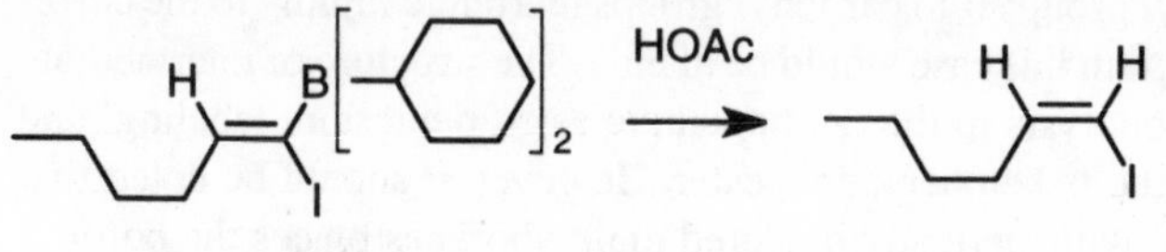

23

(Z)-Methoxyenynes (**27**) direct the boron toward the oxygen to produce 1-methoxy-3-boryl-1,3-dienes **28** [37]. Oxidation of **28** initially yielded the unstable (Z)-methoxyvinyl ketones **29**, but these isomerized to the (E)-isomers **30** during isolation. If the corresponding (E)-methoxyenynes are hydroborated, a gross mixture of regioisomers is produced, strongly suggesting that the oxygen of **27** coordinates to the boron during the hydroboration process.

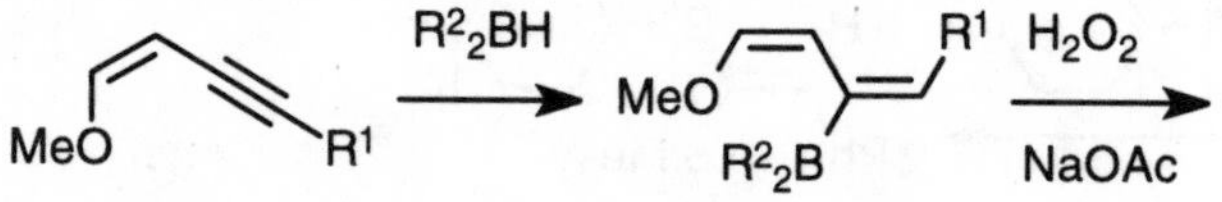

27   **28**

R^1 = cyclohexyl, siamyl; R^2 = propyl, hexyl, cyclohexyl, –CH(OSiMe$_3$)Et

29   **30**

Hydroboration of 1-iodoisopropenylacetylene with alkylthexylboranes yields intermediate boranes **31** that can be isomerized to **32** [38]. Peroxidic oxidation was carried out either under conditions that yielded the α,β-unsaturated ketone or its epoxide **33**.

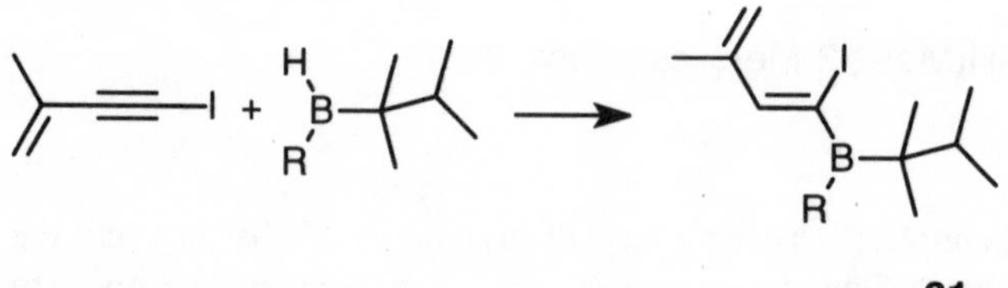

31

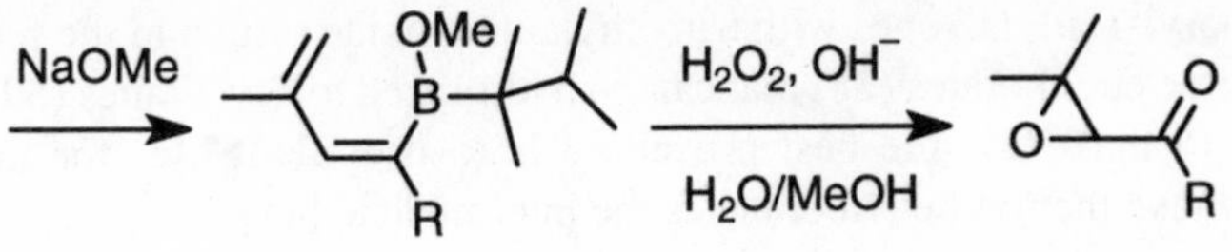

32   **33**

Zweifel and coworkers have carried out extensive investigations of transformations of alkynylsilanes. Hydroboration of a 1,3-bis(trimethylsilyl)-1-alkyne (**34**) followed by protonolysis yielded the (Z)-1,3-bis(trimethylsilyl)-1-alkene (**35**) [39]. Other similar transformations were carried out via hydroalumination.

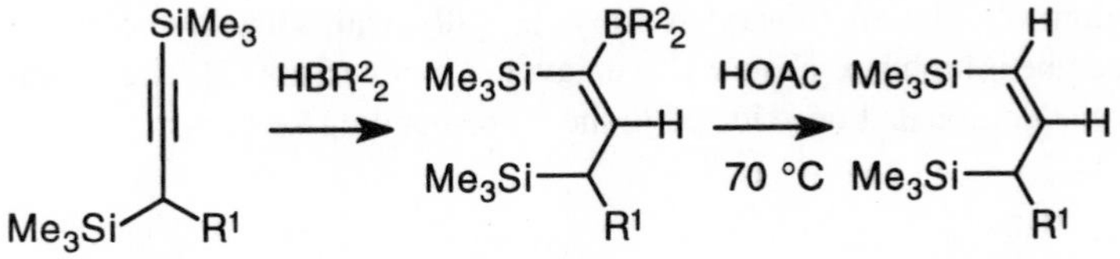

34   **35**

(R^1 = *n*-pentyl; R^2 = -CH(Me)CHMe$_2$)

Hydroboration of bis(trimethylsilyl)acetylene to **36** followed by oxidation with trimethylamine oxide and hydrolysis yields (trimethylsilylacetyl)trimethylsilane (**37**), which can undergo further transformations to provide a stereocontrolled synthesis of trisubstituted olefins [40].

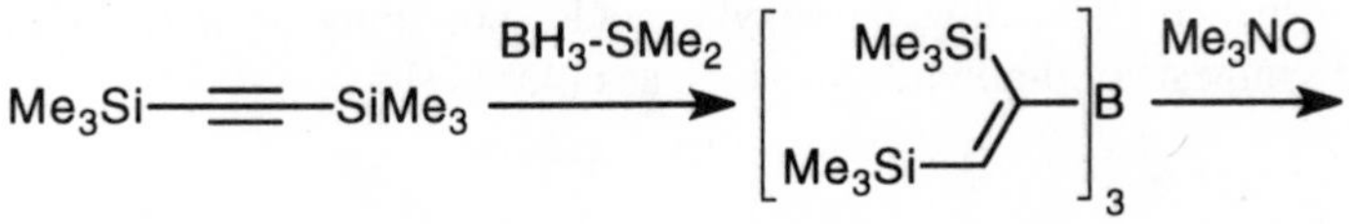

36

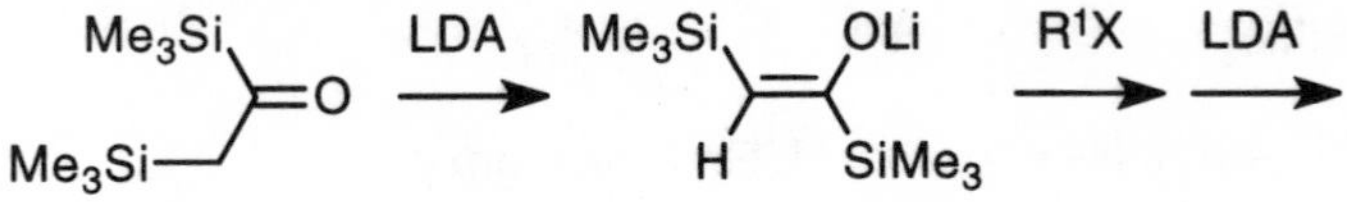

37

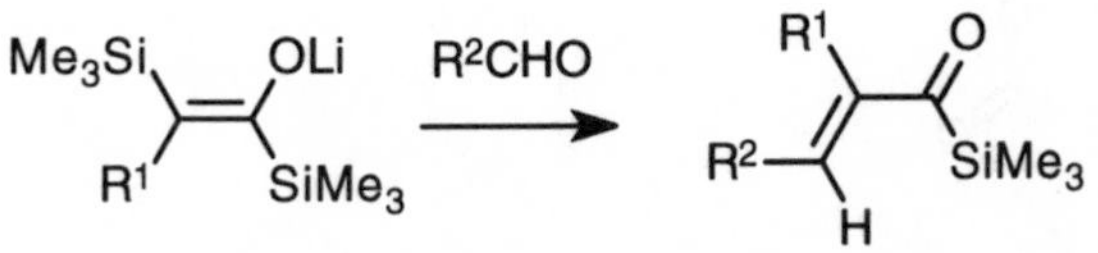

Oxidation of 1-boryl-1-silylalkenes with trimethylamine oxide results in the formation of intermediate enol borates (38) that can be hydrolyzed to acylsilanes (39). If the objective is to make 39, the best procedure is to use $BH_3\text{-}SMe_2$ for the hydroboration and make the trialkenylborane as the intermediate [41].

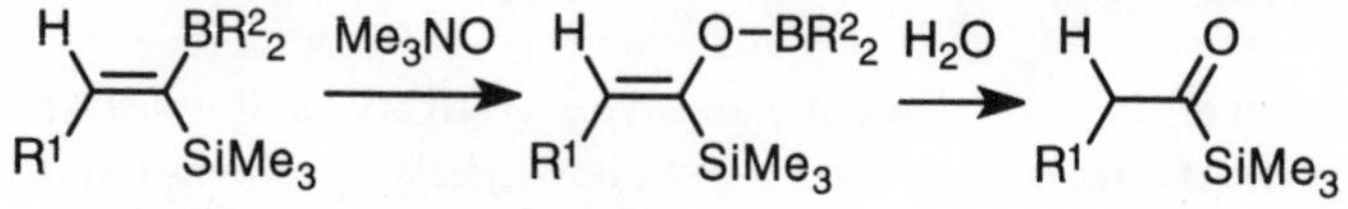

38   **39**

Hydroboration of (dimethylthexylsilyl)diynes (40) with disiamylborane or dicyclohexylborane is highly regioselective in favor of boranes 41 [42]. The boranes were either protodeboronated or oxidized to the corresponding ketones.

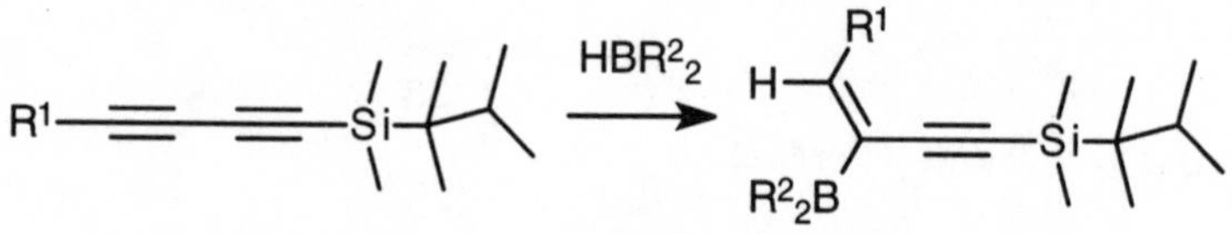

40 (R1 = *n*-hexyl, cyclohexyl)   **41**

(Trimethylsilyl)enynes 42, which were derived from (trimethylsilyl)diynes in a stereocontrolled manner via organotin/lithium chemistry, have been hydroborated and oxidized to the corresponding carboxylic acids, providing a route to the stereocontrolled synthesis of β-alkylidene-γ-lactones (43) [43].

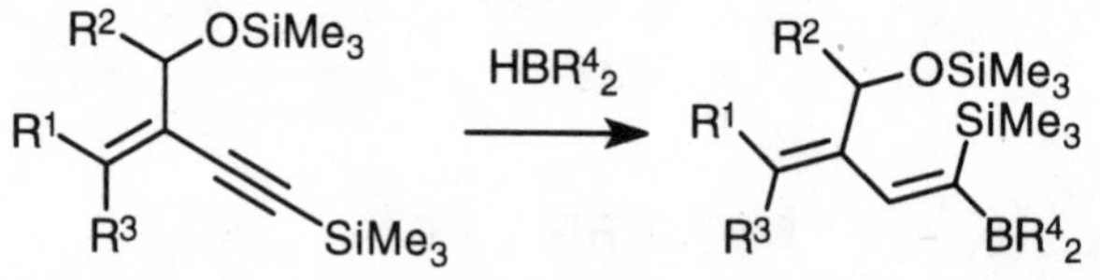

42 (R^1, R^2 = alkyl; R^3 = alkyl or H; R^4 = cyclohexyl)

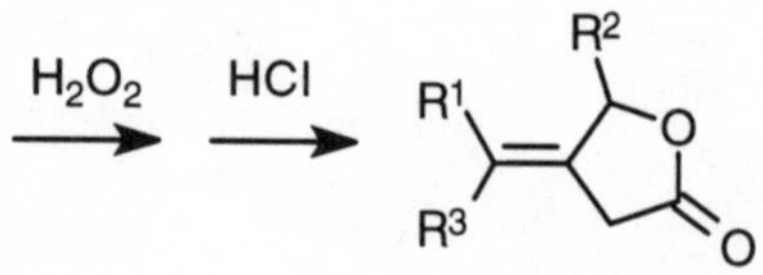

43

Steric factors affecting regioselectivity in alkynylborane hydroboration have been studied by Soderquist and coworkers. The usual preference for placing boron α to silicon can be reversed by the use of a triisopropylsilyl group, as in the formation of **44** [44]. In hydroboration of HC≡CSiR$_3$, the β-substituted borylsilane (*E*)-R$_3$SiCH=CH-9-BBN is preferred regardless of the size of R.

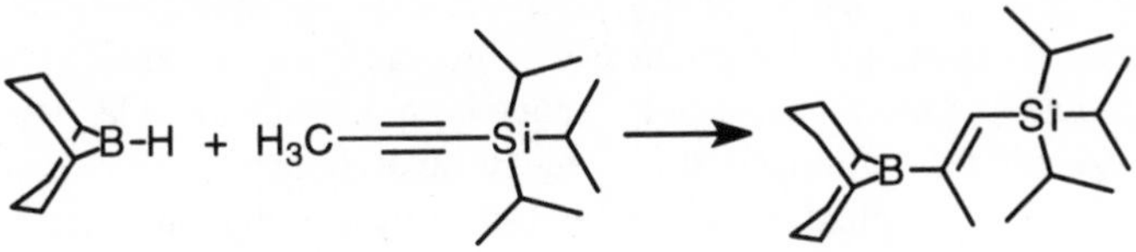

(R = *n*-alkyl)   **44**

The use of soluble nickel catalysts in hydroborations of alkoxyacetylenes or alkylthioacetylenes gives good regiocontrol, placing the boron β to the functional substituent [45]. A typical example is the preparation of **45** (92%).

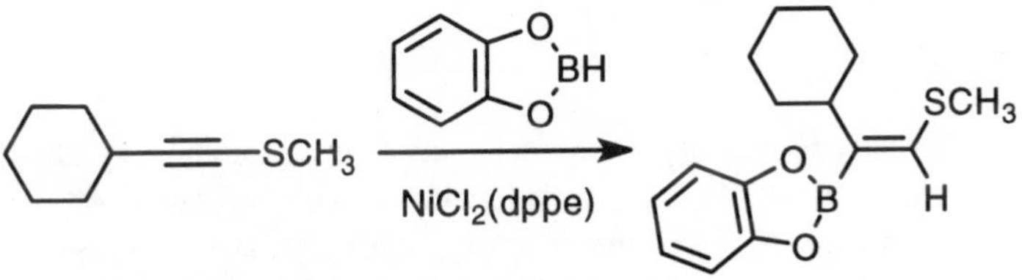

(dppe = Ph$_2$PCH$_2$CH$_2$PPh$_2$)   **45**

The reverse substitution pattern, both in the regio and steric senses as exemplified by **46**, is produced by catalyzed thioboration with *B*-phenylthio-9-BBN [46]. A wide variety of functional groups are tolerated in R, including ketone, nitrile, ester, acetal, benzyl ether, halo, and vinyl substituents, usually with three or four intervening carbon atoms. The reactive **46** were not isolated but used in catalyzed coupling with aryl, benzylic, or alkenyl halides ("Suzuki coupling, see Section 4.3) or protodeboronated to H$_2$C=C(R)SPh.

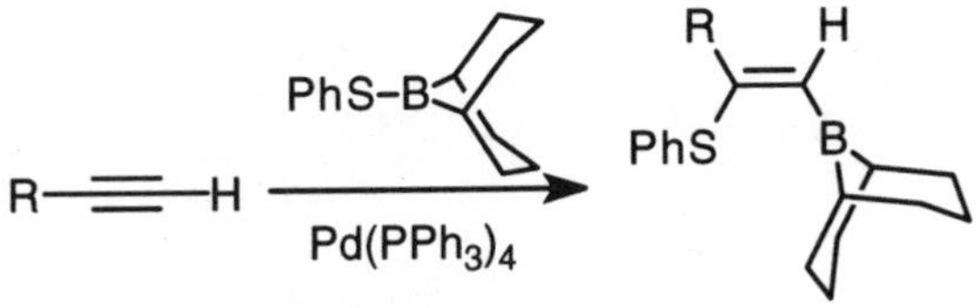

46

4.2.3 Inversion of (*E*)- to (*Z*)-Alkenylboranes

As noted in Section 3.4.3, α-haloalkylboranes react via tetracoordinate borate complexes, and this mechanism allows displacement of the halide by nucleophiles that would not ordinarily react in such a manner. The activating effect of the boron functionality is sufficient that even α-haloalkylboranes participate in nucleophilic displacements, and the alkenyl group geometry is inverted in the process. Thus, Zweifel and Arzoumanian found that the product of hydroboration of 1-iodohexyne with dicyclohexylborane (**23**) rearranges on treatment with sodium methoxide to the (*E*)-alkenylborinic ester **47** [32]. Similar rearrangements involving aryl groups were discovered in a nonsynthetic context about the same time by Köbrich and Merkle [47, 48].

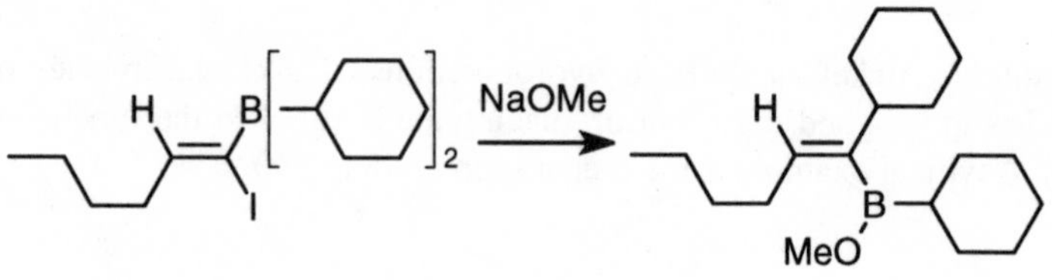

23   **47**

This process was tested successfully in a model study for prostaglandin synthesis by Corey and Ravindranathan [49]. Racemic cyclopentenol derivative **48** was hydroborated with thexylborane, and the resulting dialkylborane was used to hydroborate racemic 1-bromoalkyne **49** to form 1-(bromoalkenyl)borane **50**. Rearrangement of **50** with base to **51** was followed by protodeboronation/deprotection to the prostaglandin analogue **52**. Because the two pieces joined were both racemic, **52** was obtained in two diastereomeric forms, but only two, indicating that the hydroboration and rearrangement processes were both essentially stereospecific.

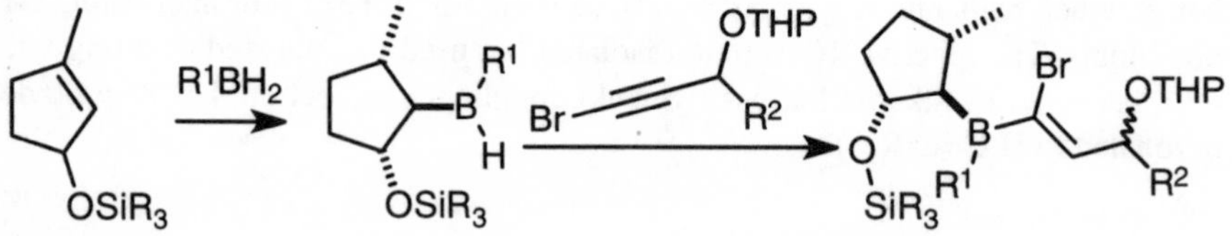

48   **49**   **50**

$SiR_3 = SiMe_2t\text{-}Bu$; $R^1 = -CMe_2CHMe_2$; $R^2 = -(CH_2)_4CH_3$

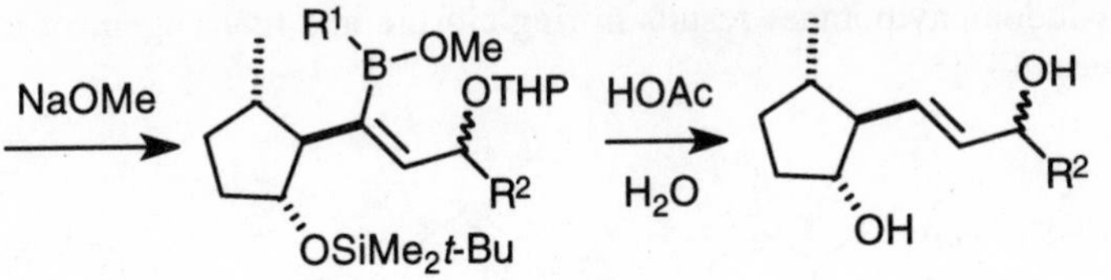

51   **52**

A stereocontrolled (*E,E*)-diene synthesis starting from a 1-halo-1-alkyne and a 1-alkyne was reported by Negishi and coworkers [50]. The coupling step yields a dienylborane **53** that can be converted to diene by protodeboronation. Two iodoalkyne fragments can be coupled to form the iododienylborane **54**, which can undergo elimination to form the labile (*E*)-butatriene **55** stereospecifically [51].

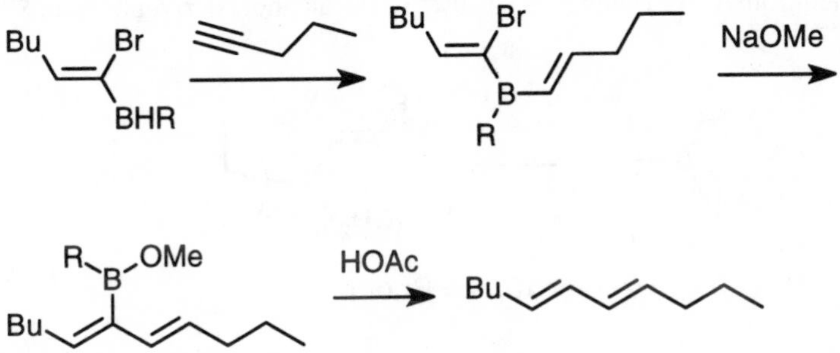

53

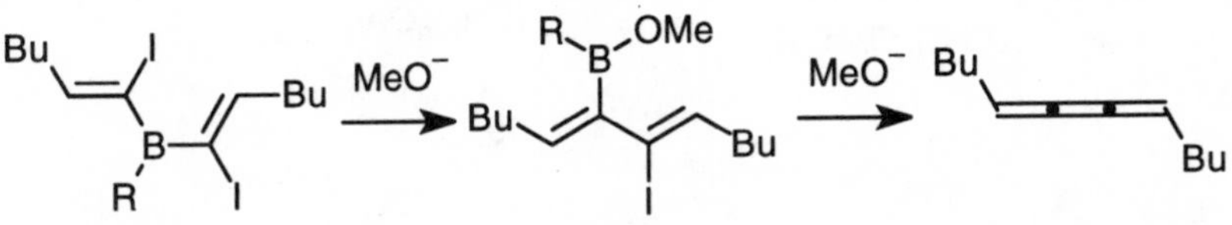

54   **55**

Lithium triethylborohydride reduces (*Z*)-1-iodo-1-alkenylboranes (**56**) to (*Z*)-alkenylboranes (**57**) with the usual inversion of the displacement site [52].

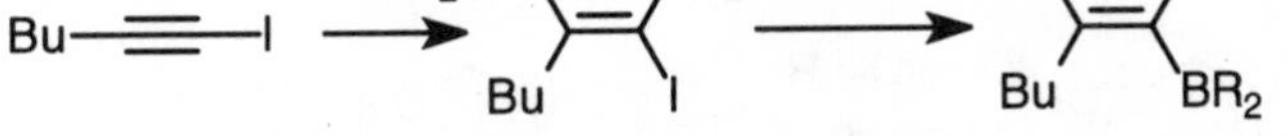

56   **57**

(R = cyclohexyl or -CHMeCHMe$_2$)

Reduction of 1-iododienylboranes results in ring closure and rearrangement to borolene derivatives (**58**) [53].

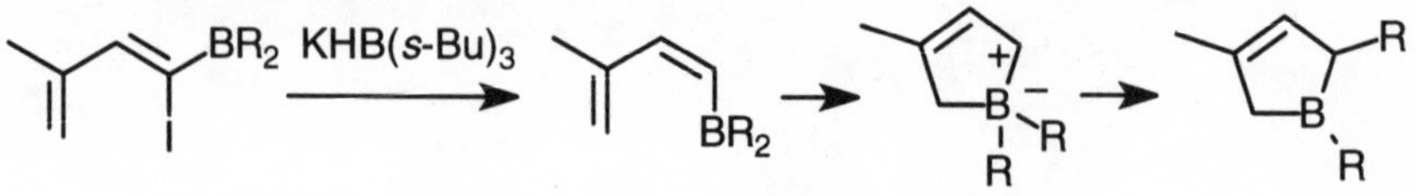

58

Preparations of boranes suitable for these reactions have been reinvestigated and improved significantly [54]. These have in turn been used for the synthesis of several examples of **59**, alcohols derived by acetate hydrolysis of **59**, and derived aldehydes, several of which are pheromones of pest species of Lepidoptera or Diptera [55]. Hydroboration of haloalkynes with borinane (**60a**) or borepane (**60b**) provides especially efficient routes to **61**, which is a subset of the alcohols derivable from **59** [56].

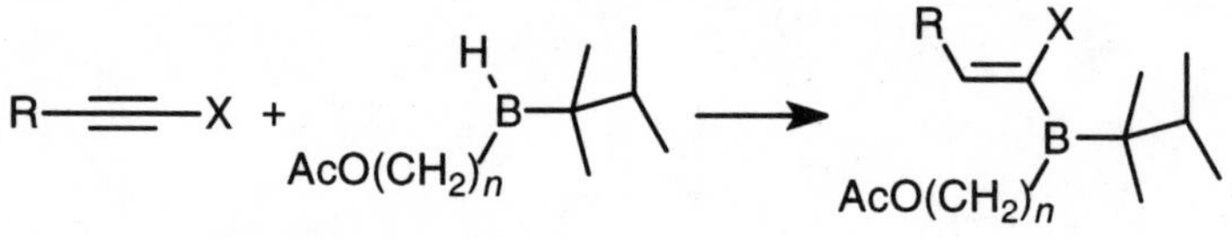

R = Et, n = 5 or 10; R = Bu, n =6 or 10; X = Br or I

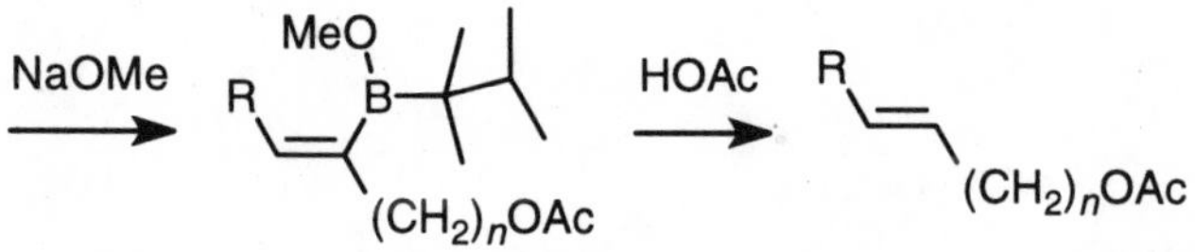

59

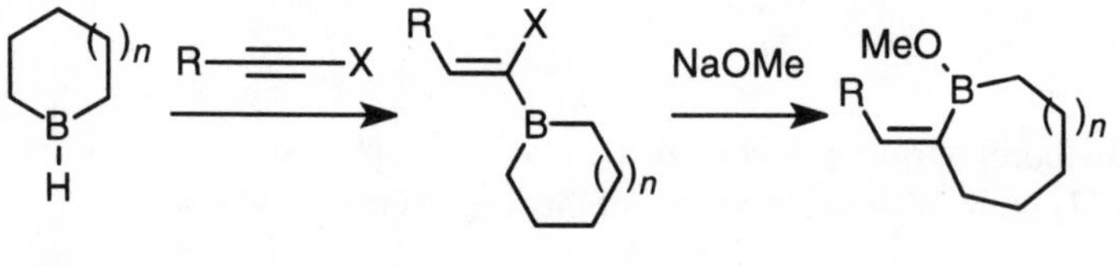

60: a, n =1; **b,** n = 2

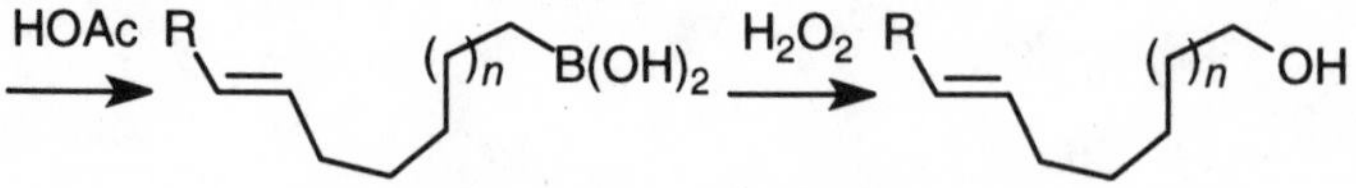

61

As noted in Section 4.2.2, alkyne hydroboration by boranes generated in situ from boron halides and silanes may offer more efficient construction of alkenylborane precursors to some of the types of intermediates described above [31].

Conversion of enantiomerically pure boronic acids to thexylborane derivatives such as **62** and the use of these in the haloalkyne reaction sequence leads to asymmetric (E)-alkenes such as **63** in high stereopurity [57].

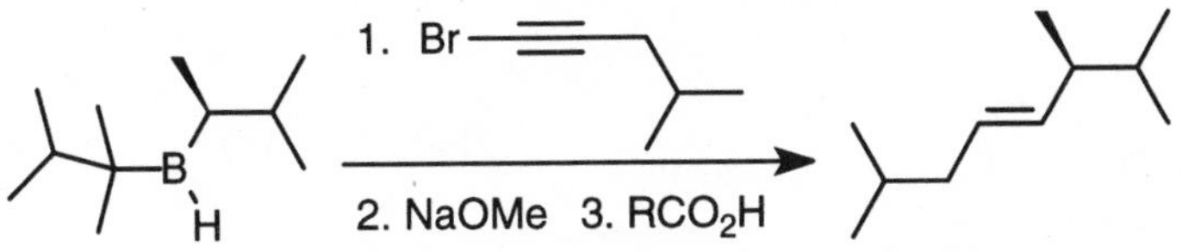

62   **63**

Stereocontrolled reactions of boronic esters have been achieved by Brown and coworkers. Preparation of (Z)-(1-bromoalkenyl)boronic esters (**65**) is easily accomplished via (Z)-1-bromo-1-(dibromoboryl)-1-alkenes (**64**) made from dibromoborane methyl sulfide and 1-bromoalkynes [58]. Reduction with potassium triisopropoxyborohydride then yields (Z)-1-alkenylboronic esters (**66**) [59].

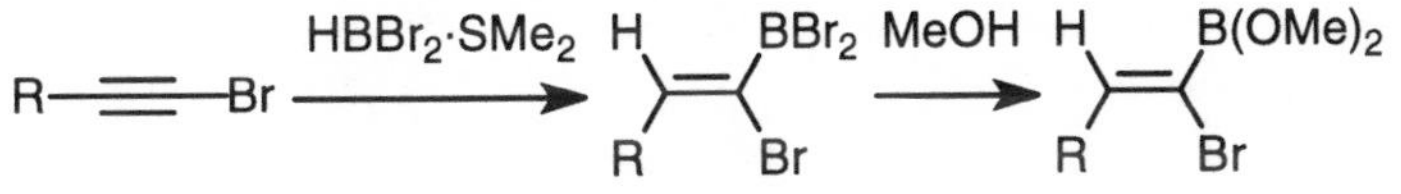

64   **65**

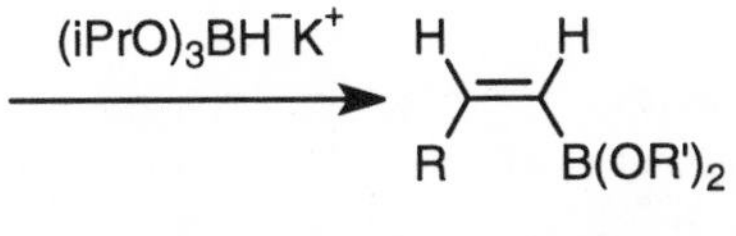

66

For replacement of bromide by alkyl, 1,3-propanediol boronic esters (**67**) were found to work better than other types [60]. Yields were higher in diethyl ether than in THF, and the rearrangement of intermediates **68** to alkenylboronic esters **69** was promoted by methanol. A side reaction was cleavage of intermediate borate **68** to alkylboronic ester [$R^2BO_2(CH_2)_3$] and, presumably, 1-bromoalkene.

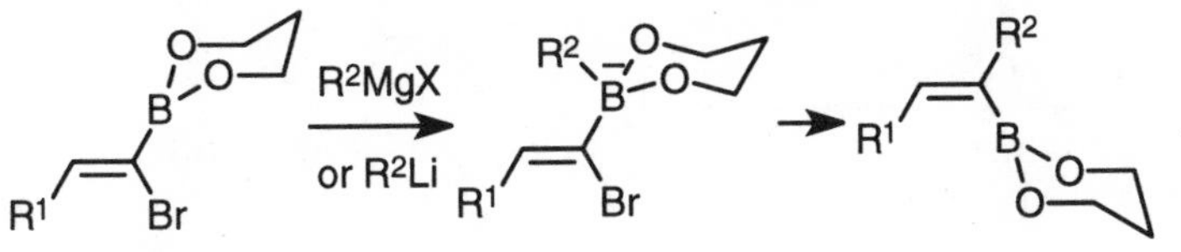

67   **68**   **69**

The group R^2 of **69** can be alkenyl with specific selected geometry, and the foregoing synthesis thus provides a general route to dienylboronic esters [61].

Hydroboration of 1-chloroalkynes with disiamylborane followed by oxidation with trimethylamine N-oxide, bromination, and base catalyzed elimination provides an efficient stereocontrolled route to (Z)-1-bromo-1-chloroalkenes (**70**) [62]. The (E)-isomers can be obtained via a similar route by starting from the 1-bromoalkyne and chlorinating the 1-bromoalkenylborinic ester intermediate.

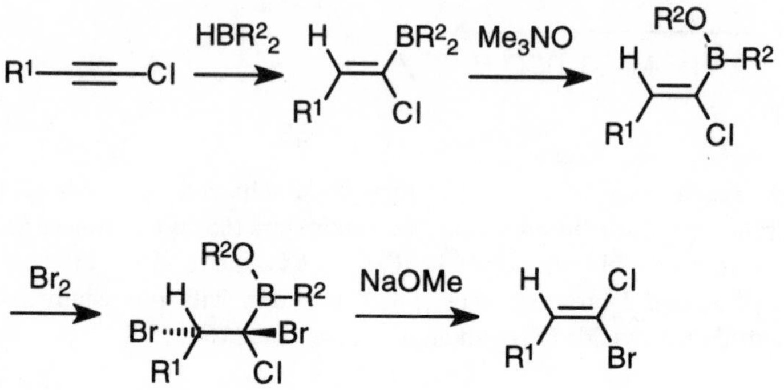

70

4.2.4 Other Routes to Alkenylboranes

Alkenylmetal compounds react with various classes of boranes with retention to produce alkenylboranes in a stereocontrolled manner. Examples of replacement of aluminum, zirconium, tin, and mercury have been mentioned in Section 2.2. Alkenyllithiums are probably the most generally useful, and were employed in the earliest stereocontrolled synthesis of alkenylboronic esters [10]. A more recent example was the use of alkenyllithiums as sources of alkenylborinic esters such as **71** for use in the Zweifel olefin synthesis (see below) [63, 64].

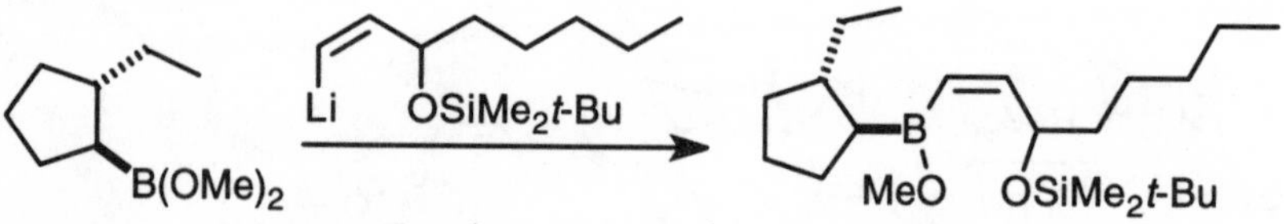

71

Internal (E)-bromoalkenes (72) are readily accessible via bromination and base catalyzed deboronobromination (Section 4.2) of internal (E)-alkenylboronic esters 69. Lithiation of 72 provides a general route to internal (Z)-alkenylboronic esters 73 [65], which are not accessible by hydroboration because of regioselectivity problems except where $R^1 = R^2$ or R^1 is much bulkier than R^2.

(Z)-1-Alkenylboronic esters (74) can be prepared by hydrogenating 1-alkynyl-boronic esters at 1 atmosphere over Lindlar's catalyst in dioxane with a small amount of pyridine [66]. These specific conditions were required in order to achieve satisfactory (Z)-isomeric purity, which ranged from 90 to >99%. The isomeric purity was readily improved via hydrolysis of the labile isopropyl esters to the recrystallizable boronic acids. The requisite alkynylboronic esters were originally made from the Grignard reagents [67], but as is often the case, it has proved more convenient to use the Brown-Cole method [68] based on the lithium reagent, triisopropyl borate, and nonaqueous work up [69].

R = n-butyl, n-hexyl, isopropyl, $tert$-butyl, cyclopentyl, phenyl

Although hydroboration of alkynes provides an almost universal solution to the synthesis of (E)-alkenylboronic esters, there is an alternative route from aldehydes and lithiated methylenediboronic esters (75) [70]. This normally less convenient alternative proved crucial in the famous synthesis of palytoxin by Kishi and coworkers, because a remote carbamate function in the very large and complex R group failed to survive hydroboration conditions [71]. The usual isomeric purity of (E)-alkenylboronic esters produced by this process was ~90-95% (60-MHz NMR estimate) in the model reactions [70], but for the valuable palytoxin intermediate this figure fell to 85% [71]. The next step in this synthesis is described in Section 4.3.2 on applications of Suzuki coupling.

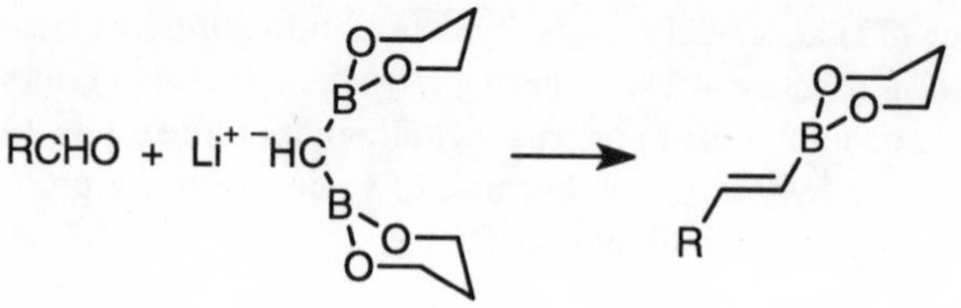

75

It is also possible to generate (Z)-alkenylboronic esters preferentially from aldehydes by the use of lithiated $Me_3SiCH_2BO_2C_2Me_4$ [72], though the (Z)/(E) ratio was
only 70:30 and the possibility of improving the selectivity of this unexpected reaction by modifying the silyl and boronic ester groups has never been explored.

4.2.5 The Zweifel Olefin Synthesis

The Zweifel synthesis in its original form provides a stereospecific synthesis of (Z)-
alkenes such as **80** from (E)-1-alkenylboron compounds such as **76**, joining the
alkenyl group and an alkyl group that were both originally linked to the boron atom.
The configuration of the alkenyl group is inverted, that of the alkyl group retained,
in this process, which involves treatment of the alkenylborane intermediate with
iodine and base [73]. It was postulated that intermediate iodonium ion **77** rearranges
with alkyl migration to the β-iodo borinic acid **78**, which is converted by base to its
anion **79** and undergoes *anti* elimination of alkylboronic acid and iodide to yield the
(Z)-alkene **80**.

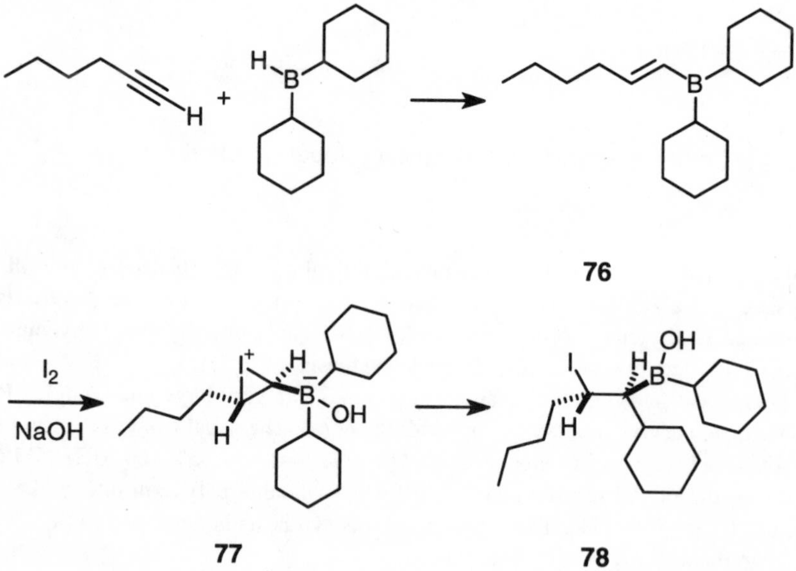

76

77   **78**

79   **80**

It is possible to achieve the synthesis of an (E)-alkene from **76** by using cyanogen bromide in dichloromethane in place of iodine and base [74], though alternatives appear to be generally more useful. If a (Z)-alkenylborane is used in the iodination route, the (E)-alkene results [32, 63]. It might also be noted that the hydroboration of haloalkynes followed by base treatment and C–B bond hydrolysis yields the same (E)-alkenes more easily than the Zweifel synthesis in many cases.

The Zweifel synthesis was tested as a potential route to prostaglandins by Evans and coworkers [63, 64], who concluded that borinic esters work better than trialkylboranes in the coupling step. The cyanogen bromide route failed with the (E)-bis(ethylcyclopentyl)alkenylborane, and the (Z)-alkenylborinic ester **71** (derived from an alkenyllithium, see above) was therefore used instead as the precursor to (E)-olefin **81**, isomeric purity >99%. The (Z)-isomer of **81** was also prepared from the (E)-isomer of **71**.

71   **81**

The Zweifel synthesis has been critically reexamined by Brown and coworkers, and has been substantially improved by the use of boronic ester intermediates, providing a wide variety of disubstituted and trisubstituted olefinic compounds in very high stereopurity. An arbitrarily selected typical example is the reaction of 1,3-propanediol propenylboronate with 2-lithiothiophene to form intermediate borate **82**, which rearranges to **83** on treatment with iodine in methanol, –78 °C → 0 °C [75]. A typical trisubstituted alkene is illustrated by **84** [76]. Borate intermediates analogous to **82** were alternatively generated from sodium methoxide and B-alkyl-B-alkenylborinic esters, which were obtained by methanolysis of bromoboranes prepared from RBHBr and alkynes [75]. The major side reaction is simple cleavage of the alkenyl group from the borate intermediate to form the vinylic iodide, and this is minimized by the use of low temperatures. Increasing steric bulk of the substituents tends to decrease the amount of iodinative cleavage.

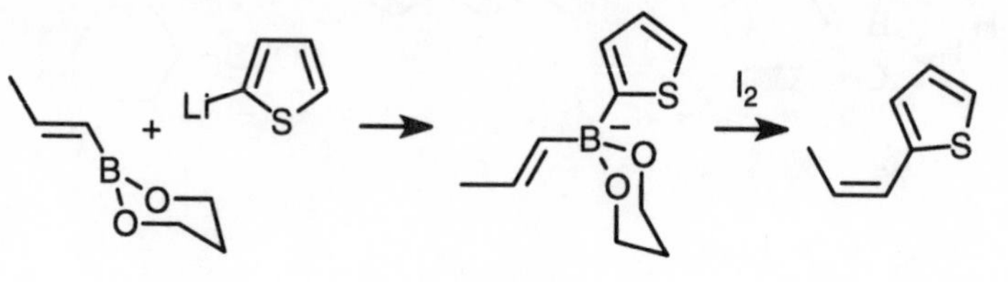

82   **83 (68%)**

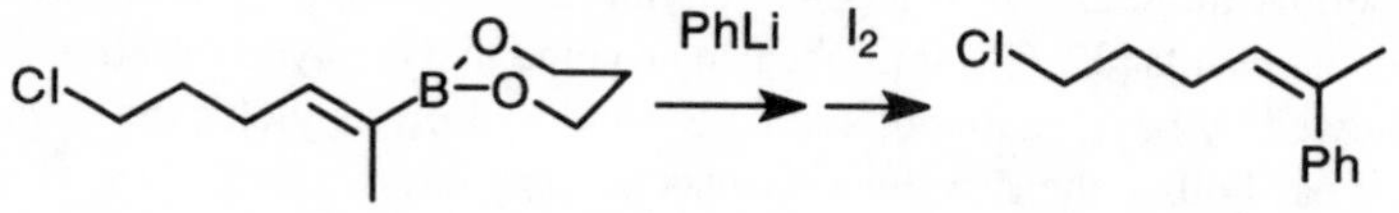

84 (76%)

Hydroboration of an alkyne with RBHBr followed by the Zweifel synthesis has been used to make the (Z)-alkene muscalure, the housefly pheromone [75], and also a (Z)-alkene that was epoxidized to racemic disparlure [77], one enantiomer of which is the gypsy moth pheromone.

Conjugated enynes **85** can be made by treatment of B-(thexyl)-B-(alkenyl)-B-(alkynyl)boranes with iodine and potassium methoxide [78]. One use of **85** is reduction via hydroboration and protodeboronation to the (Z,E)-dienes **86**. Where R^1 = Bu and R^2 = $(CH_2)_4OH$, **86** is the pheromone of the tent caterpillar *Malacosoma disstria*. The hydroxyl was protected as its THP derivative during the synthesis. As an approach to (Z,E)-diene synthesis, this method is probably inferior to routes based on the Suzuki coupling (see below).

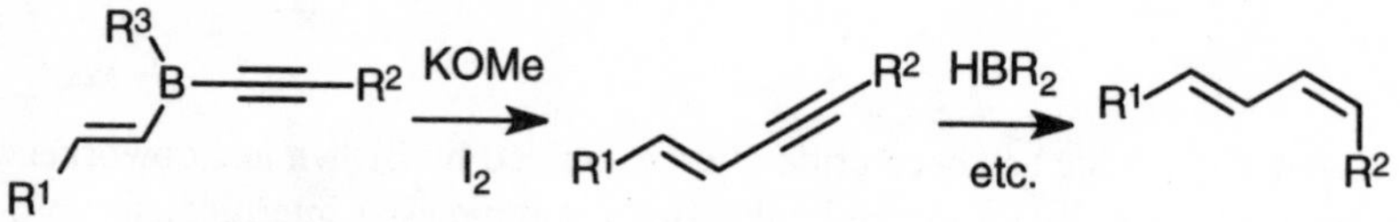

(R^3 = –CMe$_2$CHMe$_2$)   **85 (58-82%)**   **86**

The reaction of alkynylboranes with electrophiles (see also Section 3.4.3.8) bears some analogy to the Zweifel synthesis. B-Alkenyl-B-alkynylboranes with Lewis acids rearrange to intermediates that can be protodeboronated to provide (Z,E)-dienes (**87**) [79]. The pheromone of *Bombyx mori*, the silkworm moth, has R^1 = HO(CH$_2$)$_9$– and R^2 = –(CH$_2$)$_2$CH$_3$, and in this work the precursor **87** having R^1 = MeO$_2$C(CH$_2$)$_8$– was made.

87

The acetylenic analogue of the Zweifel synthesis, iodination of $R^1_3B^- - C \equiv C - R^2$, provides acetylenes $R^1 - C \equiv C - R^2$ with R^1 groups such as phenyl or cycloalkyl that cannot be introduced by ordinary displacement reactions [80]. B-Alkyl-B-alkynylborinic esters, which might provide more efficient utilization of R^1 in such couplings, are accessible from boronic esters and alkynyllithiums with nonaqueous work up [81].

4.3 The Suzuki Coupling Reaction

4.3.1 Scope of the Reaction

The nearly stereospecific coupling of organoboron compounds with alkenyl or aryl halides in the presence of a soluble palladium catalyst and base is applicable to a wide variety of structures. Most often, the boron compounds have been alkenyl- or arylboronic esters, though alkynyl- and primary alkylboranes have also been used successfully. Yields are often very high, many types of functionality are tolerated, and the procedure is often especially useful for connecting two major fragments late in a convergent synthesis.

The general cross coupling of alkenyl-, aryl-, or alkynylmetallic compounds with alkenyl or aryl halides in the presence of Ni(PPh$_3$)$_4$ or a Pd(0) catalyst prepared from Cl$_2$Pd(PPh$_3$)$_2$ and i-Bu$_2$AlH was discovered and developed by Negishi and coworkers [82, 83, 84], preceded only slightly by reports of palladium catalyzed cross couplings of Grignard reagents [85] or alkynylsodiums [86]. Negishi's early work successfully utilized aluminum, zirconium, zinc, and tin reagents, and it was observed that an alkynyltributylborate coupled with o-iodotoluene in the presence of the Pd(0) catalyst in refluxing THF, but (E)-1-hexenyldisiamylborane or its butyllithium borate complex were unreactive toward Ni(PPh$_3$)$_4$ and 1-bromo-naphthalene in THF at 25 °C.

The discovery that alkenyl- or arylboranes can generally be coupled with aryl halides with Pd(PPh$_3$)$_4$ as catalyst, provided a suitable base such as alkoxide is present, was reported by Miyaura, Suzuki, and coworkers [87, 88, 89, 90]. This reaction is often referred to as "Suzuki coupling." Several reviews of this work have appeared as it has developed [91, 92].

In its classical form, the process involved hydroboration of an alkyne to an alkenylborane (**88**), which was then coupled with an aryl bromide in the presence of 1-3 mol % of Pd(PPh$_3$)$_4$ or PdCl$_2$(PPh$_3$)$_2$ and 1 mol of sodium ethoxide to form the alkenylbenzene **89** in high isomeric purity [87, 88]. A more severe test of stereoselectivity was the coupling of the (Z)-1-hexenylboronic ester **90** with iodobenzene, which yielded 87% of **91** in 98% isomeric purity [93].

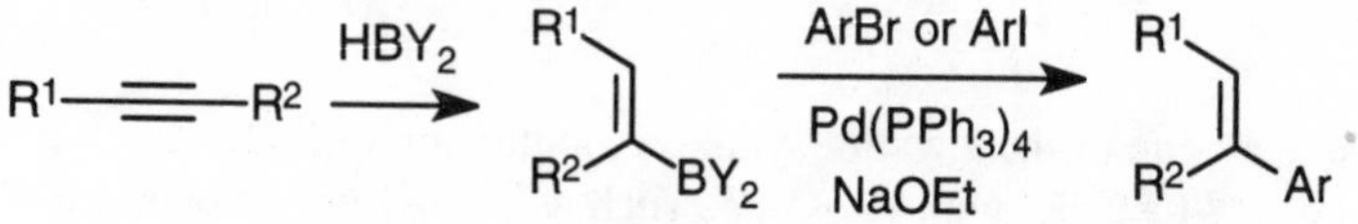

HBY_2 = disiamylborane, 9-BBN, catecholborane
R^1 = alkyl or aryl, $R^2 = R^1$ or H

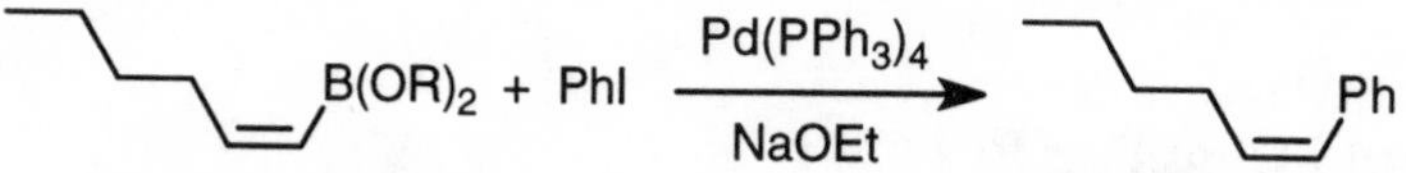

90 (R = isopropyl)   **91**

If an alkenyl halide is used in place of the aryl halide, a stereocontrolled diene synthesis results. All possible combinations of stereoisomers are possible. For example, the (Z)-1-hexenylboronic ester **90** with (Z)-1-bromooctene yields (Z,Z)-5,7-tetradecadiene (**92**), and the (Z)-1-pentenylboronic ester **93** with (E)-12-iodododecanol yields the silkworm moth pheromone bombykol (**94**), a (Z,E)-diene [94].

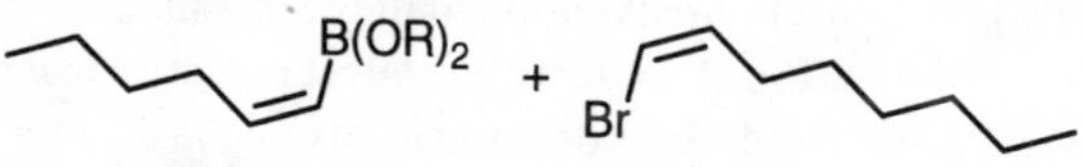

90

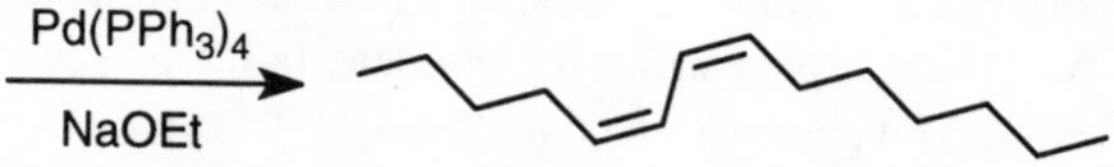

92

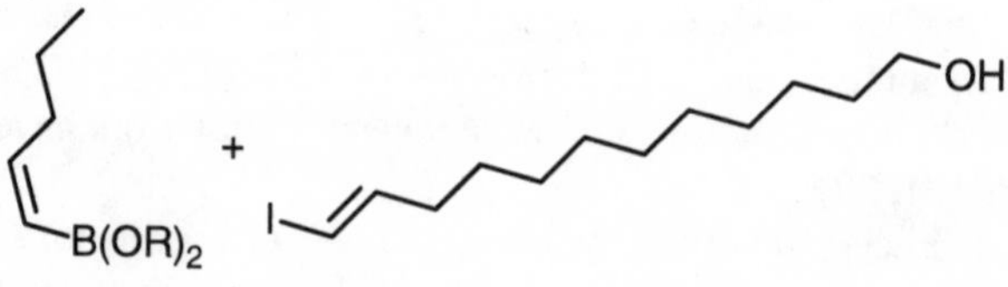

93

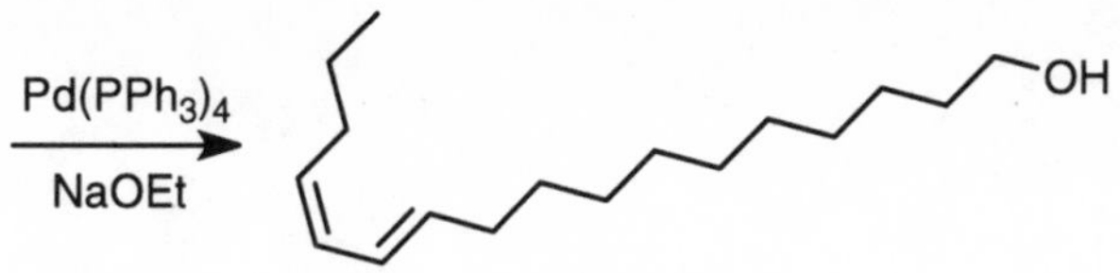

94

A typical procedure for preparation of dienes has been published in Organic Syntheses [95].

β-Halo-α,β-unsaturated ketones undergo catalyzed coupling with alkenylboronic esters in the presence of mild bases such as sodium acetate or triethylamine, as in the preparation of **95** [96]. 3-Bromochromones were coupled with sodium carbonate as the base, or following Kishi's lead (see Section 4.3.2), thallium carbonate [97].

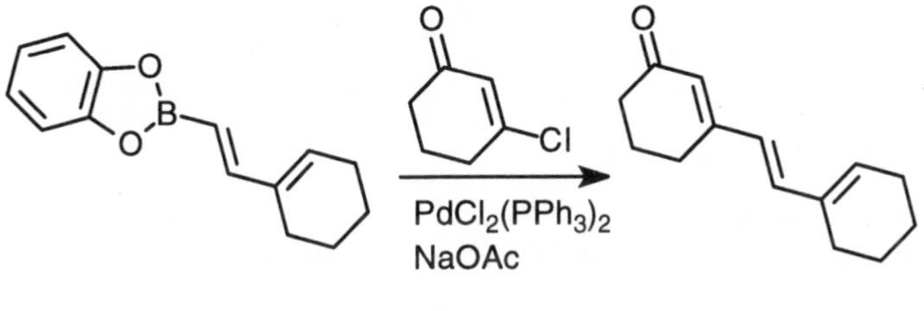

95

The coupling of 3-bromo-2-alkenoates to alkenylboronic esters requires the use of mild base such as K$_2$CO$_3$ [98]. Although **96** was obtained in >98% isomeric purity, attempted reaction of the (Z)-bromoacrylic ester **97** under similar conditions yielded a 63:37 ratio of (2Z,4E)-**99** to its (2E,4E) isomer because of isomerization of **97**. The more active catalyst PdCl$_2$(dppf) (**98**) led to **99** in >95% isomeric purity.

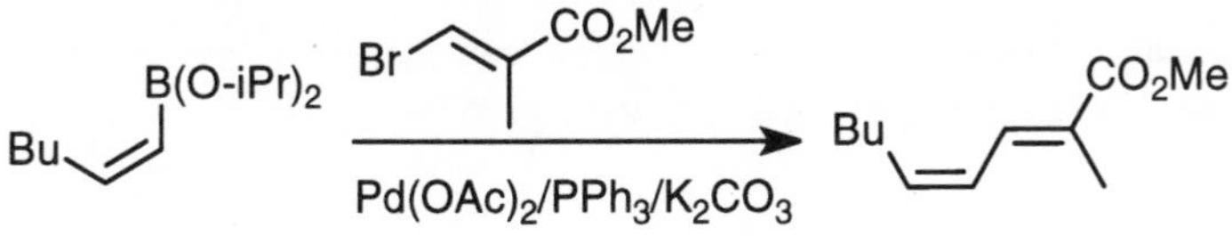

96 (95%)

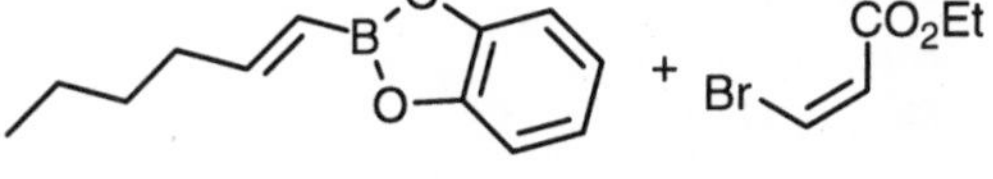

97

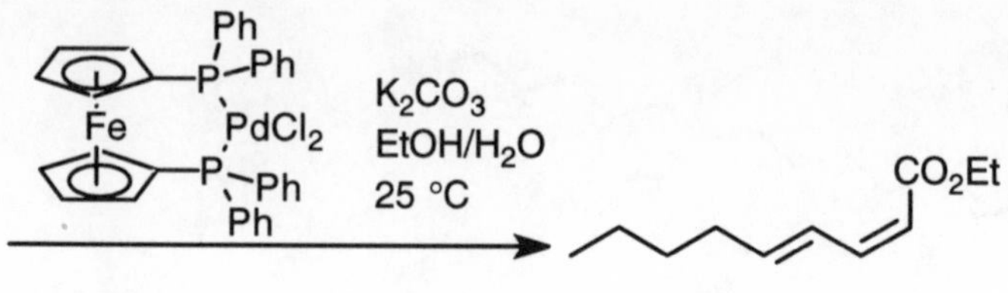

The coupling also proceeds in the presence of sensitive functionality such as enol acetates, as illustrated by the synthesis of **100** (80%) [99].

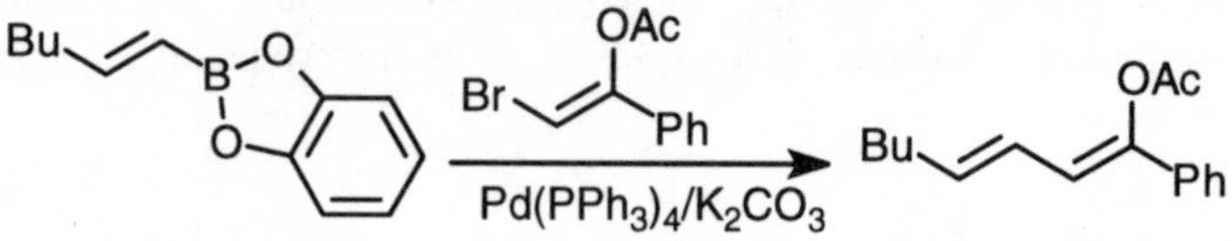

Highly hindered aryl groups can be cross coupled, for example, mesitylboronic acid with 1-bromonaphthalene, by using $Pd(PPh_3)_4$ with $Ba(OH)_2$ in water and di-methoxyethane [100].

Allylic phenoxides couple with catechol (*E*)-alkenylboronates in the presence of $Pd(PPh_3)_4$ and triethylamine [101].

Alkylboranes are less reactive than alkenylboranes, and alkylboronic esters fail to react under the usual coupling conditions. However, trialkylboranes, preferably *B*-alkyl-9-BBNs, will couple with alkenyl or aryl bromides or iodides. The alkyl group must be primary in order to avoid β-hydride elimination in the organo-palladium intermediate. Because $PdCl_2$(dppf) (**98**, above) had already been shown to catalyze alkylmagnesium coupling [102] it was tested first with a substantial series of substrates [103]. The syntheses of **101** and **102** are typical of the alkenyl halide couplings. More recently it has been found that the less expensive $Pd(PPh_3)_4$ is a perfectly adequate catalyst, as for example in the stereoselective conversion of the menthone ketal derivative **103** to the derivative **104** (83%) [104].

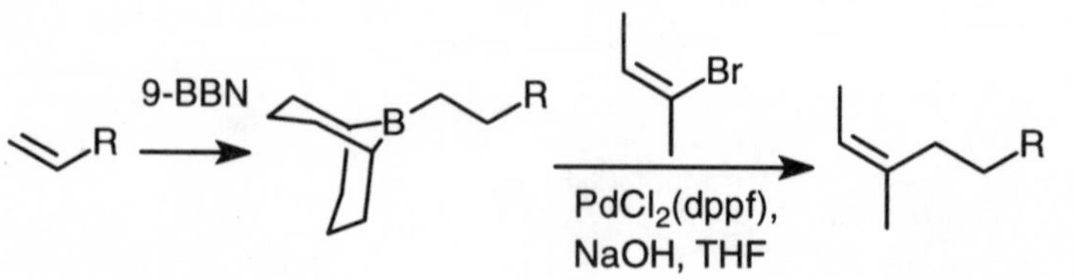

R = $(CH_2)_5CH_3$ **101** (98% gc)

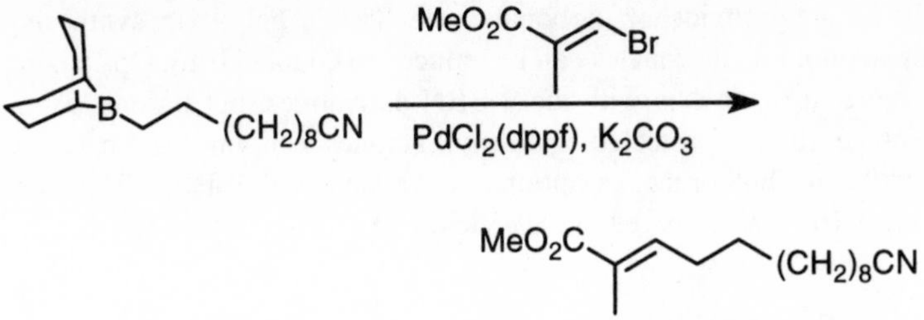

102 (80% isol)

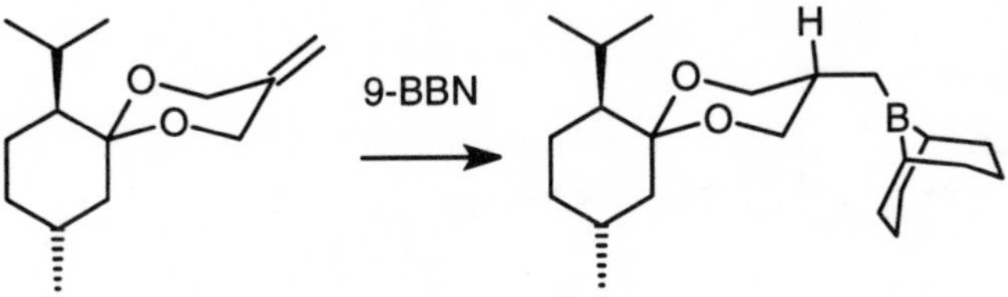

103

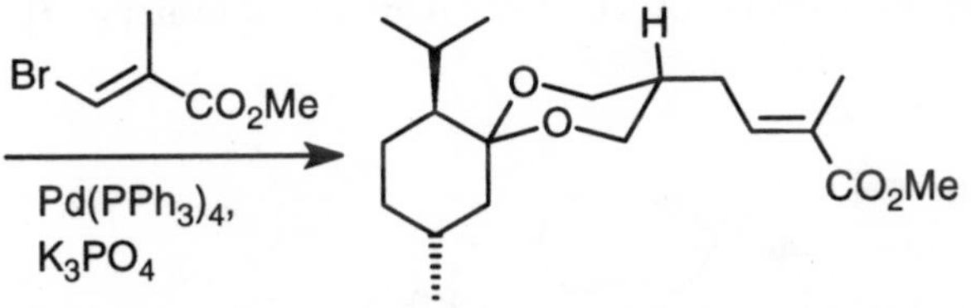

104

B-Alkyl-9-BBNs couple readily with 1-phenylthio-1- or -2-bromoalkenes in the presence of Pd(PPh₃)₄ and sodium hydroxide, as in the syntheses of **105** and **106** [105].

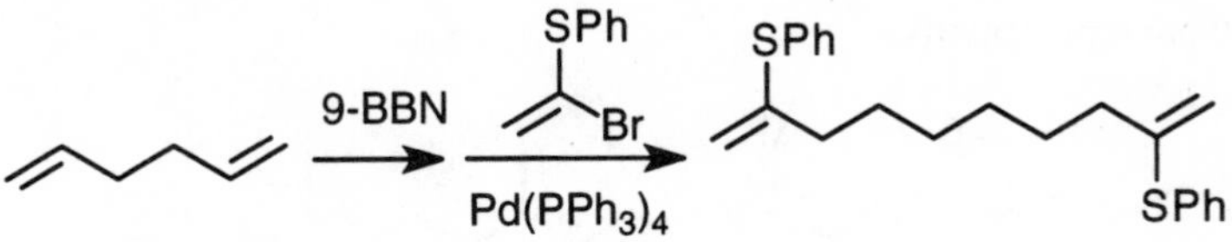

105

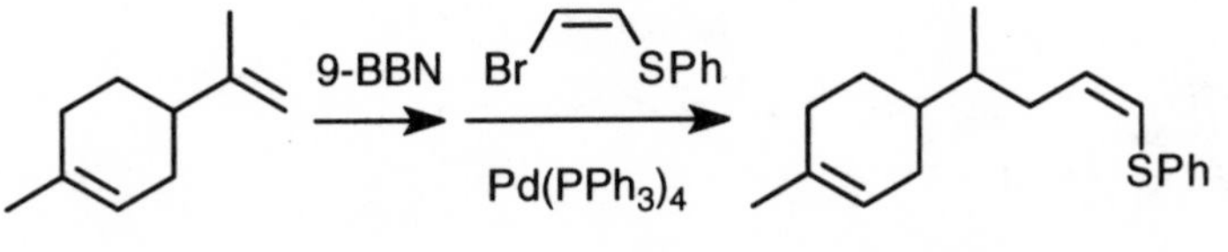

106

By using thallium hydroxide or carbonate (see Kishi's palytoxin synthesis), catechol or propanediol alkylboronates can be induced to couple [106]. Optimizing yields may be more difficult than with the 9-BBN derivatives, but boronic esters have the advantage that unprotected 1-alkenyl ketones can undergo catalytic hydroboration with catecholborane, an option not available with 9-BBN. Thus, unsaturated keto ester **107** was prepared in 68% yield [106].

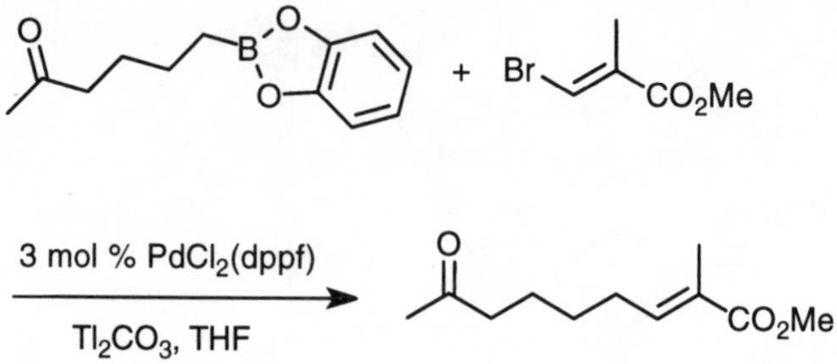

107

Cyclization of compounds having appropriately placed unsaturation can generate five- or six-membered rings having an exocyclic alkenyl group, for example, **108** [107].

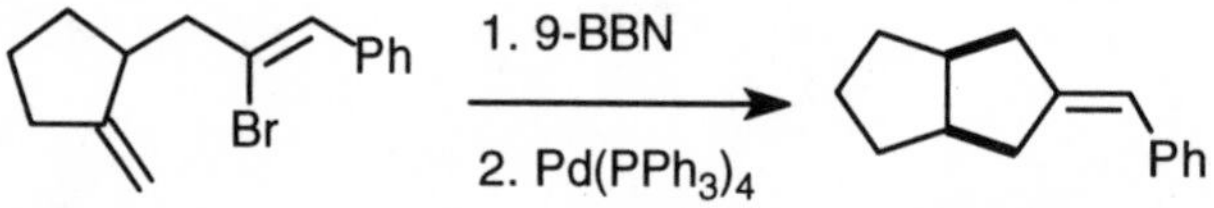

108

Alkenyl and aryl triflates are less reactive than the corresponding bromides, but are useful substrates, especially where they can be prepared easily from ketones or phenols [108, 109]. A number of examples of conversion of 1-alkenes to *B*-alkyl-9-BBNs followed by coupling with triflates were reported, of which the preparations of **109** and **110** are representative.

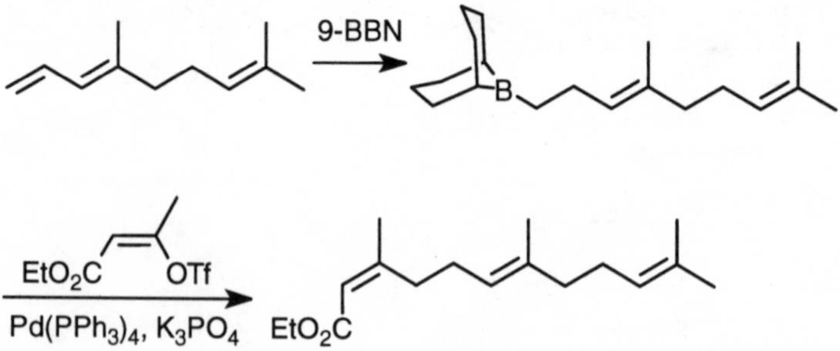

109

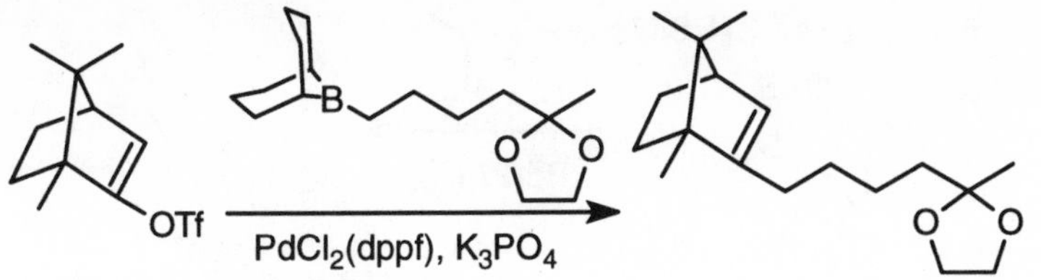

110

Alkyl-alkyl coupling between *B*-alkyl-9-BBN's and alkyl halides is possible, but only if both alkyl groups are primary, as in the preparation of **111** (57%) [110].

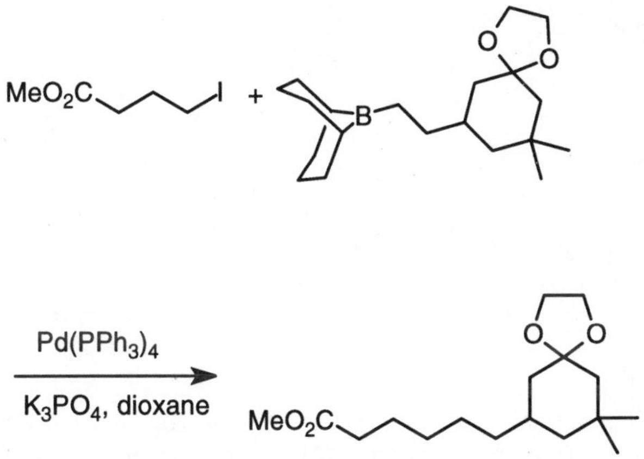

111

Soderquist and coworkers have found that silyl groups do not interfere with Suzuki coupling, as illustrated by the conversion of **44** to **112** [111] and the synthesis of **113** [112]. Similar couplings of α-trimethylsilylalkenylboranes [113] and of (*E*)-Me₃SiCH=CH-9-BBN [114] have also been reported.

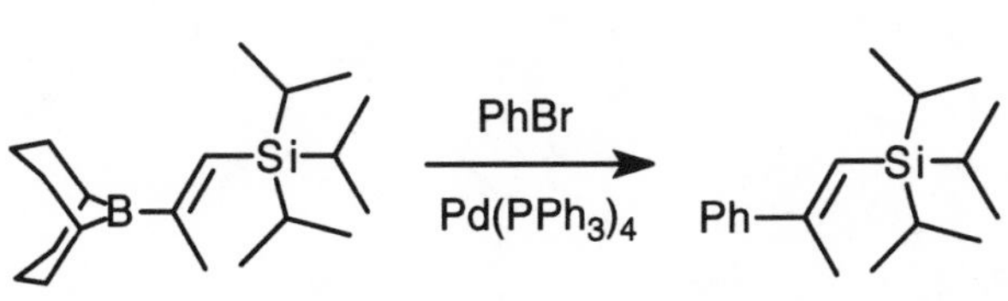

44   **112**

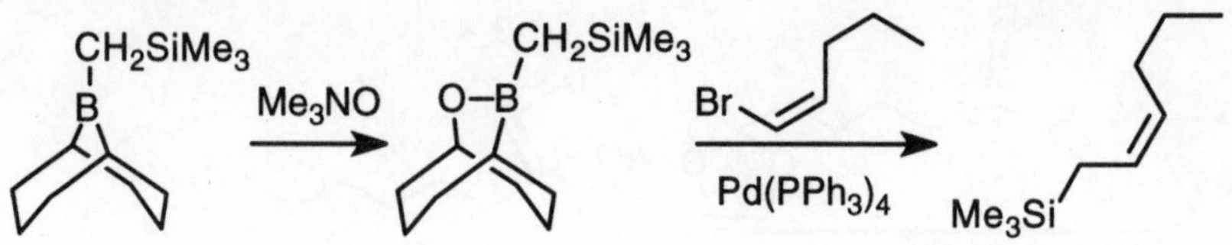

113 (99% *Z*, 97% yield)

In the presence of K_3PO_4, the *B*-alkyl-9-BBN function couples faster than trialkyltin groups, and the reaction can be used to generate functionalized organotin compounds such as **114** [115].

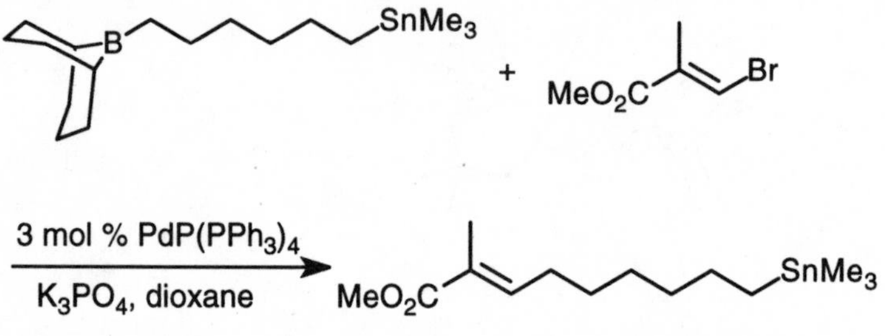

114 (83%)

Because the cross coupling of boron reagents requires base but zinc reagents do not, Knochel's (iodozincmethyl)boronic ester **115** [116] can be coupled with aryl iodides to form benzylic boronic esters **116** [117], or with alkenyl iodides or bromides to produce allylic boronic esters such as **117** with retention of configuration [118].

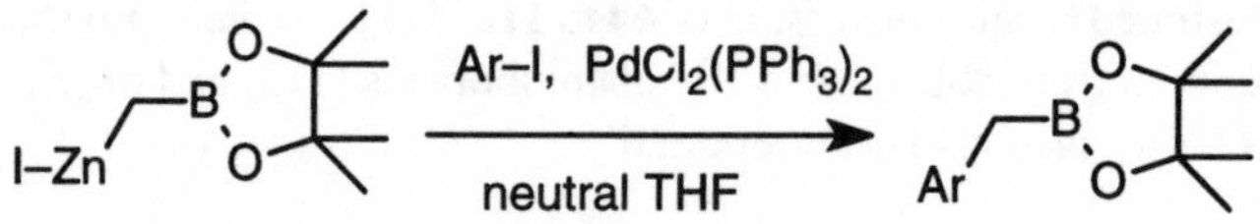

115   **116**

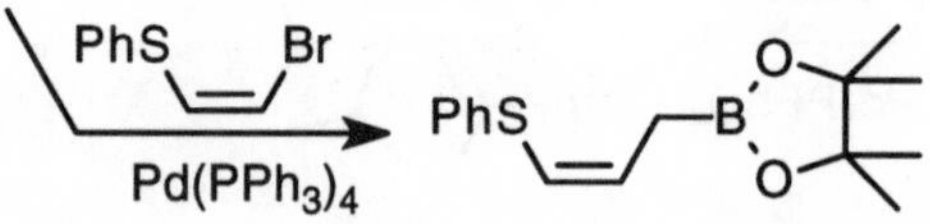

117

Also because the cross coupling of boron reagents requires base but zinc reagents do not, (*E*)-BrCH=CHBBr$_2$ (see Section 2.4) can be coupled with alkylzinc reagents, and the resulting alkenylboranes can be treated with base and an aryl halide to yield the alkene [119]. However, it is preferable to convert the dibromoborane to the diisopropyl ester (**118**) before carrying out the coupling [120, 121]. An interesting variant is the use of the 1-methoxyvinylzinc, which leads via dienylboronic ester intermediate **119** to an unsaturated ketone synthesis [122].

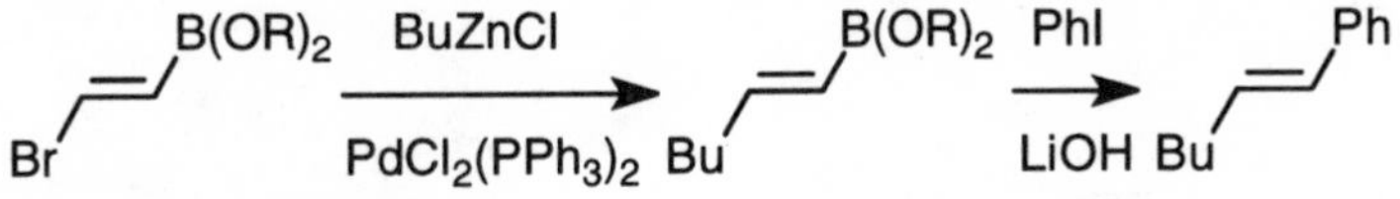

118 (R = isopropyl)

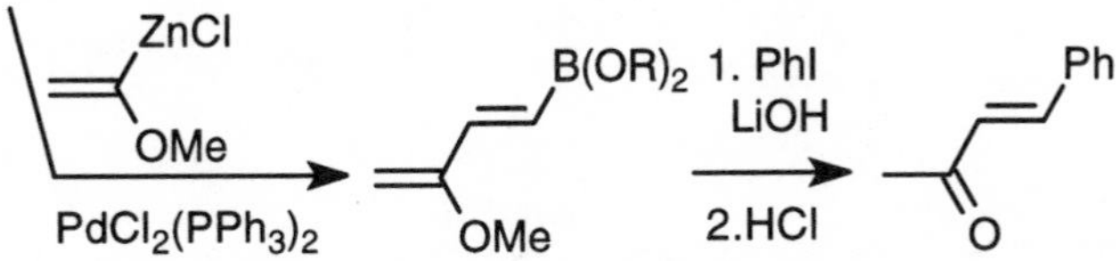

119

The (*Z*)-1-(dibromoboryl)-2-bromoalkenes **120** from bromoboration of alkynes with BBr$_3$ couple with organozinc compounds in the presence of (Ph$_3$P)$_2$PdCl$_2$ to form alkenylboron dibromides **121** in high stereopurity [123]. On addition of base and an aryl or alkenyl iodide R^{3}I, the boron reagent is cross coupled to form the trisubstituted alkene. This is one of the most general and stereoselective trisubstituted alkene syntheses known. Alternatively, **120** has been converted to the diisopropyl ester and converted by alkylzinc to the alkenylboronic esters **122**, which were carbonylated to provide unsaturated carboxylic esters in high stereopurity [121].

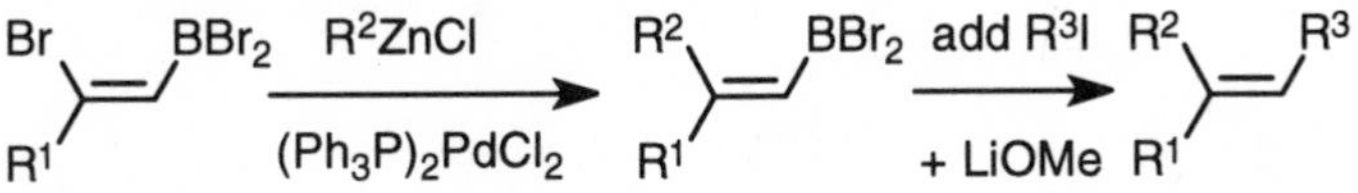

120   **121**

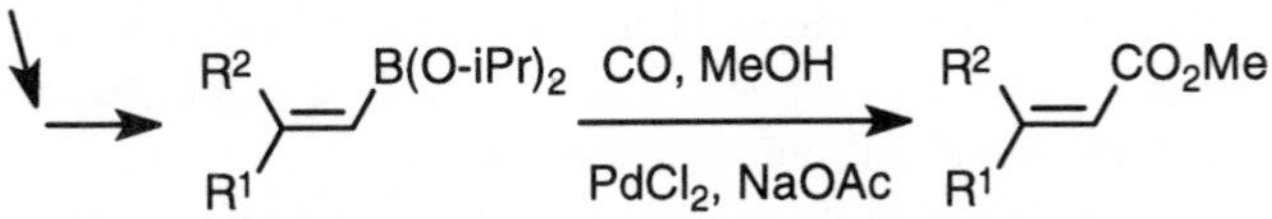

122

R^1 = alkyl; R^2 = alkyl, Ph, Me$_3$SiCH$_2$, H$_2$C=C(SiMe$_3$), BuC≡C;
R^3 = alkenyl, aryl, allyl, benzyl, H

Cross coupling in the presence of 1-3 atm of carbon monoxide results in insertion of a carbonyl group between the two coupled fragments. A variety of *B*-alkyl-9-BBN derivatives derived from 1-alkenes have been coupled with CO and aryl or alkenyl iodides in the presence of 5 mol % of $Pd(PPh_3)_4$ and excess potassium carbonate [124, 125]. The isolated yield of (*E*)-5-pentadecen-7-one (**124**) from 1-octene via 1-octyl-9-BBN (**123**) and (*E*)-1-iodohexene in benzene at 1 atm of CO was 99% [124]. A variety of functionalized substrates gave yields in the 50-75% range [124, 125], as in the preparation of the unsaturated protected hydroxy ketone **125** (53% by gc) [124].

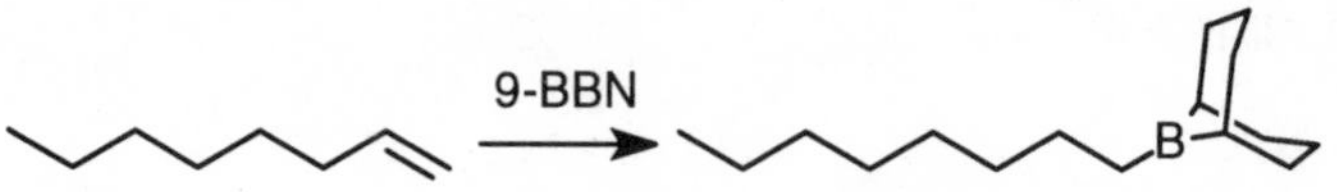

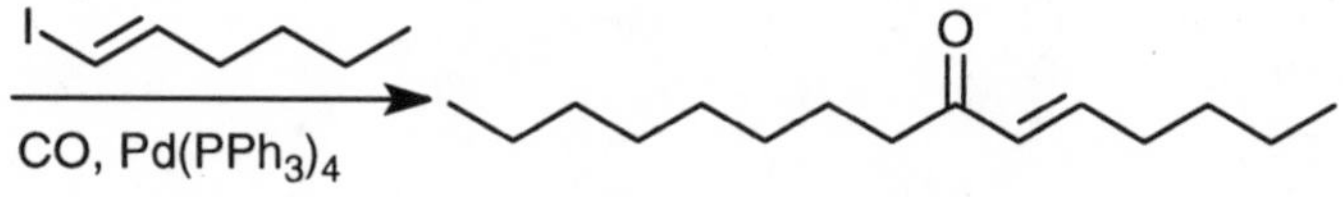

Isonitriles such as $Me_3C–NC$ can be used in place of CO, in the foregoing synthesis, though with aryl iodides only, and yield the corresponding ketones on hydrolysis [126]. It was found necessary to use less than one mole of isonitrile per mole of alkyl-9-BBN in order to avoid poisoning the catalyst.

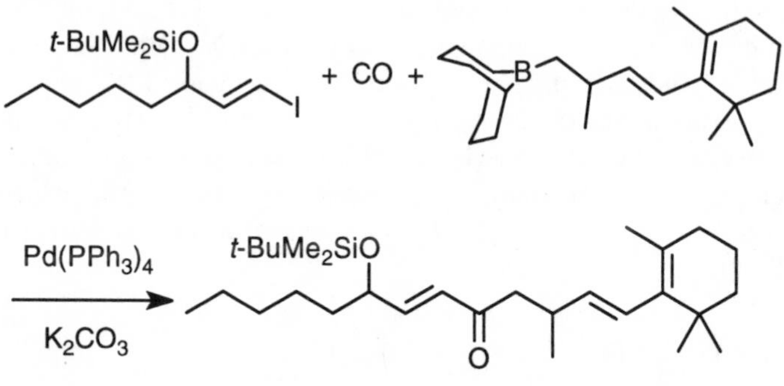

An important modification of the Suzuki conditions is provided by the water soluble catalyst **128** [127]. This catalyst couples such water soluble species as *p*-$BrC_6H_4SO_3Na$ with $ArB(OH)_2$, which is soluble in aqueous base. It also permits couplings such as that of 5-iododeoxyuridine (**126**) with styreneboronic acid (**127**) to form the corresponding substituted uridine (**129**).

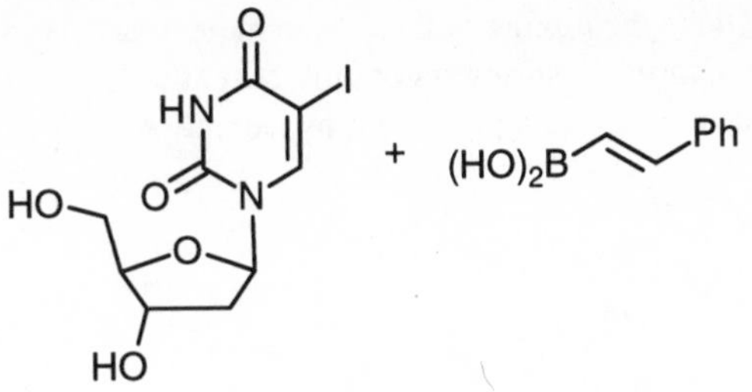

126   **127**

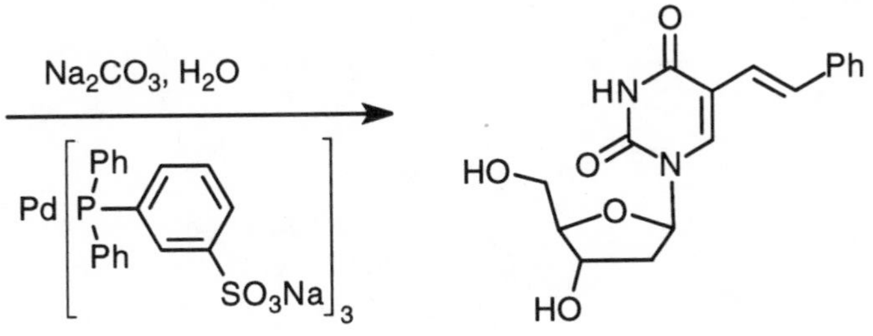

128   **129**

The 1,2-diborylalkenes obtained by catalyzed diborylation of alkynes readily undergo catalyzed coupling to form the corresponding alkenes, as in the conversion of **130** to **131** [128].

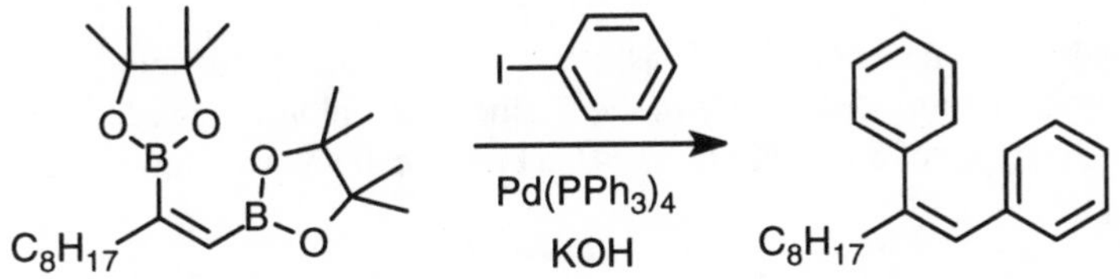

130   **131**

4.3.2 Synthetic Applications

The Suzuki coupling is especially valuable because it tolerates the presence of other functionality. A particularly striking example is its use in a key step to join two large fragments to form (Z,E)-diene **132** in Kishi's synthesis of palytoxin [71]. In order to achieve a sufficient reaction rate to make the joining of these high molecular weight fragments practical, it was necessary to use a strong halophilic base, thallium hydroxide, which yielded 1000-fold faster rates than potassium hydroxide.

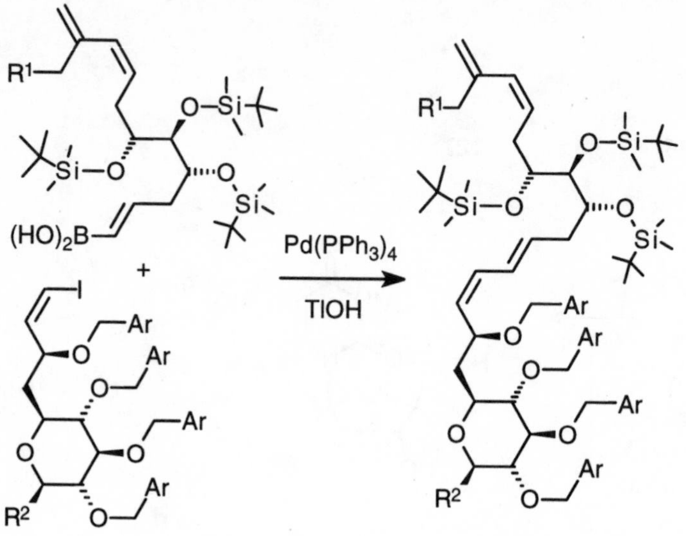

132

R^1 is a functionalized 29-carbon chain, R^2 is a functionalized 15-carbon chain, Ar is *p*-methoxyphenyl

Thallium hydroxide has been used as the base for the preparation of a functionalized triene intermediate (**133**) in the synthesis of chlorothricolide [129], and also for the synthesis of 5,6-DiHETE (**134**), a biologically active metabolite of leukotriene A_4 [130]

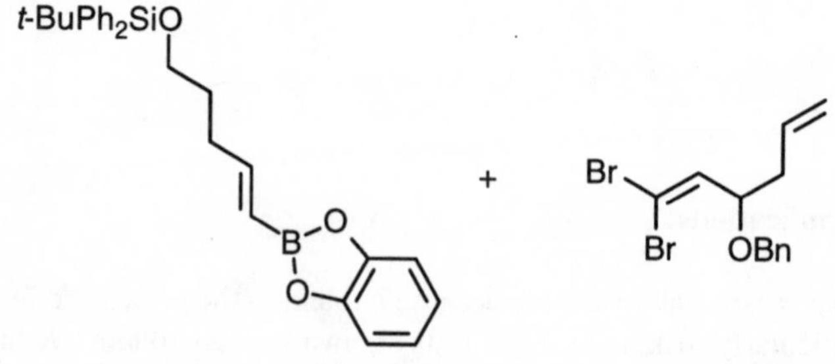

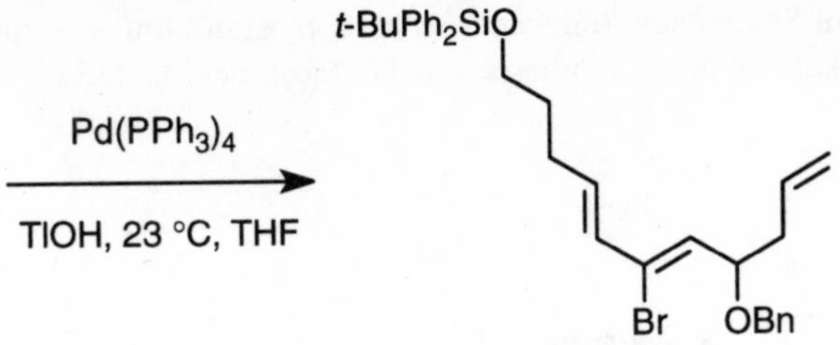

133

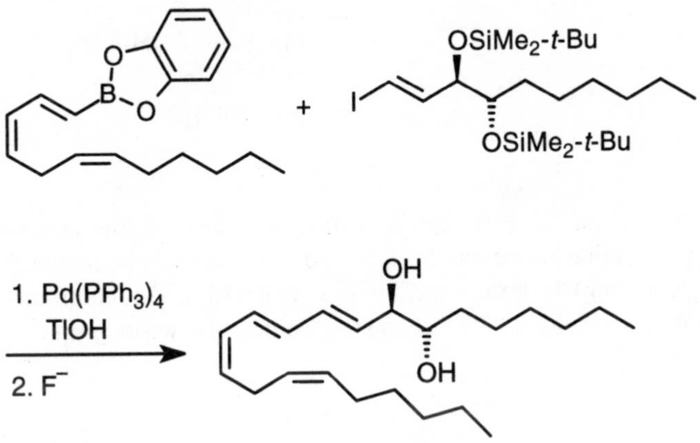

134

Leukotriene B$_4$ (**135**) has been made via a disiamylborane intermediate [131].

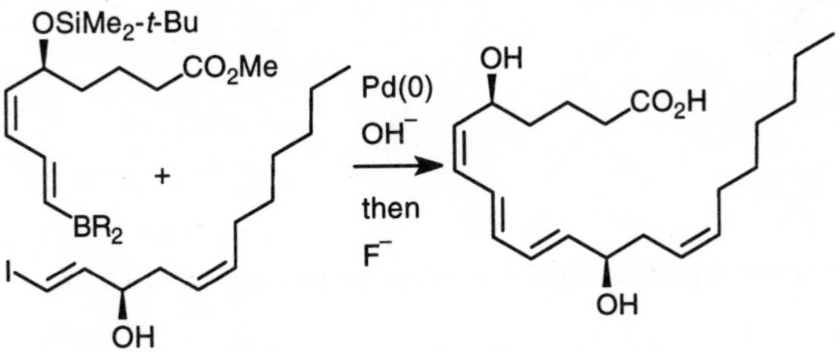

135

The Kishi modification of Suzuki coupling has been used in assembling a major portion **136** of the carbon skeleton in the synthesis of (–)-chlorothricolide [132].

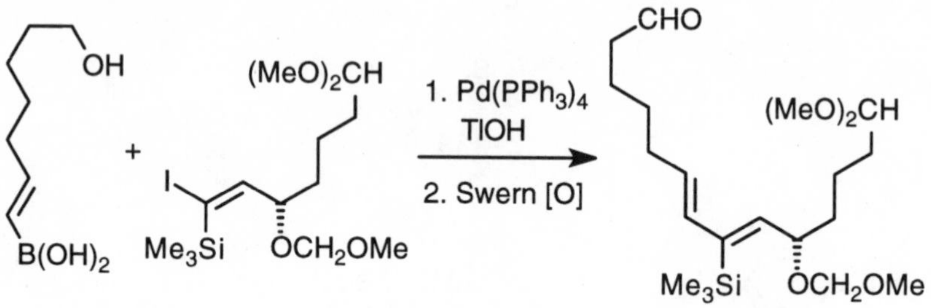

136

Tamoxifen (**137**, Y = NMe$_2$, Z = H) can be made efficiently via the original Negishi coupling of phenylzinc halide, but for a bromo substituted derivative (**137**, Z = Br, Y = Cl during coupling) the boronic acid works better [133]. This coupling is not 100% stereospecific, but yields a ~20:1 geometric isomer ratio when optimized.

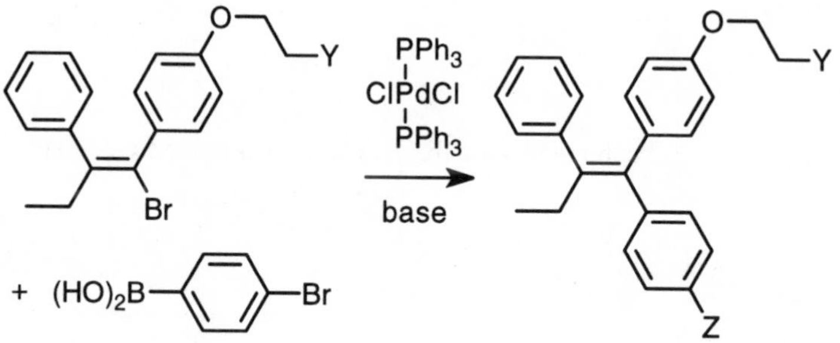

137 (20:1 isomer ratio)

Dihydroxyserrulatic acid (**138**) has been prepared via the coupling of a B-alkyl-9-BBN [134].

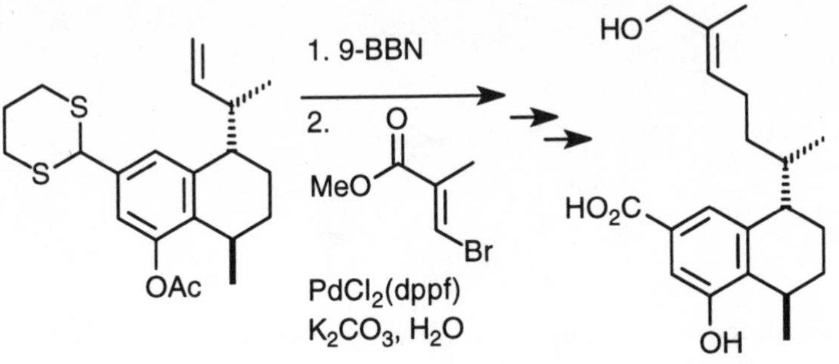
1. 9-BBN
2.
MeO
Br
PdCl2(dppf)
K2CO3, H2O
OAc
HO
HO2C
OH

138

An efficient prostaglandin synthesis utilizes the coupling of iodocyclopentenone **139** with borane **140** to form the intermediate **141** [135].

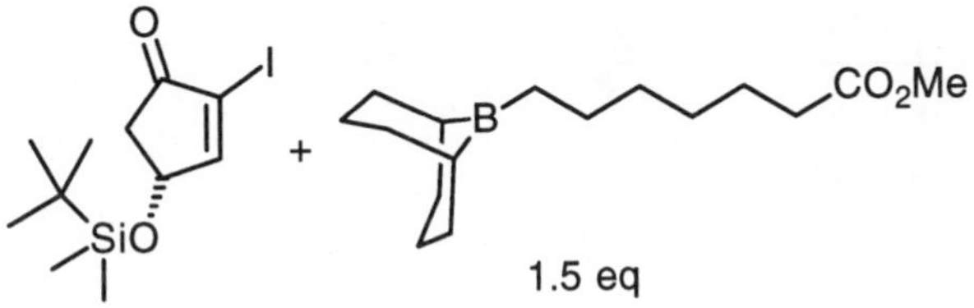
I
SiO
B
CO2Me
1.5 eq

139 **140**

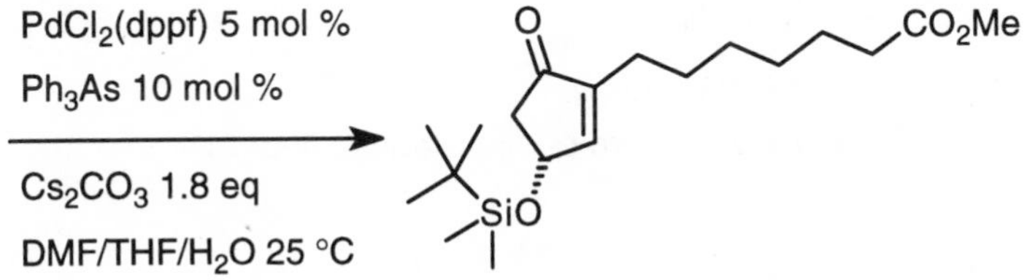
PdCl2(dppf) 5 mol %
Ph3As 10 mol %
Cs2CO3 1.8 eq
DMF/THF/H2O 25 °C
CO2Me
SiO

141

Another type of prostaglandin synthesis catalytically joins boronic ester **143** and organozinc reagent **144** under neutral conditions, then under basic conditions connects bromo ketone **142** in the same pot, resulting in a ketone that is reduced with sodium borohydride to the methyl ester of PGB_1 **145** [136].

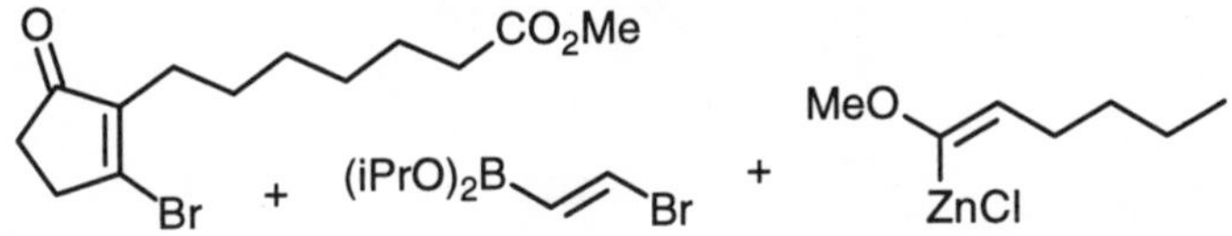
CO2Me
Br
(iPrO)2B
Br
MeO
ZnCl

142 **143** **144**

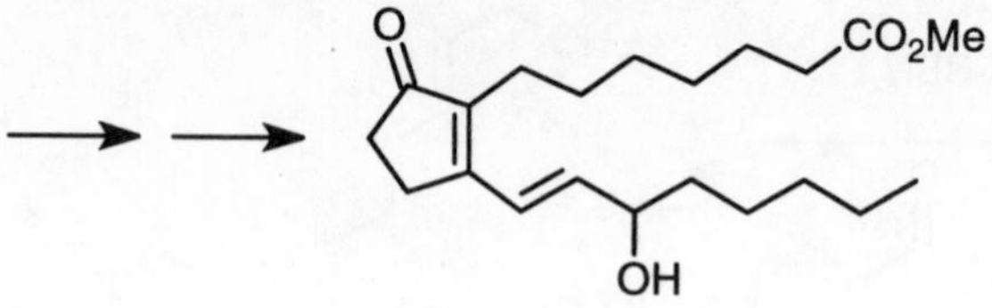

145

An efficient construction of the pine beetle pheromone brevicomin (**147**) utilizes Suzuki coupling to form the immediate precursor **146** [137]. The boron chemistry worked as well for a related synthesis of the pheromone frontalin, but the olefin did not have the right substitution pattern to permit Sharpless dihydroxylation with good enantioselectivity [138].

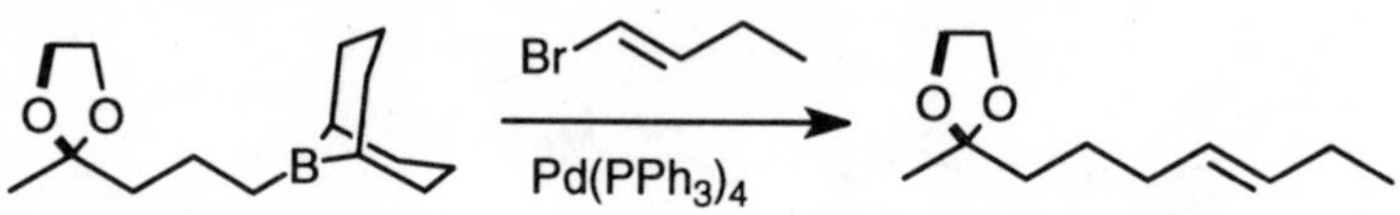

146

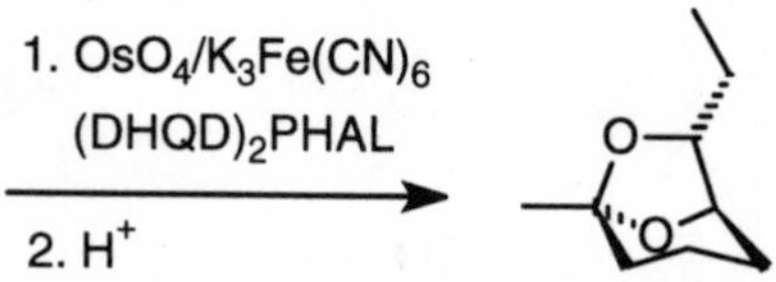

147 (95% ee)

(DHQD)₂PHAL = Phthalazine derivative of dihydroquinidine [2]

Several types of polyphenylenes have been made via Suzuki coupling [139, 140, 141, 142]. Because these are not stereodirected, only one example is illustrated, the coupling of water soluble 4,4'-dibromodiphenyl-2,2'-dicarboxylate **148** with diphenyl-4,4'-diboronic ester **149** to make a water soluble polyphenylene **150** [141]

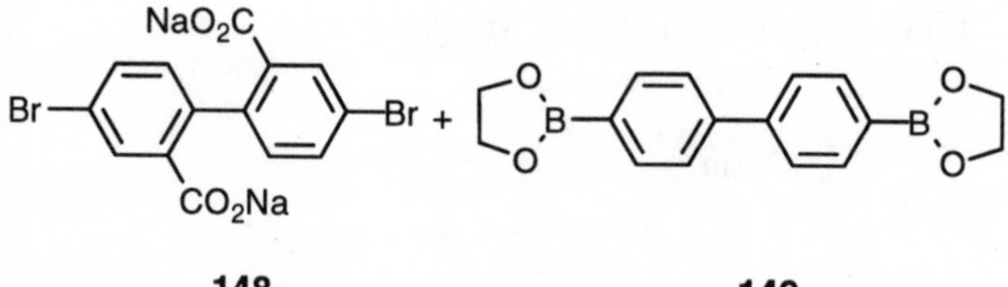

148   **149**

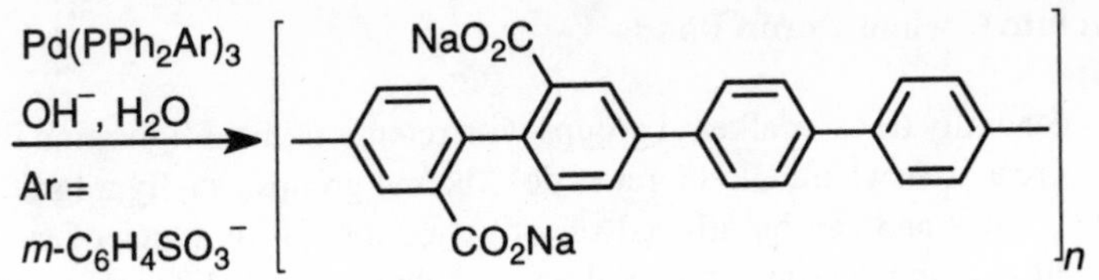

150

4.4 Other Stereoselective Routes to Alkenes

4.4.1 Alkenylpropargylic and Alkenylallenic Boranes

The borate anions from trialkylboranes and 1-lithio-3-chloropropyne rearrange to allenic boranes at –78 °C [143], and these rearrange to propargylic boranes at 25 °C [144]. Thexyl chloroborane is a useful hydroborating agent for producing borane intermediates such as **151** [145], which with 1-lithio-3-chloropropyne leads to borate intermediate **152**. The alkenyl group retains its geometry as it migrates from boron to propargylic carbon with intramolecular S$_N$2′ displacement of chloride ion to form allenylborane **153**. Allylic rearrangement occurs in the formation of **154** at 25 °C, as well as the reactions of **153** and **154** with acetaldehyde to form **155** and **156**, respectively, but the alkenyl group is unaffected [146]. This chemistry is reviewed in another context in Section 3.4.5.7.

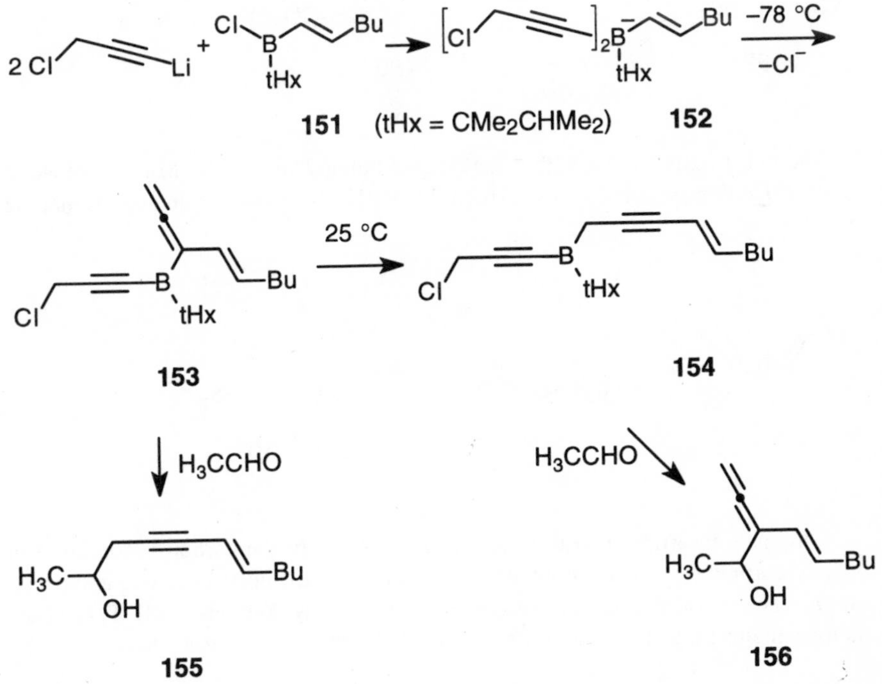

151   (tHx = CMe$_2$CHMe$_2$)   **152**

153

154

155

156

4.4.2 Insertion into Carbon-Boron Bonds

Alkenylboranes generally transfer alkenyl groups with retention of configuration in the same kinds of reactions where alkylboranes do. Alkenyl groups usually migrate faster than alkyl groups and can be utilized selectively. Most of the types of carbon—carbon bond forming rearrangements described in Section 3.4.3, including haloborate rearrangements, work well with alkenylboranes. (Dichloromethyl)-lithium insertions with control of chirality are discussed in Chapter 5.

Ethyl diazoacetate inserts into the boron—carbon bond of alkenylboron dichlorides to form β,γ-unsaturated carboxylic esters of controlled geometry [147]. Examples are provided by the reactions of (E)-2-cyclopentyl-1-ethenyldichloroborane (**157**) with ethyl diazoacetate to form (E)-alkenylacetic acid **158**, and the similar reaction of the (Z)-isomer **159** to produce the corresponding (Z)-alkenyl-acetic acid **160**, both in ≥99% isomeric purity by gc analysis, yields 63-68% [147].

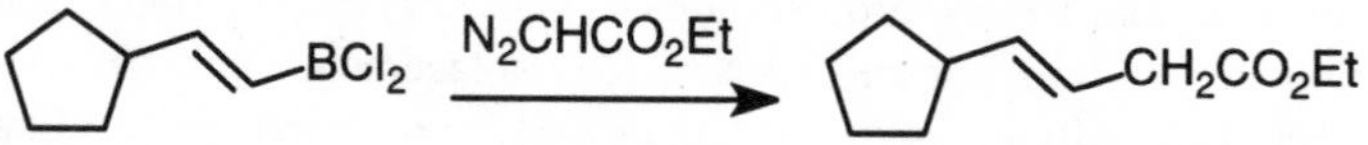

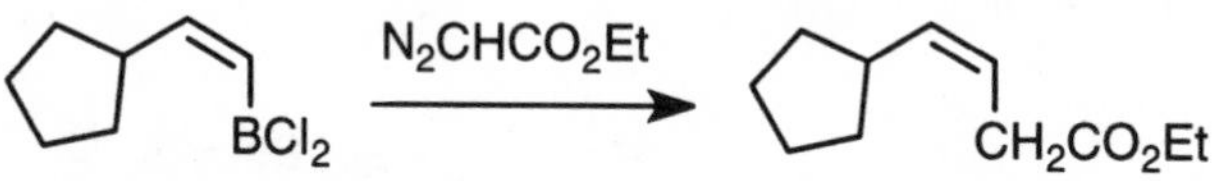

Alkenyl-9-BBN's react with 4-methoxy-3-buten-2-one in the Michael sense to form (E,E)-dienones (**161**) [149]. Alkynyl-9-BBN's behave in an analogous manner [150].

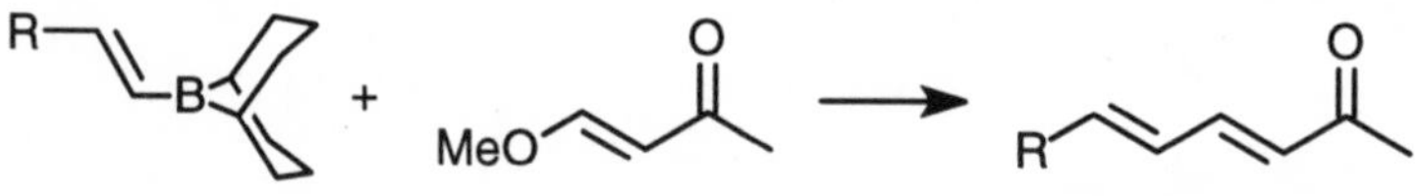

Conjugate insertion of α,β-unsaturated ketones into the carbon-boron bond of alkenylboronic esters occurs in the presence of boron trifluoride etherate and provides a very general stereoselective synthesis of γ-alkenyl ketones [148]. Typical examples include the preparations of **162** and **163** in ≥98% isomeric purity.

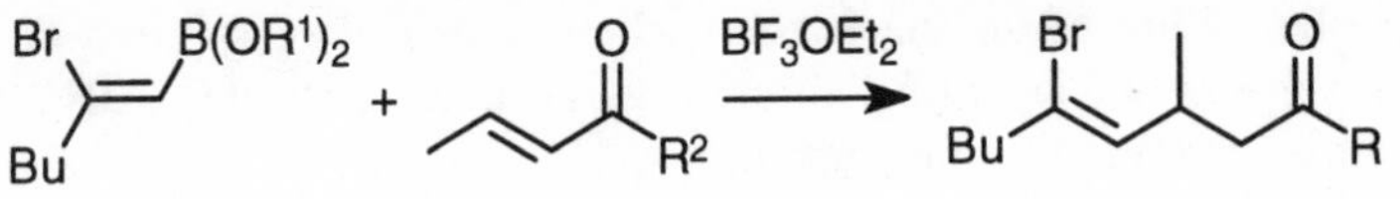

$R^1 = CHMe_2$; $R^2 = (CH_2)_5CH_3$ **162 (84%)**

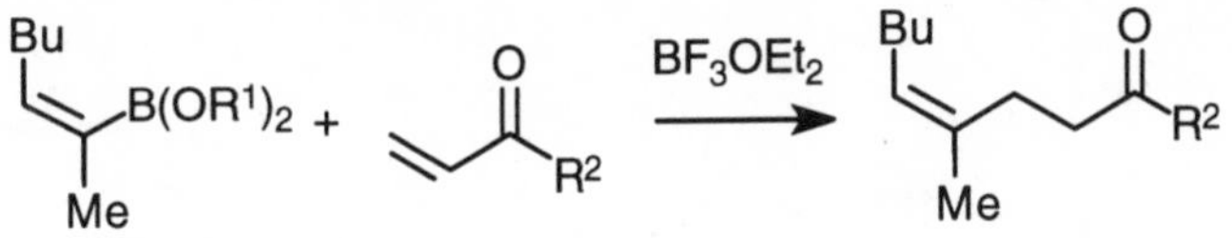

$R^1 = CHMe_2$; $R^2 = (CH_2)_5CH_3$ **163 (60%)**

If a β-methoxy-α,β-unsaturated ketone is used with a (Z)-B-iodoalkenyl-9-BBN, iododienones **164** result [151]. These were unstable and were converted to ethylene ketals for isolation (52-56%). The diastereoselectivity was ≥95%.

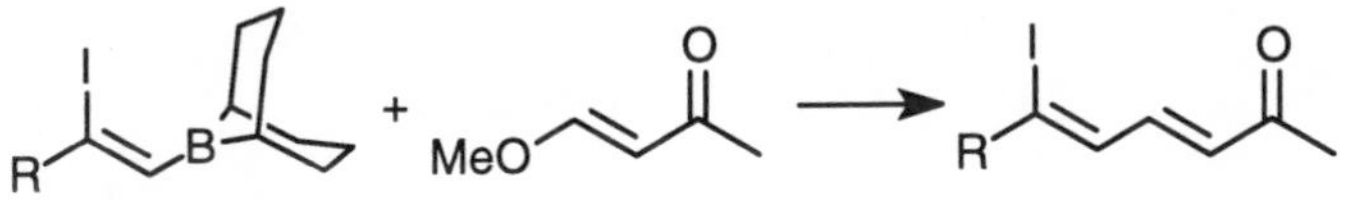

$R = CH_3(CH_2)_n$; n = 3, 4, 5, 7   **164**

Reaction of **44** (Section 4.2.2.2) with benzaldehyde yields **165** [152]. Precedent for such substitutions is described in Section 3.3.4.1.

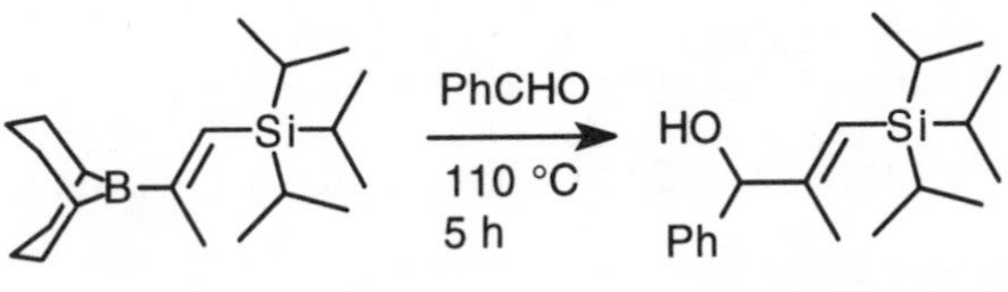

44   **165**

4.4.3 α,β-Unsaturated Ketones from Boron Enolates

Addition of 9-BBN bromide to vinyl ketones yields bromomethyl enolates (**166**) that condense with aldehydes to yield α,β-unsaturated ketones (**167**), which from NMR data appear to consist of a single isomer in each case [153]. Where R^1 = Me and R^2 =Ph, the geometry was shown to be that indicated. The essential distinction in this process from aldol condensations leading to β-hydroxy ketones (Section 7.4) appears to lie in the acidity of the medium, which consists of benzene and dichloromethane as solvents and includes no base. Thus, small amounts of 9-BBN or adventitious HBr may catalyze the elimination.

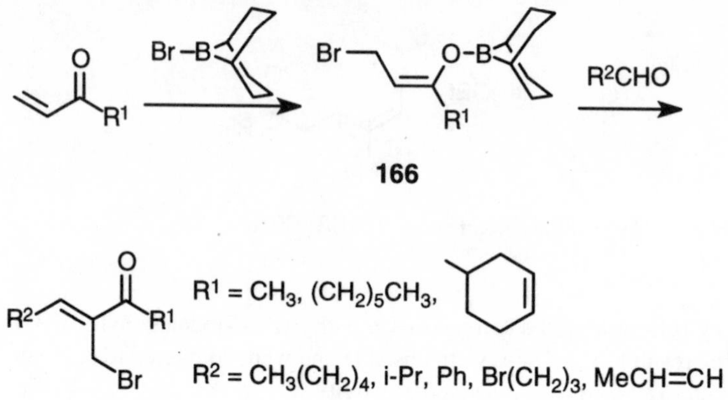

166

$R^1 = CH_3, (CH_2)_5CH_3,$

$R^2 = CH_3(CH_2)_4$, i-Pr, Ph, $Br(CH_2)_3$, MeCH=CH

167 (50-62%)

4.5 References

1. Kolb HC, Andersson PG, Bennani YL, Crispino GA, Jeong KS, Kwong HL, Sharpless KB (1993) J. Am. Chem. Soc. 115:12226
2. Wang ZM, Sharpless KB (1994) J. Org. Chem. 59:8302
3. Katsuki T, Sharpless KB (1980) J. Am. Chem. Soc. 102:5974.
4. Sharpless KB, Behrens CH, Katsuki T, Lee AWM, Martin VS, Takatani M, Viti SM, Walker FJ, Woodard SS (1983) Pure Appl. Chem. 55:589
5. Johnson RA, Sharpless KB (1993) in Catalytic Asymmetric Synthesis, Ojima I, Ed, VCH Publishers: 103-158
6. Brandes BD, Jacobsen EN (1994) J. Org. Chem. 59:4378
7. Jacobsen EN (1993) in Catalytic Asymmetric Synthesis, Ojima I, Ed, VCH Publishers: 159-202
8. Hawthorne MF, Dupont JA (1958) J. Am. Chem. Soc. 80:5830.
9. Matteson DS, Liedtke JD (1963) J. Org. Chem. 28:1924.
10. Matteson DS, Liedtke JD (1965) J. Am. Chem. Soc. 87:1526
11. Brown HC, Subrahmanyam C, Hamaoka T, Ravindran N, Bowman DH, Misumi S, Unni MK, Somayaji V, Bhat NG (1989) J. Org. Chem. 54:6068
12. Brown HC, Hamaoka T, Ravindran N (1973) J. Am. Chem. Soc. 95:6456
13. Brown HC, Bhat NG (1988) Tetrahedron Lett 29:21
14. Kunda SA, Smith TL, Hylarides MD, Kabalka GW (1985) Tetrahedron Lett. 26:279
15. Brown HC, Somayaji V (1984) Synthesis 919
16. Brown HC, Hamaoka T, Ravindran N, Subrahmanyam C, Somayaji V, Bhat NG (1989) J. Org. Chem. 54:6075

17. Brown HC, Bhat NG, Rajagopalan S (1986) Synthesis 480
18. Marshall JA, Bundy GL (1966) J. Am. Chem. Soc. 88:4291
19. Marshall JA, Bundy GL (1967) J. Chem. Soc., Chem. Commun. 854
20. Marshall JA, Huffman WF (1970) J. Am. Chem. Soc. 92:6359
21. Goralski CT, Singaram B, Brown HC (1987) J. Org. Chem. 52:4014
22. Singaram B, Goralski CT, Rangaishenvi MV, Brown HC (1989) J. Am. Chem. Soc. 111:384
23. Singaram B, Rangaishenvi MV, Brown HC, Goralski CT, Hasha DL (1991) J. Org. Chem. 56:1543
24. Brown HC, Zweifel G (1961) J. Am. Chem. Soc. 83:3834
25. Zweifel G, Clark GM, Polston NL (1971) J. Am. Chem. Soc. 93:3395
26. Brown HC, Chandrasekharan J (1983) J. Org. Chem. 48:5080
27. Wang KK, Scouten CG, Brown HC (1982) J. Am. Chem. Soc. 104:531
28. Zweifel G, Arzoumanian H (1966) Tetrahedron Lett. 2535
29. Brown HC, Basavaiah D, Kulkarni SU (1982) J. Organomet. Chem. 225:63
30. Colberg JC, Rane A, Vaquer J, Soderquist JA (1993) J. Am. Chem. Soc. 115:6065
31. Soundararajan R, Matteson DS (1990) J. Org. Chem. 55:2274
32. Zweifel G, Arzoumanian H (1967) J. Am. Chem. Soc. 89:5086
33. Zweifel G, Backlund SJ (1977) J. Am. Chem. Soc. 99:3184
34. Plamondon J, Snow JT, Zweifel G (1971) Organomet. Chem. Syn. 1:249
35. Negishi E, Yoshida T (1973) J. Am. Chem. Soc. 95:6837
36. Brown HC, Blue CD, Nelson DJ, Bhat NG (1989) J. Org. Chem. 54:6064
37. Zweifel G, Najafi MR, Rajagopalan S (1988) Tetrahedron Lett. 29:1895
38. Zweifel G, Shoup TM (1988) Synthesis 130
39. Rajagopalan S, Zweifel G (1984) Synthesis 113
40. Miller JA, Zweifel G (1981) J. Am. Chem. Soc. 103:6217
41. Miller JA, Zweifel G (1981) Synthesis 288
42. Stracker EC, Leong W, Miller JA, Shoup TM, Zweifel G (1989) Tetrahedron Lett. 30:6487
43. Lee Y, Leong W, Zweifel G (1992) Heteroatom Chem. 3:227
44. Soderquist JA, Colberg JC. Del Valle L (1989) J. Am. Chem. Soc. 111:4873
45. Gridnev ID, Miyaura N, Suzuki A (1993) Organometallics 12:589
46. Ishiyama T, Nishijima K, Miyaura N, Suzuki A (1993) J. Am. Chem. Soc. 115:7219
47. Köbrich G, Merkle HR (1967) Angew. Chem. 79:50; Angew. Chem. Int. Ed. Engl. 6:74
48. Köbrich G, Merkle HR (1967) Chem. Ber. 100:3371
49. Corey EJ, Ravindranathan T (1972) J. Am. Chem. Soc. 94:4013
50. Negishi E, Yoshida T (1973) J. Chem. Soc., Chem. Commun. 606
51. Yoshida T, Williams RM, Negishi E (1974) J. Am. Chem. Soc. 96:3688
52. Negishi E, Williams RM, Lew G, Yoshida T (1975) J. Organomet. Chem. 92:C4
53. Zweifel G, Backlund SJ, Leung T (1977) J. Am. Chem. Soc. 99:5192
54. Brown HC, Basavaiah D, Kulkarni SU, Lee HD, Negishi E, Katz JJ (1986) J. Org. Chem. 51:5270
55. Brown HC, Lee HD, Kulkarni SU (1986) J. Org. Chem. 51:5282
56. Brown HC, Basavaiah D, Singh SM, Bhat NG (1988) J. Org. Chem. 53:246
57. Brown HC, Bakshi RK, Singaram B (1988) J. Am. Chem. Soc. 110:1529
58. Brown HC, Bhat NG, Somayaji V (1983) Organometallics 2:1311
59. Brown HC, Imai T (1984) Organometallics 3:1392
60. Brown HC, Imai T, Bhat NG (1986) J. Org. Chem. 51:5277
61. Brown HC, Bhat NG, Iyer RR (1991) Tetrahedron Lett. 32:3655
62. Fisher RP, On HP, Snow JT, Zweifel G (1982) Synthesis 127
63. Evans DA, Crawford TC, Thomas RC, Walker JA (1976) J. Org. Chem. 41:3947
64. Evans DA, Thomas RC, Walker JA (1976) Tetrahedron Lett. 1427
65. Brown HC, Bhat NG (1988) Tetrahedron Lett. 29:21
66. Srebnik M, Bhat NG, Brown HC (1988) Tetrahedron Lett. 29:2635

67. Matteson DS, Peacock K (1963) J. Org. Chem. 28:369
68. Brown HC, Cole TE (1985) Organometallics 4:816
69. Brown HC, Bhat NG, Srebnik M (1988) Tetrahedron Lett. 29:2631
70. Matteson DS, Moody RJ (1982) Organometallics 1:20
71. Uenishi J, Beau JM, Armstrong RW, Kishi Y (1987) J. Am. Chem. Soc. 109:4756
72. Matteson DS, Majumdar D (1983) Organometallics 2:230
73. Zweifel G, Arzoumanian H, Whitney CC (1967) J. Am. Chem. Soc. 89:3652
74. Zweifel G, Fisher RP, Snow JT, Whitney CC (1972) J. Am. Chem. Soc. 94:6560
75. Brown HC, Basavaiah D, Kulkarni SU, Bhat NG, Vara Prasad JVN (1988) J. Org. Chem. 53:239
76. Brown HC, Bhat NG (1988) J. Org. Chem. 53:6009
77. Brown HC, Basavaiah D (1983) Synthesis 283
78. Brown HC, Bhat NG, Basavaiah D (1986) Synthesis 674
79. Zweifel G, Backlund SJ (1978) J. Organomet. Chem. 156:159
80. Suzuki A, Miyaura N, Abiko S, Itoh M, Midland MM, Sinclair JA, Brown HC (1986) J. Org. Chem. 51:4507
81. Brown HC, Srebnik M (1987) Organometallics 6:629
82. Negishi E, Baba S (1976) J. Chem. Soc., Chem. Commun. 596
83. Baba S, Negishi E (1976) J. Am. Chem. Soc. 98:6729
84. Negishi E (1982) Acc. Chem. Res. 15:340
85. Yamamura M, Moritani I, Murahashi SI (1975) J. Organomet. Chem. 91:C39
86. Cassar L (1975) J. Organomet. Chem. 93:253
87. Miyaura N, Yamada K, Suzuki A (1979) Tetrahedron Lett. 3437
88. Miyaura N, Suzuki A (1979) J. Chem. Soc., Chem. Commun. 866
89. Miyaura N, Suginome H, Suzuki A (1981) Tetrahedron Lett. 22:127
90. Miyaura N, Suzuki A (1981) J. Organomet. Chem. 213:C53
91. Suzuki A (1982) Acc. Chem. Res. 15:178
92. Suzuki A (1991) Pure Appl. Chem. 63:419
93. Miyaura N, Yamada K, Suginome H, Suzuki A (1985) J. Am. Chem. Soc. 107:972
94. Miyaura N, Satoh M, Suzuki A (1986) Tetrahedron Lett. 27:3745
95. Miyaura N, Suzuki A (1989) Org. Syn. 68:130
96. Satoh N, Ishiyama T, Miyaura N, Suzuki A (1987) Bull. Chem. Soc. Jpn. 60:3471
97. Hoshino Y, Miyaura N, Suzuki A (1988) Bull. Chem. Soc. Jpn. 61:3008
98. Yanagi T, Oh-e T, Miyaura N, Suzuki A (1989) Bull. Chem. Soc. Jpn. 62:3892
99. Abe S, Miyaura N, Suzuki A (1992) Bull. Chem. Soc. Jpn. 65:2863
100. Watanabe T, Miyaura N, Suzuki A (1992) Synlett 207
101. Sasaya F, Miyaura N, Suzuki A (1987) Bull. Korean Chem. Soc. 8:329
102. Castle PL, Widdowson DA (1986) Tetrahedron Lett. 52:6013
103. Miyaura N, Ishiyama T, Sasaki H, Ishikawa M, Satoh M, Suzuki A (1989) J. Am. Chem. Soc. 111:314
104. Nomoto Y, Miyaura N, Suzuki A (1992) Synlett 727
105. Hoshino Y, Ishiyama T, Miyaura N, Suzuki A (1988) Tetrahedron Lett. 29:3983
106. Sato M, Miyaura N, Suzuki A (1989) Chemistry Lett. 1405
107. Miyaura N, Ishikawa M, Suzuki A (1992) Tetrahedron Lett. 33:2571
108. Oh-e T, Miyaura N, Suzuki A (1990) Synlett 221
109. Oh-e T, Miyaura N, Suzuki A (1993) J. Org. Chem. 58:2201
110. Ishiyama T, Abe S, Miyaura N, Suzuki A (1992) Chemistry Lett. 691
111. Soderquist JA, Colberg JC (1989) Synlett 25
112. Soderquist JA, Santiago B, Rivera I (1990) Tetrahedron Lett. 31:2311
113. Soderquist JA, Leon-Colon G (1991) Tetrahedron Lett. 32:43
114. Soderquist JA, Colberg JC (1994) Tetrahedron Lett. 35:27
115. Ishiyama T, Miyaura N, Suzuki A (1991) Synlett 687
116. Knochel P (1990) J. Am. Chem. Soc. 112:7431
117. Kanai G, Miyaura N, Suzuki A (1993) Chemistry Lett. 845

118. Watanabe T, Miyaura N, Suzuki A (1993) J. Organomet. Chem. 444:C1
119. Hyuga S, Chiba Y, Yamashina N, Hara S, Suzuki A (1987) Chemistry Lett. 1757
120. Hyuga S, Yamashina N, Hara S, Suzuki A (1988) Chemistry Lett. 809
121. Yamashina N, Hyuga S, Hara S, Suzuki A (1989) Tetrahedron Lett. 30:6555
122. Ogima M, Hyuga S, Hara S, Suzuki A (1989) Bull. Chem. Soc. Jpn. 1959
123. Satoh Y, Serizawa H, Miyaura N, Hara S, Suzuki A (1988) Tetrahedron Lett. 29:1811
124. Ishiyama T, Miyaura N, Suzuki A (1991) Bull. Chem. Soc. Jpn. 64:1999
125. Ishiyama T, Miyaura N, Suzuki A (1991) Tetrahedron Lett. 32:6923
126. Ishiyama T, Oh-e T, Miyaura N, Suzuki A (1992) Tetrahedron Lett. 33:4465
127. Castelnuovo AL, Calabrese JC (1990) J. Am. Chem. Soc. 112:4324
128. Ishiyama T, Matsuda N, Miyaura N, Suzuki A (1993) J. Am. Chem. Soc. 115:11018
129. Roush WR, Riva R (1988) J. Org. Chem. 53:710
130. Nicolaou KC, Ramphal JY, Palazon JM, Spanvello RA (1989) Angew. Chem., Int. Ed. Engl. 28:587
131. Kobayashi Y, Shimazaki T, Taguchi H, Sato F (1990) J. Org. Chem. 55:5324
132. Roush WR, Sciotti RJ (1994) J. Am. Chem. Soc. 116:6457
133. Potter GA, McCague R (1990) J. Org. Chem. 55:6184
134. Uemura M, Nishimura H, Minami T, Hayashi Y (1991) J. Am. Chem. Soc. 113:5402
135. Johnson CR, Braun MP (1993) J. Am. Chem. Soc. 115:11014
136. Hyuga S, Hara S, Suzuki A (1992) Bull. Chem. Soc. Jpn. 65:2303
137. Soderquist JA, Rane AM (1993) Tetrahedron Lett. 34:5031
138. Santiago B, Soderquist JA (1992) J. Org. Chem. 57:5844
139. Rehahn M, Schluter AD, Wegner G, Feast WJ (1989) Polymer 30:1060
140. Kim YH, Webster OW (1990) J. Am. Chem. Soc. 112:4592
141. Wallow TI, Novak BM (1991) J. Am. Chem. Soc. 113:7411
142. Miller TM, Neena TX, Zayas R, Bair HE (1992) J. Am. Chem. Soc. 114:1018
143. Zweifel G, Leung T (1974) J. Am. Chem. Soc. 96:7520
144. Zweifel G, Backlund SJ, Leung T (1978) J. Am. Chem. Soc. 100:5561
145. Zweifel G, Pearson NR (1980) J. Am. Chem. Soc. 102:5919
146. Zweifel G, Pearson NR (1981) J. Org. Chem. 46:829
147. Brown HC, Salunkhe AM (1991) Synlett 684
148. Hara S, Hyuga S, Aoyama M, Sato M, Suzuki A (1990) Tetrahedron Lett. 31:247
149. Molander GA, Singaram B, Brown HC (1984) J. Org. Chem. 49:5024
150. Molander GA, Brown HC (1977) J. Org. Chem. 42:3106
151. Tayano T, Satoh Y, Hara S, Suzuki A (1990) Main Group Metal Chemistry 13:211
152. Soderquist JA, Vaquer J (1990) Tetrahedron Lett. 31:4545
153. Shimizu H, Hara S, Suzuki A (1990) Synthetic Commun. 20:549

5 Asymmetric Synthesis via (α-Haloalkyl)boronic Esters

5.1 Introduction

5.1.1 Overview

The discovery of a practical synthesis of (α-haloalkyl)boronic esters in very high enantiomeric and diastereomeric purity has resulted in a novel method of asymmetric synthesis of broad scope [1]. For example, this chemistry provides the most general known synthesis of chiral secondary alcohols in >99% enantiomeric excess. Several adjacent chiral centers and a wide variety of functional groups can be incorporated into the target compounds. The only restrictions are those imposed by high steric hindrance and reactive functionality.

The asymmetric insertion of a CHCl group into a boron—carbon bond can be controlled with very high precision if the chiral ligand on the boron atom is appropriately chosen. Typical diastereomer ratios with good chiral directors and optimized reaction conditions are in the 100:1 range. Displacement of the chloride by a wide variety of nucleophiles under mild conditions, together with the stability of the resulting substituted boronic esters and the possibility of repeating the chain extension process to provide additional chiral centers, make this a very versatile approach to asymmetric organic synthesis.

The first successful chiral director chosen for this work was the diol derived by osmium tetraoxide catalyzed oxidation of α-pinene [1]. This choice was the result of chemist's intuition, inspired by the historic success achieved by Brown and Zweifel using α-pinene as a chiral director [2]. The diastereoselectivity is further enhanced when the chloride is displaced if the chiral director has C_2 symmetry, leading to products having ultrahigh diastereomeric and enantiomeric purity even before final purification.

After a brief discussion of conventions for describing chirality (Section 5.1.2), this chapter begins with a description of the fundamental chemistry, beginning with the results obtained with pinanediol (Section 5.2), the diol resulting from osmium tetraoxide catalyzed oxidation of α-pinene. Chiral directors of C_2 symmetry are then described (Section 5.3). An extensive description of applications to synthesis of natural products and other molecules of biological interest concludes the chapter (Section 5.4).

5.1.2 Nomenclature Notes

5.1.2.1 Chirality descriptors. The systematic names for pinanediol and its boronic esters are more intricate than the conjugation of Basque verbs [3] and intelligible to fewer people. Accordingly, informal names will be used for these compounds.

"(S)-Pinanediol" (**1**), also designated "(s)-pinanediol" or "(+)-pinanediol" in earlier publications, is systematically named [1S-(1α,2β,3β,5α)]-2,6,6-trimethyl-bicyclo[3.1.1]heptane-2,3-diol. This name was imported directly from *Chemical Abstracts* by computer, and the index order of word segments was rearranged to normal order. Although "(S)-pinanediol" could be ambiguous in some contexts, if there were other diastereomers under discussion, the informal name might be extended to [1S-(1α,2β,3β,5α)]-pinanediol, (1S,2S,3R,5S)-pinanediol, or (S)-αββα-2,3-pinanediol in order to distinguish it [1].

1   **2**

The pinanediol moiety is completely renamed by *Chemical Abstracts* when it is esterified with a boronic acid. For example, (S)-pinanediol (S)-[(chloro)(phenyl)-methyl]boronate (**2**) is named {3aS-[2(R*),3aα,4β,6β,7aα]}-2-(chlorophenylmeth-yl)hexahydro-3a,5,5-trimethyl-4,6-methano-1,3,2-benzodioxaborole. The permuted numbering makes the ring chiral centers (3aS,4S,6S,7aR) if they are designated according to the Cahn-Ingold-Prelog system.

Chemical Abstracts nomenclature chooses a single chiral center to specify absolute configuration and defines all the rest in a relative sense by descriptors α and β for rings, R* and S* for open chains. This nomenclature is used because it correctly groups pairs of enantiomers under a single name. The designation of the side chain chirality as "2(R*)" means that if this were the enantiomer in which index carbon 3a is (R), the 2-(chloro)(phenyl)methyl substituent would also be (R), but in this case carbon 3a is (S), therefore the side chain at position 2 is (S). The logic is equivalent to regarding the relative configuration descriptors (R*, S*, α, β) as a set of vectors, which are to be multiplied by the scalar +1 if the absolute configuration is (R) or the scalar −1 if the absolute configuration is (S). The net result is that all of the relative configuration indicators (R*, S*, α, β) are to be taken literally if the overall chirality indicator is (R), but every one of them becomes its opposite when the overall chirality indicator is (S).

For some purposes the *Chemical Abstracts* nomenclature can be useful, but in most situations it is easier to communicate the chemistry if each chiral center is designated by its own Cahn-Ingold-Prelog (R) or (S) absolute configuration. Chemists are usually trying to follow what is happening to one or two chiral centers at a time, and having to keep track of all of the steric relationships all of the time in a series of complex molecules is likely to cause total confusion in anyone whose brain has not been replaced by a silicon chip transplant.

5.1.2.2 Bond illustration conventions. Three basic conventions regarding the illustration of chiral centers will be followed as consistently as possible in this book. (1) A chiral center takes priority over surrounding groups, so that a bold wedge bond always projects in front of the chiral center, and a hashed bond projects behind the chiral center. (2) The broad end of a wedge bond is always toward the viewer, regardless of whether the bond is bold or hashed. If two chiral centers are joined by a wedge bond, the broad end of the bond is in front, regardless of whether the bond is wedged or hashed. (3) Racemic mixtures or meso structures will be illustrated with bold or hashed parallel-sided bonds, and wedge bonds will be used in such structures only if application of rule (1) might be inherently ambiguous.

Rule (3), that wedge bonds be reserved for truly asymmetric compounds, is a convention that has gained some acceptance [4]. The use of wedge bonds is applicable to all compounds that are "scalemic" [5], that is, asymmetric compounds that have one enantiomer in significant excess over the other, even though they are not necessarily enantiomerically pure. To illustrate, **1** represents (*S*)-pinanediol, **3** is (*R*)-pinanediol, and **4** is racemic (1α,2β,3β,5α)]-pinanediol. Materials represented by **1** or **3** might be "enantiomerically pure" or merely 92% ee or some lower value.

1   **3**   **4**

Rule (2) is contrary to majority practice regarding hashed wedge bonds projecting behind a chiral center. These are most often written with the pointed end of the wedge bond at the chiral center and the broad end meant to be farther away, perhaps because of the esthetics of joining a chiral center indicated by a line junction point and a substituent indicated by an alphabetic symbol. Usually such structure drawings are unambiguous because the bond is by convention meant to extend behind the chiral center, but it makes no sense at all in a perspective drawing. The lack of consensus and the violation of perspective by the more common convention has led to a suggestion that all hashed wedge bonds should simply be avoided in favor of dashed lines [6]. However, ambiguities can arise, especially if the bond joins two chiral centers in a complex structure, and it is useful to have a convention which defines the near end of the bond. Furthermore, where mechanisms are discussed, dashed lines suggest partial bonds.

In this book the broad end of a hashed wedge bond, like the broad edge of a bold wedge bond, is always closer to the viewer. A glance at the "railroad tracks on LSD" versus "down to earth railroad tracks" illustrated should be convincing.

The nonperspective heavy or hashed bonds used for meso or racemic structures such as **4** are, of course, ambiguous if they happen to join two chiral centers. In that event, a single enantiomer would have to be chosen for representation, but few structures of such complexity are encountered as racemates.

5.1.2.3 Drawings of bicyclic structures. Relatively rigid structures such as those derived from α-pinene are represented by plane projections mechanically traced from structures computed by the program "PC Model", Serena Software Company, Bloomington, Indiana. This software is based on W. C. Still's MMX modeling program. The structures have been turned so as to provide as clear as possible illustration. Where feasible, attached groups are oriented in a realistic low energy direction, but bond lengths are not scaled and structures are not minimized for zigzag chains or simple rings.

5.2 Synthetic Methodology: Pinanediol Esters

5.2.1 Original Process

5.2.1.1 Preparation of pinanediol. The first highly stereoselective preparations of α-chloro boronic esters reported by Matteson and Ray [7] utilized the chiral director (S)-pinanediol (**1**), which was obtained by the osmium tetraoxide catalyzed oxidation of (+)-α-pinene with trimethylamine N-oxide [8, 9]. The preparation of **1** was based on a published procedure that utilized N-methylmorpholine N-oxide [10], but in this case trimethylamine N-oxide was required for satisfactory results. The (–)-enantiomer of α-pinene is also readily available.

1

Common natural sources do not supply either enantiomer of α-pinene in high enantiomeric purity, but recrystallization of a sodium or potassium borate salt of pinanediol can lead to high enantiomeric purity [11, 12]. Pinanediol is most easily recoverable from the salt as its boric acid ester, which on treatment with acid and a boronic acid or ester readily yields the corresponding pinanediol boronic ester. However, α-pinene is available commercially in laboratory quantities [13] from a process consisting of hydroboration, crystallization of the resulting borane, and cleavage back to α-pinene by treatment with an aldehyde [14]. This chemistry is described in more detail in Section 6.2.1.

A kinetic study of the reaction of α-pinene with osmium tetraoxide and tri-methylamine N-oxide has revealed that the reaction is first-order in total osmium species and first-order in amine oxide, but zero-order in α-pinene [15]. The osmium tetraoxide catalyzed dihydroxylation of α-pinene is much slower than that of cyclohexene, but the presence of a small amount of α-pinene greatly retards the reaction. The kinetics are in accord with a mechanism in which essentially all of the osmium is present as the ester of one mole of osmium(VI) with two moles of pinanediol, and the rate-limiting step is the oxidation of this ester by the amine oxide to form free pinanediol and the osmium(VIII) ester with one mole of pinanediol, which rapidly reacts irreversibly with more α-pinene.

5.2.1.2 Discovery of pinanediol boronic ester chemistry. The first work showed that (S)-pinanediol boronic esters with (dichloromethyl)lithium reliably yielded pre-dominantly α(S)-(α-chloroalkyl)boronic esters [7, 11], as illustrated by the conver-sion of (S)-pinanediol butylboronate (5) to (S)-pinanediol 1(S)-(1-chloropentyl)-boronate (7) followed by treatment with methyllithium to form 9 and oxidation with hydrogen peroxide to (S)-2-hexanol (10) in 79% ee. This is far below the stereo-selectivity achieved later. Evidence for the stereochemistry of the postulated borate intermediates 6 and 8 is presented in Section 5.2.6.

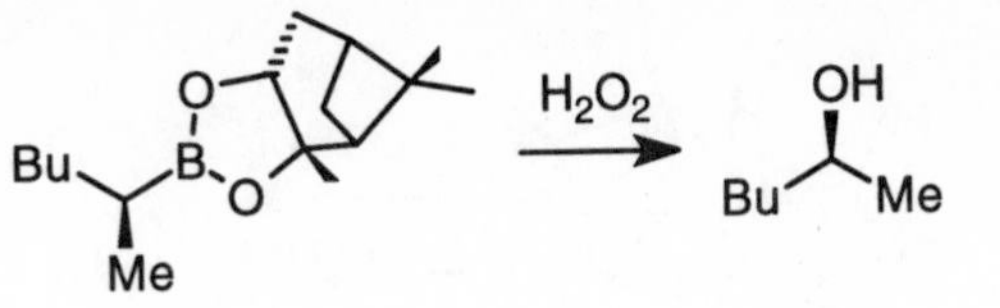

9   **10**

The synthetic potential of this chemistry was first demonstrated with a synthesis of a single enantiomer of each of the two diastereomers of 3-phenyl-2-butanol [7, 11]. One reason for this choice was that earlier work by Cram had unambiguously defined the structures [16]. The choice proved fortuitous in other ways, as the migration of aryl and benzylic groups proved particularly facile and provided better than average stereocontrol in the unimproved process. The crude "*erythro*" isomer, (*S*)-(2*R**,3*R**)-3-phenyl-2-butanol (**13**) was obtained in 88% de and the overall yield from (*S*)-pinanediol phenylboronate (**11**) of this isomer contained in the mixture was 67% by ^{1}H-NMR analysis.

To make the "*threo*" isomer, it might be imagined that one could start with a methylboronic ester, extend the chain with (dichloromethyl)lithium, and phenylate the resulting chloroethylboronic ester. This approach was defeated by the low diastereoselectivity of the methylboronic ester under the conditions used at that time, but is no longer a problem. Another possibility, conversion of a halo boronic ester to a hydroxy boronic ester and then to a mesylate for double inversion, was unsuccessful at that time but has since been achieved. The remaining option was to remove the chiral director from the intermediate (methyl)(phenyl)methylboronic ester **12** and replace it by its enantiomer to form the diastereomeric boronic ester **14**. With the extra steps, the resulting overall yield of (*R*)-(2*R**,3*S**)-3-phenyl-2-butanol (**15**) from **11** was 40%, de 92%.

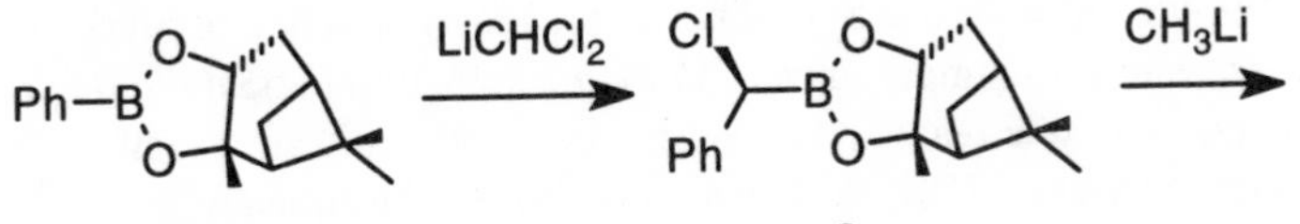

11   **2**

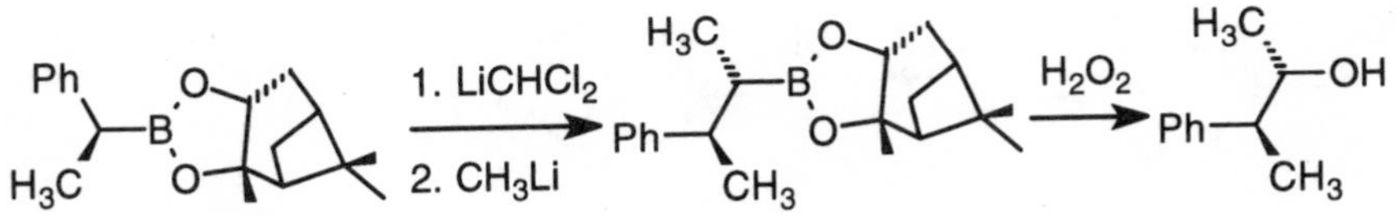

12   **13**

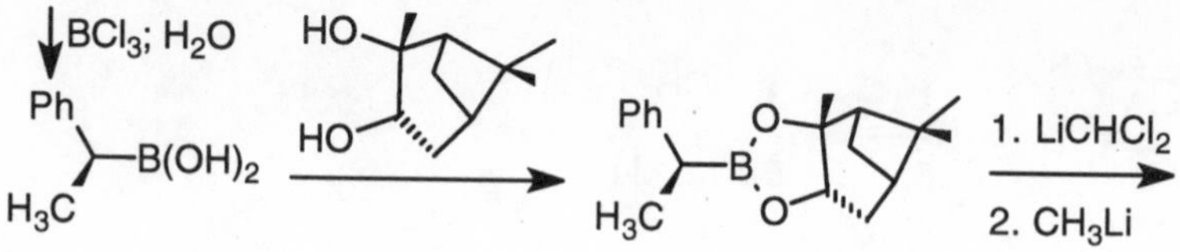

14

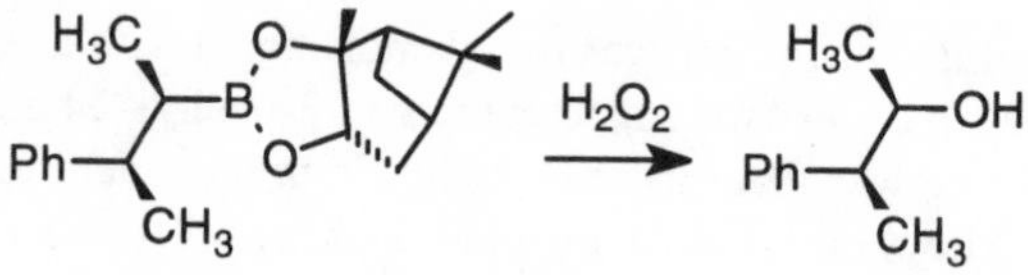

15

The de's achieved in the second chain extension step do not differ much in view of the similar de's obtained in both crude final products **13** and **15**, and it may be concluded that the diastereomer fraction is nearly independent of the relative configuration of the starting boronic ester. Accordingly, the amount of the enantiomer (R)-(2R*,3R*)-3-phenyl-2-butanol present in the (S)-enantiomer **12** is essentially equal to the product of the fractions of minor diastereomer formed in each of the two steps. This quantity is maximized if the amounts are equal, ~3%, in each step. Then the product $(0.03) \times (0.03) = 0.0009$, which corresponds to an ee of 99.8%. Whether the enantiomeric purity of the pinanediol might have been this high cannot be proved but does not seem likely.

5.2.1.3 Epimerization of (α-chloroalkyl)boronic esters. While this topic might seem a diversion from the main goal of asymmetric synthesis, it provides an example of how an understanding of reaction mechanisms and unwanted side reactions can lead to revolutionary improvement of a synthetic method.

Halide exchange in α-haloalkylboronic esters has been known since this class of compounds was first discovered [17]. Since chloride ion exchange inverts the chiral center, and since lithium chloride is a byproduct of the reaction which produces α-(chloroalkyl)boronic esters, the importance of knowing something about the rate of this potentially deleterious side reaction was recognized early, and an investigation of the kinetics was undertaken [18]. (S)-Pinanediol (S)-(α-chlorobenzyl)boronate (**2**) and its epimer (S)-pinanediol (R)-(α-chlorobenzyl)boronate (**16**) were chosen for detailed study.

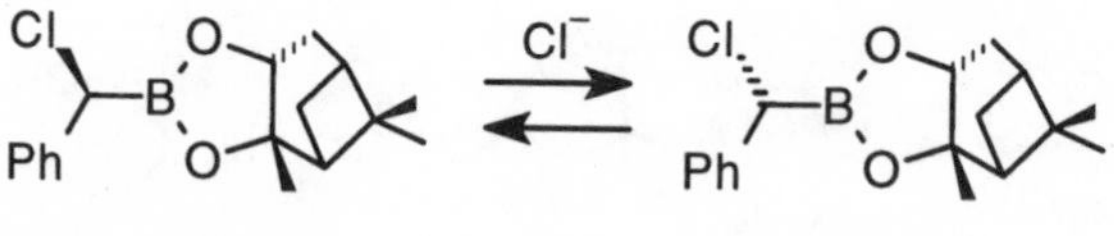

2   **16**

The rate of epimerization of 2 in THF was followed polarimetrically and was found to be first-order in 2 and approximately 0.75-order in lithium chloride, which exists largely as ion tetramers in THF solution. The results were interpreted as indicating that free chloride ion is the active species, and the kinetics are compatible with the dissociation of Li_2Cl_2 to Li_2Cl^+ and Cl^-, which would result in half-order kinetics, but which would be subject to a strong positive salt effect.

The rate constant k_1 for epimerization of 2 by 0.45 M lithium chloride (nearly saturated) in THF at 25 °C is 5.7×10^{-5} sec^{-1}, which corresponds to a half-life of ~3.4 h or a loss of stereopurity of about 0.35% per minute. The k_1 for epimerization of (S)-pinanediol (S)-[(1-chloro)pentyl]boronate, a typical saturated α-chloro boronic ester, is 2.5×10^{-6} sec^{-1} under the same conditions, half-life 50 h, ~1% loss of stereopurity per hour.

The addition of 0.1% water to the THF solution doubled the rate of epimerization of 2. Dimethyl sulfoxide also had a strongly accelerating effect, precipitating most of the lithium chloride while quadrupling the pseudo-first-order k_1.

Metal cations that complex with chloride ion greatly retard epimerization. An equivalent amount of mercuric chloride reduced the rate below the limit of observability, at least two orders of magnitude. However, mercuric chloride is not a satisfactory additive for synthetic operations, as it is very difficult to remove from the products in addition to being a notorious toxic environmental hazard. Zinc chloride was therefore examined as an alternative.

In the $LiCl/ZnCl_2$ system, the minimum epimerization rate is an order of magnitude lower than that for a similar amount of LiCl alone. The minimum occurs near the stoichiometry $LiZnCl_3$. At the higher concentrations of salts, Li_2ZnCl_4 does not cause nearly as much increase in rate as does $ZnCl_2$, and rates are low throughout the range of $Li_2ZnCl_4/LiZnCl_3$ compositions. Dilute $ZnCl_2$ does not cause epimerization, but there is a term in the rate law that is first-order each in [$ZnCl_2$] and [$LiZnCl_3$]. This second-order process becomes very fast as these concentrations approach 1 M. Accordingly, in reactions catalyzed by zinc chloride it is inadvisable to use an excess of zinc chloride beyond what will ultimately be complexed by nitrogen or oxygen functionality plus what will be converted to $LiZnCl_3$ by the chloride ion released in the reaction.

5.2.2 Zinc Chloride Promotion of α-Halo Boronic Ester Formation

As the results of the foregoing studies became apparent, zinc chloride was tested as an inhibitor of epimerization and possible promoter for the rearrangement of the borate complexes formed from boronic esters and (dichloromethyl)lithium. The improvement in the results was dramatically successful, and a new method of asymmetric synthesis was launched.

In its original form, the α-chloro boronic ester synthesis was only moderately stereoselective with alkylboronic esters, worked poorly with pinanediol methylboronate or isobutylboronate, and failed altogether with pinanediol (benzyloxymethyl)boronate [11]. The addition of 0.5-1.0 molar equivalent of zinc chloride to the cold solution of borate complex generally improves the isolable yields of α-chloro boronic esters to 80-95% and the de's to 98-99%, with a few exceptions.

The first example tested, (S)-pinanediol isobutylboronate (**17**), only yielded 30% of (S)-pinanediol 1(S)-(1-chloro-3-methylbutyl)boronate (**19**), de 76%, under the original reaction conditions. When 0.5 molar equivalent of zinc chloride was added to the (dichloromethyl)borate complex (**18**), the yield of **19** rose to 89% and the de to 99%. The de was measured on the α-acetamido derivative (**21**), which showed well separated NH peaks for the two diastereomers in the 200 MHz ¹H-NMR spectrum of material prepared without the zinc chloride promoter. The 0.5% of minor diastereomer was barely detectable in **21** prepared via the zinc chloride promoted route. The roundabout route to **21** via the silylated amino boronic ester (**20**) is discussed in Section 5.4.2 [19, 20].

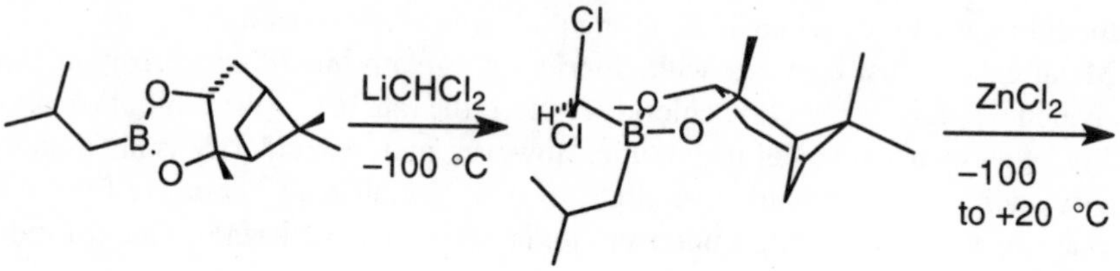

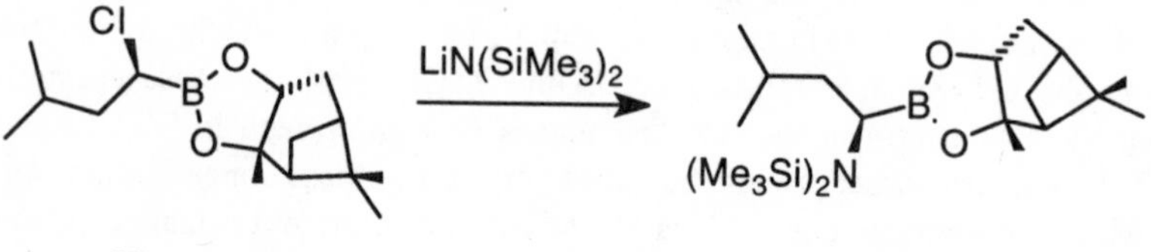

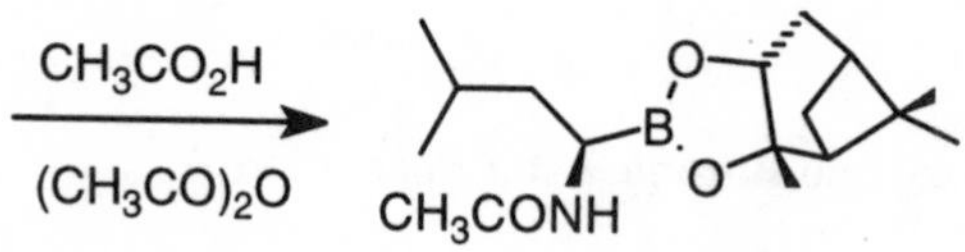

Further exploration showed that the benefits of zinc chloride promotion are general [12]. Compounds containing oxygen functionality, most notably R = PhCH₂OCH₂, fail to undergo chain extension with (dichloromethyl)lithium unless zinc chloride is added. Even then, the yields and diastereoselection are merely satisfactory, though the diastereoselection improves when the oxygenated side chain is more sterically demanding [21]. On the other hand, the improvement with pinanediol phenylboronate (**11**), which reacts efficiently in the absence of zinc chloride, is marginal or nonexistent. It was believed that the crude (S)-pinanediol (S)-(chloro)(phenyl)methylboronate (**2**) was of higher chemical purity when prepared with the aid of zinc chloride, but no rigorous comparison was ever made [18]. A comparison of results with and without zinc chloride is shown in Table 4-1.

Table 4-1. Yields and diastereomeric excesses (de's) in reactions of pinanediol boronates [RB(O$_2$C$_{10}$H$_{16}$)] with (dichloromethyl)lithium with and without zinc chloride.

R of RB(O$_2$C$_{10}$H$_{16}$)	without ZnCl$_2$		with ZnCl$_2$		Refs.
	de, %	yield, %	de	yield, %	
CH$_3$	48	57	89	83	[11, 12]
CH$_3$CH$_2$CH$_2$			98	87	[11, 12]
CH$_3$CH$_2$CH$_2$CH$_2$	79	61	97	86	[11, 12]
(CH$_3$)$_2$CHCH$_2$	77	15-33	99	89	[12, 20]
C$_6$H$_5$CH$_2$	85	75	99	99	[12, 22]
C$_6$H$_5$	96[a]	87[a]	(88)[b]	(84)[b]	[11, 18]
PhCH$_2$OCH$_2$	—	0	80[c]	70-83[c]	[11, 21]

[a] Chloro compound not isolated, analyzed crude by NMR, converted in situ to PhCH(Me)B(O$_2$C$_{10}$H$_{16}$) (**12**), for which yield is given. [b] Yield and purity after chromatography on silica, which may have caused some epimerization. However, subsequent experience has indicated that PhCHClB(O$_2$C$_{10}$H$_{16}$) (**2**) prepared in this way does not need to be chromatographed, and although it contains some epimer, it is easily recrystallized to high enantiomeric purity. [c] Required 1.7 molar equivalents of ZnCl$_2$.

5.2.3 Functional Group Compatibility

5.2.3.1 Alkoxy substituents. Among the first questions was whether the synthesis could accommodate functionality. An early set of examples was provided by transformations of (S)-pinanediol (S)-(1-chloropentyl)boronate **7**, which was readily available from the butylboronate **5** [12]. Lithium benzyl oxide converted **5** to (S)-pinanediol (R)-[1-(benzyloxy)pentyl]boronate (**22**), and a second chain extension yielded **23**.

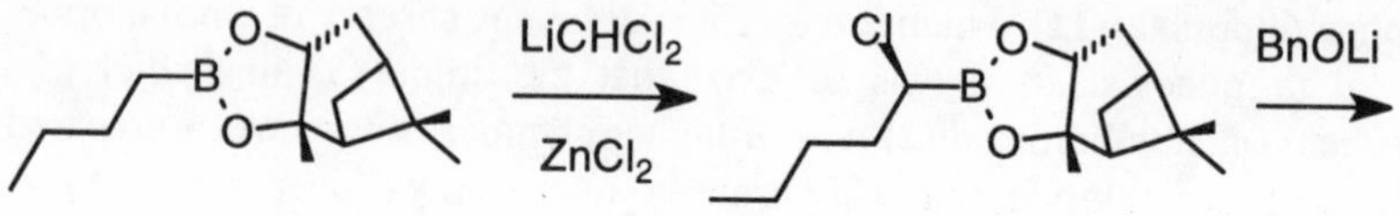

5 →(LiCHCl₂ / ZnCl₂)→ **7** →(BnOLi)→

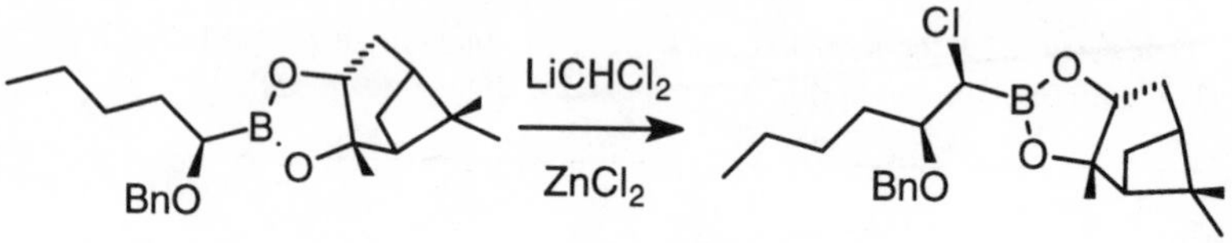

22 →(LiCHCl₂ / ZnCl₂)→ **23**

Butylmagnesium chloride converted **23** to boronic ester **24**, which was deboronated with hydrogen peroxide to **25** and debenzylated with hydrogen and palladium to (–)-(*S,S*)-5,6-decanediol (**26**), de 97%. The *meso* isomer was synthesized by another route for comparison. Benzyloxy boronic esters have subsequently been used as intermediates in the synthesis of insect pheromones and carbohydrates.

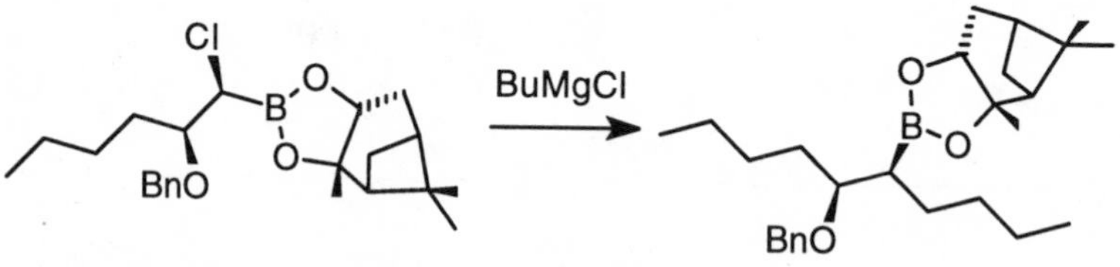

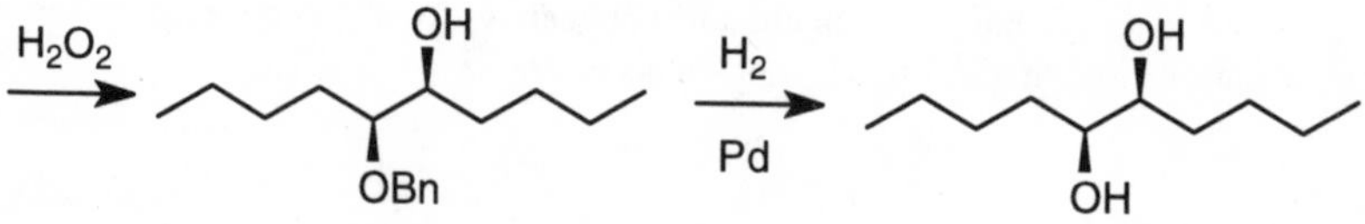

23 →(BuMgCl)→ **24**

25 →(H₂ / Pd)→ **26**

Treatment of intermediate **23** with methylmagnesium bromide followed by chain extension, butylmagnesium chloride treatment to form **28**, deboronation to **29**, and hydrogenolysis led to (*S,S*)-6-methyl-5,7-undecanediol (**30**). The reaction of **27** with (dichloromethyl)lithium was best carried out in the presence of 1.7 moles of zinc chloride, sufficient to complex with the benzyloxy group and have enough left over to promote the rearrangement of the borate intermediate. Independent synthesis of a mixture of all of the possible diastereomers of **30** revealed that no diastereomer was detectable in the crude **30** above the 0.5% limit of detectability. Since there was at least 1-2% diastereomer in the precursor **22**, it may be presumed that chromatography of intermediates resulted in purification, but none of the intermediates after **27** were purified, and the last chiral carbon therefore must have been introduced in ≥99% de.

Benzyloxy alcohol **29** provided the first example of three adjacent chiral centers prepared via this chemistry, but it may be noted in passing that the central carbon of the 1,3-diol **30** is not a chiral center. A C_2 rotation about the vertical axis in the plane of the page interchanges the positions of the two hydroxyl groups but not their appearance, and moves the center methyl group from in back to in front. Thus, the molecule is unchanged by formal inversion of the methyl group. Similar symmetry types are, of course, well known in sugar chemistry.

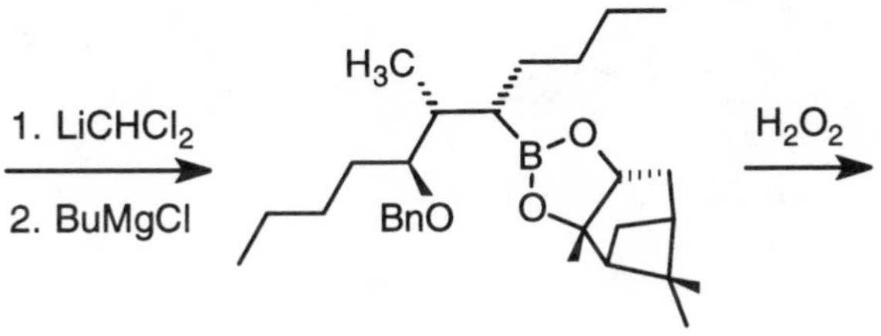

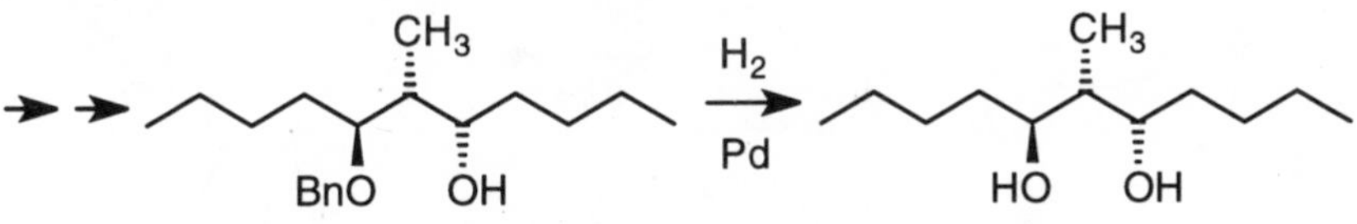

5.2.3.2 An azido substituent. Compatibility with an azido substituent was not anticipated, and the success of this exploratory work revealed new possibilities for asymmetric synthesis.

(*S,S*)-6-Amino-5-decanol (**35**) was prepared from (1-chloropentyl)boronate **7** via a reaction with sodium azide to form (*S*)-pinanediol (*R*)-(1-azidopentyl)boronate (**31**). The reaction with sodium azide was carried out in a two-phase system, dichloromethane and water, with tetrabutylammonium ion as phase transfer catalyst. Chloride ion liberated in the displacement competes with chloride as a nucleophile, epimerizing the chloro boronic ester **7**, and therefore azide was used in 10-fold excess. The reaction of **31** with (dichloromethyl)lithium proceeded normally, but (*S*)-pinanediol (1-chloro-2-azidohexyl)boronate (**32**) with butylmagnesium chloride under normal conditions yielded only butylboronate **5** and an unsaturated product, presumably 1-chloro-1-hexene from elimination of the boronic ester and azido groups. However, the addition of 4 moles of zinc chloride before the rearrangement of the borate complex promoted the formation of azido boronic ester **33**, with only 5% elimination to **5**, and deboronation led to azido alcohol **34** in 95.5% de by gc analysis, 71% yield based on **32**. Reduction with lithium aluminum hydride furnished (*S,S*)-6-amino-5-decanol (**35**), de 96% by NMR analysis.

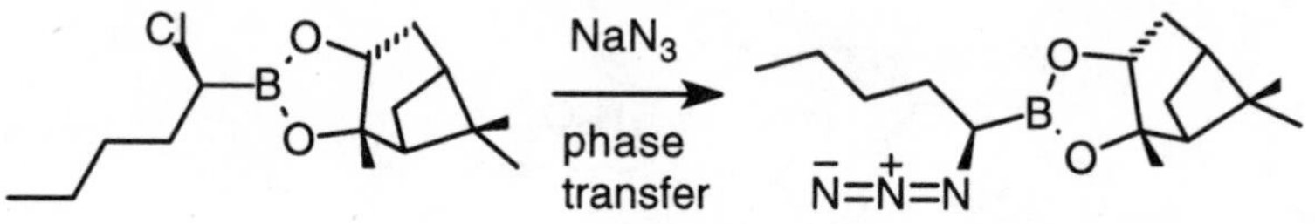

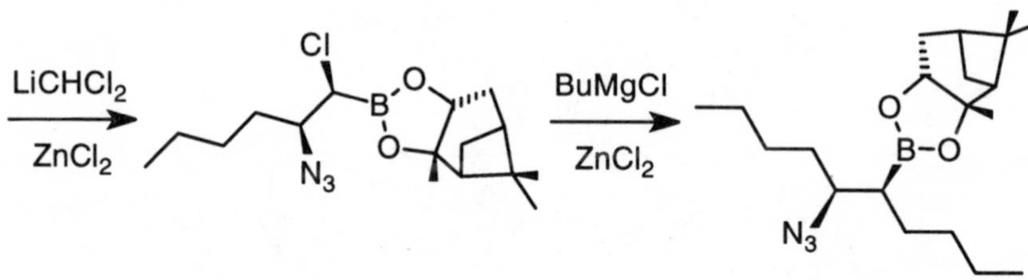

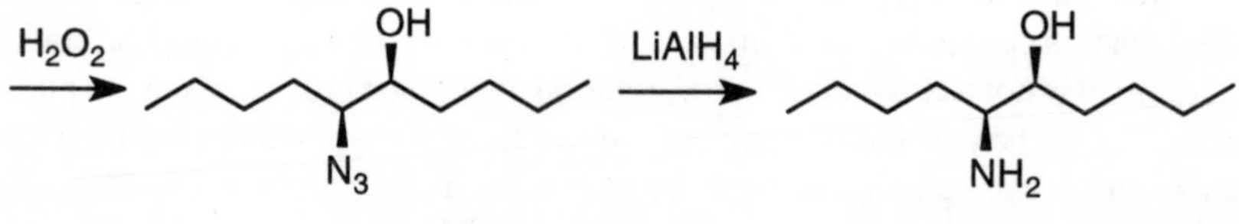

The stability of the azido boronic ester **31** was unexpected, inasmuch as it would be possible to eliminate molecular nitrogen and the boronic ester group to form a carbon—nitrogen double bond, but this reaction has not been observed. Azido boronic esters have subsequently been used in an asymmetric synthesis of amino acids (see Section 5.4.5).

5.2.3.3 Thioether substituents. The chain extension process was found to be incompatible with an α-phenylthio group, **36** apparently undergoing cleavage of the (phenylthio)methyl anion, but a β-hexylthio group did not interfere with chain extension of **37** to **38** [12]. However, it was subsequently found that the analogue of **37** having a methylthio group in place of the hexylthio group failed to undergo the chain extension reaction [23]. Perhaps the failure resulted from the greater acidity of the methylthio group compared to the hexylthio group.

5.2.3.4 Halogen substituents. β-Halogens are strictly incompatible with the homologation reaction because of β-elimination of boron and halide, but with three intervening carbon atoms the chain extension can be accomplished without difficulty [20]. An example is illustrated in Section 5.4.2 on α-amido boronic acids.

5.2.3.5 Carboxylic ester substituents. Carboxylic ester groups are compatible with the chain extension chemistry (see Sections 5.4.3.3 and 5.4.5). However, there must be at least two carbon atoms between the carbonyl group and the boronic ester function. Thermodynamics favors B—O bonds over B—C bonds by more than 30 kcal/mol (Section 1.2.3), and if the boron is tricoordinate there is no orbital symmetry barrier to a 1,3-shift of boron from carbon to oxygen in a C-bonded boron enolate, because the boron has a vacant orbital available for bonding. In accord with this expectation, reactions which acylate boron substituted carbanions and would theoretically give rise to β-boryl carbonyl compounds as intermediates yield deboronated products [24, 25].

No boronic esters are known in which the carbonyl group is bonded directly to boron. The enol ether (S)-pinanediol (1-methoxyvinyl)boronate (**39**) has been prepared in a straightforward manner from (1-methoxyvinyl)lithium and triisopropyl borate followed by acidification with acetic acid and transesterification with pinanediol [26]. However, attempts to hydrolyze **39** with acid to the acetylboronic ester led only to decomposition. Attempts to extend the chain of **39** with (dichloromethyl)lithium were also unsuccessful. It was found possible to alkylate **39** with butylmagnesium chloride and rearrange the resulting borate with zinc chloride followed by acetic acid to produce a ~1:1 mixture of diastereomeric pinanediol (1-methoxy-1-methylpentyl)boronates (**40**) in 89% yield.

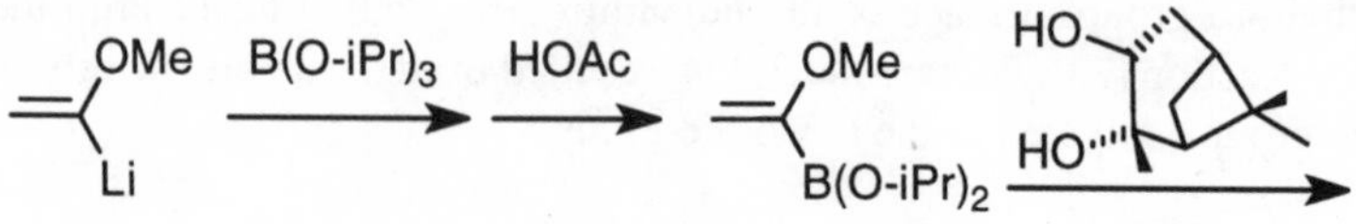

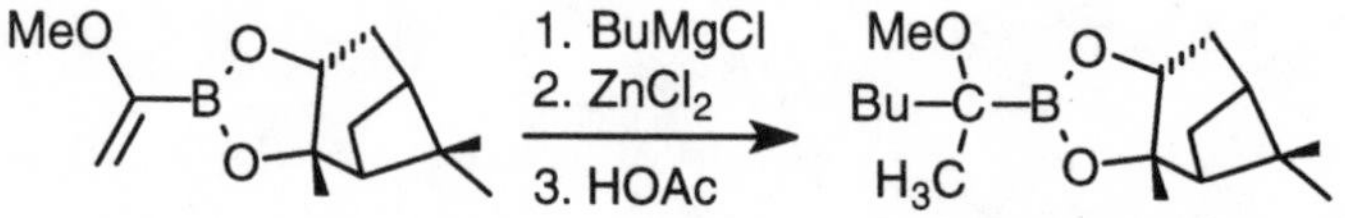

39   **40** (diastereomeric mixture)

5.2.3.6 Unsaturated boronic esters. The usual chain extension works well with pinanediol vinylboronate (**41**). The diastereomeric purity of the resulting (*S*)-pinanediol (α-chloroallyl)boronate (**42**) is high even without zinc chloride, which in this case appears to be of little if any benefit. Normal reactions have been observed with **42** and butylmagnesium chloride [11] or lithiohexamethyldisilazane [23], but attempted substitutions by alkoxides or even carboxylate salts have led only to decomposition products [27].

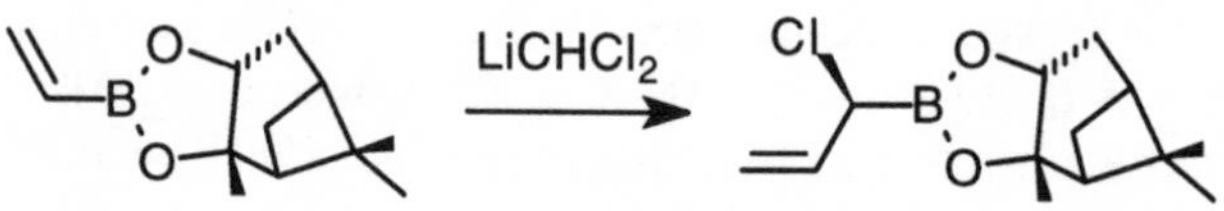

41   **42**

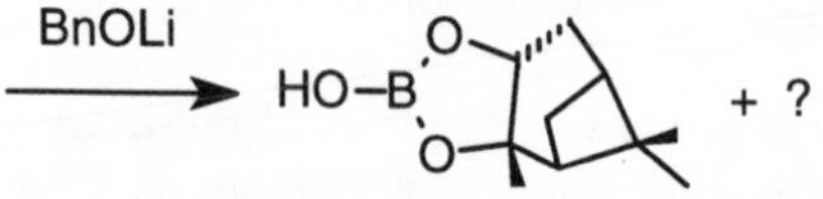

(*S*)-Pinanediol allylboronate (**43**) behaves normally in the reaction with (dichloromethyl)lithium and yielded the expected chain extension product (**44**), which underwent several normal further reactions [28]. However, some problems have been encountered in similar chain extensions of other boronic esters (Section 5.3.7).

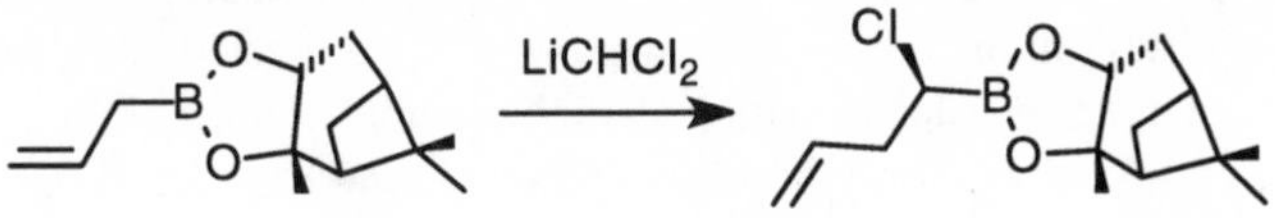

43   **44**

5.2.4 Quaternary Chiral Centers

In all of the foregoing, each chiral carbon has had one hydrogen and three other atoms attached. The possibility of introducing chiral centers having all four atoms not hydrogen has been explored briefly with mixed results [29].

Reaction of (S)-pinanediol phenylboronate (**11**) with (1,1-dichloroethyl)lithium, prepared in situ by addition of LDA to a mixture of **11** and 1,1-dichloroethane in THF at −78 °C, resulted in formation of (S)-pinanediol (S)-(1-chloro-1-phenyl-ethyl)boronate (**45**). Note that this is opposite the configuration that would be expected based on the behavior of (dichloromethyl)lithium. Ethylmagnesium bromide converted **45** to boronic ester **48**, which with hydrogen peroxide yielded (R)-2-phe-nyl-2-butanol (**49**), 85% ee. The de of the intermediate **45** was estimated to be 92%, and it appeared that some loss occurred during the reaction with ethylmagnesium bromide to form **46** [29].

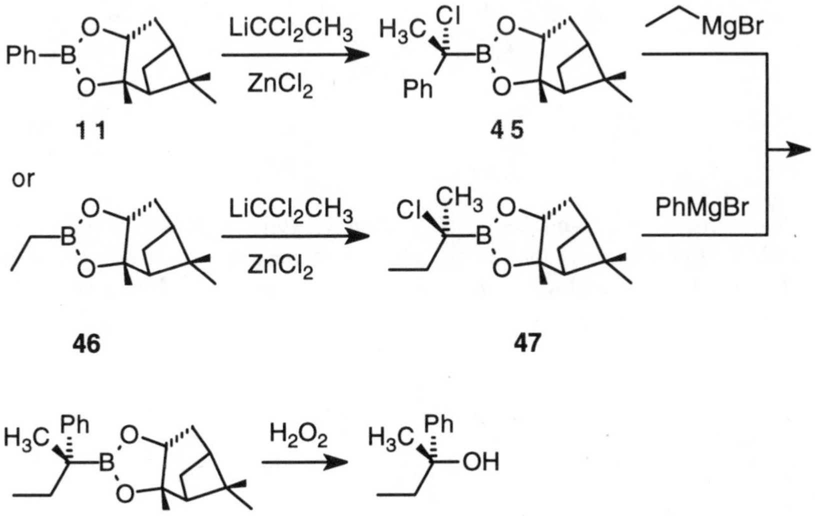

11   **45**

or

46   **47**

48   **49**

Starting from (S)-pinanediol ethylboronate (46) with (1,1-dichloroethyl)lithium led to the same ultimate product, (R)-2-phenyl-2-butanol (49), 77% ee. Thus, the intermediate chloro boronic ester is the (S)-(1-chloro-1-methylpropyl)boronate (47), which has the same configuration that would be expected by analogy to the behavior of (dichloromethyl)lithium, but opposite what would be expected based on the behavior of the phenylboronic ester 11 with (1,1-dichloroethyl)lithium [29].

Chain extension of (S)-pinanediol butylboronate (5) was found to proceed with essentially no diastereoselection at all. Ethylation of the intermediate chloroboronic ester and deboronation yielded ~2% ee (S)-3-methyl-3-heptanol. The functionalized boronic ester 50 leaned toward the pattern of the ethylboronic ester, and led to 54, the enantiomer of the insect pheromone frontalin, in only 22% ee. Thus, the slightly favored intermediates are 51 and 52. Completion of the synthesis via 53 included conventional peroxidic deboronation, hydrogenolysis of the benzyl ether, and acid catalyzed hydrolysis of the ethylene ketal with ring closure to the bicyclic ketal 54 [29].

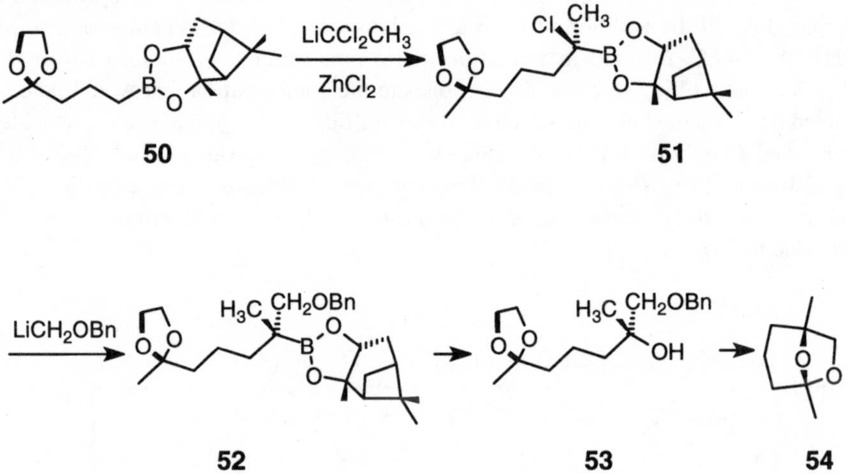

Chain extension of (S)-pinanediol (R)-(1-benzyloxypentyl)boronate (55) with (1,1-dichloroethyl)lithium yielded a 10:1 mixture of separable diastereomers (56) and (57), and although analogy to the phenylboronic ester suggests that 56 might predominate, this was not proved.

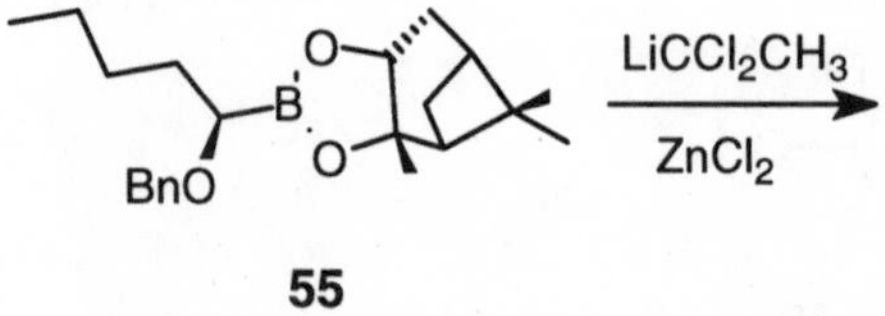

56   **57**

It may be concluded from the foregoing results that the chiral direction in chain extensions of boronic esters with (dichloromethyl)lithium and related compounds depends on the relative sizes of the substituents in the vicinity of the boron atom rather than their polarity or ability to coordinate with metal cations. If a specific interaction of the chlorine atom were the determining factor, the (1,1-dichloro-ethyl)lithium reactions should show a preference for the same direction of asymmetric induction as does (dichloromethyl)lithium, instead of the mixed results actually observed.

5.2.5 Pinanediol Recovery

A perennial problem with pinanediol boronic esters has been cleavage of the pinanediol without destroying either the pinanediol or the boronic acid moiety. In the syntheses described so far, where the free boronic acid was needed, the pinanediol was destroyed with boron trichloride.

For synthetic operations where oxidation of the boronic ester to an alcohol is the final step, there is no real problem. The recovered pinanediol is in the form of its boric acid ester or a borate salt, and crystallization of the borate salt improves its enantiomeric purity. Pinanediol boric acid ester, freed from the pinanediol borate salt by treatment with aqueous acid, reacts with boronic esters to form pinanediol boronic esters. Even so, it can be desirable to free the pinanediol from its boric acid ester.

Brown and Rangaishenvi have found that hydrofluoric acid, in the form of triethylamine trihydrofluoride, will react with sodium bis(pinanediol) borate in ether and water to form sodium tetrafluoroborate and free pinanediol [30].

Hydrofluoric acid does not cleave pinanediol from boronic esters. Neither do a wide variety of boron chelating agents, including diethanolamine and other diols. Reduction with lithium aluminum hydride does work, provided of course that any functionality present in the boronic acid moiety tolerates that reagent, and the product is a useful alkylborohydride. An example is the conversion of pinanediol cyclohexylboronate (**58**) to lithium cyclohexylborohydride (**59**). (Pinanediol)(hydrido)-aluminum methoxide (**60**) is precipitated from ether as its sodium salt, and pinanediol can be recovered by treatment of **60** with aqueous base and ether extraction [30].

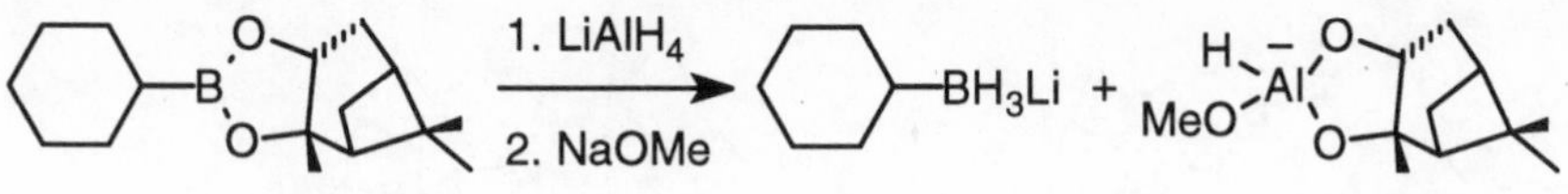

58   **59**   **60**

Another reaction that releases pinanediol is methylation with methyllithium and silylation of the resulting borinic ester anion. The borinic ester can be separated by conversion to the ethanolamine ester [30]. The use of this reaction in synthesis is detailed in Section 6.3.2.3.

5.2.6 Nonequivalent Faces of Boron in Pinanediol Esters

The nonequivalence of the two faces of trigonal boron in pinanediol boronic esters is revealed by the contrast between the reactions of (S)-pinanediol (dichloromethyl)-boronate (**61**) with isobutylmagnesium chloride [31] and of (S)-pinanediol isobutylboronate (**17**) with (dichloromethyl)lithium [20]. Without zinc chloride promotion, **61** yields a 2:1 mixture of diastereomeric (α-chloroalkyl)boronic esters **63** and **64**, reversing the diastereoselection found in the reaction of isobutylboronic ester **17** with (dichloromethyl)lithium, ~8:1 in favor of **64** over **63** [12]. These results suggest that the organometallic reagent adds to the less hindered face of the boron atom of the boronic ester in each case. (Dichloromethyl)boronic ester **61** and isobutylmagnesium chloride yield intermediate borate **62**, which is diastereomeric to **18** formed from isobutylboronic ester **17** and (dichloromethyl)lithium.

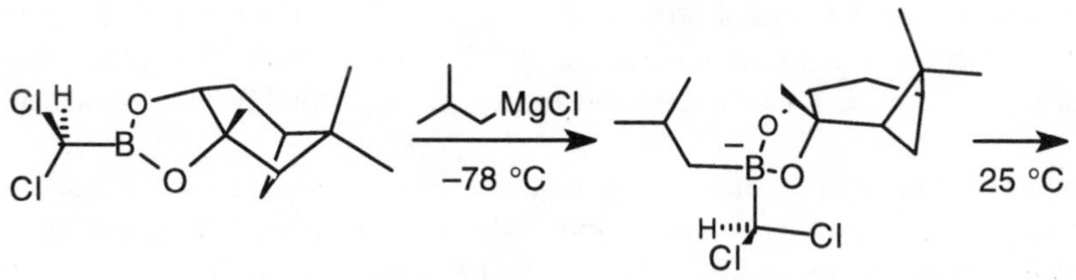

61   **62**

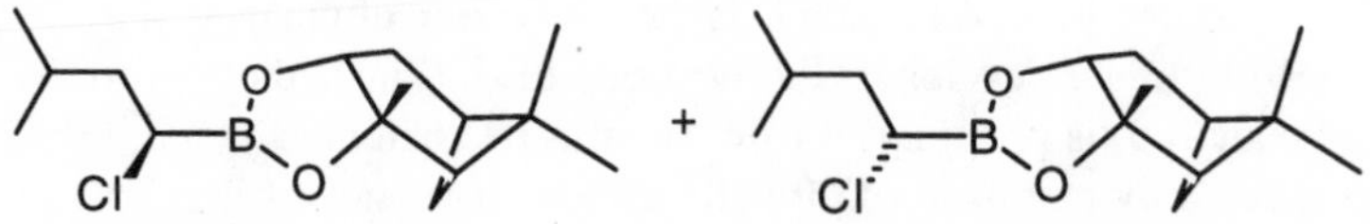

63 (66%)   **64** (34%)

17   **18**

64 (89%)   **63 (11%)**

Zinc chloride promotion shifts the isomer ratios strongly toward **64** in either case. From **61** and isobutylmagnesium chloride, the **64**:**63** ratio was 92:8 [31], and from **17** and (dichloromethyl)lithium, the **64**:**63** ratio was 200:1 [12].

(Dichloromethyl)boronic ester **61** with butyllithium and no zinc chloride yielded a 35:65 ratio of the diastereomers analogous to **63** and **64** [31]. The analogous ratio with phenyllithium was 32:68, and with methylmagnesium bromide, 78:22, which was shifted to 49:51 by zinc chloride.

In keeping with the foregoing results, (S)-pinanediol (1,1-dichloroethyl)boronate yielded a 2:1 mixture of diastereomers, favoring (S)-(1-chloro-1-methylpropyl)-boronate, on treatment with ethylmagnesium bromide [29].

Clearly **61** is not a useful synthetic intermediate. The other major conclusion to be drawn from this study is that reaction of (dichloromethyl)lithium occurs at least 91-94% at the preferred face of typical boronic esters such as **17**. Since it is unlikely that rearrangements of borate **18** or its analogues are stereospecific, the actual face selectivity is probably higher than this calculated minimum and may well approach 100%. The results of this study prompted the investigation of boronic esters of chiral directors of C_2 symmetry, in which the two faces of the boron atom are equivalent.

5.3   Synthetic Methodology: Chiral Directors Having C_2 Symmetry

5.3.1 Butanediol Esters

5.3.1.1 Synthetic routes made possible by C_2 symmetry. The first chiral director of C_2 symmetry to be tested was (R,R)-2,3-butanediol, which is a fermentation product. Because of the C_2 symmetry, both faces of the boron atom are equivalent in boronic esters of this diol. Thus, both the reaction of (dichloromethyl)lithium with

(R,R)-2,3-butanediol butylboronate (**65**) and the reaction of butyllithium with (R,R)-2,3-butanediol (dichloromethyl)boronate (**66**) must yield the same borate intermediate **67**. Rearrangement of **67** in the presence of zinc chloride yields (R,R)-2,3-butanediol (1S)-1-chloropentylboronate (**68**) in 90-92% de, regardless of the source of **67** [32]. Several other simple alkyl and aryl groups yielded similar diastereoselection.

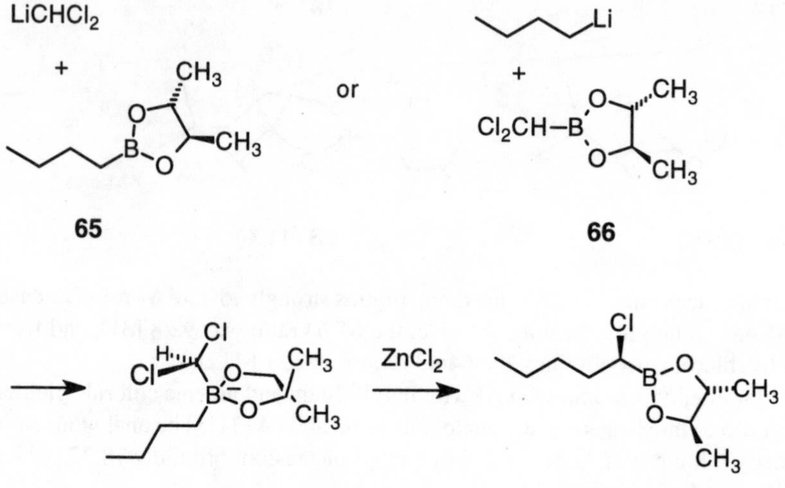

65   **66**

67   **68** (90-92% de)

Although historically important for demonstrating the benefits of C_2 symmetry, (R,R)-2,3-butanediol suffers from several deficiencies and has had little further development as a chiral director. It could be, but has not been, produced industrially on a large scale, and the cost is high. The (S,S)-enantiomer is accessible only via a multistep synthesis from tartaric acid. It is easier to make a much more effective chiral director, (R)- or (S)-(3R*,4R*)-2,5-dimethyl-3,4-hexanediol, from this source, and easier still to make (R)- or (S)-(3R*,4R*)-1,2-dicyclohexyl-1,2-ethanediol from *trans*-stilbene via Sharpless' chemistry.

(R,R)-2,3-Butanediol did prove useful for chain extension of a *tert*-butylboronic ester, the pinanediol ester having proved to be too hindered to react with (dichloromethyl)lithium. (R,R)-2,3-Butanediol ester **71** was converted via the α-chloro boronic ester to the (1,2,2-trimethylpropyl)boronic ester **72** (94% de) [33].

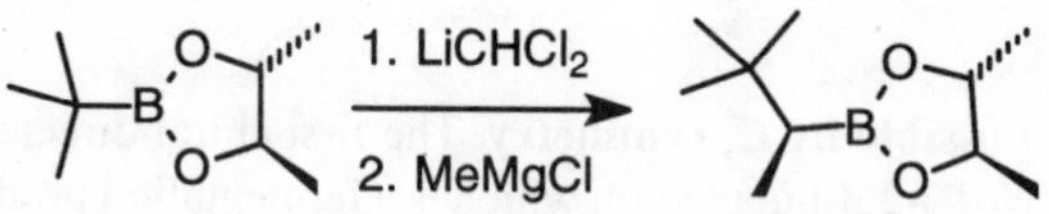

71   **72**

5.3.1.2 Nomenclature of cyclic boronic esters. Representative *Chemical Abstracts* names for (*R,R*)-2,3-butanediol boronic esters include [4*R*-(4α,5β)]-2-butyl-4,5-dimethyl-1,3,2-dioxaborolane) (**65**); [4*R*-(4α,5β)]-2-(dichloromethyl)-4,5-dimethyl-1,3,2-dioxaborolane) (**66**); and {4*R*-[2(*S**),4α,5β]}-(1-chloropentyl)-4,5-dimethyl-1,3,2-dioxaborolane (**68**).

5.3.2 Diisopropylethanediol (DIPED) Esters

5.3.2.1 Introduction. (*S*)-(*R**,*R**)-Diisopropylethanediol (**70**), systematically named (*S*)-(3*R**,4*R**)-2,5-dimethyl-3,4-hexanediol, will be abbreviated as "(*S*)-DIPED". This chiral director provides excellent stereoselection, with isomer ratios having reached more than 1000:1 after chain extension and substitution in a carefully studied system [34]. DIPED has been superseded by (*R**,*R**)-1,2-dicyclohexylethanediol ("DICHED") for most purposes because the latter is easier to make. However, the chemical and physical properties of DIPED as a chiral director have not been surpassed, and the fundamental chemistry of this class of chiral directors was established with DIPED.

5.3.2.2 Synthesis of DIPED. The first synthesis of (*S*)-DIPED (**70**) was accomplished from (*S*)-pinanediol isopropylboronate (**69**) via two chain extensions with (dichloromethyl)lithium [35]. This route to DIPED has become obsolete, but it does illustrate the applicability of the general method to the rapid laboratory solution of a chiral synthesis problem.

69

70

The initial testing of (*S*)-DIPED boronic esters indicated that DIPED is a highly effective chiral director [35], and a better route for large scale synthesis was sought. Tartaric acid can be converted via **73** to enantiomerically pure (*S*)-DIPED (**70**) via the sequence outlined [36]. The only nonroutine step was the acetylation of the bis(tertiary) alcohol **74** to diacetate **75**, for which trimethylsilyl chloride proved to be an effective catalyst. Gas phase ester pyrolysis to the diene **76** was efficient in a narrow range of temperature/flow rate combinations. The sequence outlined allows the synthesis of half-mole batches of DIPED in a few days of work.

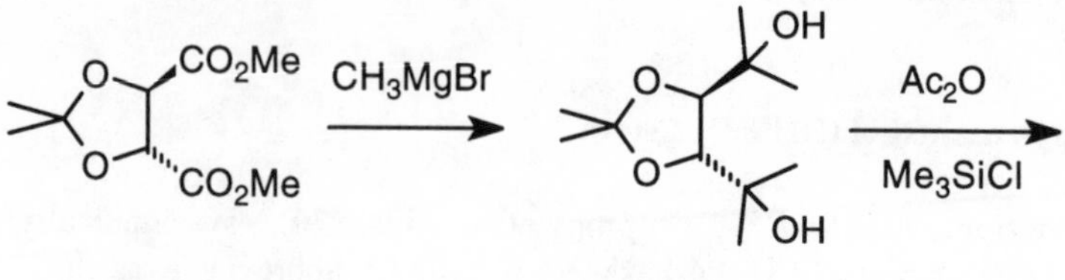

5.3.3 Dicyclohexylethanediol (DICHED) Esters

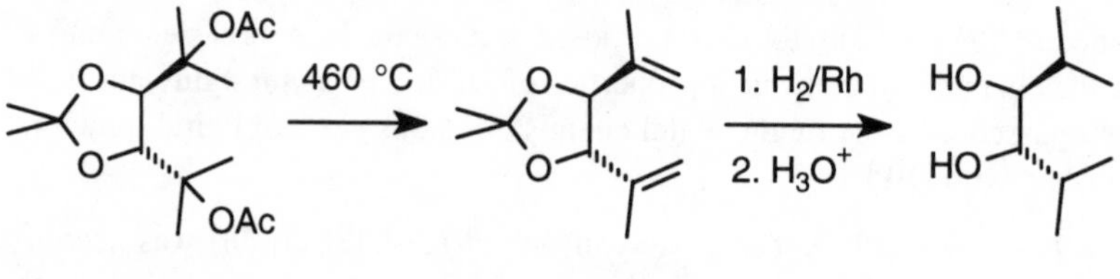

5.3.3.1 Introduction. These were first used by Hoffmann and coworkers [37, 38]. The chiral directing properties of DICHED are similar to those of DIPED, and recent developments in asymmetric synthesis of diols by Sharpless and coworkers have made it easier to make DICHED [39, 40, 41]. The only apparent disadvantage of DICHED is its higher molecular weight, which may require higher dilutions for reaction mixtures. Because DICHED became easily available only after the fundamental chemistry of the asymmetric synthesis had been explored, specific examples of the use of DICHED in reactions are deferred to Section 5.4 on applications.

5.3.3.2 Synthesis of DICHED. Dihydroxylation of *trans*-stilbene in the presence of a catalytic amount of a quinine or quinidine derivative yields (*R*)- or (*S*)-(*R**,*R**)-1,2-diphenylethanediol [39, 40, 41]. This is one of those relatively rare organic compounds that forms separate (*R*)- and (*S*)-isomer crystals, not racemic crystals, and is thus resolvable by Pasteur's first method. More importantly, the lack of a racemic crystalline phase makes purification to very high ee particularly straightforward. Unfortunately, (*R**,*R**)-1,2-diphenylethanediol as chiral director for boronic ester synthesis produces poor ee's, and the rings must be hydrogenated over rhodium on alumina [38]. The best conditions appear to involve methanol as a solvent and a small amount of water and acetic acid [42]. The synthesis of (*S*)-DICHED (**77**) is illustrated.

77

(DHQ)₂PHAL = Phthalazine derivative of quinine [41]

5.3.4 Ultrahigh Stereoselectivity with C_2 Symmetry

5.3.4.1 Introduction. The two-step sequence of adding (dichloromethyl)lithium to a DIPED boronic ester to form an (α-haloalkyl)boronic ester followed by nucleophilic replacement of the remaining chlorine atom has been shown to lead to diastereomer ratios >1000:1, with enantiomeric purity essentially absolute because of the origin of the DIPED from a natural chiron. This ultrahigh stereocontrol was discovered and proved in several stages.

5.3.4.2 Background. The only carefully documented attempt to estimate the stereoselection in the first stage of the two-step sequence was made during the first study of DIPED [35]. It was not generally possible to distinguish diastereomeric DIPED (α-chloroalkyl)boronic esters by 200-MHz ¹H-NMR, but pinanediol esters nearly always show one well separated pair of doublets [12]. Therefore, after conversion of (S)-DIPED (dichloromethyl)boronate and Grignard or lithium reagents to (S)-DIPED (αR)-(α-chloroalkyl)boronates, the (αS)-isomer content was estimated by transesterification with (S)-pinanediol and ¹H-NMR analysis. The apparent (αS)-isomer content for six widely different alkyl groups (methyl, n-butyl, isopropyl, benzyl, vinyl, phenyl) was an anomalously uniform 3-4% (de 92-94%).

The flaw in the analytical method was that the enantiomeric purity of the pinanediol was perhaps 98-99%. If the pinanediol contained 1% enantiomer, the DIPED esters made from it would contain the same. Transesterification with similarly impure pinanediol would randomly introduce a second 1% of diastereomeric impurity, for an apparent maximum 96% de. Thus, the measured 94% de might reflect a true diastereomeric ratio of 50:1, 100:1, or more. This general magnitude appears plausible based on subsequent nonsystematic inspection of spectra, but subsequent findings also make the exact purity of the (α-chloroalkyl)boronic esters irrelevant.

The first evidence that C_2 symmetrical chiral directors could lead to very high enantiomeric excesses in products was reported by Hoffmann and coworkers, who obtained a product in >99% ee from a 1,2-dicyclohexyl-1,2-ethanediol boronic ester [43]. Although this ee seemed anomalously high, the reason for it was not recognized.

5.3.4.3 Discovery of enhanced selectivity. The sequential double diastereoselection arising from C_2 symmetry was recognized only when a synthetic scheme involving a change of chiral directors failed. Treatment of (R,R)-2,3-butanediol (S)-(1-chlorobutyl)boronate (**78**) with diethanolamine yielded crystalline chelate **79**, which with aqueous acid and (S)-DIPED formed (S)-DIPED (S)-(1-chlorobutyl)-boronate (**80**), which is not the diastereomer that would be obtained from the usual reaction of (S)-DIPED propylboronate with (dichloromethyl)lithium [34]. Treatment of **80** with methylmagnesium bromide yielded only an estimated 6% of the expected methyl migration product **81**. The remainder underwent oxygen migration/ ring expansion to form the air-sensitive cyclic borinic ester **82**, which could be isolated by distillation. Normal work up with exposure to air yielded butyraldehyde and (S)-DIPED methylboronate (**83**).

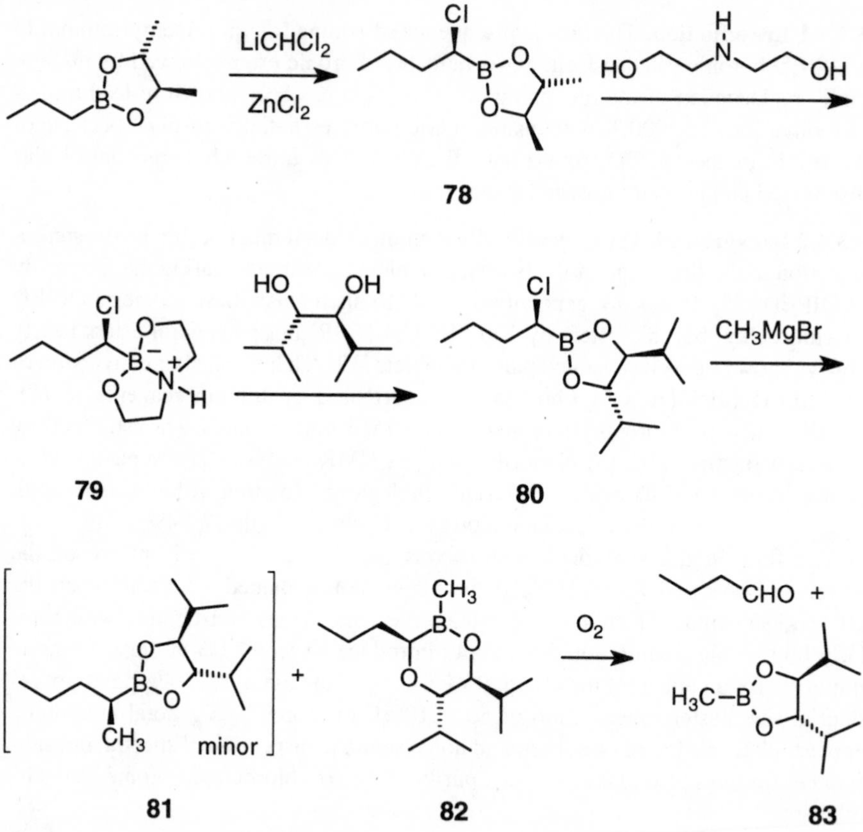

5.3.4.4 Kinetic resolution. This result suggested the possibility of kinetic resolution. Such resolution was demonstrated by the conversion of racemic ethylene glycol (1-chloro-2-phenylethyl)boronate (**84**) to the diastereomeric mixture of (S)-DIPED esters **85** and **86**, which was converted by methylmagnesium bromide to a mixture consisting mainly of (S)-DIPED (R)-(1-methyl-2-phenylethyl)boronate

(87) derived from (R)-α-chloro boronic ester 85 and the dioxaborin 89 derived from ring expansion of (S)-α-chloro boronic ester 86 [34]. Work up in air converted 88 to a lower boiling fraction consisting of phenylacetaldehyde and methylboronic ester 83, leaving a higher boiling fraction of boronic ester 87, which was oxidized to (–)-(R)-1-phenyl-2-propanol (88), ee 95% (after correction for the presence of 6% benzyl alcohol measured by ^{1}H-NMR, derived from the benzylboronic ester precursor to 84).

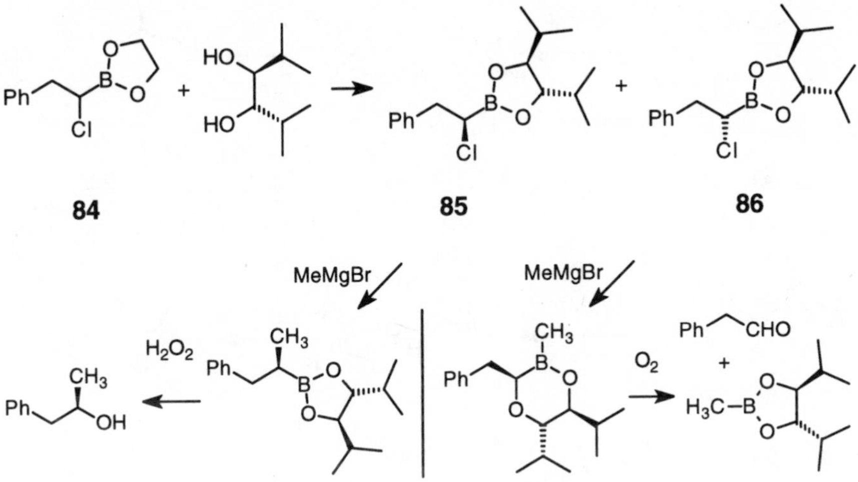

The minor diastereomers formed from (S)-DIPED alkylboronates and (dichloromethyl)lithium are (S)-DIPED (S)-(1-chloroalkyl)boronates, exemplified by 80 and 86. Thus, the diastereomer that is only present in a very small amount to begin with is virtually eradicated by conversion to unrelated products when the (1-chloroalkyl)boronic ester reacts with a Grignard reagent.

5.3.4.5 Mechanistic interpretation. In view of the thermodynamic preference for carbon migration over oxygen migration (Section 3.4.3.3), the foregoing results were unexpected. What steric factors might apply? Based on the observed stereochemical outcome and the need to invert the carbon at which S_N2 displacement occurs, the reactive conformation for borate complex 90 must be that illustrated. After inversion to form 91, the remaining chlorine is placed in the position vulnerable to displacement by the migrating B-methyl group of the derived borate 92. In contrast, B-methylation of the minor diastereomer 93 derived from 90 produces a borate complex 94 in which the favored conformer illustrated is oriented for oxygen migration and ring expansion to 96, assuming that the favored conformation for displacement depends on the relative positions of the larger group (Cl or R^1) versus hydrogen.

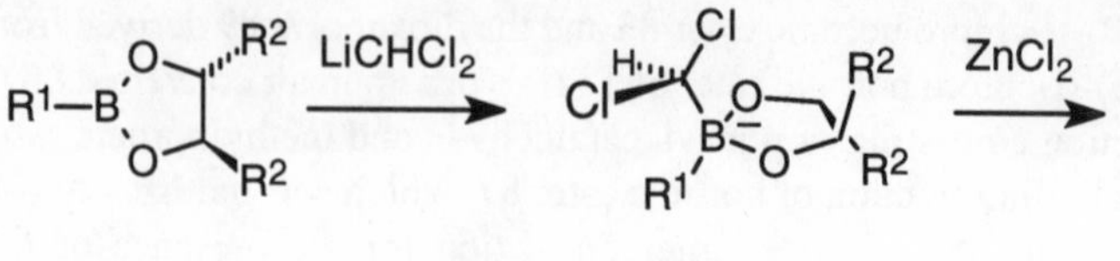

90

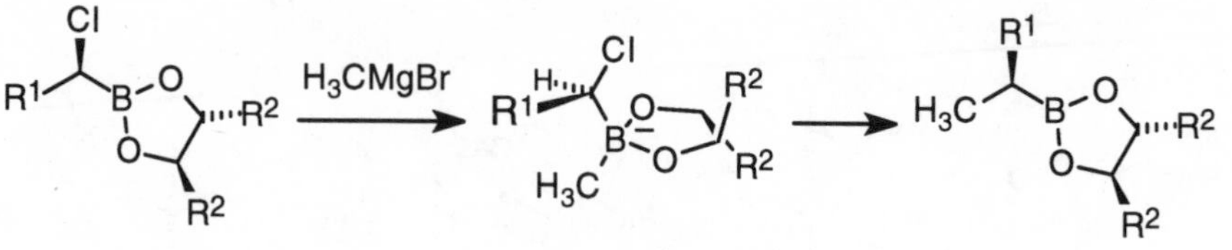

91   **92**   **93**

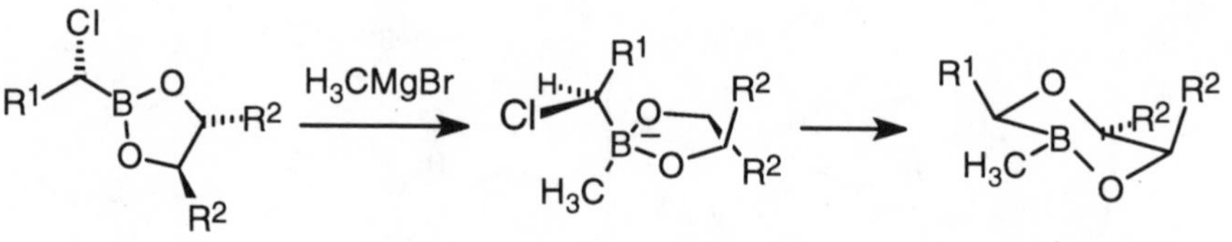

94   **95**   **96**

The reason for the preference for conformer **90** over the rotamer which exposes the other chlorine to displacement is not known. However, it seems likely that a metal cation is involved in all of the rearrangements of this type. Combining **90** with zinc chloride might lead to intermediate **97**, in which the zinc coordinates both with the departing chloride and one of the ring oxygens. Modeling with MMX calculations suggests that structure **97** is more stable than the alternative **98** by about 1 kcal/mol.

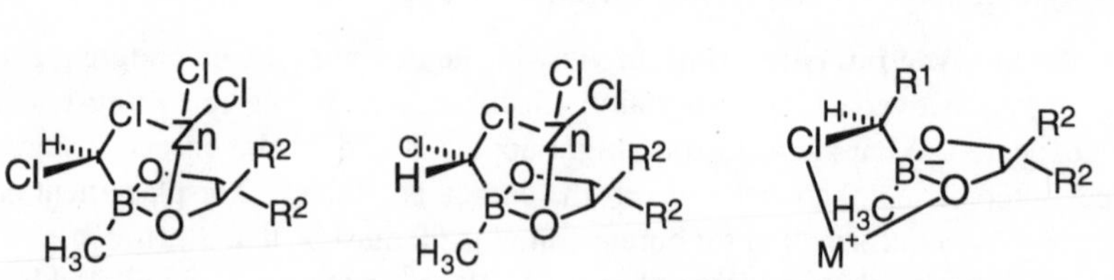

97   **98**   **99**

Coordination of **95** with a metal cation leads to **99**, which (using $R^1 = Cl$, $M^+ = ZnCl_2$) is calculated to be several kcal/mole more stable than **97**, and would allow oxygen migration and ring expansion to form **96**. This calculation provides a rationale for allowing the thermodynamically disfavored oxygen migration to gain sterically directed kinetic preference. On the other hand, these simple model calculations do not account for all of the subtleties of the system. Analogues of **92** having an alkoxy substituent in place of the *B*-methyl group undergo migration of the substituent alkoxy group in preference to ring expansion.

5.3.5 Convergent Connection of Two Asymmetric Centers

α-Halo boronic esters provide an asymmetric electrophilic carbon atom, which offers the possibility of connection to an asymmetric nucleophilic carbon atom to join two asymmetric molecular fragments in a convergent synthesis. α-Lithio ethers had been previously known to be enantiomerically stable [44], and had been synthesized asymmetrically via acylstannane reduction [45, 46], though the binaphthol chiral director required would not be cheap for scaling up. Accordingly, a major challenge in this work was to find an efficient route to enantiomerically pure α-lithio ethers via boronic ester chemistry, and this finding may be the most important successful outcome of this work.

(*S*)-DIPED α-chloro boronic esters (**100**) with (tributylstannyl)lithium efficiently yielded α-(tributylstannyl)boronic esters (**101**) [47]. These were easily deboronated by alkaline hydrogen peroxide to the acid sensitive α-tributylstannyl alcohols (**102**). If pinanediol esters were used in place of DIPED esters, this deboronation step failed. The labile tin substituted alcohols (**102**) were promptly converted to stable methoxymethyl ethers (**103**) or benzyloxymethyl ethers by the previously established method [44]. Treatment with butyllithium at low temperatures yielded the α-lithio ethers (**104**), which presumably owe their configurational stability to internal coordination [44]. Evidence that the formation of **104** and tetrabutyltin from **103** and butyllithium is reversible was found later in the course of the boronic ester work.

100   **101**   **102**

103   **104**

The reaction of (S)-DIPED (R)-(1-chloropentyl)boronate (**105**) with the lithio ether **106** proceeded easily to form the coupled boronic ester **107**, which was further transformed by conventional deboronation and methoxymethyl ether cleavage to (S,S)-5,6-decanediol (**108**) in 96% de [47].

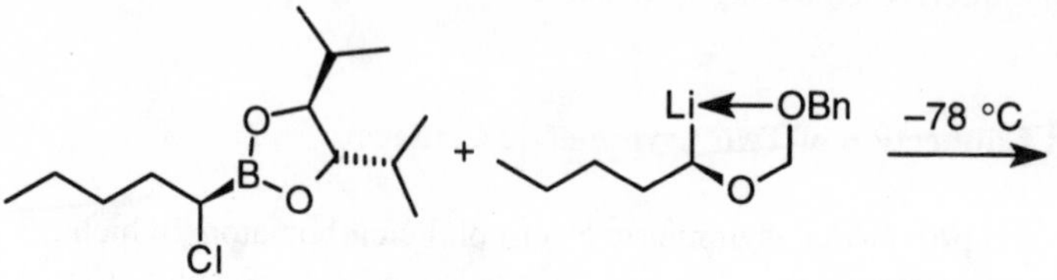

105   **106**

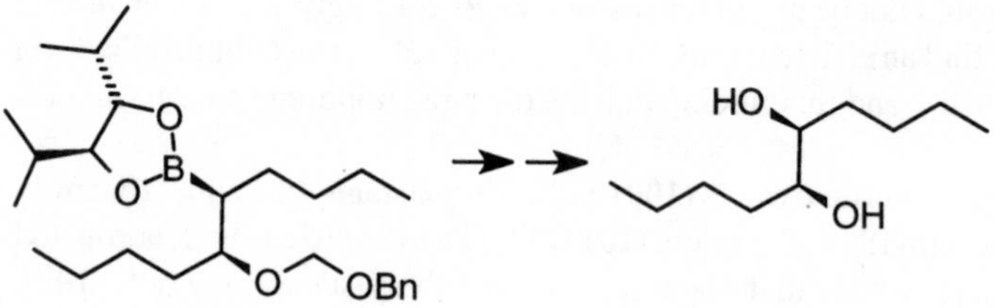

107   **108**

The possibility of using a branched R group in the couplings of **100** with **101** was tested with R = isopropyl on both sides, chloro boronic ester **109** and lithio ether **110**. Initially, the reaction was carried out at −78 °C, but the yield of coupling product **113** was only ~20%. It appeared that the major pathway was reaction of **110** with tetrabutyltin, the reverse of the formation of **110**, to form an equilibrium level of precursor tin compound **111** and butyllithium, which then reacted rapidly with chloro boronic ester **109** to form butyl substitution product **112**. When the reaction temperature was lowered to −100 °C, the rate of reversal of formation of **110** was evidently reduced more than the rate of reaction of **110** with **109**, and the major product (~70%) became the expected coupling product **113**, 98% de. For proof of structure, **113** was converted to (S)-DIPED (**70**).

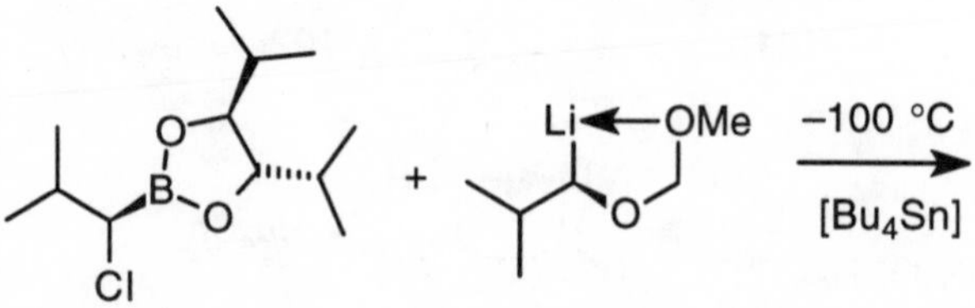

109   **110**

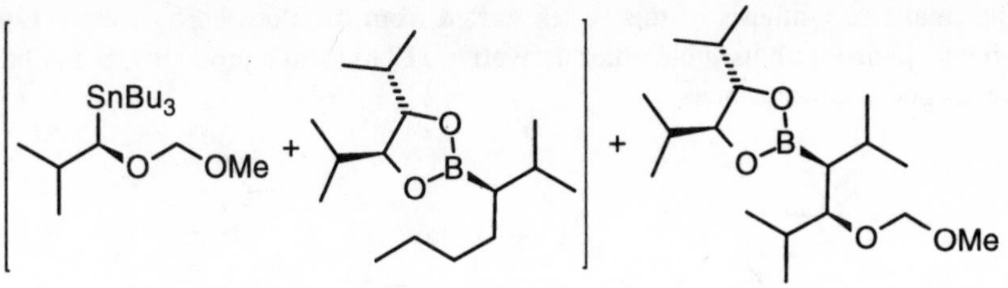

111   (byproducts) **112**   **113**

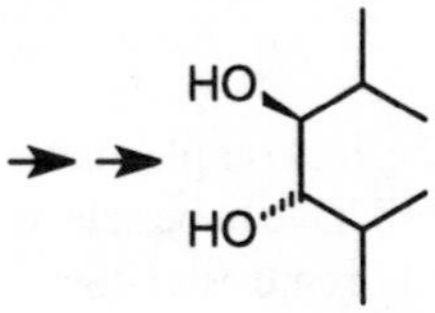

70

The only reason for using (S)-DIPED as the chiral director for its own synthesis was to test the stereoselectivity and feasibility of the synthetic method in an easy way. The 90% recovery of the original (S)-DIPED together with the newly synthesized material did bring the total output to 109% of the starting amount. This synthesis serves as a reminder that a chiral director can be used to increase the supply of itself in an exponential fashion via a repeated cycle, though there are more efficient processes than this one that could be chosen for such a purpose.

To prove that the coupling was independent of the relative configurations of the two chiral centers being coupled, the reaction of **114** (the enantiomer of **110**) was shown to react with **109** to produce **115**, the meso diastereomer of DIPED (**70**).

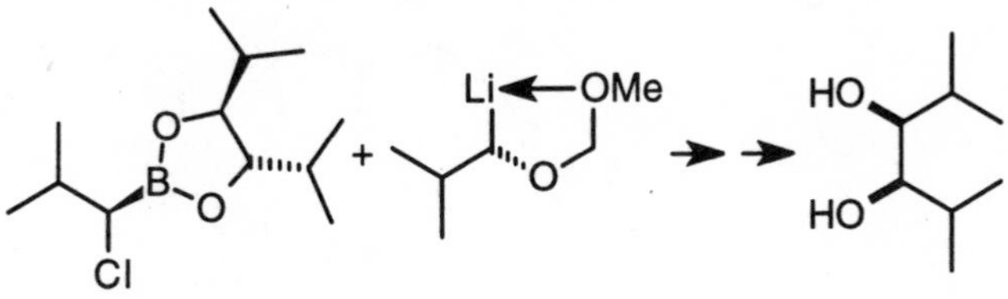

109   **114**   **115**

The stereospecificity and controllability of the coupling of **110** or **114** with α-chloro boronic ester **109** may be contrasted with the reported reaction of **110/114** with propionaldehyde, which resulted in a 77:23 diastereomer mixture [48].

The final test synthesis of this series started from α-chloro boronic ester **116**, which was joined with its lithio ether derivative **117** to form a product **118** that has four contiguous chiral centers [47].

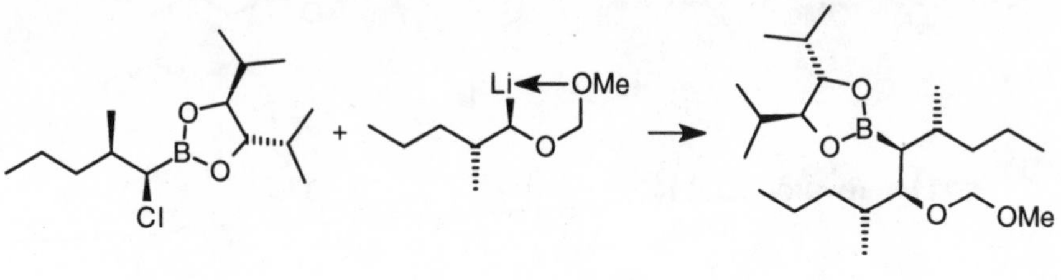

116   **117**   **118**

It had been hoped that this chemistry might be used to synthesize complex natural products, but a different target from the one that was chosen will have to be selected. Attempted coupling of α-lithio ether **110** to a model α-chloro boronic ester containing a β-alkoxy substituent failed, evidently because the alkoxy group is too deactivating for the displacement process to proceed [49].

5.3.6 Reaction of Enolates

The reaction of (α-bromoalkyl)boronic esters with the lithium enolate of *tert*-butyl propionate provides threo α-methyl-β-boryl carboxylic esters in high diastereomeric purity [50], but not in useful enantiomeric purity if extraordinary precautions are not taken [51]. An example is the conversion of **119** to **120**. Peroxidic deboronation yields the β-hydroxy ester **121**.

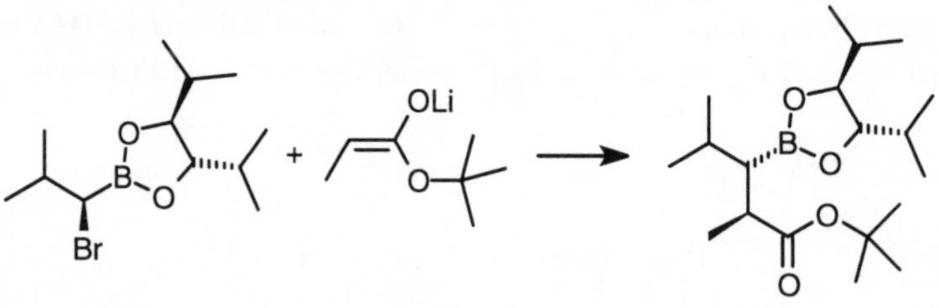

119   **120**

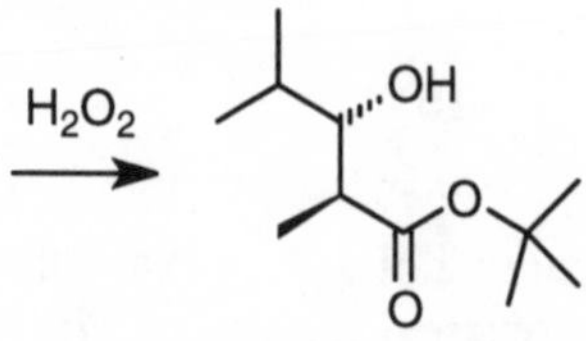

121

If chirality is controlled by the enolate instead of the boronic ester, as in the reaction of racemic pinacol 1-bromoethylboronate (**123/124**, X = Br) with the chiral enolate **122**, it is possible in favorable cases to consume all of the racemic boronic ester via a "retroracemization" sequence in the presence of iodide ion so that a single product **125** results in high diastereomeric and enantiomeric purity [51].

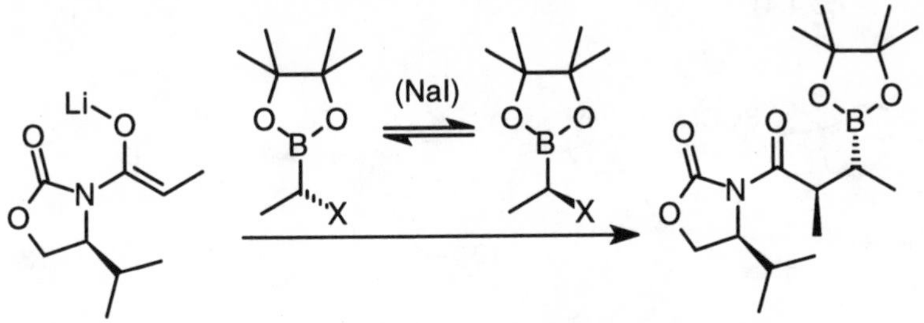

122   **123**   **124**   **125**

5.3.7 Allylic Boronic Esters

5.3.7.1 (α-Haloallyl)boronic esters. Allylic boronic esters have useful properties but have also caused some special problems. These do not fit neatly into boronic ester types, and some of the work done on pinanediol esters is described in Section 5.2.3.6. Other types of esters, including achiral pinacol esters, are included here.

The most serious difficulties have been encountered with simple (α-chloroallyl)boronic esters. For example, pinacol (α-chloroallyl)boronate (**126**) failed to yield substitution products with alkoxides, and although use of the bromine analogue **127** proved successful [52], allylic bromo boronic esters epimerize very readily and have not proved useful in asymmetric synthesis.

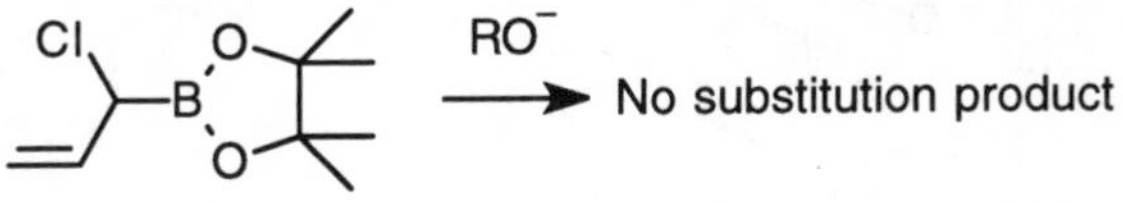

126

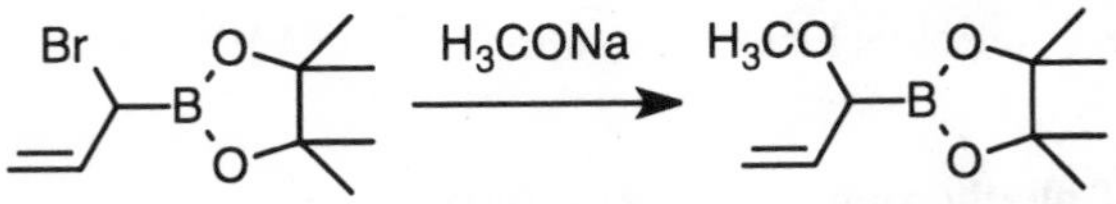

127

In contrast to the simple (α-chloroallyl)boronic esters, longer chain allylic boronic esters such as pinacol (α-chlorocrotyl)boronate (**129**) do undergo straightforward substitution by alkoxides [53, 54], but when an attempt was made to utilize this reaction in asymmetric synthesis, the chain extension reaction proceeded with substantial epimerization [55]. Instead, a roundabout route to **129** via **128** was found, in which the asymmetric synthesis of **128** was accomplished by Midland's method, reduction of an acetylenic ketone with a chiral pinylborane [55]. (α-Methoxycrotyl)boronic esters such as **130** are particularly useful in the reaction with aldehydes (Section 7.2.4.4).

128   **129**   **130**

5.3.7.2 Asymmetric synthesis of an allylic boronic ester.

The epimerization problem described in Section 5.3.7.1 was encountered during attempted synthesis of an (α-methylcrotyl)boronic ester from the corresponding (α-chlorocrotyl)boronic ester. The difficulty was avoided by the simple expedient of making DICHED (1-chloroethyl)boronate (**132**) by treatment of DICHED (dichloromethyl)boronate (**131**) with methyllithium followed by zinc chloride, then reacting **132** with (Z)-1-propenyllithium to generate **133**, which proved to be >99% diastereomerically pure [43]. Allylic boronic ester **133** then served as a key intermediate in the synthesis of the aglycon of mycinolide V [38]. This work was done before the sequential double diastereoselection [34] described above in the Section 5.3.4 had been discovered, but the excellent results are in accord with those more recent findings.

131   **132**   **133**

A simpler though less stereoselective kinetic resolution route to **133** is described in Section 7.2.2.2.

5.3.7.3 Chain extension of allylboronic esters.

A different kind of problem has been encountered in the attempted chain extension of a simple allyl group. Allylmagnesium chloride added to a stoichiometric amount of (R,R)-2,3-butanediol (dichloromethyl)boronate (**134**) yielded not only the expected (1-chloro-3-butenyl)-boronic ester (**136**) but also substantial amounts of byproduct (**137**) that must arise by transfer of allyl groups from one boron atom to another, presumably by reaction of the borate intermediate **135** with the initial product **136** [28]. The ratio of **136** to **137** produced varied between 3:1 and 1:3 in different batches, with no clue regarding what the significant variable might be.

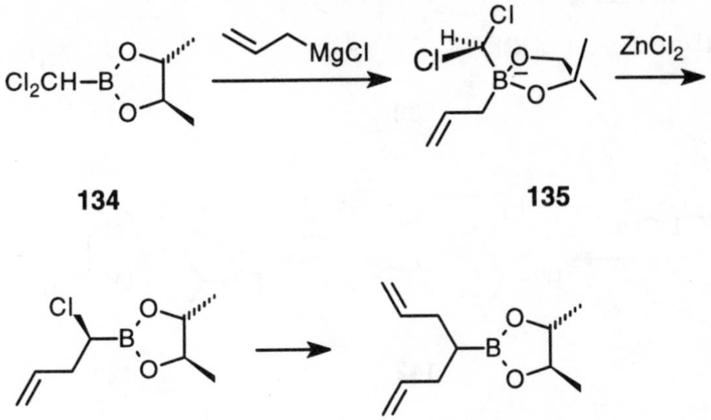

Recent work with DIPED esters has encountered the same phenomenon, though the amount of allyl transfer appeared to be smaller. It does not matter whether the allyl group is attached to boron before or after the dichloromethyl group [56]. As noted in Section 5.2.3.6, chain extension of pinanediol allylboronate proceeded without difficulty.

5.4 Synthetic Applications

5.4.1 A Useful Mnemonic for Chiral Direction

For planning syntheses, it is useful to have a mnemonic for remembering the correlation between the chiral director and the products. In halo boronic ester chemistry, "products point parallel" to the chiral directing groups. The context of that mnemonic requires that the structures be written in zigzag line style. Then introduced substituents point to the front on the side of the molecule where the chiral director bond points to the front, and to the back on the side where the chiral director points back. The opposite rule, "intermediate inverted," applies to the C-X bond of an α-halo intermediate, which undergo inversion on substitution. Structures **138**, **139**, and **140**, in which R^1 groups are the chiral directors and Y and Z represent substituents introduced by nucleophilic substitution of α-chlorine along a chain being grown from R^0, illustrate this point. Drawings **141**, **142**, and **143** represent the identical set of structures turned over.

138   **139**   **140**

141   **142**   **143**

Though pinanediol does not have the simple C_2 symmetry of the group depicted, it does contain a methyl substituent and a bond in the 6-membered ring that correspond to the mnemonic, that presumably determine the chiral direction, and that are always marked accordingly in this book with a wedge bond and a hashed bond, respectively.

Zigzag structure depictions will be used throughout except for the description of carbohydrate syntheses, where Fischer projections are more familiar. Fischer projections turn everything backwards. They are pointed opposite the direction used for defining (R) or (S) configuration, and in boronic ester chemistry they point all substituents in front on the side opposite the front group of the chiral director.

5.4.2 Amido Boronic Esters

This group of unnatural products includes compounds that act as peptide mimics capable of inhibiting certain enzymes. It also illustrates some of the functional group compatibilities of the boronic ester synthesis.

The first α-amido boronic ester studied was (S)-N-acetylboraphenylalanine (**147**), which is a good inhibitor of chymotrypsin, dissociation constant 2.1×10^{-6} M at pH 7.5, 25 °C [57]. (S)-Pinanediol (S)-(1-chloro-2-phenylethyl)boronate (**144**) is was prepared in high yield and diastereomeric purity from the reaction of (S)-pinanediol benzylboronate with (dichloromethyl)lithium [22]. Displacement of chloride ion from **144** was accomplished with an unusual nucleophile, lithiohexamethyldisilazane, to provide **145**. Treatment with acetic acid and acetic anhydride yielded the acetamido boronic ester **146**, which was cleaved with boron trichloride to yield

(S)-N-acetylboraphenylalanine (**147**). The coordination of boron to oxygen was indicated by X-ray crystallography of a subsequently prepared amido boronic ester [23].

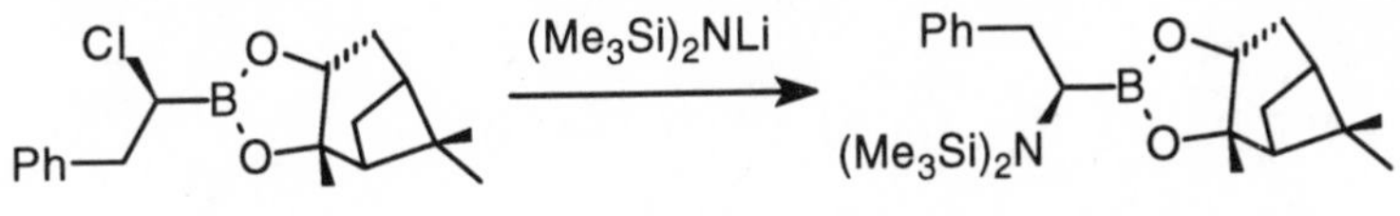

144   **145**

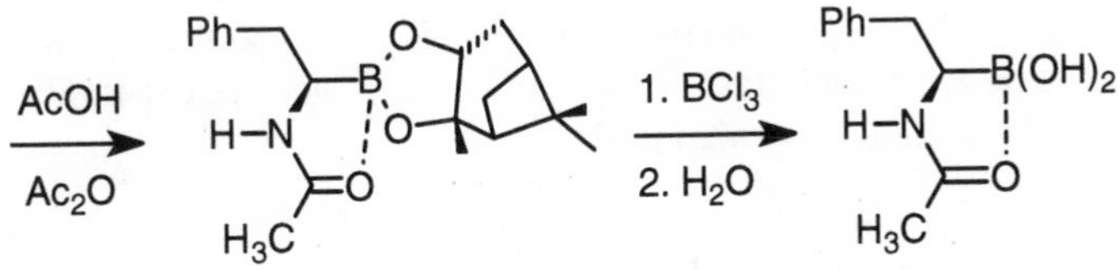

146   **147**

Once the route to **147** had been discovered and its expected enzyme inhibiting activity confirmed, there was immediate interest in synthesizing and testing a variety of other amido boronic acids or esters. One side chain of particular interest was 3-halopropyl, since it might serve as a source of arginine or proline analogues. The illustrated synthesis of (S)-pinanediol (R)-(1-acetamido-4-bromobutyl)boronate (**151**) began with hydroboration of allyl bromide by catecholborane to form **148**, transesterification with (S)-pinanediol to **149**, conversion to the silylated amino boronic ester **150**, and treatment with acetic acid and acetic anhydride to form **151** [20].

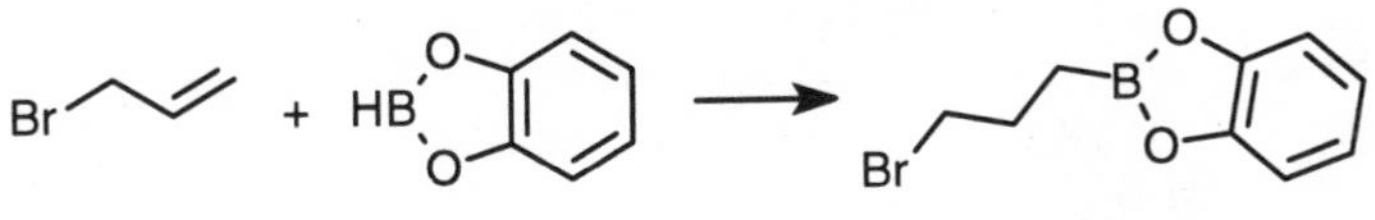

148

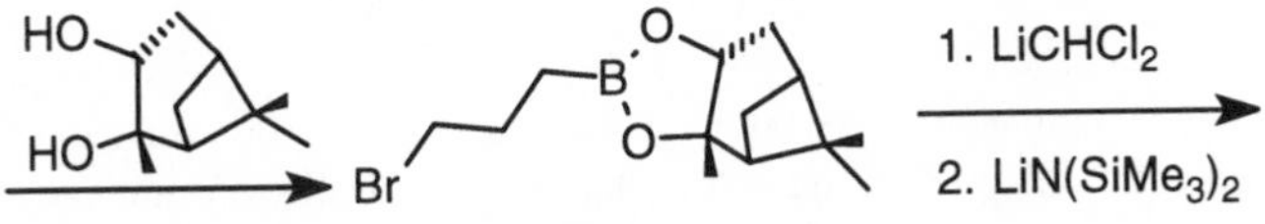

149

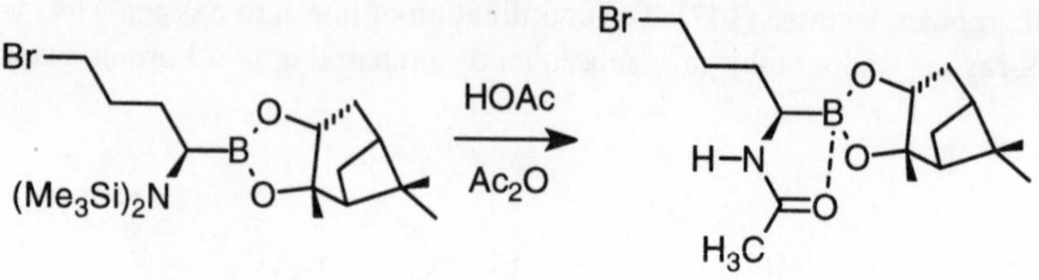

150   **151**

Several other pinanediol amido boronic esters summarized by structures **152**, **153**, and **154** were synthesized by using similar chemistry [20]. Enzyme inhibition studies were carried out in borate buffer, which equilibrate with pinanediol and release the free amido boronic acid in solution. These studies showed the D-amino acid analogues **153** to be active inhibitors of *Bacillus cereus* β-lactamase, with $K_i =$ 44 μM at pH 7 for **153** having R = isobutyl [58].

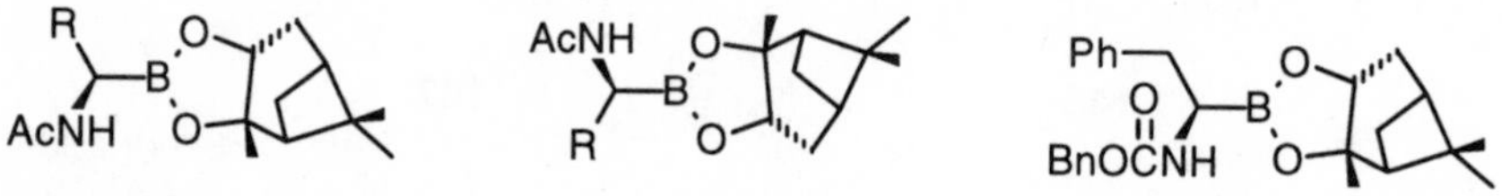

152   **153**   **154**

(R = Me, Me₂CH, Me₂CHCH₂) (R = Me₂CH, Me₂CHCH₂)

Several racemic α-acetamido boronic acids (**156**) were synthesized by similar chemistry, starting from the corresponding *meso*-butanediol esters (**155**), and were found to inhibit elastase and chymotrypsin [59]. An interesting finding was that stable crystalline difluoro derivatives **157** could be prepared easily by treatment of **156** with aqueous hydrofluoric acid.

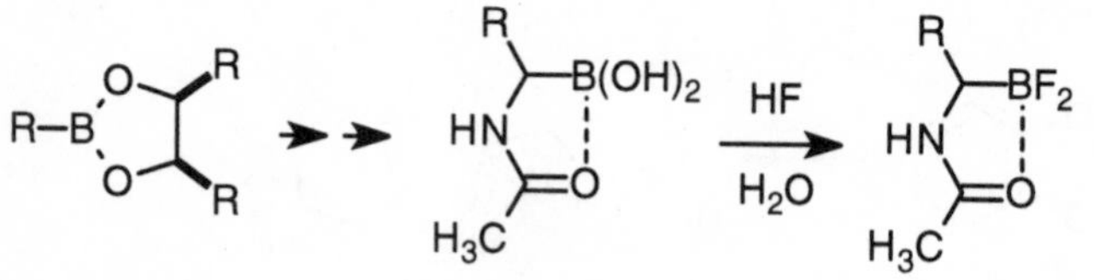

155   **156**   **157**

R = (*S*)-EtMeCH, other alkyl, Ph

Another amido boronic acid synthesis which revealed more of the functional group compatibilities of the chloro boronic ester chemistry, and allowed an X-ray structure determination as well, was that of the *N*-acetyl-L-methionine analogue **161** [23]. Reaction of (*S*)-pinanediol (*S*)-1-chloroallyl)boronate (**42**) (see Section

5.2.3.6) with lithiohexamethyldisilazane to form **158** and desilylation/acetylation to (*S*)-pinanediol (*R*)-(1-acetamidoallyl)boronate (**159**) were routine. Radical addition of methyl mercaptan to the double bond of **159** yielded crystalline boronic ester **160**, which proved unsuitable for X-ray analysis. Treatment of **160** with boron trichloride led to the *N*-acetyl-L-methionine analogue **161**, which was not isolated but esterified with ethylene glycol to ester **162**, which did form crystals suitable for X-ray crystallography. An unexpected feature of the crystal structure was that the boron was linked to the amide oxygen in a ring, B—O distance 1.64 Å [23]. Some of the other bond lengths from this study have been listed in Section 1.2.2, Table 1-1.

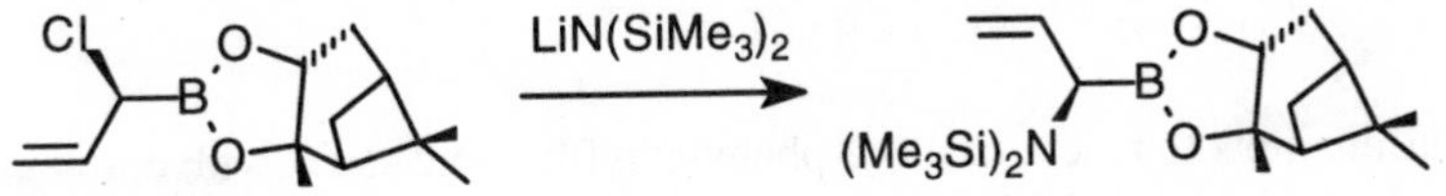

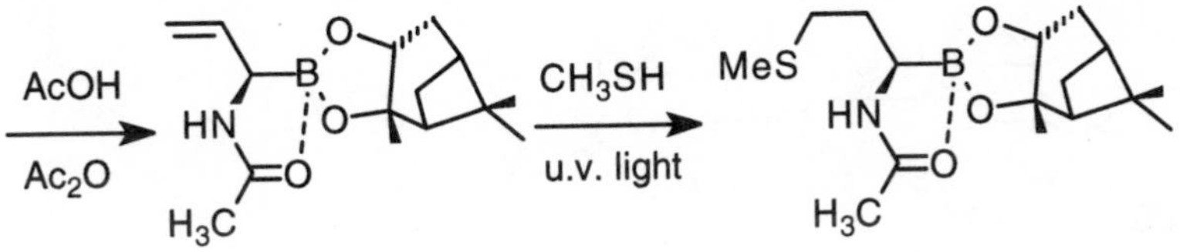

The synthesis of amido boronic esters was arrived at only after a long series of failed attempts by other routes. Replacement of the iodide of dibutyl (iodomethyl)-boronate by piperidine or dimethylamine was achieved long ago [60]. Several reactions of pinacol (iodomethyl)boronate (**163**) with tertiary amines to form quaternary ammonium salts (**164**) were found [61]. Products from **163** and piperidine or morpholine proved hard to purify, but dibutyl (iodomethyl)boronate caused no problem, for example, in the reaction with morpholine to form **165** [61]. However, dibutyl (iodomethyl)boronate failed to yield an isolable amino boronic ester product with benzylamine.

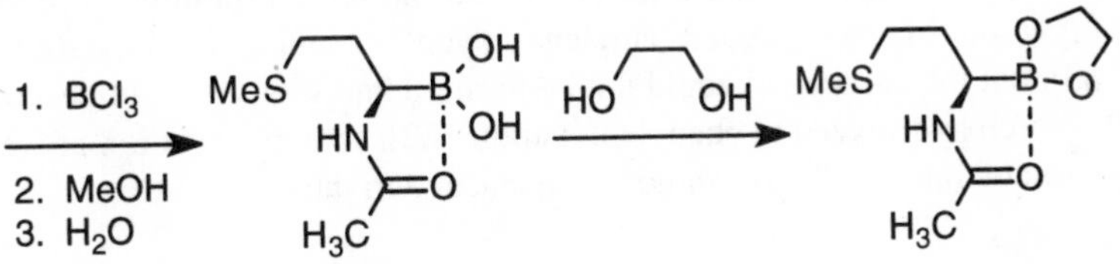

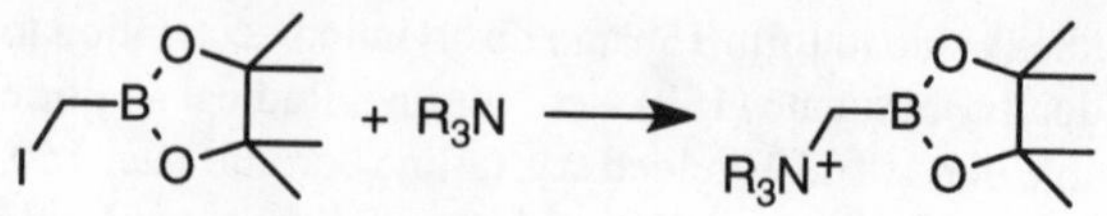

163   (R$_3$ = Et$_3$, Bu$_3$, PhMe$_2$)   **164**

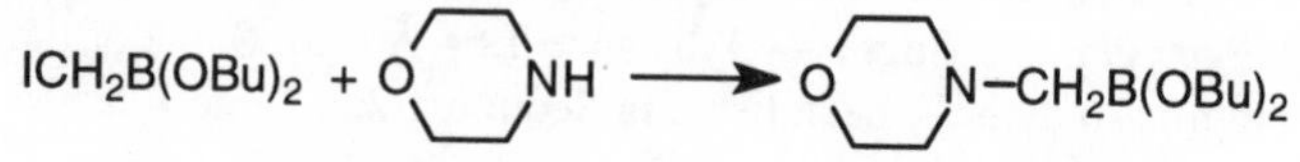

165

Numerous attempts to react (1-halo-2-phenylethyl)boronic esters with ammonia or ammonia derivatives failed. The consistent product from pinacol (1-iodo-2-phenylethyl)boronate (**166**) with ammonia after aqueous work up was 2-phenylethylamine [61].

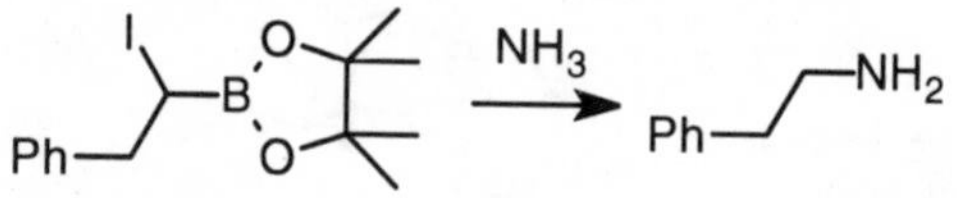

166

The problem turned out to be that α-amino boronic acids are unstable if there is a proton on the nitrogen. The boronic ester group and the proton on nitrogen trade places. Desilylation of **167** yielded ethylene glycol (1-amino-2-phenylethyl)-boronate (**168**), but the compound could not be purified and was shown to decompose to **169**, which hydrolyzed to phenylethylamine [22]. Prompt acylation proved to be the key to obtaining stable α-amino boronic acid derivatives.

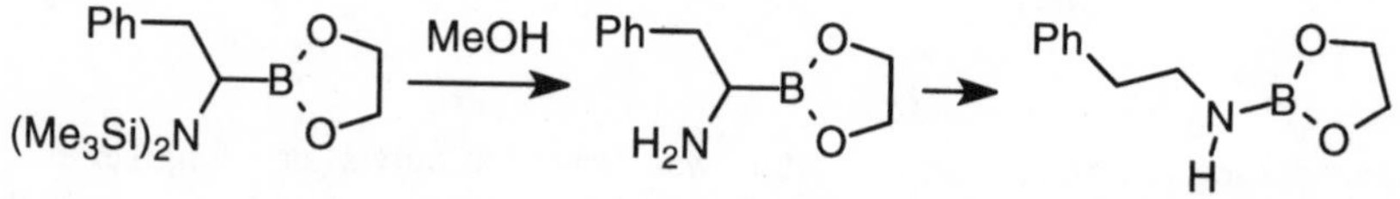

167   **168**   **169**

An alternative to prompt acylation is isolation of the free amino boronic acid or ester as a trifluoroacetate salt [62]. It was claimed that such salts can survive several days in water at pH 7 with negligible loss [62], though that may well depend on the particular structure, as other experience has been contrary.

A free α-amino boronic acid that has been tested as an enzyme inhibitor is racemic boraalanine (**171**), which was obtained in solution by hydrolysis of boronic ester **170**, which proved to be a good inhibitor of alanine racemase from *Bacillus stearothermophilus*, K_i = 20 mM. It was slow binding, k_{inact} = 0.15-0.35 min^{-1}, and the complexation was reversible. Slow binding of **171** to D-alanine:D-alanine ligase from *Salmonella typhimurium* was also observed and showed two binding constants for different enzyme sites, K_i = 35 μM and K_i' = 18 μM [63]. Solutions of **171** were found to have largely decomposed and lost activity after 24 h at 37 °C.

Cl
H3C
B
O
2
(Me3Si)2NLi
H3C
(Me3Si)2N
B
O
2
H2O
H3C
H2N
B(OH)2

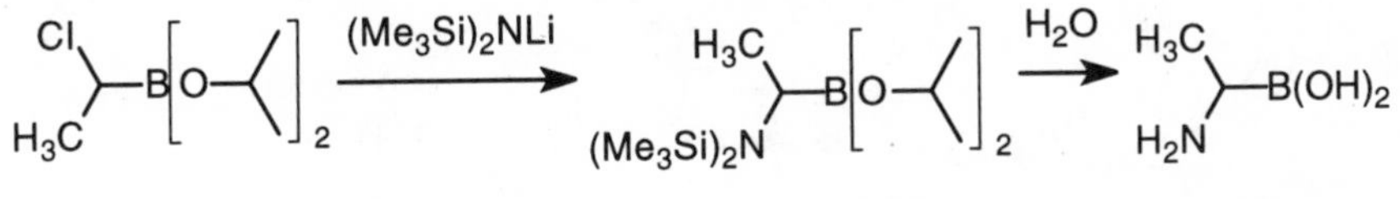

170 **171**

Reaction of sodiobenzamide with dibutyl (iodomethyl)boronate followed by hydrolysis was thought at first to yield (benzamidomethyl)boronic acid [64]. However, (benzamidomethyl)boronic acid was synthesized unambiguously from dibutyl (iodomethyl)boronate and lithiohexamethyldisilazane followed by benzoylation and shown to be different (**172**). The product from sodiobenzamide must be the imido ester isomer **173** [65]. Both **172** and **173** have proved to be good inhibitors of chymotrypsin. The K_i for **172** is 8.1 μM at pH 7.5, which is 400 times stronger binding than that by methyl hippurate. The K_i for **173** is ~2.5 μM at pH 7.5, a little stronger binding than that shown by amido boronic acid **172**.

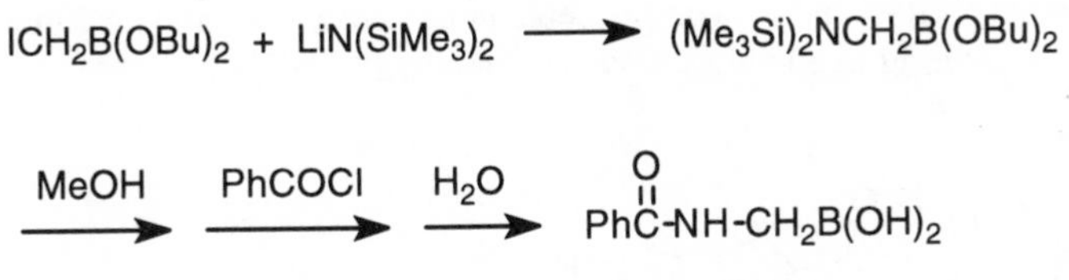

ICH2B(OBu)2 + LiN(SiMe3)2
(Me3Si)2NCH2B(OBu)2
MeOH
PhCOCl
H2O
O
PhC-NH-CH2B(OH)2

172

ICH2B(OBu)2 + PhCONHNa
H2O
H
N
B(OH)2
Ph
O

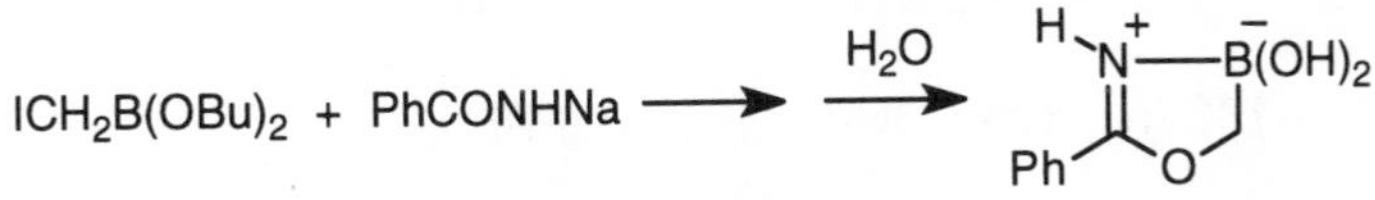

173

Treatment of **172** with methanol and water yielded (aminomethyl)boronic acid, but this compound decomposed in solution at 20-25 °C within 24 h to methylamine and boric acid, and partially decomposed at 5 °C in the solid state within a week.

Perhaps the most interesting enzyme inhibitor in this series reported to date is "du Pont 714" (**174**), a thrombin inhibitor that is active in rabbits at dose levels of 0.1 mg kg^{-1} h^{-1} [66, 67]. The synthesis proceeds via the (3-bromopropyl)boronic ester **149** and the silylated amino boronic ester **150**, described above, and elaboration to the α-amido boronic ester **174** is by straightforward transformations [68].

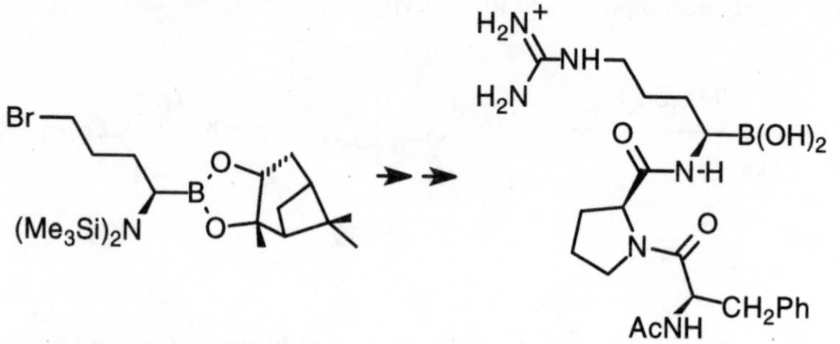

150   **174**

A methoxy group in place of guanidino on **174** also provides a potent thrombin inhibitor [1, 69].

Amido boronic acids having the general structure MeOSucc-Ala-Ala-Pro-CH(R)B(OH)$_2$ (**175**) have proved to be excellent inhibitors of leucocyte elastase and pancreatic elastase, with dissociation constants in the nanomolar range, as well as strong inhibitors of cathepsin G and chymotrypsin [70]. Such compounds have been found effective against elastase induced emphysema in hamsters [71], and the kinetics and mode of their binding to enzymes have been studied [72, 73].

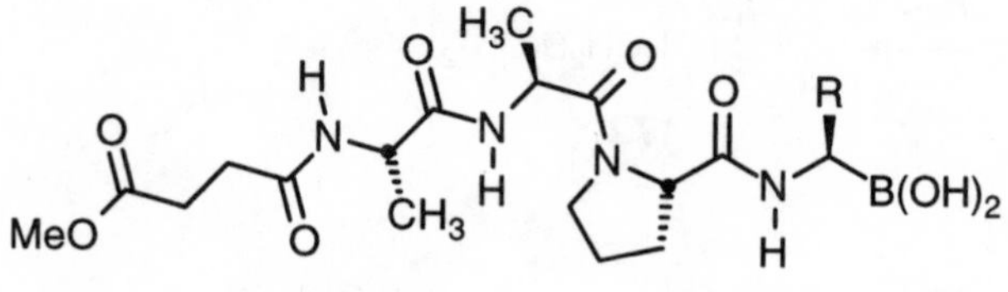

175 (R = CH$_3$, CH(CH$_3$)$_2$, CH$_2$Ph)

Several other studies of peptidyl boronic acids and their mode of bonding to protease enzymes, which is often accessible to detailed NMR analysis, have been reported [74-78]

5.4.3 Insect Pheromones

5.4.3.1 Introduction. Some of the most promising results from asymmetric synthesis via (α-haloalkyl)boronic esters have been with insect pheromones. This extremely varied class of natural products contains a number of examples of volatile oily substances having one or two asymmetric centers. In general, only one stereoisomer of a chiral substance is attractive to the particular type of insect that utilizes it. Most often the other stereoisomers are inert, but occasionally a stereoisomer has turned out to be repellent, so that impure samples of pheromone are not useful as attractants. The very high degree of stereocontrol achievable with the (α-haloalkyl)-boronic ester route is particularly useful in situations where a stereoisomer is hard to separate and has a deleterious effect on activity.

All of the pheromones described in this section were synthesized first in other ways, most often from natural chirons, and these syntheses will not be reviewed here. However, there is one case where boronic ester chemistry has solved an important problem that had been intractable previously. Previous syntheses of stegobiol and stegobinone had not yielded pure samples. The impure samples had only a small fraction of the biological activity of the natural pheromone, and impure stegobinone underwent epimerization to a repellent diastereomer in a few hours or days.

5.4.3.2 *exo*-Brevicomin. *exo*-Brevicomin (**181**) is the aggregation pheromone of the western pine beetle, *Dendroctonus brevicomis* [79, 80]. Although **181** had previously been synthesized in several ways, it was chosen as an early test of the functional group compatibility of the boronic ester chemistry [12]. The problem amounts to a 1,2-diol synthesis, that of (6R,7R)-6,7-dihydroxynonan-2-one, which is illustrated as its 6-benzyl ether **180**. The Grignard reagent **176** was converted to (R)-pinanediol boronic ester **177** by well established means. The first chain extension with (dichloromethyl)lithium followed by treatment with lithium benzyl oxide yielded **178**. The second chain extension followed by ethylmagnesium bromide led to **179**, which was deboronated to **180** and debenzylated to *exo*-brevicomin (**181**).

| | 176 | (R)-pinanediol | 177 |

178

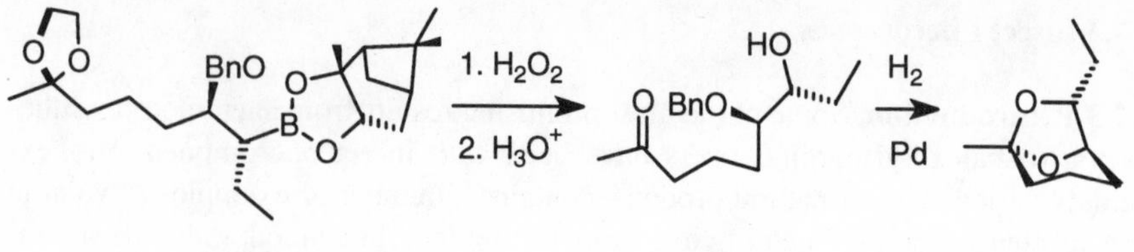

179   **180**   **181**

It was estimated from 200-MHz ^{1}H-NMR spectra that the de of the (α-chloro-alkyl)boronic ester intermediate between **177** and **178** was 99%, but apparent epimerization during the benzyloxide displacement reduced the de of **178** to ~95%. Assuming that this epimerization was real and not an artifact of inadequate NMR data, it may be noted that this is a problem peculiar to pinanediol esters, as a chiral director of C_2 symmetry would result in destruction of the minor epimer. The de of the brevicomin (**181**) was found to be 95% (±1%) by gas chromatography and ^{1}H-NMR. Intermediate **180** crystallized and could have been purified further if there had been any incentive to make pure **181**.

5.4.3.3 Eldanolide. Eldanolide (**187**), the wing gland pheromone of the male African sugar cane borer moth, *Eldana saccharina* [81], provided another test of functional group compatibility of the boronic ester chemistry [12]. (*R*)-Pinanediol methylboronate (**182**) was converted to the (*R*)-(1-chloroethyl)boronate **183**, which with lithio *tert*-butyl acetate formed **184**. Chain extension of **184** proceeded normally, as did substitution with prenylmagnesium chloride to form the boronic ester **185** containing the complete carbon skeleton. Deboronation to **186** and acid catalyzed lactonization yielded eldanolide (**187**) [12].

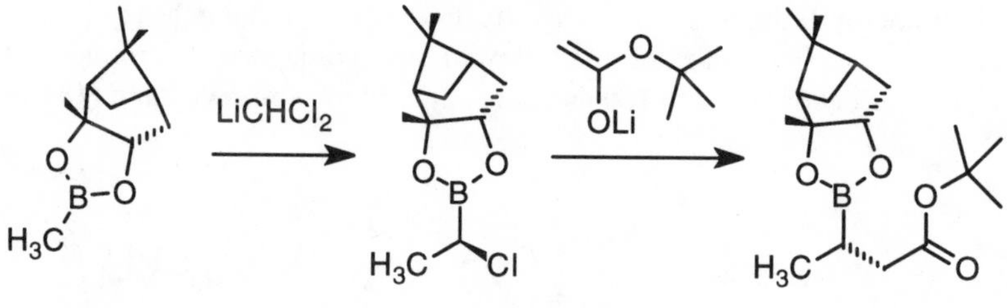

182   **183**   **184**

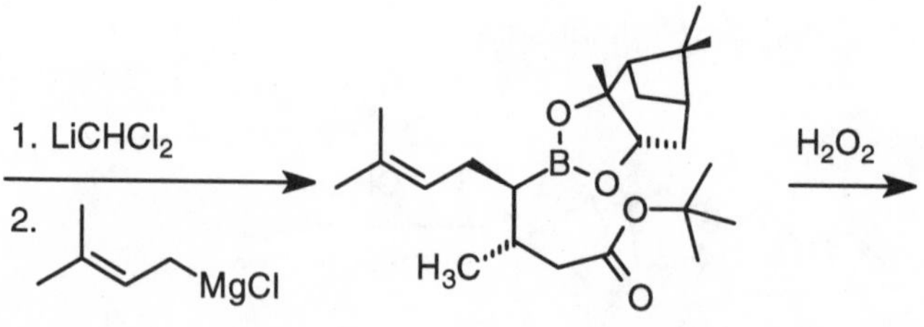

185

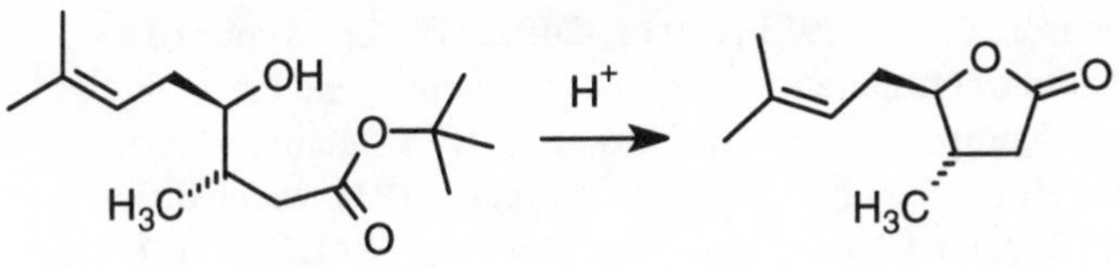

186   **187**

5.4.3.4 (3S,4S)- and (3R,4S)-4-Methyl-3-heptanol. The minor component of the aggregation pheromone of the elm bark beetle, *Scolytus multistriatus*, is (3S,4S)-4-methyl-3-heptanol (**191**) [82, 83, 84]. This simple compound was chosen as an early test of the efficacy of the new synthetic method [12, 19]. The stereoselectivity was not rigorously measured in the original synthesis based on pinanediol esters [12], but in any event this route has been superseded by the achievement of ultrahigh stereoselectivity with DIPED boronic esters [34]. The two-step double diastereo-differentiation which leads to ultrahigh stereoselectivity has been described in Section 5.3.4.

The correct relationship of asymmetric centers follows naturally if (R)-DIPED propylboronate (**188**) is chosen as the starting material. Chain extension with (di-chloromethyl)lithium and followed by treatment with methylmagnesium bromide yielded (R)-DIPED (S)-(1-methylbutyl)boronate (**189**). A second chain extension and reaction with ethylmagnesium bromide led to (R)-DIPED (1S,2S)-(1-ethyl-2-methylpentyl)boronate (**190**), which was deboronated with hydrogen peroxide to yield **191** [34]. The overall yield of **191** from **188** was 60-65%.

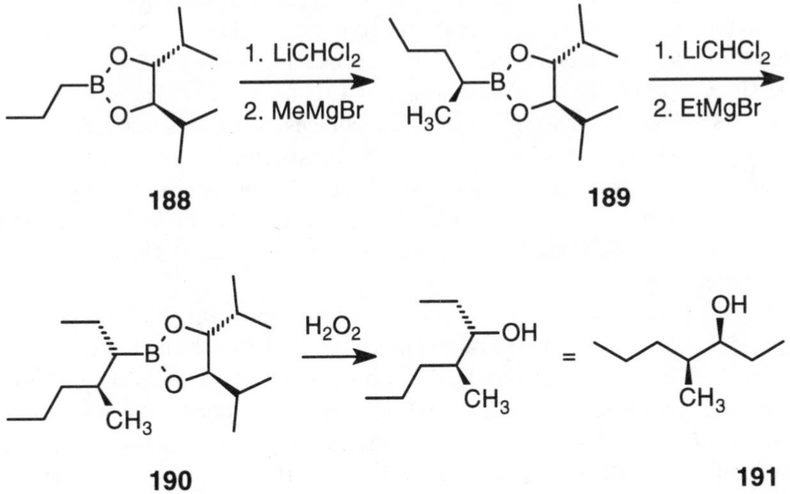

188   **189**

190   **191**

(3R,4S)-4-Methyl-3-heptanol (**195**), the diastereomer of **191**, is the trail pheromone of *Leptogenys diminuta*, a southeast Asian ponerine ant [85]. The synthetic strategy differs from that for **191** in that (S)-DIPED is used as chiral director to obtain the correct absolute configuration at the alcohol carbon, and the methyl group is introduced before the propyl group [34]. Thus, (S)-DIPED methylboronate (**192**) was converted to (S)-DIPED (S)-(1-methylbutyl)boronate (**193**), which differs from **189** only in the chiral director. Chain extension of **193** led to (S)-DIPED (1R,2S)-(1-ethyl-2-methylpentyl)boronate (**194**), which was deboronated to **195**, overall yield 57-58% [34].

The pair of diastereomers **191** and **195** had well separated peaks in the ^{13}C-NMR spectra, making it possible to estimate diastereomeric purities with high precision. One sample showed the ratio of **191** to **195** (and enantiomer of **195**) to be ~700:1, the other showed **195** to **191** (and enantiomer of **191**) to be 500:1. The enantiomers of **191** and **195** were also made by appropriate choices of chiral director, and were obtained in 300:1 and 180:1 diastereomeric ratios, respectively.

With all four stereoisomers available, it was easily shown that the (3R,4S)-isomer **195** is the only active trail pheromone [86]. The slight activity of other isomers was in accord with expectation based on the presence of traces of active **195** contaminating the samples.

It is not usual for organic chemists to be concerned about 0.1-0.5% impurity in starting materials, but in syntheses of this precision such minor and not easily detectable levels of impurities can significantly affect the results. Thus, if only 0.2 mol % of the starting boronic esters **188** or **192** were an achiral boronic ester contaminant, then 0.2% of the product would be a racemic mixture of diastereomers, in accord with the levels of impurities actually detected. If the chiral starting material were free from such contaminants, random production of a 1000:1 isomer ratio at each of two chiral centers would result in a 1000:2 diastereomer ratio, with only 1 part per million double error to form the enantiomer of the major product. Obviously, the current limitations on the isomeric purities that can be produced by this synthesis have to do with purification and analysis of starting materials rather than the limitations of the process itself.

5.4.3.5 Stegobiol and stegobinone. The principal attractant pheromone of the drugstore beetle *Stegobium paniceum* (Anobiidae), an economically important pest of stored foodstuffs, is (2S,3R,1'R)-stegobinone (**196**) [87, 88]. The natural pheromone also contains a few percent of (2S,3R,1'S,2'S)-stegobiol (**197**), which by itself causes a slower response from the insects [89]. (2S,3R,1'R)-Stegobinone (**196**) is also the attractant of the furniture beetle, *Anobium punctatum*, a wood eating pest [90].

196   **197**   **198**

Stegobinone (**196**) readily epimerizes to (**198**), which is repellent to *S. paniceum*. Stegobinone (**196**) from the natural source has crystallized, mp 52.5-53.5 °C [87], but previous synthetic samples of stegobinone (**196**) were at best partially crystalline [88, 91] or oily [92], and were unstable to prolonged storage. Synthetic racemate of (**196**) was also repellent, and samples of (**196**) at room temperature lost attractiveness in two weeks [93]. Ironically, the repellent epimer (**198**) crystallizes easily and its structure was readily established by X-ray crystallography [91].

The successful boronic ester strategy utilized (R)-DICHED boronic ester **202** as a common intermediate to make two segments of stegobiol (**197**) to be joined in an aldol condensation [42]. The synthesis of **202** from (R)-DICHED ethylboronate (**199**) via **200** and **201** was straightforward. Hydrogen peroxide at pH 9 converted **202** to the labile aldehyde **203** (99% de), which was isolated by distillation in 65% yield based on (R)-DICHED ethylboronate (**199**). Methylmagnesium bromide converted another portion of **202** to boronic ester **204**, which was deboronated to **205**, silylated to **206**, and debenzylated with palladium hydroxide and calcium carbonate and oxidized with pyridinium chromate to **207**, which was distilled in 28% yield overall from **199** [42].

199   **200**

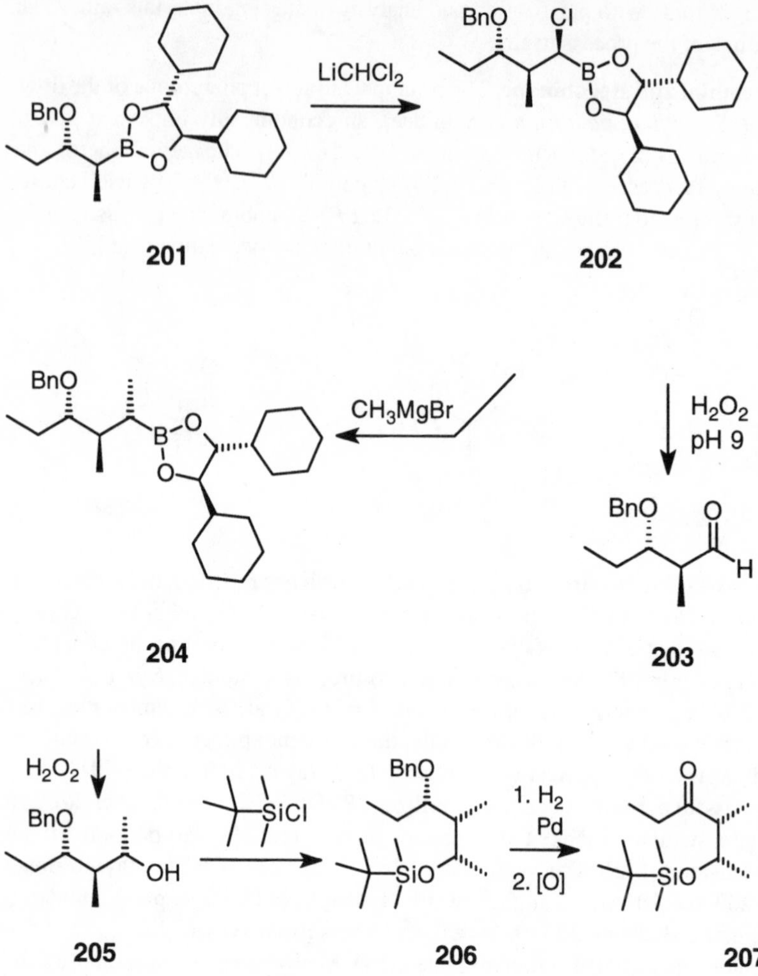

The synthesis was completed by conversion of **207** to the boron enolate **208**, which with aldehyde **203** yielded aldol **209**. Oxidation of **209** with pyridinium dichromate to **210** and desilylation with acid completed the synthesis of benzyl-stegobiol (**211**). Debenzylation yielded stegobiol (**197**), which was the first compound in the entire sequence to require chromatographic purification. Stegobiol (**197**) had been reported previously only as an oil from either natural [89] or synthetic [94] sources, but **197** prepared via the boronic ester chemistry crystallized, mp 73-74 °C. Oxidation of pure **197** with N-methylmorpholine-N-oxide catalyzed by tetrapropylammonium perruthenate readily yielded pure stegobinone (**196**), which crystallized when the solvent was evaporated [42].

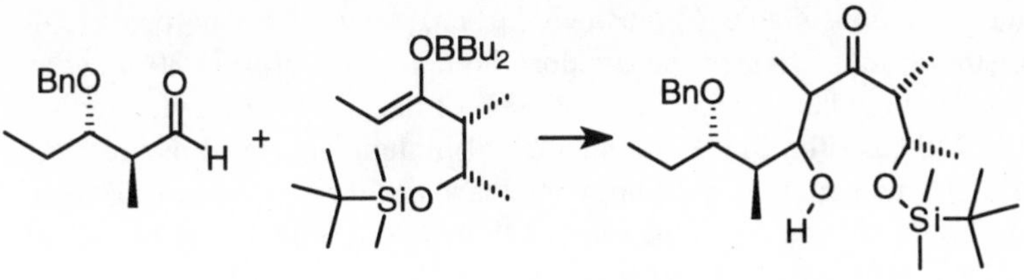

203   **208**   **209**

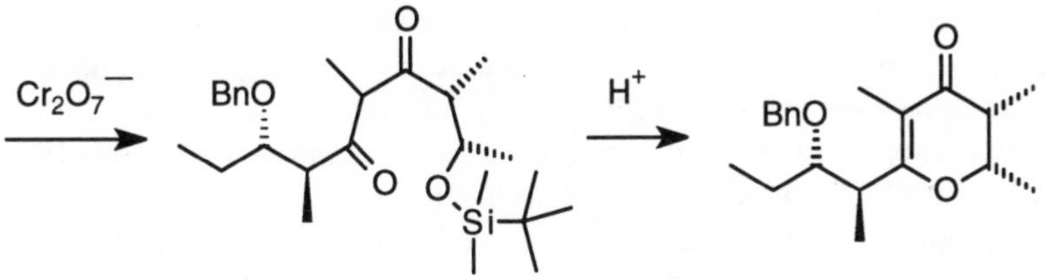

210   **211**

Also tested with some success was the alternative in which the boronic ester group of **204** was left in place until after conversion to a ketone, aldol condensation, and oxidation to the analog of **210** having the boronic ester group in place of the silyloxy [42]. A lithium enolate in place of boron enolate **208** resulted in lower yields [42].

Serricorole, the homologue of stegobiol (**197**) bearing an ethyl group instead of the methyl at C(2), has recently been prepared via an unrelated route [95].

As an example of systematic names, the key intermediate **202** is [4R-[2(1S*,2S*,3S*),4α,5β]]-2-[1-chloro-2-methyl-3-(phenylmethoxy)pentyl]-4,5-dicyclohexyl-1,3,2-dioxaborolane. As noted in Section 5.1.2, such systematic names are useful for information storage and retrieval purposes, provided one has a computer to keep track of it all.

5.4.4 Polyols

5.4.4.1 L-Ribose. In view of the extreme ease of elimination of ethylene from dibutyl (2-bromoethyl)boronate [96], it was surprising to find that a β-alkoxy boronic ester was stable until temperatures approached 100 °C [97]. Success with a β-benzyloxy boronic ester intermediate in the asymmetric synthesis of brevicomin encouraged attempts to extend the synthetic method to carbohydrates. L-Ribose (**221**) was chosen as a synthetic target which could be reached efficiently by using (S)-pinanediol, the less expensive chiral director.

(S)-Pinanediol (chloromethyl)boronate (**212**) was prepared by transesterification of the diisopropyl ester (preparation: Section 2.2.1.1) [21]. Treatment with lithium benzyl oxide yielded distillable (S)-pinanediol [(benzyloxy)methyl]boronate (**213**). (Dichloromethyl)lithium yielded the α-chloro boronic ester **214a** in 80% de, and treatment with lithium benzyl oxide yielded 68% of the [1,2-bis(benzyloxy)ethyl]-boronic ester **215** based on **213**. (Dibromomethyl)lithium, generated in situ by the addition of LDA to a mixture of dibromomethane and the substrate in THF at – 78 °C, provided the α-bromo boronic ester **214b** in 94% de and the bis(benzyloxy) derivative **215** in 82% overall yield from **213**.

The bromo boronic ester intermediates proved necessary for further chain extensions, as the overall yield of the [1,2,3-tris(benzyloxy)propyl]boronic ester **216** was only 10% if chloro boronic ester intermediates were used but a useful 64% via the bromo boronic esters. The yield fell to 58% for the chain extension of **216** to **217**. After that, yields fell precipitously. Further chain extension with (dibromomethyl)-lithium failed altogether, and (dichloromethyl)lithium yielded only 13%. (Chloromethyl)lithium, prepared in situ from iodochloromethane and *n*-butyllithium [98], yielded 36% of the [2,3,4,5-tetrakis(benzyloxy)pentyl]boronic ester **218** from **217**. Deboronation with hydrogen peroxide and Swern oxidation with dimethyl sulfoxide/oxalyl chloride/triethylamine yielded tetrabenzyl ribose (**220**) which was debenzylated with hydrogen and palladium to L-ribose (**221**) in an overall yield of 13% from **213** [21].

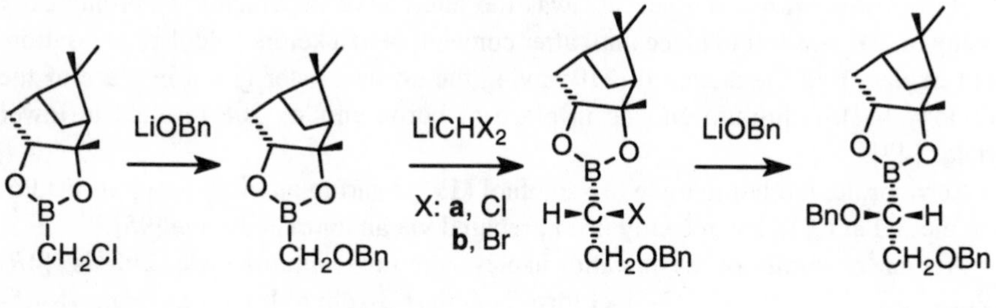

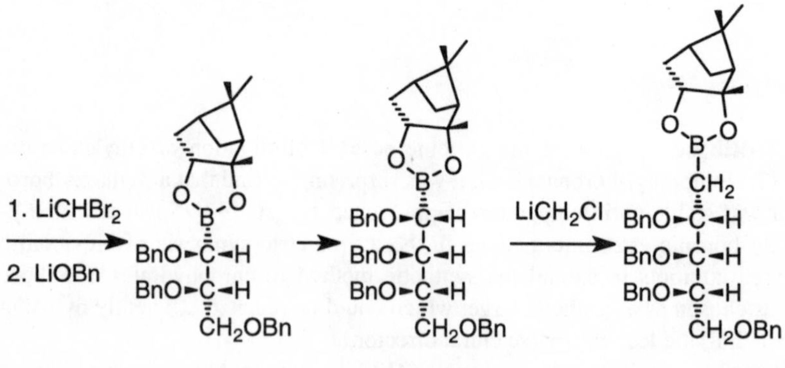

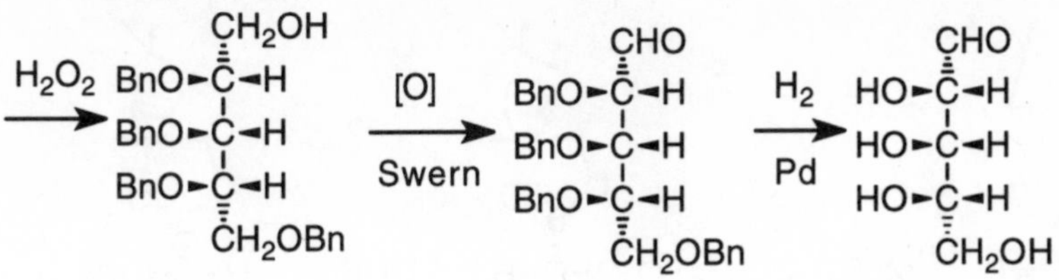

219   **220**   **221**

The reason for the difficulty in extending the chain beyond four benzyloxy substituted carbons is not understood. However, there is some evidence in less highly substituted systems to suggest that a 4-benzyloxy substituent, which is in a position to form a six-membered ring if the oxygen coordinates to the boron, undergoes unexpectedly facile debenzylation [99].

The L-ribose synthesis introduces each carbon as an (R)-(benzyloxy substituted) chiral center. In order to establish a truly general polyol synthesis, there has to be a way to switch to an (S) chirality at any desired point. The known way to cleave a pinanediol boronic ester, treatment with boron trichloride [11], was known to be incompatible with retention of benzyloxy groups [20]. Therefore a double inversion sequence was undertaken [100]. The bromo boronic ester **222** from chain extension of the ribose intermediate **216** was treated with lithium dimethoxybenzyloxide (instead of benzyloxide) to form **223**, then cleaved by dichlorodicyanoquinone (DDQ) [101] to the α-hydroxy boronic ester **224**. Conversion to the mesylate **225** then provided the synthetic equivalent of the diastereomer of the halide **222**. Lithium benzyl oxide converted **225** to benzyloxy derivative **226**, a diastereomer of the ribose intermediate **217**. Diastereomers **217** and **226** were easily distinguishable by 200-MHz ^{1}H-NMR spectra.

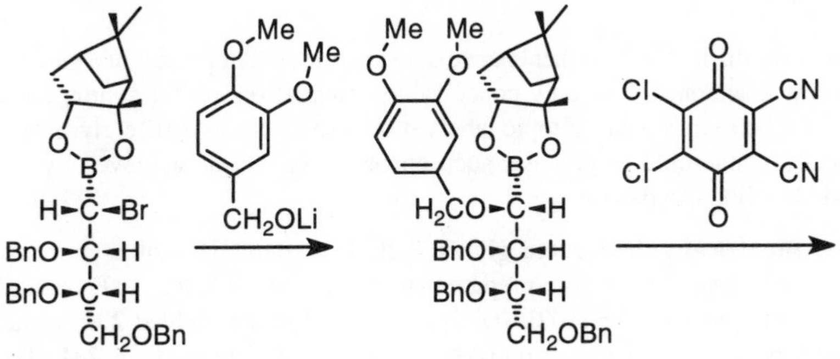

222   **223**

224 **225** **226**

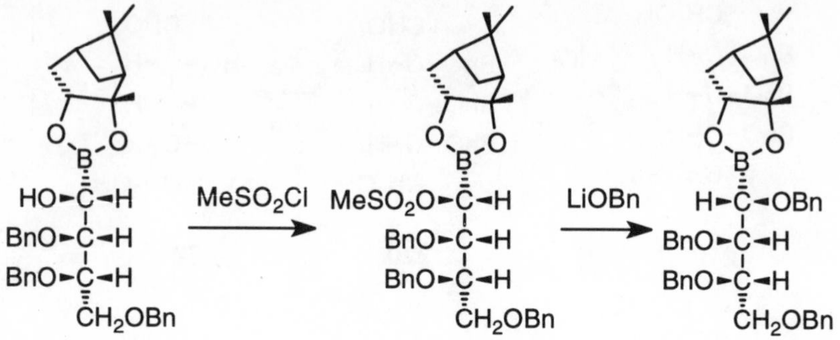

The use of lithium dimethoxybenzyloxide as the nucleophile for achieving hydroxyl substitution in the synthesis of **224** might seem to be roundabout, but model studies on chloro boronic ester **227** indicated that direct substitution by hydroxide ion or trimethylsilyloxide ion led to other products, possibly migration of either of the oxygens of the pinanediol from boron to carbon with ring expansion to **228** or **229** [100]. It is not understood why such ring expansion is so seldom a problem with alkoxy substitution.

227 **228** (NMR evidence only) **229**

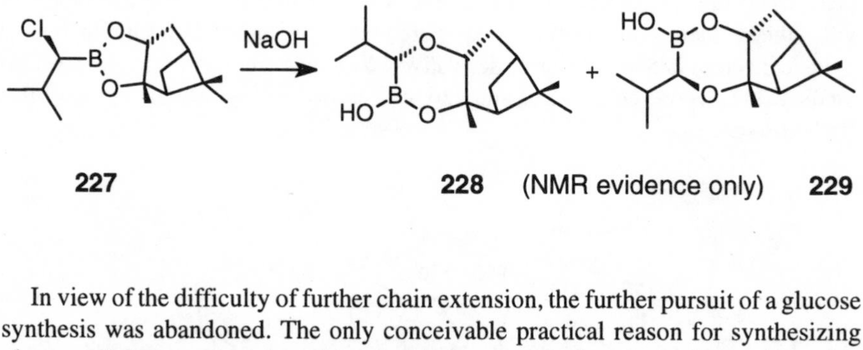

In view of the difficulty of further chain extension, the further pursuit of a glucose synthesis was abandoned. The only conceivable practical reason for synthesizing glucose might be to introduce isotopic labels at precise locations. Since glycerol is enzymatically convertible to glucose, such an objective can be achieved by the asymmetric labeling of glycerol.

5.4.4.2 Asymmetrically deuterated glycerol. Both diastereomers of asymmetrically deuterated glycerol have been prepared via straightforward chain extension of the L-ribose intermediate **215** [102]. (Dibromomethyl)lithium yielded **230**, which was reduced by potassium triisopropoxyborodeuteride to intermediate **231**. The usual peroxidic deboronation to **232** and reductive debenzylation to (*S*)-(1*R**,2*R**)-1-deuteroglycerol (**233**) followed.

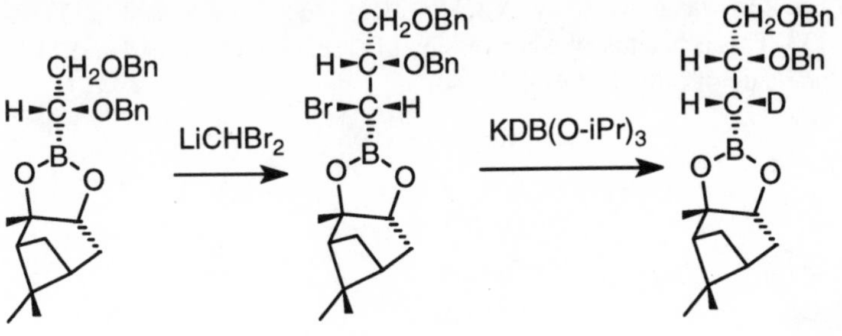

215   **230**   **231**

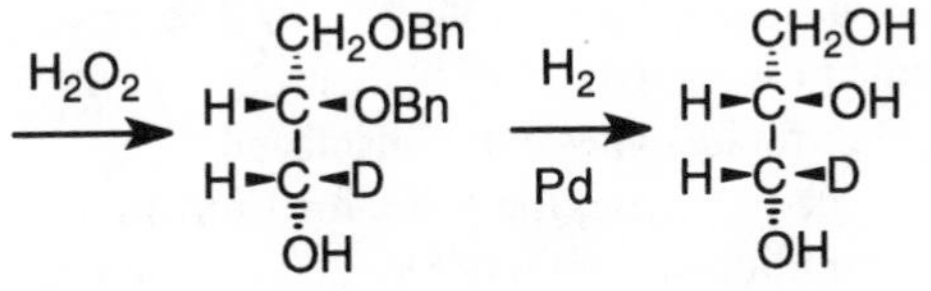

232   **233**

The synthesis of (R)-(1R*,2S*)-1-deuteroglycerol (**236**) was achieved by the use of dideuterodibromomethane to generate (deuterodibromomethyl)lithium and make **234**, which was reduced to **235** with lithium triethylborohydride, then deboronated and debenzylated.

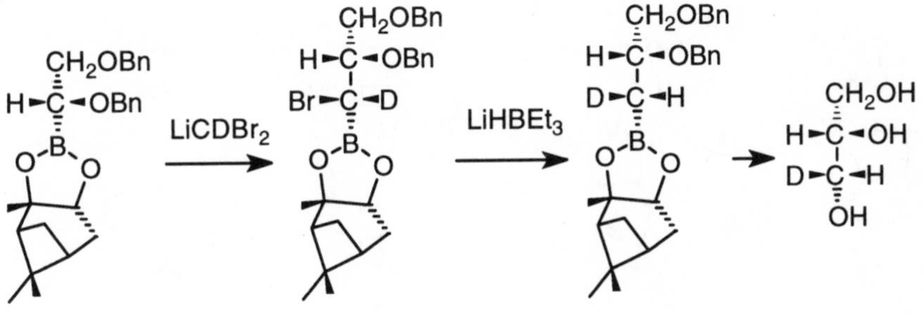

215   **234**   **235**   **236**

Estimation of the purity (S)-(1R*,2R*)-1-deuteroglycerol (**233**) and (R)-(1R*,2S*)-1-deuteroglycerol (**236**) was accomplished by 200-MHz ^{1}H-NMR analysis, which was based on the chemical shift difference of ~0.1 δ between the diastereotopic protons of the CH$_2$ groups of glycerol. The measured de's of **233** and **236** were only 92-94%. However, most of what appeared to be **236** in the sample of **233** was probably the enantiomer, (S)-(1R*,2S*)-1-deuteroglycerol (**238**), and most of the impurity in **236** was conversely (R)-$(1R^*,2R^*)$-1-deuteroglycerol (**237**), the enantiomer of **233**. This is because the conversion of precursor **213** to **214** and **215** is known to produce a de of only 92-94%.

237   **238**

Biological oxidation to glyceraldehyde would involve the top carbon of any of the four isomers **233**, **236**, **237**, or **238** as written. Thus, the de of label placement would turn out to be very high, probably ≥98%, but there would be 3-4% of missing label or regioisomerically misplaced label.

Glyceraldehyde, the basis for Emil Fischer's system of chirality nomenclature and a traditional building block of the training of organic chemists, was the original target of the foregoing work. Unfortunately, glyceraldehyde is not a well defined molecular species but a gross mixture of acetals from self-condensation [103], and the 200-MHz ^{1}H-NMR spectrum of a commercial sample of glyceraldehyde showed a forest of peaks. However, boronic ester chemistry provides routes to compounds that could be used as glyceraldehyde synthons. Chain extension of (S)-pinanediol [(p-methoxybenzyl)methyl]boronate (**239**) yielded the deuterated boronic ester intermediate **240**. Deprotection with DDQ yielded the hydroxy substituted boronic

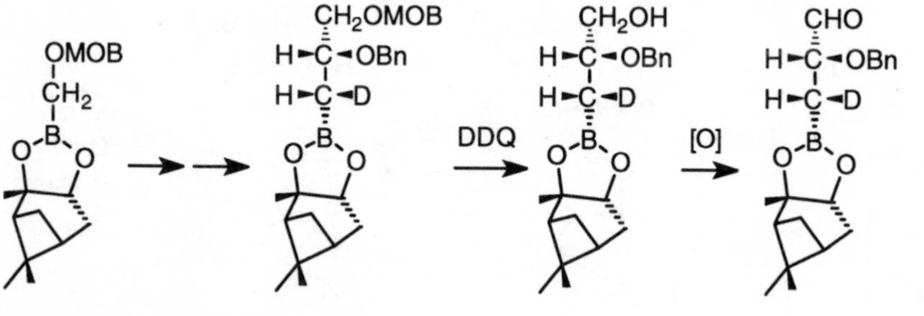

239 (MOB = p-MeOC$_6$H$_4$) **240**   **241**   **242**

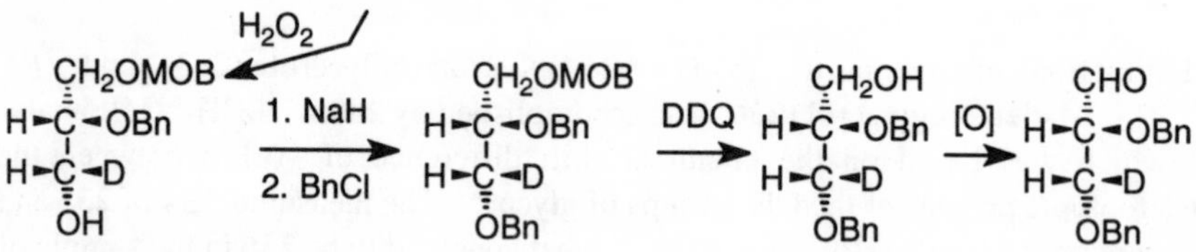

243   **244**   **245**   **246**

ester **241**, which on Swern oxidation was converted to the boronic ester aldehyde **242**, an obvious glyceraldehyde synthon. Alternatively. **240** was deboronated to **243**, benzylated to **244**, deprotected with DDQ to the bis(benzyloxy) alcohol **245**, and oxidized by Swern's method to dibenzylglyceraldehyde (**246**). Attempted debenzylation of **246** resulted in partial reduction of the aldehyde group to glycerol, and it was impossible to say whether glyceraldehyde was a major component of the resulting mixture because of the complexity of the NMR spectrum and the likelihood that glycerol also forms acetals with glyceraldehyde.

5.4.5 Amino Acids

α-Halo boronic ester chemistry provides a straightforward asymmetric synthesis of amino acids [104]. A simple example is the synthesis of valine (**251**). Pinanediol isopropylboronate (**69**) was converted to the (α-bromoisobutyl)boronic ester **247**, chosen after it was found that the chloro analog reacted so slowly with sodium azide in a water/dichloromethane phase transfer system [12] that formation of explosive diazidomethane [105] was a hazard. The reaction of **247** with azide ion differs from reactions with more basic nucleophiles in that the boron atom is not deactivated as a borate complex prior to rearrangement, and as a result the bromide ion generated during the course of the reaction can epimerize the remaining **247**. The ratio of azide to substrate **247** was increased to 50:1 in order to obtain azide **248** with minimal epimerization. Chain extension to chloro boronic ester **249** was straightforward, but a means of oxidizing the α-chloro boronic ester to carboxylic acid was required. Sodium chlorite, a reagent known to oxidize aldehydes to carboxylic acids [106], was found to oxidize **249** directly to the α-azido acid **250**, which was hydrogenated to valine (**251**), 57% based on **69**, ee ~96%, estimated by 200-MHz ^{1}H-NMR analysis of the methyl ester in the presence of a chiral shift reagent [107].

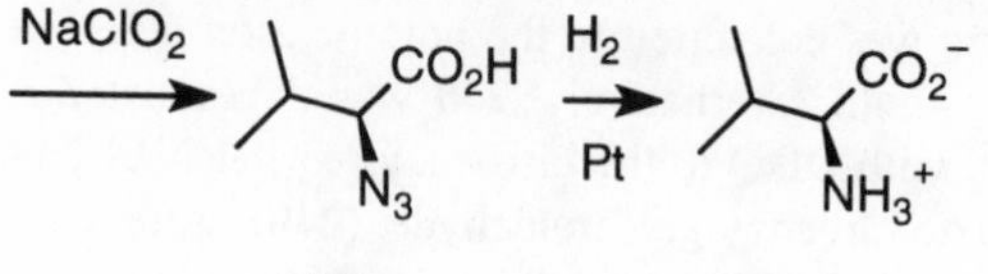

250   **251**

The synthesis of glutamic acid (255) began from a (halomethyl)boronic ester and *tert*-butyl lithioacetate. (Chloromethyl)boronic esters are insufficiently reactive, and the acetate ester is partially lost to Claisen condensation. Pinacol (iodomethyl)-boronate (252) was therefore prepared from the chloro precursor and sodium iodide. *tert*-Butyl lithioacetate with 252 yielded 253, which was transesterified with pinanediol and converted via the azido boronic ester 254 to glutamic acid (255). The final step was cleavage of the *tert*-butyl ester with trifluoroacetic acid, and the over all yield of 255 from 253 was 32% [104].

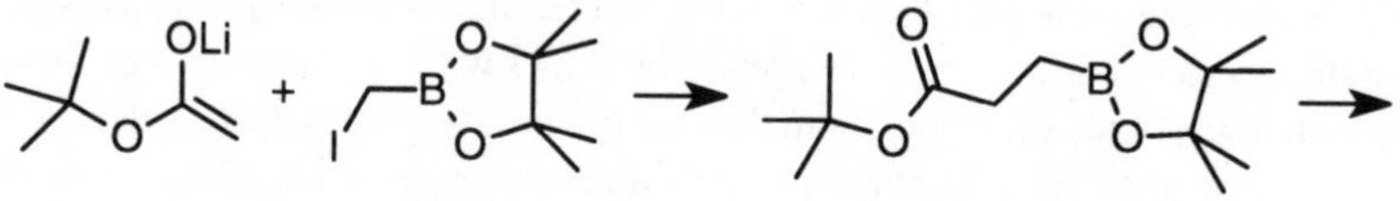

252   **253**

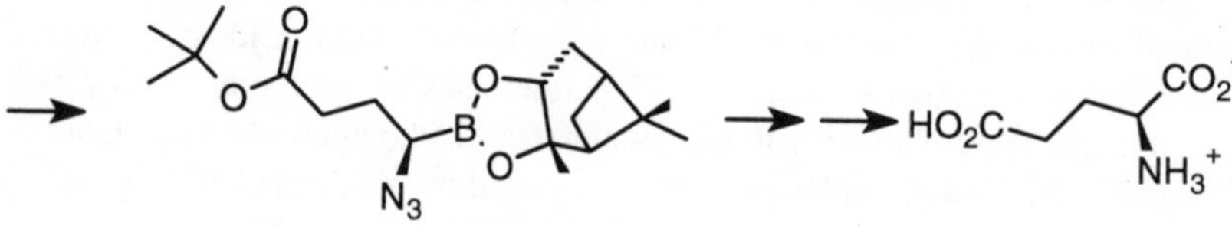

254   **255**

Serine (257) was synthesized from (S)-pinanediol [(benzyloxy)methyl]boronate (213) in a similar manner. Hydrogenation of the azide intermediate 256 over palladium accomplished removal of the benzyl protecting group in the same step. The over all yield of 257 from 213 after chromatography was 39%, de 92% [104].

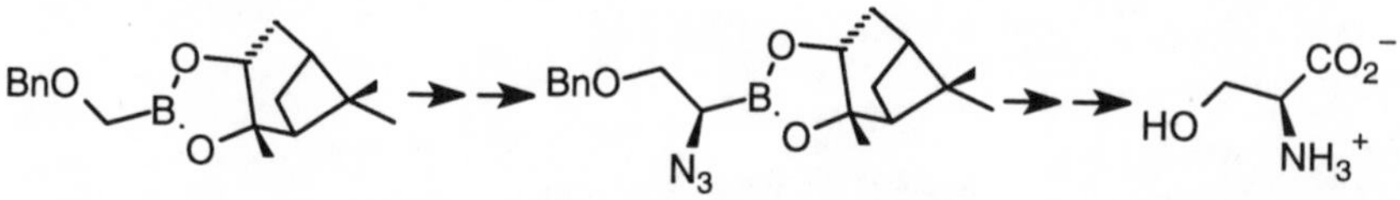

213   **256**   **257**

Phenylalanine was made in a manner similar to the others, except that the (1-chloro-2-phenylethyl)boronic ester 259 was used in preference to its bromo analog

[104]. Not only was **259** sufficiently reactive, but the bromo analog epimerized to a significant extent. The phenylalanine synthesis was then repeated with introduction of a chirally specific deuterium label [108].

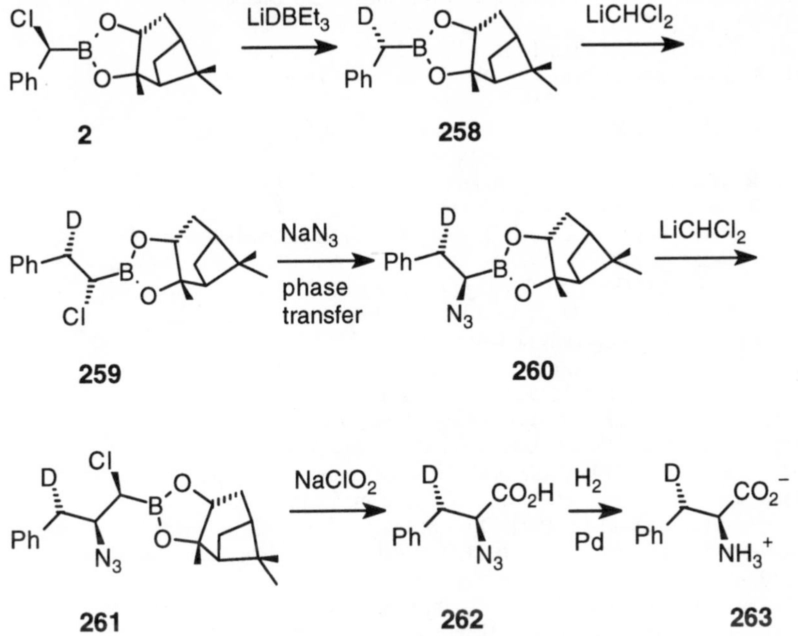

The conversion of (S)-pinanediol phenylboronate (**11**) to the (S)-(chlorobenzyl)boronate (**2**) is less stereoselective than most, usually yielding only about a 95:5 diastereomer ratio, perhaps because (**2**) epimerizes so readily [18]. However, **2** was purified easily by recrystallization before triethylborodeuteride reduction to **258**. Conversion of **258** to (S)-benzyl alcohol-d and NMR analysis with a chiral shift reagent indicated that the **258** contained no more than 2% of isomeric or undeuterated material. Conversion of **258** to chloro boronic ester **259** by the usual method was followed by treatment with a large excess of sodium azide under phase transfer conditions to form **260**. Chain extension of the azido boronic ester **260** to chloro boronic ester **261** was followed by oxidation of the chloro boronic ester with sodium chlorite to form the azido carboxylic acid **262**, which was hydrogenated to (2S,3S)-phenylalanine-3-d (**263**).

5.5 References

1.   Matteson DS (1989) Chem. Rev. 89:1535
2.   Brown HC, Zweifel G (1961) J. Am. Chem. Soc. 83:2544
3.   Houghton HP (1944) "The Basque Verb, Guipuzcoan Dialect", Mohn Printing Co.,
     Northfield MN, 58 pp
4.   Nakai T (1991) Chem. Eng. News 69:July 1:p 2
5.   Heathcock C (1991) Chem. Eng. News 69:Feb 4:p 3
6.   Testa B (1991) Chirality 3:159
7.   Matteson DS, Ray R (1980) J. Am. Chem. Soc. 102:7590
8   Ray R, Matteson DS (1980) Tetrahedron Lett. 21:449
9.   Ray R, Matteson DS (1982) J. Indian Chem. Soc. 59:119
10. Van Rheenen V, Kelly RC, Cha DY (1976) Tetrahedron Lett. 1973
11. Matteson DS, Ray R, Rocks RR, Tsai DJS (1983) Organometallics 2:1536
12   Matteson DS, Sadhu KM, Peterson ML (1986) J. Am. Chem. Soc. 108:812
13. Aldrich Chemical Company Catalog (1994)
14. Brown HC, Jadhav PK, Desai MC (1982) J. Org. Chem. 47:4583
15. Erdik E, Matteson DS (1989) J. Org. Chem. 54:2742
16. Cram DJ (1952) J. Am. Chem. Soc. 74:2149
17. Matteson DS, Mah, RWH (1963) J. Am. Chem. Soc. 85:2599
18. Matteson DS, Erdik E (1983) Organometallics 2:1083
19. Matteson DS, Sadhu KM (1983) J. Am. Chem. Soc. 105:2077
20. Matteson DS, Jesthi PK, Sadhu KM (1984) Organometallics 3:1284
21. Matteson DS, Peterson ML (1987) J. Org. Chem. 52:5116
22. Matteson DS, Sadhu KM (1984) Organometallics 3:614
23. Matteson DS, Michnick TJ, Willett RD, Patterson CD (1989) Organometallics 8:726
24. Matteson DS, Moody RJ (1982) Organometallics 1:20
25. Matteson DS, Arne KH (1982) Organometallics 1:280
26. Matteson DS, Beedle EC (1990) Heteroatom Chemistry 1:135
27. Michnick TJ (1991) Ph. D. Thesis, Washington State University
28. Matteson DS, Campbell JD (1990) Heteroatom Chemistry 1:109
29. Matteson DS, Hurst GD (1990) Heteroatom Chemistry 1:65
30. Brown HC, Rangaishenvi MV (1988) J. Organomet. Chem. 358:15
31. Tsai DJS, Jesthi PK, Matteson DS (1983) Organometallics 2:1543
32. Sadhu KM, Matteson DS, Hurst GD, Kurosky JM (1984) Organometallics 3:804
33. Rangaishenvi MV, Singaram B, Brown HC (1991) J. Org. Chem. 56:3286
34. Tripathy PB, Matteson DS (1990) Synthesis 200
35. Matteson DS, Kandil AA (1986) Tetrahedron Lett. 27:3831
36. Matteson DS, Beedle EC, Kandil AA (1987) J. Org. Chem. 52:5034
37. Ditrich K, Bube T, Stürmer R, Hoffmann RW (1986) Angew. Chem. 98:1016; Angew.
     Chem., Internat. Ed. 25:1028
38. Hoffmann RW, Ditrich K, Köster G, Stürmer R (1989) Chem. Ber. 122:1783
39. Jacobsen EN, Markó I, Mungall WS, Schröder G, Sharpless KB (1988) J. Am. Chem.
     Soc. 110:1968
40. Sharpless KB, Amberg W, Bennani YL, Crispino GA, Hartung J, Jeong KS, Kwong
     HL, Morikawa K, Wang ZM, Xu D, Zhang XL (1992) J. Org. Chem. 57:2768
41. Wang ZM, Sharpless KB (1994) J. Org. Chem. 59:8302
42. Matteson DS, Man HW (1993) J. Org. Chem. 58:6545
43. Hoffmann RW, Ditrich K, Köster G, Stürmer R (1989) Chem. Ber. 122:1783
44. Still WC, Sreekumar C (1980) J. Am. Chem. Soc. 102:1201
45. Marshall JA, Gung WY (1988) Tetrahedron Lett. 29:1657
46. Chan, PCM, Chong JM (1988) J. Org. Chem. 53:5584
47. Matteson DS, Tripathy PB, Sarkar A, Sadhu KM (1989) J. Am. Chem. Soc. 111:4399

48. McGarvey GJ, Kimura M (1982) J. Org. Chem. 47:5420
49. Soundararajan R, Matteson DS (1991) Unpublished results
50. Matteson DS, Michnick TJ (1990) Organometallics 9:3171
51. Matteson DS, Man HW (1994) J. Org. Chem. 59:5734
52. Hoffmann RW, Landmann B (1986) Chem. Ber. 119:1039
53. Hoffmann RW, Dresely S (1987) Tetrahedron Lett. 28:5303
54. Hoffmann RW, Dresely S (1988) Synthesis 103
55. Hoffmann RW, Dresely S, Lanz JW (1988) Chem. Ber. 121:1501
56. Soundararajan R, Matteson DS (1991) Unpublished results
57. Matteson DS, Sadhu KM, Lienhard GE (1981) J. Am. Chem. Soc. 103:5241
58. Philipp M, Maripuri S, Matteson DS, Jesthi PK, Sadhu KM (1983) Biochemistry (Washington, DC) 22:A13
59. Kinder DH, Katzenellenbogen JA (1985) J. Med. Chem. 28:1917
60. Matteson DS, Cheng TC (1968) J. Org. Chem. 33:3055
61. Matteson DS, Majumdar D (1979) J. Organomet. Chem. 170:259
62. Shenvi A (1986) Biochemistry (Washington, D. C.) 25:1286
63. Duncan K, Faraci SW, Matteson DS, Walsh CT (1989) Biochemistry 28:3541
64. Lindquist RN, Nguyen A (1977) J. Am. Chem. Soc. 99:6435
65. Amiri P, Lindquist RN, Matteson DS, Sadhu KM (1984) Arch. Biochem. Biophys. 234:531
66. Knabb RM, Kettner CA, Timmermans PBM, Reilly TM (1992) Thromb. Haemostasis 67:56; Chem. Abstr. 116:166038g
67. Hussain MA, Knabb R, Aungst BJ, Kettner C (1991) Peptides (Fayetteville, NY) 12:1153
68. Wityak J, Chorvat RJ, Earl RA, Kettner CA, Pierce ME (1992) 203rd Am. Chem. Soc. Meeting, San Francisco, California, Abstract MEDI 160
69. Philipp M, Claeson G, Matteson DS, deSoyza T, Agner E, Sadhu, KM (1987) *Federation Proc.*, 46:2223
70. Kettner CA, Shenvi AB (1984) J. Biol. Chem. 259:15106
71. Soskel NT, Watanabe S, Hardie R, Shenvi AB, Punt JA, Kettner CA (1986) Am. Rev. Resp. Dis. 133:635; 133:639
72. Kettner CA, Bone R, Agard DA, Bachovchin WW (1988) Biochemistry (Washington, DC) 27:7682
73. Bachovchin WW, Wong WYL, Farr-Jones S, Shenvi AB, Kettner CA (1988) Biochemistry (Washington, DC) 27:7689
74. Tsilikounis E, Kettner CA, Bachovchin WW (1992) Biochemistry 31:12839
75. Flentke GR, Munoz E, Huber BT, Plaut AG, Kettner CA, Bachovchin WW (1991) Proc. Natl. Acad. Sci. U.S.A. 88:1556
76. Bachovchin WW, Plaut AG, Flentke GR, Lynch M, Kettner CA (1990) J. Biol. Chem. 265:3783
77. Kettner CA, Mersinger L, Knabb R (1990) J. Biol. Chem. 265:18289
78. Zhong S, Jordan F, Kettner C, Polgar L (1991) J. Am. Chem. Soc. 113:9429
79. Silverstein RM, Brownlee RG, Bellas TE, Wood DL, Browne LE (1968) Science 159:889
80. Bellas TE, Brownlee RG, Silverstein RM (1969) Tetrahedron 25:5149
81. Kunesh G, Zagatti P, Lallemand JY, Debal A, Méric R, Vigneron JP (1981) Tetrahedron Lett. 22:5271
82. Pearce GT, Grove WE, Silverstein RM, Peacock JW, Cuthbert RA, Lanier GN, Simeone JB (1975) J. Chem. Ecol. 1:115
83. Mori K (1977) Tetrahedron 33:289
84. Mori K, Isawa H (1980) Tetrahedron 36:2209
85. Attygalle AB, Vostrowsky O, Bestmann HJ, Steghaus-Kovâc S, Maschwitz U (1988) Naturwissenschaften 75:315

86. Steghaus-Kovâc S, Maschwitz U, Attygalle AB, Frighetto RTS, Vostrowsky O, Bestmann HJ (1992) Experientia 48:690
87. Kuwahara Y, Fukami H, Howard R, Ishii, S, Matsumura F, Burkholder WE (1978) Tetrahedron 34:1769
88. Hoffmann RW, Ladner W (1979) Tetrahedron Lett. 4653
89. Kodama H, Ono M, Kohno M, Ohnishi A (1987) J. Chem. Ecol. 13:1871
90. White PR, Birch MC (1987) J. Chem. Ecol. 13:1695
91. Hoffmann RW, Ladner W, Steinbach K, Massa W, Schmidt R, Snatzke G (1981) Chem. Ber. 114:2786
92. Mori K, Ebata T (1986) Tetrahedron 42:4413
93. Kodama H, Mochizuki K, Kohno M, Ohnishi A, Kuwahara Y (1987) J. Chem. Ecol. 13:1859
94. Mori K, Ebata T (1986) Tetrahedron 42:4685
95. Oppolzer W, Rodriguez I (1993) Helv. Chim. Acta 76:1275
96. Matteson DS, Liedtke JD (1963) J. Org. Chem. 28:1924
97. Matteson DS, Majumdar D (1983) Organometallics 2:1529
98. Sadhu KM, Matteson DS (1985) Organometallics 4:1687
99. Ho O, Matteson DS (1993) unpublished results
100. Matteson DS, Kandil AA (1987) J. Org. Chem. 52:5121
101. Oikawa Y, Yoshioka T, Yonemitsu O (1982) Tetrahedron Lett. 23:885
102. Matteson DS, Kandil AA, Soundararajan R (1990) J. Am. Chem. Soc. 112:3964
103. Duke CC, MacLeod JK, Williams JF (1981) Carbohydrate Research 95:1
104. Matteson DS, Beedle EC (1987) Tetrahedron Lett. 28:4499
105. (a) Bretherick L (1986) Chem. Eng. News 64:Dec 22:2; (b) Hassner A (1986) Angew. Chem., Internat. Ed. 25:478
106. Lindgren BO, Nilsson T (1973) Acta Chem. Scand. 27:888
107. Ajisaka K, Kamisaku M, Kainosho M (1972) Chem. Lett. (Jpn.) 857
108. Matteson DS, Beedle EC, Christenson E., Dewey MA, Peterson ML (1988) J. Labelled Compd. Radiopharm. 25:675

6 Asymmetric Hydroboration Chemistry

6.1 Introduction

Hydroboration provided the first nonenzymatic asymmetric synthesis of any kind to have resulted in a truly high enantioselection. The hydroboration of *cis*-2-butene with (–)-diisopinocampheylborane (**1**) derived from (+)-α-pinene proceeded with high diastereoselectivity to form the intermediate trialkylborane **2** [1, 2]. When **1** of high enantiomeric purity [3] is used and the procedure is otherwise optimized, this process yields 98% ee in the final product, (*R*)-(–)-2-butanol (**3**), which was obtained via the usual alkaline hydrogen peroxide oxidation of **2** [4]. Both enantiomers of α-pinene are readily available, and it is therefore equally simple to prepare (*S*)-(+)-2-butanol by this method.

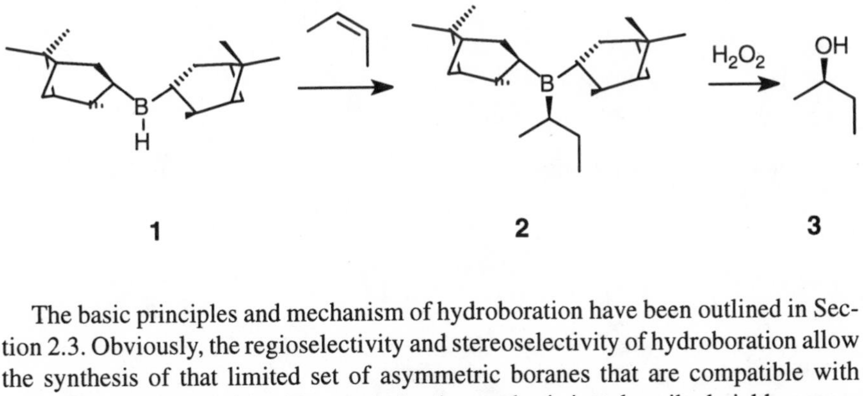

1	**2**	**3**

The basic principles and mechanism of hydroboration have been outlined in Section 2.3. Obviously, the regioselectivity and stereoselectivity of hydroboration allow the synthesis of that limited set of asymmetric boranes that are compatible with these inherent selectivities. For example, the synthesis just described yields a gross regioisomer mixture if *cis*-2-pentene is used as the substrate. From *cis*-3-hexene, the ee of the resulting (*R*)-3-hexanol was only 93%, and from norbornene, the ee of the (*S*)-2-*exo*-norborneol was only 83% [4, 5]. Even so, this chemistry can be used to provide a considerable variety of chiral building blocks in good to excellent enantiomeric purity. Brown and Singaram have reviewed this chemistry recently [6], and there are also earlier reviews [7, 8].

In this chapter, the hydroborations which have given useful enantioselection will be reviewed, followed by applications of the resulting boranes in asymmetric synthesis. Hydroborations which are diastereoselective with chiral substrates, including stereoselective catalytic hydroborations, will also be reviewed.

6.2 Asymmetric Hydroboration

6.2.1 Reagents

6.2.1.1 Purification of pinylboranes. Because α-pinene is cheap, often gives good diastereoselection, and is available in both enantiomeric forms, boranes derived from α-pinene have been studied extensively. However, α-pinene from practical commercial sources is not enantiomerically pure. The best commercial (+)-α-pinene is ~92% ee [6], and (–)-α-pinene is easily made in ~95% ee by isomerization of (–)-β-pinene [9]. Fortunately, (–)-diisopinocampheylborane (**1**) from (+)-α-pinene of 92% ee can easily be upgraded to high enantiomeric purity by crystallization from THF [10, 11]. The ~4% of (–)-α-pinene will form mainly the diastereomeric *meso*-diisopinocampheylborane **4** in combination with the majority (+)-α-pinene, but the system equilibrates within a few days at 0 °C, and if a slight excess of α-pinene is used, the (–)-α-pinene is left behind in the solution as the chiral isomer **1** crystallizes.

(+)-α-pinene	(–)-α-pinene		
(96%)	(4%)	**1**	**4 (disappears)**

Monoisopinocampheylborane (**6**) is also a useful hydroborating agent. It cannot be prepared directly from α-pinene, but is easily obtained by reaction of **1** with tetramethylethylenediamine (TMEDA), which liberates α-pinene and forms the crystalline TMEDA adduct **5**, which is readily converted to **6** by boron trifluoride [12]. It is particularly valuable for hydroborating trisubstituted olefins, especially cycloolefins. Although the resulting de's are generally only fair to good, ranging from 53-72% with a β-methyl or 82-97% with a β-phenyl substituent (see below), many of the boranes can be crystallized from the THF solution in which they are formed and consequently obtained in high enantiomeric purity.

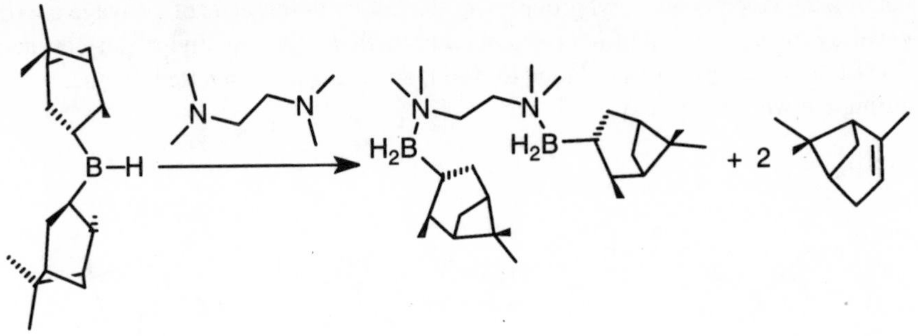

1   **TMEDA**   **5**

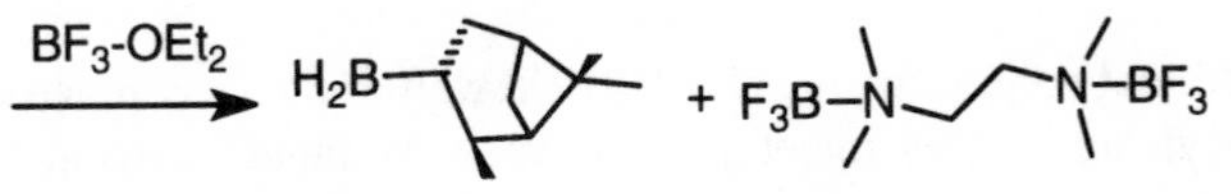

6

The reaction of enantiomerically pure **1** with TMEDA regenerates enantiomerically pure (+)-α-pinene. However, if recovery of enantiomerically pure (+)-α-pinene is the objective, heating pure **1** with benzaldehyde results in cleavage to benzyl borate and (+)-α-pinene [13]. Both pure enantiomers of α-pinene, 99% ee, are now commercially available in laboratory quantities [14].

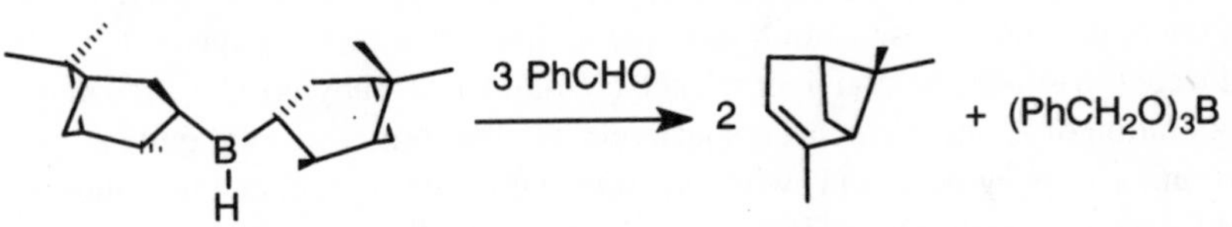

1   (+)-α-pinene

6.2.1.2 Recovery of α-pinene. The cleavage of (isopinocampheyl)(alkyl)boranes with acetaldehyde to form boronic esters provides a key step in many of the processes which utilize these asymmetric hydroboration products in synthesis. This reaction appears to be very general, because if an olefin is not too hindered to be hydroborated by diisopinocampheylborane or monoisopinocampheylborane, then α-pinene will be cleaved far more rapidly than that olefin from the resulting trialkylborane or dialkylborane [15]. Its importance lies not only in the recovery of the valuable chiral director, but also in the fact that boronic ester chemistry can utilize the chiral alkyl group in further transformations without side reactions involving the isopinocampheyl groups. An illustration of this type of cleavage is provided by the conversion of intermediate **2** to diethyl (*R*)-(1-methylpropyl)boronate (**7**). The ethyl esters are very labile to water, and are often converted to other esters during the work up process [15].

It may also be noted that isopinocampheyl-9-BBN, a well known asymmetric reducing agent for acetylenic ketones sometimes known as Midland's reagent or Alpine Borane, eliminates α-pinene in the reduction process (see Section 9.2).

6.2.1.3 Alternative chiral directors. The 2-ethyl homologue of **6**, mono(2-ethylapoisopinocampheyl)borane (**8**), has been tested as a hydroborating agent [16]. Hydroborations of trisubstituted olefins with **8** usually gave somewhat higher ee's than hydroborations with monoisopinocampheylborane (**6**), though typical numbers would be 78% ee for hydroboration of 1-methylcyclohexene with **8** and 72% with **6** [16]. Inasmuch as the utility of **6** for hydroboration of such substrates depends on crystallization of the borane intermediates from the reaction medium with consequent upgrading of the de to nearly 100% [6], the extra effort of preparing **8** does not appear justified. The analogue of **8** having a phenyl group in place of ethyl yielded much lower ee's than those from **8** or **6** and is thus not a useful reagent [17]. Also tested were (1*S*)-di-2-isocaranylborane and (1*S*)-di-4-isocaranylborane, prepared by hydroboration of 2- and 3-carene, respectively, but the former was not quite as good as diisopinocampheylborane (**1**) with any of the substrates tested, and the latter was generally poor, though it gave 75% and 80% ee, respectively, with norbornene and norbornadiene [18].

(*R*)-(*R**,*R**)-2,5-Dimethylborolane (**14**) and its (*S*)-enantiomer are chiral hydroborating agents that give exceptionally high asymmetric induction and are of considerable theoretical interest [19]. Unfortunately, the reported synthesis first yields a 53:47 mixture of racemic borinic ester (R,S)-**9** and meso diastereomer **10**, which first have to be separated by selective reaction of the meso isomer with N,N-dimethylethanolamine to form **11**. Then L-prolinol forms a stable, crystalline derivative **12** with the (R)-enantiomer of **9**, leaving (S)-**9** in the form of the volatile borinic ester. After separation, **12** is converted back to (R)-**9** with acid and methanol, then reduced by lithium aluminum hydride to the borohydride (R)-**13**, and oxidized with methyl iodide to the free borane (R)-**14**. The (S)-borinic ester (S)-**9** is purified via the crystalline L-valinol derivative, then regenerated and transformed to the borane (S)-**14** in a similar manner (not illustrated) [19].

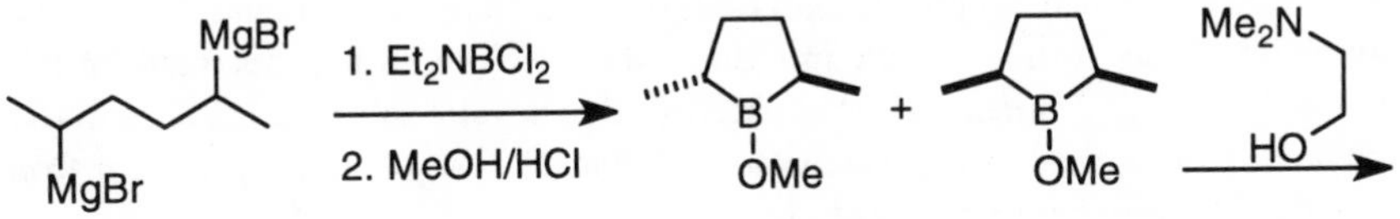

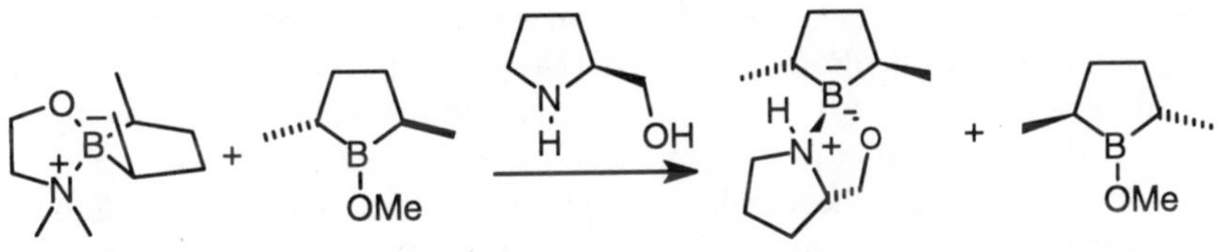

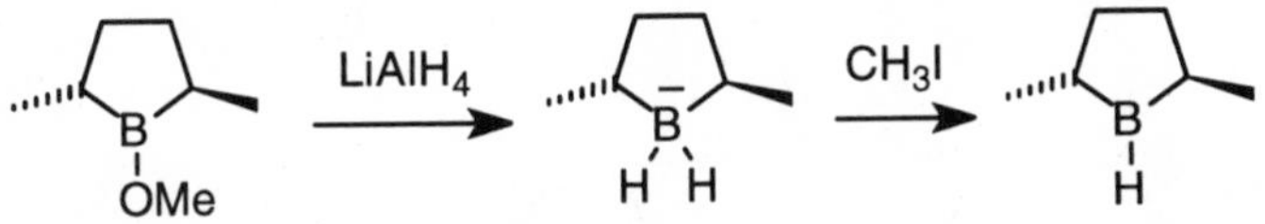

As if the foregoing were not enough obstacle to the practical use of (R)- or (S)-**14** as a hydroborating agent, the compound is unstable to storage and has to be generated fresh. Dissociable and reactive dimer (R)-**15** isomerizes via the monomer (R)-**14** to the unreactive dimer (R)-**16** with a half-life of several days at 0.5 M in THF at 25 °C [19].

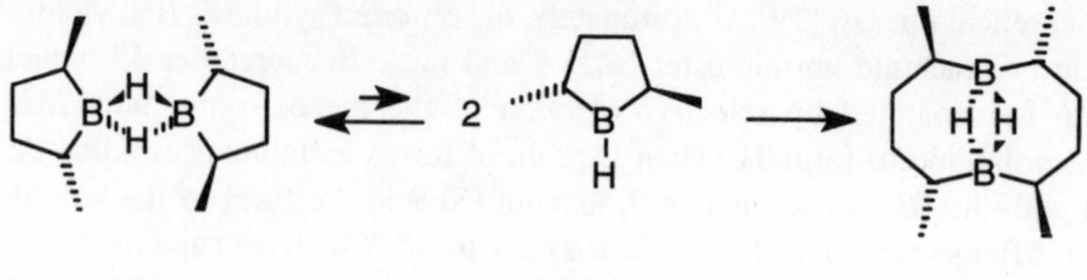

(R)-15   (R)-14   (R)-16

6.2.2 Hydroboration with Diisopinocampheylborane

Purified (−)-diisopinocampheylborane (1) made from (+)-α-pinene [3] is an effective reagent for the asymmetric hydroboration of *cis* olefins that are unsubstituted at both olefinic sites, as noted in the introduction [2, 4]. The 98% ee obtained for (R)-2-butanol from *cis*-2-butene is not equaled by other acyclic substrates. The ee of (R)-3-hexanol from *cis*-3-hexene was 93%, and that of (S)-1-phenyl-1-propanol from *cis*-1-phenylpropene only 63% [4].

In contrast to the mediocre results with acyclic olefins, 1 with several mono-unsaturated five-membered heterocycles has yielded enantiomerically pure products described as "100% ee" [20]. For example, treatment of 2,3-dihydrofuran with 1 leads to intermediate borane (17), which with acetaldehyde liberates (+)-α-pinene and diethyl (R)-(3-tetrahydrofuranyl)borate (18) [15], which is converted by alkaline hydrogen peroxide to (−)-(R)-3-hydroxytetrahydrofuran (19). Similar treatment of 2,5-dihydrofuran results in the same stereoselectivity of attack on the double bond, with boron going to the front left site, but the different regioisomer leads to the enantiomer of 19, (+)-(S)-3-hydroxytetrahydrofuran (20). Similar transformations have yielded (+)-(R)-3-hydroxytetrahydrothiophene (21) from 2,3-dihydrothiophene and (1R,2S,4R)-1,4-epoxy-2-hydroxy-1,2,3,4-tetrahydronaphthalene (22) from 1,4-epoxy-1,4-dihydronaphthalene [20]. Product alcohols were isolated in 68-92% yields

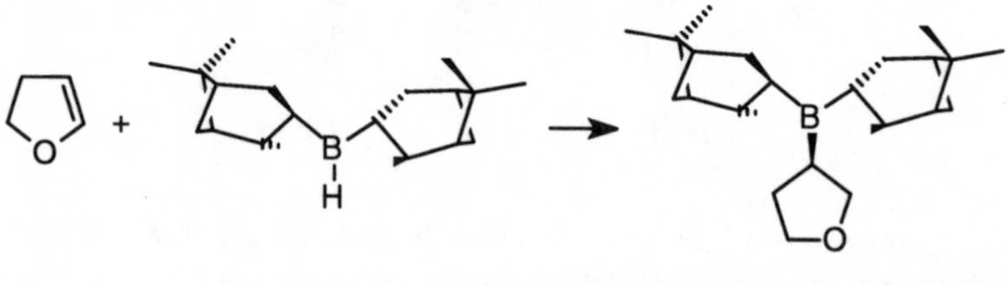

1   17

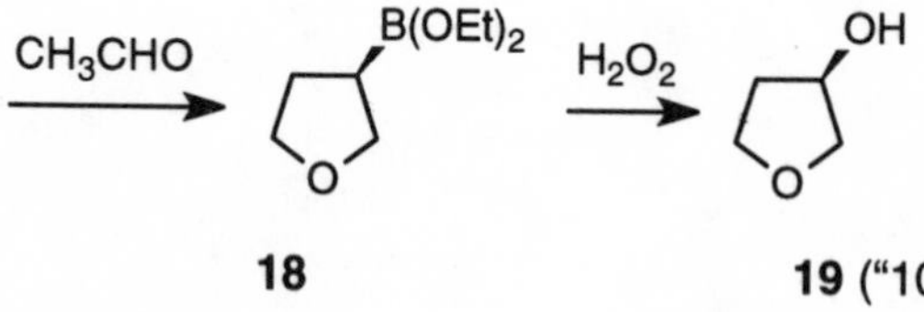

18   19 ("100% ee")

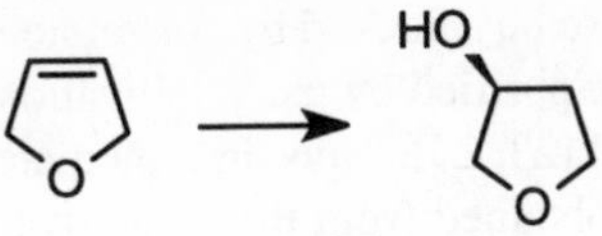

20 ("100% ee") 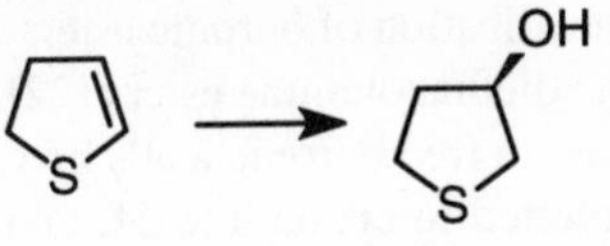   **21** ("100% ee")

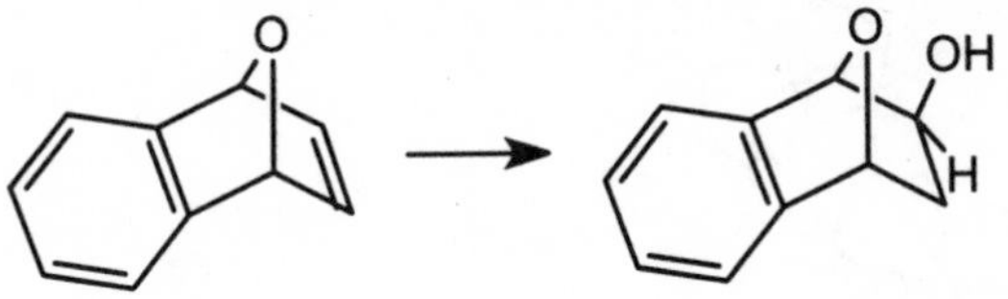

22 ("100% ee")

2,5-Dihydropyrrole fails to undergo hydroboration with **1** because the basic nitrogen complexes with the borane and the double bond is deactivated. This problem was overcome by making the *N*-carbobenzyloxy derivative, which yielded the corresponding (*S*)-alcohol (**23**) in 89% ee [20].

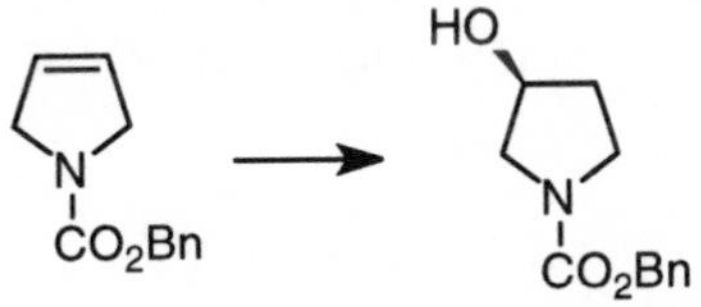

23 (89% ee)

All of the foregoing reactions were also carried out with the enantiomer of **1** to provide the enantiomers of the listed products. The enantiomeric purities were determined from ^{19}F-NMR spectra of the (methoxy)(trifluoromethyl)(phenyl)acetates (Mosher esters) taken on a 200-MHz instrument. Rotations were also taken, but because they were higher than literature values in several cases, they are uninformative. Several of the ee's were listed as "100%", but no critical evidence was presented regarding to how many significant figures this approximation was valid.

Recently it has been found that aldehydes, preferably benzaldehyde, cleave the major diastereomer of the (diisopinocampheyl)boranes significantly faster than they attack the minor isomer, and that this difference in rates provides additional upgrading by kinetic resolution [21]. Thus, propanediol esters of (2-butyl)-, (3-hexyl)-, and (3-cyclohexen-1-yl)boronic esters were obtained in ≥99% ee, 60-74% yields. The (*exo*-2-norbornyl)boronic ester was obtained in 97% ee, upgraded to ≥99% by recrystallization of the boronic acid, 54% yield.

Further purification of boronic esters can usually also be achieved by conversion to crystalline diethanolamine esters [22], which can be purified by recrystallization and converted to free boronic acids by cold dilute acid [23, 24]. Thus, intermediate **18** was converted to crystalline **24**, and solids were obtained from the other five-membered ring heterocyclic boronic esters in the series as well [20].

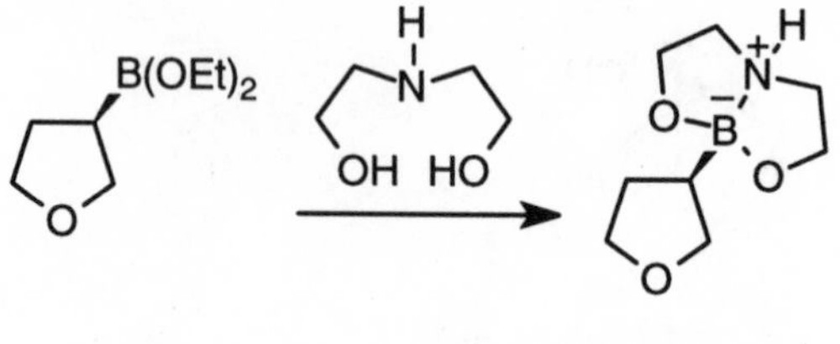

18   **24**

Several amino alcohols and amino acids have also been tested for making chelates of boronic or borinic acids for purposes of upgrading enantiomeric purity [25, 26].

Six-membered heterocycles gave only fair to good stereoselection in hydroboration with **1** [20]. The ee of (R)-3-hydroxytetrahydropyran (**25a**) from 3,4-dihydropyran was 83%, and that of (R)-3-hydroxytetrahydrothiapyran (**25b**) from 3,4-dihydrothiapyran was 66%. N-Carbobenzyloxy-1,2,3,6-tetrahydropyridine was converted to a mixture of 85% 3-alcohol **26**, 70% ee, and 15% achiral 4-alcohol **27**.

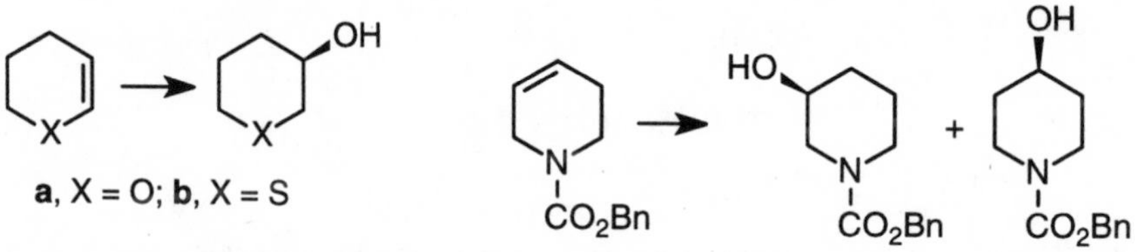

a, X = O; b, X = S

25   **26**   **27**

Useful intermediates for synthesis have been obtained in high ee's from **1** and carbocycles. Examples include **28** from 5-methylcyclopentadiene, used in a synthesis of loganin [27], prostaglandin intermediate **29** from methyl cyclopentadiene-5-acetate [28], **30**, a precursor to zeaxanthin, from safranol ether [29], and **31**, used in a synthesis of capsorubicin [30].

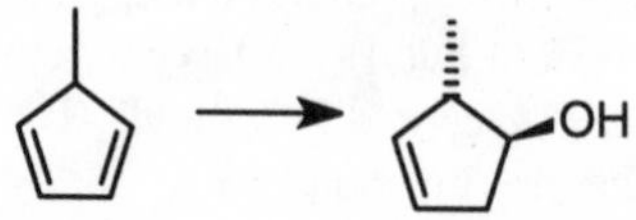

28 (96% ee)

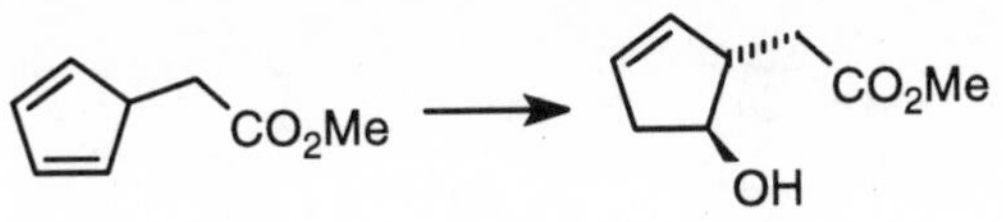

29 (92% ee)

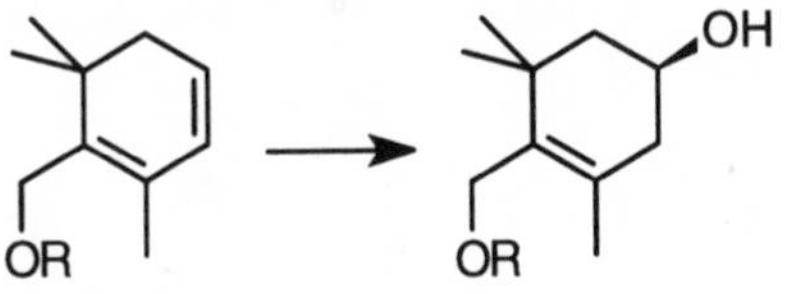

30 (>95% ee)

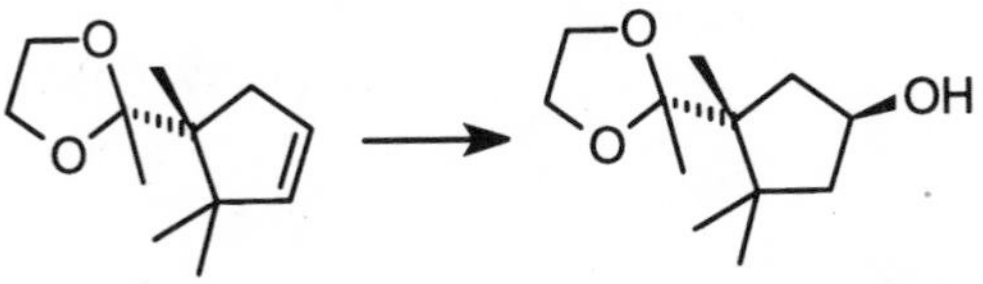

31 (>95% de)

In contrast to the behavior of cyclopentadienes, hydroboration of cyclohexadiene with **1** places the boron atom mainly in the allylic position, and the oxidation product is 92% **32**, 93% ee, with only 8% homoallylic isomer [31]. The allylic borane intermediate does not undergo allylic rearrangement and consequent epimerization as long as it is kept below –25 °C.

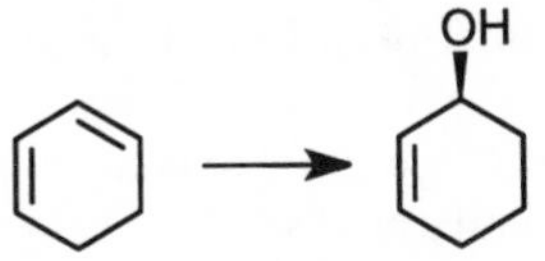

32

Although (–)-diisopinocampheylborane (**1**) provides almost no asymmetric induction in the hydroboration of simple 2-methyl-1-alkenes, it becomes an effective reagent when a sufficiently large alkyl group is present [32]. Thus, tylonolide intermediate **33** has been converted to the next intermediate **34** in at least a 50:1 diastereomeric ratio by hydroboration with **1**. The use of the (+)-enantiomer of **1** yielded the diastereomer of **34**, again >50:1, having the CH₃ group in front [32]. Thus, the chiral director **1** or its enantiomer and not the chirality of the substrate governs the stereoselection in the hydroboration process. Incidentally, it was observed that the achiral hydroborating agent 9-BBN yielded a 2:1 ratio of **34** to **35**.

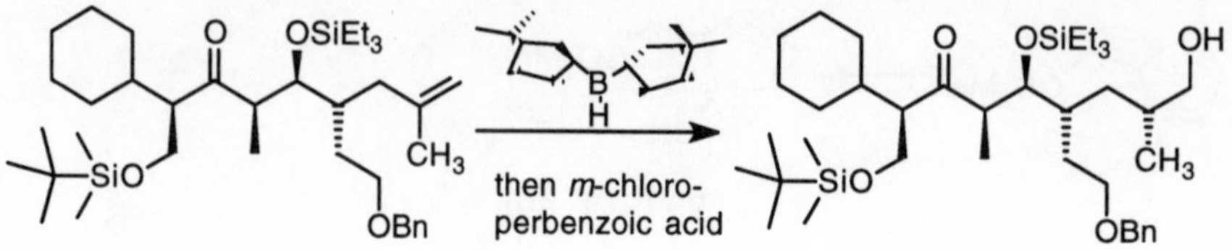

33

34 (>96% de)

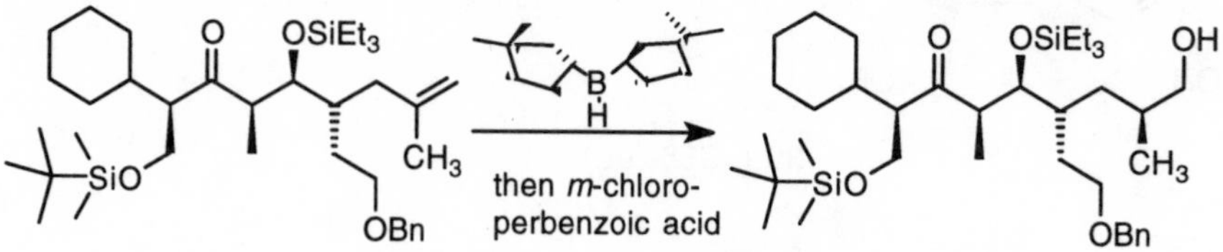

33

35 (>96% de)

6.2.3 Hydroboration with Monoisopinocampheylborane

Hydroboration of *trans*-alkenes and trisubstituted alkenes, including cyclic examples, has been accomplished with fair to excellent enantioselection by monoisopinocampheylborane (**6**). The enantiomer derived from (+)-α-pinene places the boron on the front right site of double bonds, opposite the preference of diisopinocampheylborane (**1**). Trisubstituted olefins also require that the boron add to the less hindered end, so that only one enantioface is attacked.

The best results have been obtained with phenyl substituted olefins. For example, (1S,2S)-2-phenylcyclopentanol (**38**) has been obtained in 85% ee from 1-phenylcyclopentene via the borane **36** and the boronic ester **37** [33]. The conversion to the boronic ester prior to peroxidic oxidation allows recovery of the chiral director, (+)-α-pinene, and simplifies purification of the product **38**. Similarly, (1S,2S)-2-phenylcyclohexanol (**40**) was obtained in 97% ee from 1-phenylcyclohexene via the borane **39** [33], and (2S,3S)-3-phenyl-2-butanol (**41**) in 82% ee from 2-phenyl-2-butene [12].

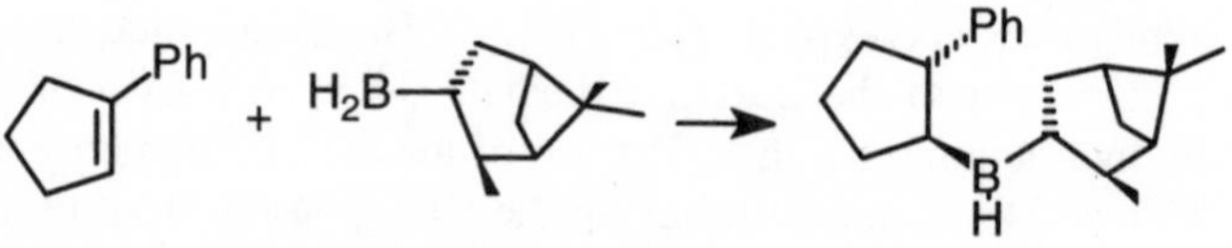

6   **36**

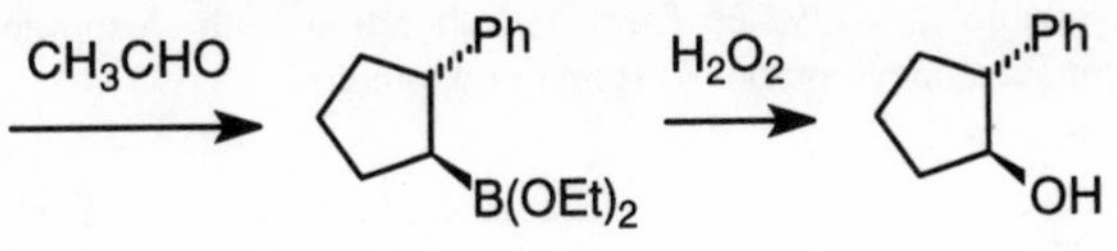

37   **38**

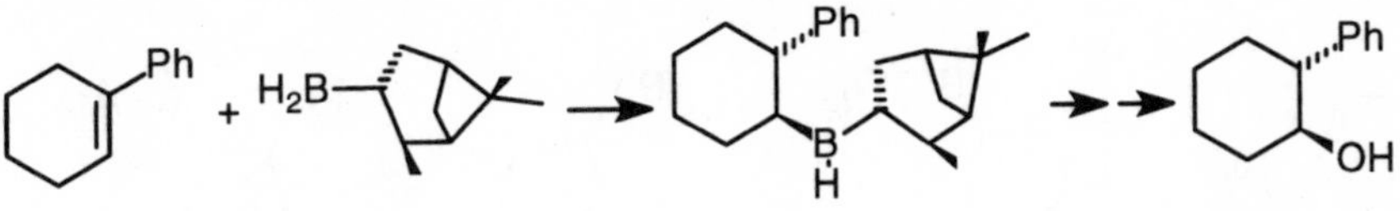

6   **39**   **40**

41

Some borane intermediates such as **36** crystallize from the reaction mixture, permitting upgrading of the purity of the product to nearly 100% ee. Others such as **39** have failed to crystallize, which merely illustrates the well known limits of crystallization as a general technique for purifying diastereomers.

Hydroboration of several methyl substituted olefins with monoisopinocampheylborane (**6**) initially gave the following alcohols in mediocre ee's: **42**, 53%; **43**, 66%; **44**, 72% [34, 35]. However, in these cases the intermediate boranes can be crystallized and the ee's of the resulting alcohols upgraded to essentially 100% ee [35].

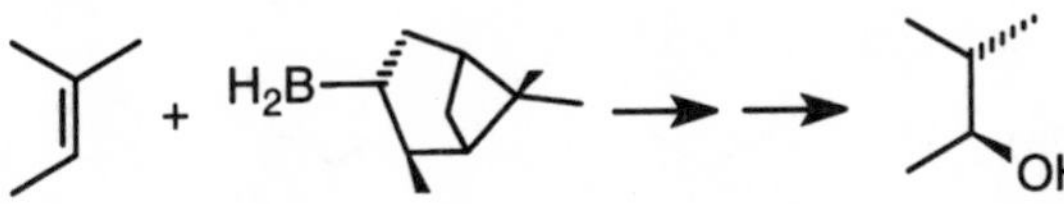

42

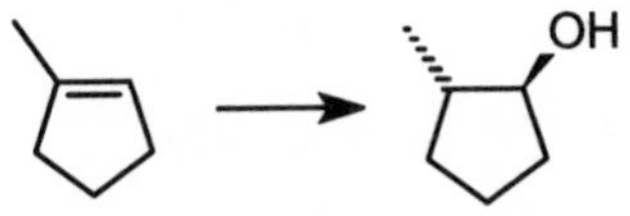

43 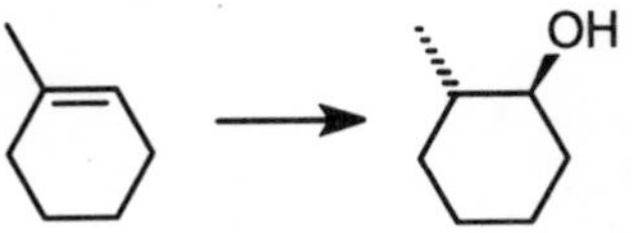   **44**

Table 6-1. Alkylboronates available in >99% ee from hydroboration with diisopino-campheylborane (**1**) ormonoisopinocampheylborane (**6**) from (+)-α-pinene.

Olefin + **1** →	Boronic Ester	Olefin + **6**	Boronic Ester

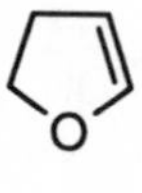

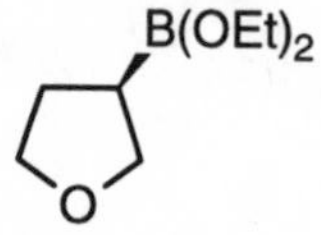

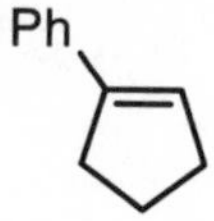

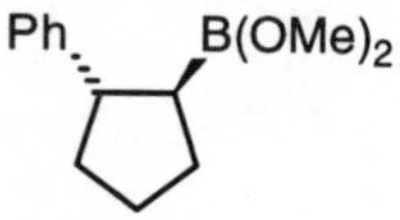

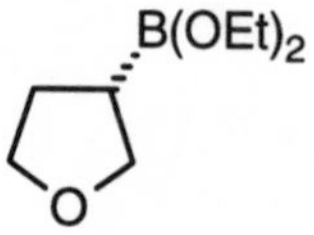

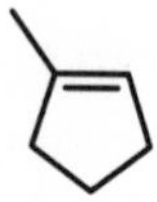

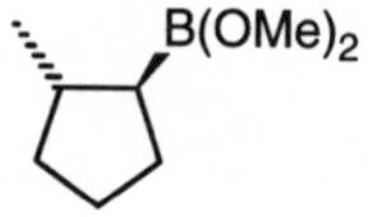

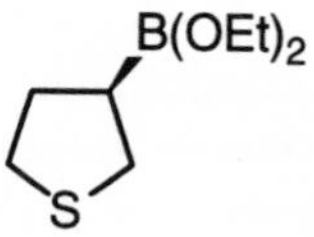

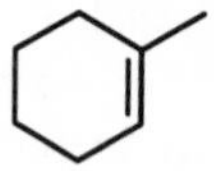

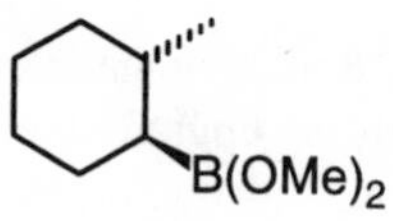

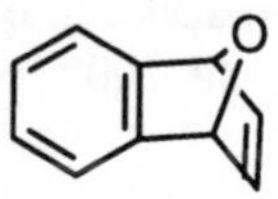

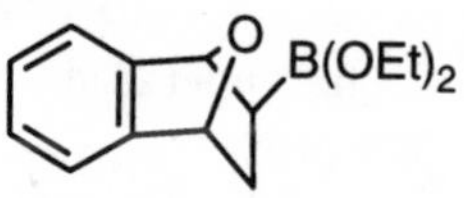

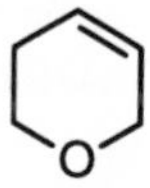

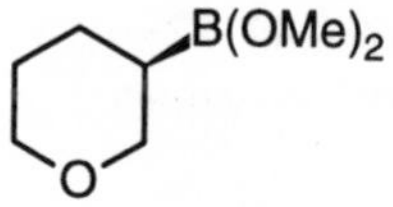

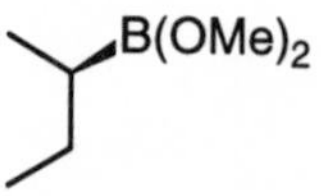

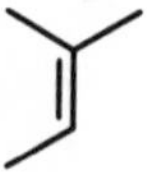

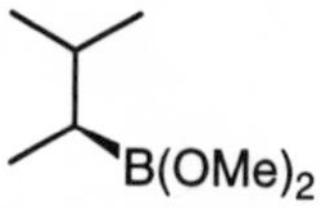

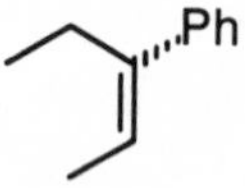

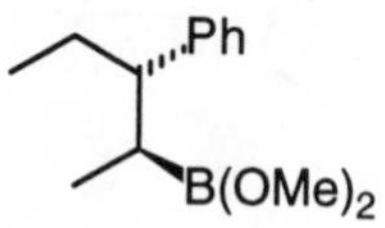

The diethyl boronate intermediates in the foregoing sequences may be isolated directly or converted to diethanolamine esters [20]. Alternatively, extraction into aqueous base and regeneration with acid separates the boronic acids, which are converted by methanol/pentane partitioning [36] to methyl esters [35]. Some of the typical boronic esters that have been obtained in essentially 100% ee from this chemistry are summarized in Table 6-1 [6]. Although three examples of acyclic chiral alkylboronic esters are included, it should be noted that these, as well as many other acyclic chiral alkylboronic esters inaccessible by hydroboration, could be prepared more easily in very high enantiomeric and diastereomeric purity via the reaction of asymmetric boronic esters with (dichloromethyl)lithium described in detail in Chapter 5. The hydroboration technique is valuable for generating certain specific types of chiral cycloalkylboronic esters, which are otherwise inaccessible.

Diethyl boronates listed in Table 6-1 were obtained from (diisopinocampheyl)(alkyl)boranes of high diastereomeric purity by cleavage with acetaldehyde [15] and removal of the solvents and (+)-α-pinene under vacuum [20]. These were also converted to crystalline diethanolamine esters [20]. Dimethyl (R)-2-butylboronate was obtained via a roundabout route involving conversion of (diisopinocampheyl)(2-butyl)borane of 93-97% diastereomeric purity to ethyl ester, then via borohydride to (monoisopinocampheyl)(2-butyl)borane, which was cleaved and converted to the methyl ester [35]. Dimethyl boronates were obtained from purified (monoisopinocampheyl)(alkyl)boranes via cleavage with acetaldehyde, base extraction to separate boronic acids from (+)-α-pinene, regeneration of the boronic acids, and conversion to methyl esters [35].

In addition to the tabulated examples, a series of nine cycloalkenylheterocyclic compounds have been hydroborated with monoisopinocampheylborane (6), initially in 83-90% ee's, and were upgraded by the various recrystallization techniques to [trans-2-(heteroaryl)-1-cycloalkyl]boronic esters of ≥98-99% ee [37]. Ring sizes included 5, 6, and 7 carbons. Heterocyclic groups included 2- and 3-furyl, 2-benzofuryl, and 2- and 3-thienyl.

6.3 Transformations of Asymmetric Organoboranes

6.3.1 Overview

6.3.1.1 General summary of transformations. Most of the organoborane transformations summarized briefly in Section 3.2 and 3.4 are stereospecific, and once the borane has been made in high enantiomeric purity, all of its derivatives can be made in high enantiomeric purity as well. The chemistry described in this section has all been reported first in the context of diastereoselective reactions. Only the applications to asymmetric substrates of high enantiomeric purity are recorded here. The model syntheses performed have generally utilized just those asymmetric boranes that are accessible via hydroboration, but the chemistry would often be equally applicable to the broader range of asymmetric boronic esters provided by the more general route described in Chapter 5.

Table 6-2. Structures accessible stereospecifically from boronic esters

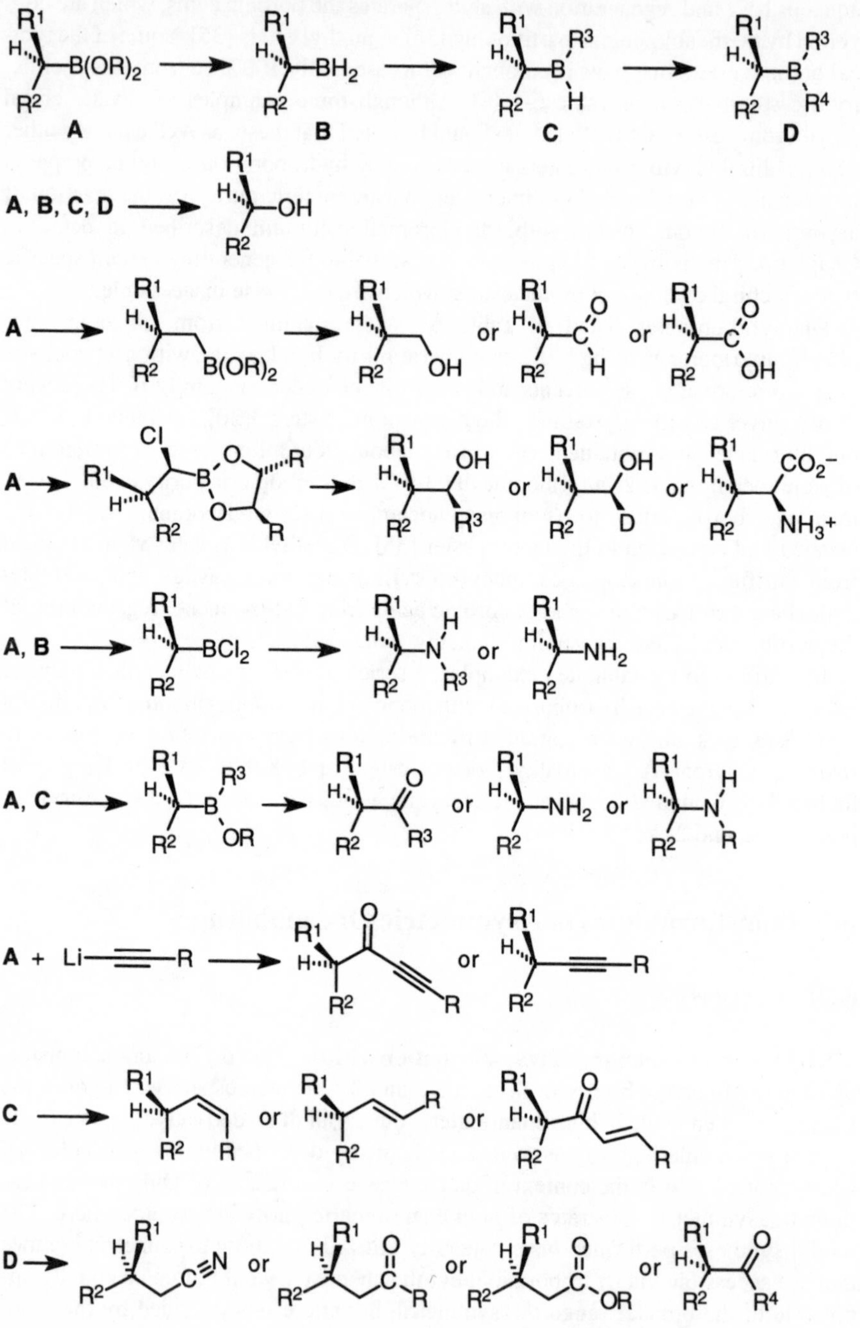

A brief summary of some of the known stereospecific transformations of chiral R*BY$_2$ that can generally be carried out without isolation and purification of intermediates is provided in Table 6-2. Included is an indication of which borane intermediate is required as an immediate precursor to each type of product. Transformations are selected from Brown's circular chart of twenty-four replacements of boron that retain configuration [6], with some pruning of redundant items and reactions not yet carried out with pure enantiomers, plus a few of the more significant transformations of boronic esters from Chapter 5 to provide perspective. Such a summary is inherently an arbitrary selection, and is only meant to suggest the wide variety of asymmetric products that can be made via borane chemistry.

6.3.1.2 Interconversions of boron oxidation states.

Borinic esters can be made by adding alkyllithium reagents to 2-alkyl-1,3,2-dioxaborinanes and quenching with trimethylsilyl chloride [38, 39]. An example is the [(R)-2-butyl](phenyl)borinic ester **46**, which is derived from the boronic ester **7** via (R)-2-(2-butyl)-1,3,2-dioxaborinane (**45**) [39].

| 7 | 45 | 46 |

The reduction of alkylboronic esters to alkylborohydrides can be accomplished with lithium aluminum hydride [40], and the resulting alkylborohydrides have been converted to alkylborane dimers, which have been tested as hydroborating agents [41]. The method reduces boronic esters of high enantiomeric purity to alkylborohydrides without loss of configuration [42]. Chlorotrimethylsilane is the preferred reagent for regenerating the alkylborane. An example is the conversion of diethyl *trans*-(1S,2S)-2-phenylcyclopentylboronate (**37**) via the propanediol ester **47** to the insoluble propanediol aluminum hydride **48** and the soluble borohydride **49**. Conversion of **49** to the alkylborane dimer **50** is followed by hydroboration of 1,5-cyclooctadiene by **50** to form the 9-BBN derivative **51** [42]. Alternatively, **50** can be used to hydroborate tetramethylethylene to form the thexylborane **52**, which will hydroborate terminal olefins to trialkylboranes **53** [43]. This chemistry is, of course, applicable to the entire series of boronic esters prepared in high enantiomeric purity via hydroboration chemistry, as well as much of the wider variety of asymmetric boronic esters accessible by other means.

| 37 | 47 | 48 | 49 |

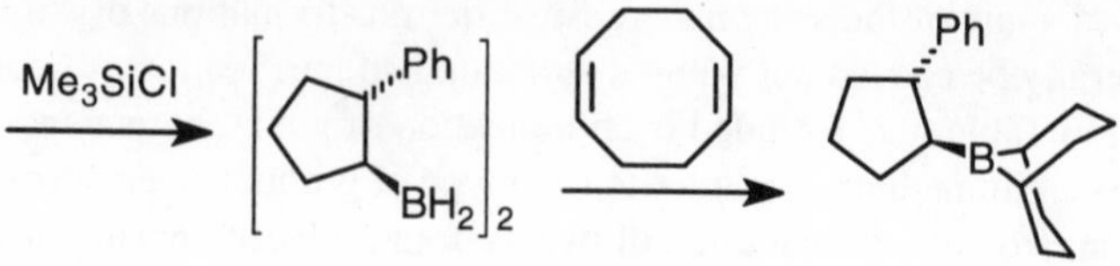

50   **51**

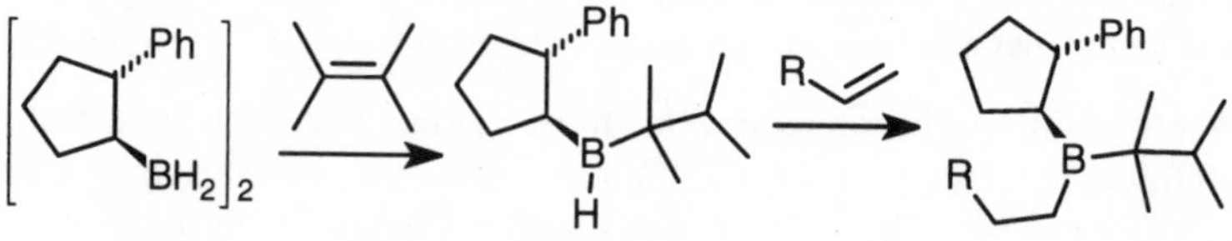

50   **52**   **53**

6.3.2 Carbon—Carbon Bond Formation

6.3.2.1 Introduction. These reactions were all run first in other contexts. In the asymmetric hydroboration context, they are useful for stereospecific conversion of the boron substituted chiral center to one bearing only C and H substituents. The most generally useful carbon insertions are those of (dichloromethyl)lithium described in Chapter 5.

6.3.2.2 Methylene insertion routes. The stereospecific insertion of the unsubstituted methylene group derived from (chloromethyl)lithium into the carbon—boron bond of an asymmetric boronic ester (**54**) was first noted with the original discovery of the efficient in situ generation of the reagent [44]. Preparation of **54** was carried out via reaction of recrystallized (S)-pinanediol (S)-(α-chlorobenzyl)-boronate with ethylmagnesium bromide. The high diastereomeric and enantiomeric purity of **55** was proved by peroxidic oxidation to (+)-(S)-2-phenylbutanol (**56**), apparent ee 100.3% based on comparison with the literature rotations [44]. Rotations can be unreliable guides to enantiomeric purity, but this is well within experimental error of the expected value, >99%. A similar methylene insertion was used in the last stage of the ribose synthesis described in Section 5.4.4.1.

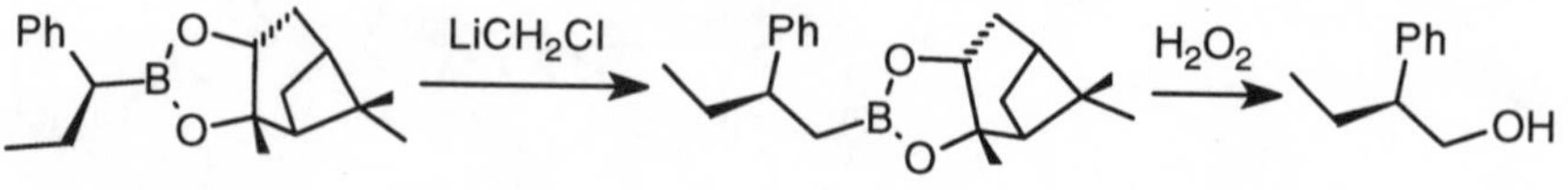

54   **55**   **56**

The synthetic need for such methylene insertions arises in the context of hydroboration chemistry because enantioface selectivity has proved to be poor to mediocre for most combinations of terminal alkene and hydroborating agent [6], with the exception of olefins bearing one very bulky substituent, such as **33** described in Section 6.2.2. For example, alcohol **56** cannot be made in high enantiomeric purity via hydroboration of the corresponding alkene, $PhEtC=CH_2$. Hydroborations of such terminal alkenes being the most regioselective of all, the lack of stereoselectivity is disappointing. However, the steric interactions between the attacking borane and two different alkyl groups on the β-carbon are too weak to provide direction in most instances, even with Masamune's 2,5-dimethylborolane as chiral director [19]. Mechanistic reasons for this failure, and the contrast with the useful stereoselectivity seen in ketone reductions, have been discussed [45].

As part of the program to make as wide a variety of enantiomerically pure model compounds as possible by hydroboration, an investigation of the possible methods for converting boronic esters, $RB(OR')_2$, to their homologues, $RCH_2B(OR')_2$, was carried out by Brown and coworkers. The first route found was to use (dichloromethyl)lithium to make the (α-chloroalkyl)boronic ester (see Section 3.4.3.5 and Chapter 5). The CHCl group is reduced to CH_2 by potassium triisopropoxyborohydride [46, 47, 48]. An example is the conversion of **47** via **57** to **58** [47].

47   57   58

The previously reported use of in situ generation of (dichloromethyl)lithium from dichloromethane and LDA in the presence of the boronic ester [49] has been reexamined and extended [50]. It has been found that the (dichloromethyl)lithium reaction with added methoxide will transfer both alkyl groups of borinic esters (made by hydroboration in this case) from boron to carbon, and that trialkylboranes react well with (dichloromethyl)lithium only if they are sterically unhindered [46]. In addition to (dichloromethyl)lithium, the only other known haloalkyllithiums that react with all three oxidation states of boron are $LiCHClSiMe_3$ and $LiCH(OMe)SPh$ [51].

A critical comparison of the various methods of chain extension has been made with 1,3,2-dioxaborins, for example, conversion of **59** to **60** [52]. The route via the α-chloro boronic ester and reduction with potassium triisopropoxyborohydride was found to be equally efficient (90-92%) with preformed (dichloromethyl)lithium and with (dichloromethyl)lithium generated in situ from dichloromethane and LDA. The in situ method worked as well at 0 °C as it did at −78 °C, though it should be noted that in previous work on this chemistry, which involved possibly somewhat less reactive substrates, a sharp drop in yield was noted as the internal temperature of the reaction was increased to 0 °C and above [53]. An interesting innovation, which has evidently not been exploited subsequently, was the finding that *sec*-butyllithium and dichloromethane provided a satisfactory means for generating (dichloromethyl)lithium in situ, and produced only a marginally lower yield (87%) [52]. (Chloromethyl)lithium generated from bromochloromethane and butyllithium was found to be almost as efficient (89%) as that generated from iodochloromethane (96%) [52]. Although bromochloromethane is cheaper than iodochloromethane, the similarity of starting material and product and consequent separation problem, as well as the probable cost of any substrate of real interest in this type of reaction, would make this a dubious savings at best.

(a) LiCHCl$_2$, then KHB(OR)$_3$: 90-92%
(b) LiCHCl$_2$ from *sec*-BuLi, etc.: 87%

(c) LiCH$_2$Cl from ICH$_2$Cl: 96%
(d) LiCH$_2$Cl from BrCH$_2$Cl:89%

59   **60**

(Chloromethyl)lithium from bromochloromethane has been used for chain extension of several of the heterocyclic boronic esters listed in Table 6-1, as well as (1*S*,2*S*)-*trans*-(2-furylcyclopentyl)boronic ester and its 2- and 3-thiophenyl analogues [54].

6.3.2.3 Borinic ester intermediates. Borinic ester **46** (see Section 6.3.1) has been converted to (*R*)-2-butyl phenyl ketone (**61**) in high enantiomeric purity by the use of the reaction with lithiated dichloromethyl methyl ether [39]. This is but one of a series of twelve similar transformations reported, which included several of the chiral groups derived from boronic esters in Table 6-1 paired with such achiral groups as methyl, *tert*-butyl, and 2- and 3-furyl [39]. Yields were 68-92%, ee's ≥99%, except that **61** from the route illustrated was only 97% ee and was brought to ≥99% by the use of trimethylamine oxide in place of hydrogen peroxide. (Also, one member of the reported series was 2-norbornylboronic ester, not available at that time in enantiomerically pure form.)

Cl$_2$CHOMe, LiOCMe$_3$ H$_2$O$_2$

[LiCCl$_2$OMe]   pH 8

46   **61**

Further transformations of pinanediol boronic esters **62** made by chain extension with (dichloromethyl)lithium and methylation have been reported [55]. The use of methyllithium and chloro(trimethyl)silane to form the borinic esters **63** was followed by purification via the ethanolamine esters **64** and conversion to the reactive methyl borinates **65**. The in situ formation of MeOCCl$_2$Li from (dichloro)methyl methyl ether and lithium *tert*-butoxide led to the methyl ketone derivatives **66**, and hydroxylaminesulfonic acid yielded the primary amines **67**.

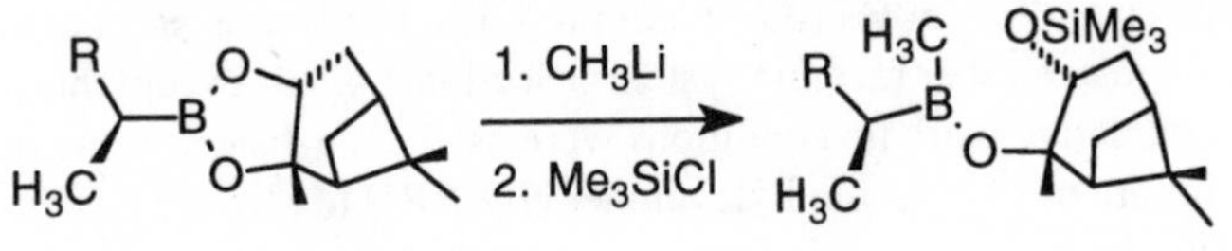

62 (R = *tert*-butyl, cyclohexyl, benzyl)   **63**

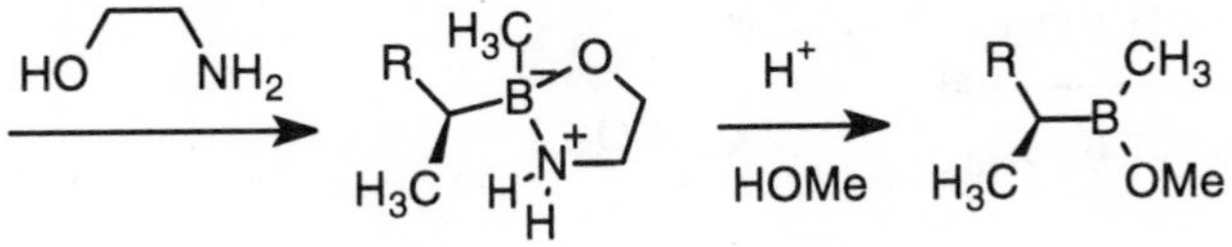

64   **65**

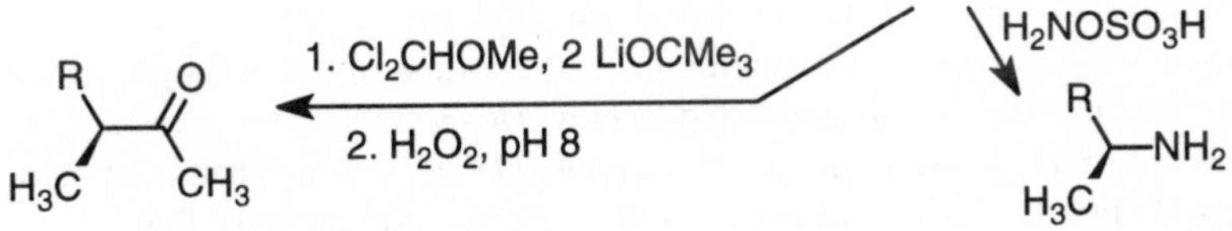

66   **67**

The pinanediol boronic esters (**62**) were also converted to their lithium borohydride derivatives **68**, which were converted to boronic esters **69**. The aldehydes **70** were made via the synthesis with boronic esters and PhSCH(OMe)Li described in Section 3.4.3.10, and these were converted by aqueous dichromate oxidation to the carboxylic acids **71**. The ee's of the products were found to be the same as those of the boronic ester starting materials [55].

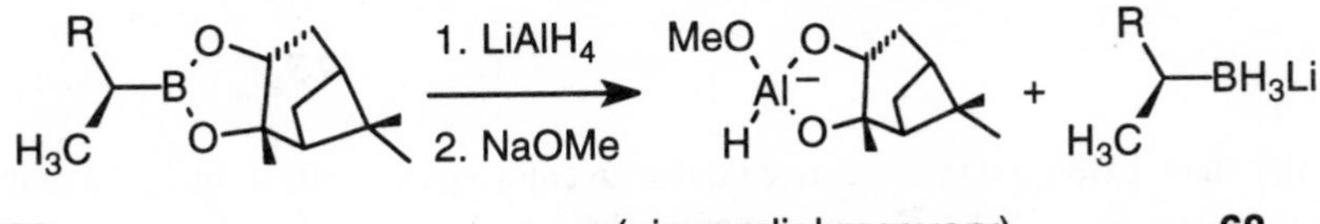

62   (pinanediol recovery)   **68**

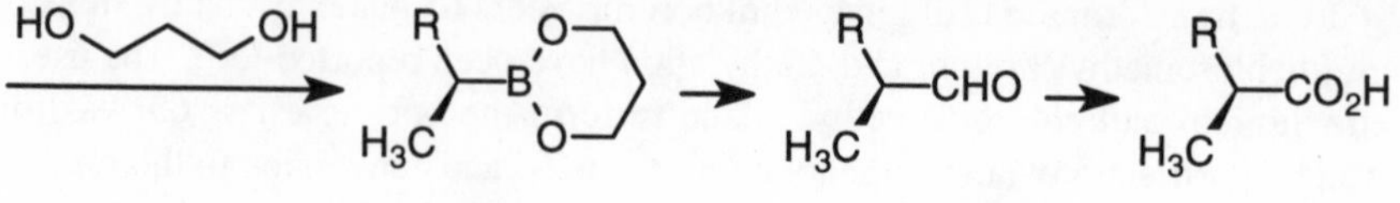
HO
OH
R
O
B
O
H3C
R
CHO
H3C
R
CO2H
H3C

69 70 71

6.3.2.4 Enolate insertions. Treatment of asymmetric alkyl-9-BBN's such as **72** with ethyl bromoacetate and sodium *tert*-butoxide yielded (+)-ethyl [(1*R*,2*S*)-*trans*-2-methylcyclopentyl]acetate (**73**) (50%) (≥99% ee) [42]. The last step is similar to reactions run previously in achiral or racemic systems (Section 3.4.3.4) except that a different base and nonoxidative work up conditions were used. One disappointment encountered was the failure of *trans*-2-methylcyclohexyl-9-BBN to provide a satisfactory yield of ester product.

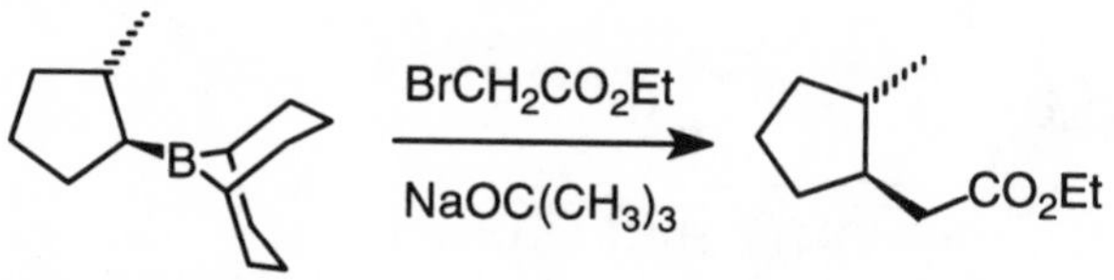
B
BrCH2CO2Et
NaOC(CH3)3
CO2Et

72 73

In case there was any room for doubt that the alkyl migration from boron to the acetate α-carbon proceeds with retention of configuration of the alkyl group, the (*R*)-2-butylboronic ester **7** was converted to ethyl (*R*)-3-methylpentylacetate (**74**), which was reduced to (–)-(*R*)-3-methyl-1-pentanol (**75**) [42], which had been made previously via two successive chain extensions with (dichloromethyl)lithium [47].

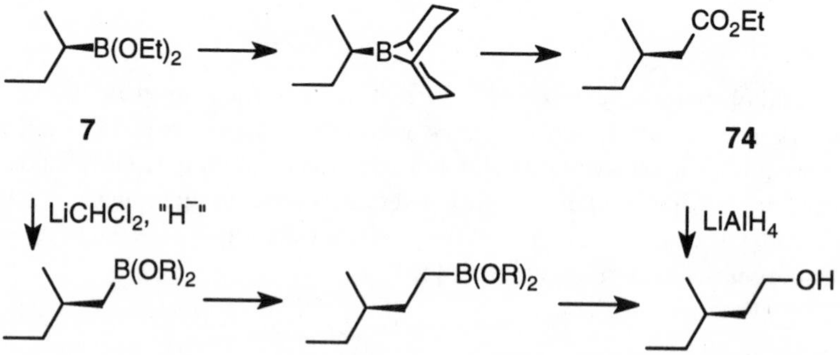
B(OEt)2
B
CO2Et
LiCHCl2, "H⁻"
LiAlH4
B(OR)2
B(OR)2
OH

7 74 75

Similar chain extensions were carried out with chloroacetonitrile to make **76** and with bromoacetone to make **77**, and several other similar examples were reported [42]. Yields were in the 45-66% range, and enantiomeric purities were all very high.

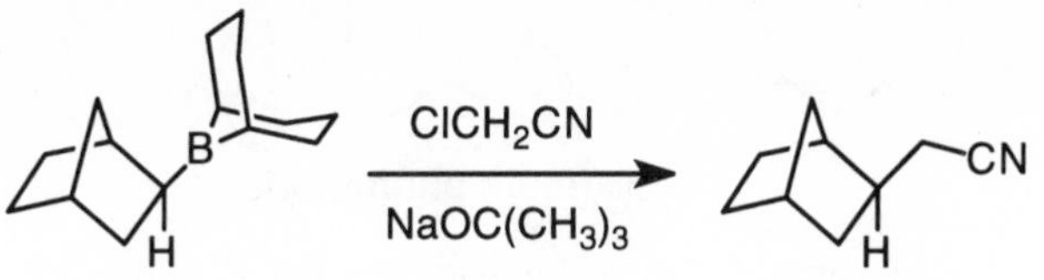

76

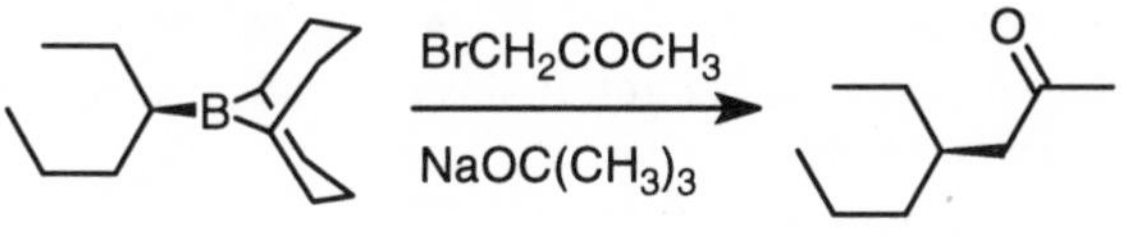

77

6.3.2.5 Thexylborane and pinylborane intermediates. Although RR'BH where R is isopinocampheyl can sometimes be used directly as hydroborating agents, they are often too sterically hindered, and indirect routes via thexylborane intermediates (see **52** in Section 6.3.1.2) are useful. (*R*)-(2-Butyl)(thexyl)borane (**78**) has been converted by methanol to the borinic ester **79**. Treatment of **79** with a lithium acetylide yielded the borate salt **80**, which rearranged on treatment with iodine to yield the acetylenic product **81** [43]. For **81**, R = $(CH_2)_7OSiMe_2CMe_3$, after conversion to **81**, R = $(CH_2)_7OAc$, reduction to the (*Z*)-alkene **82** was accomplished by treatment with disiamylborane followed by acetic acid, and reduction with diimide yielded the saturated compound **83**, which is the sex pheromone of the lesser tea tortrix moth (genus *Adoxopheyes*) [43]. The yield based on **79** was 70%, and the rotation was in good agreement with that of the natural product.

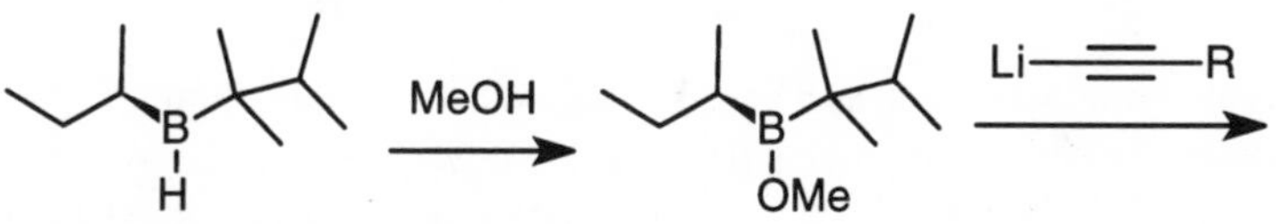

78   **79**

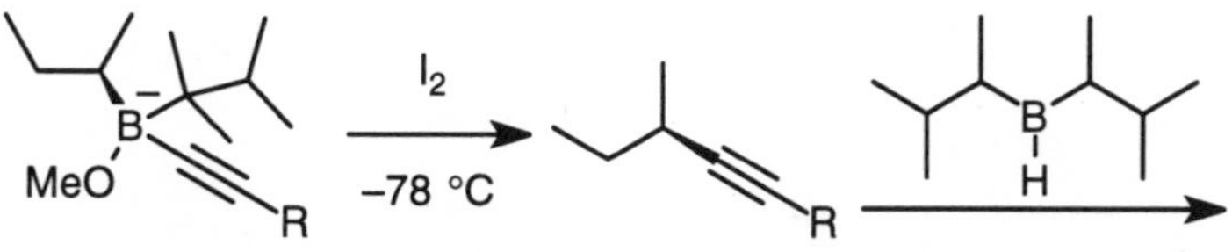

80   **81**

[R = $(CH_2)_7OSiMe_2CMe_3$, **81** converted to $(CH_2)_7OAc$; also R = $SiMe_3$]

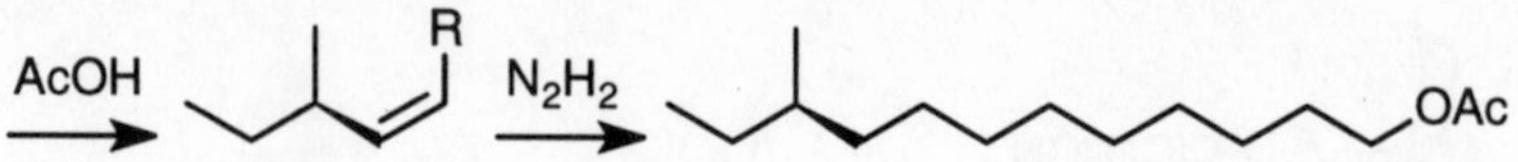

82   83   (moth pheromone)

Preparation of **81**, R = SiMe$_3$, and related compounds has been followed by desilylation to provide the terminal acetylenes, for example, **81**, R = H [56].

Thexylborane **45** (prepared in the same manner as **78**) hydroborates a 1-bromoalkyne to form the α-bromoalkenylborane **84**. Sodium methoxide causes rearrangement of **84** to the borinic ester **85**, in which the carbon chain has been constructed with both asymmetric and geometric control. In this case, **85** was merely protodeboronated to the hydrocarbon **86** [43].

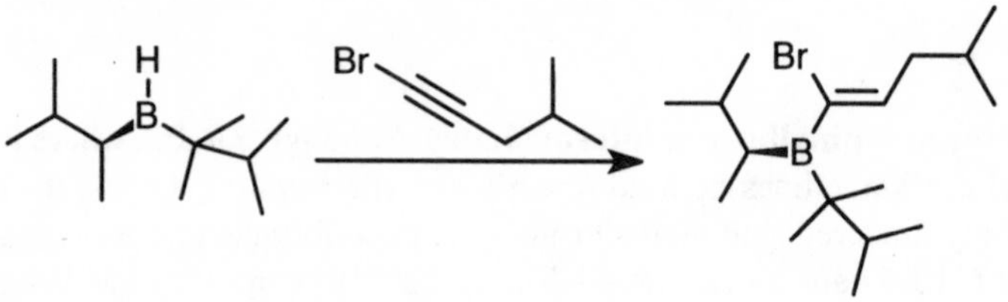

45   84

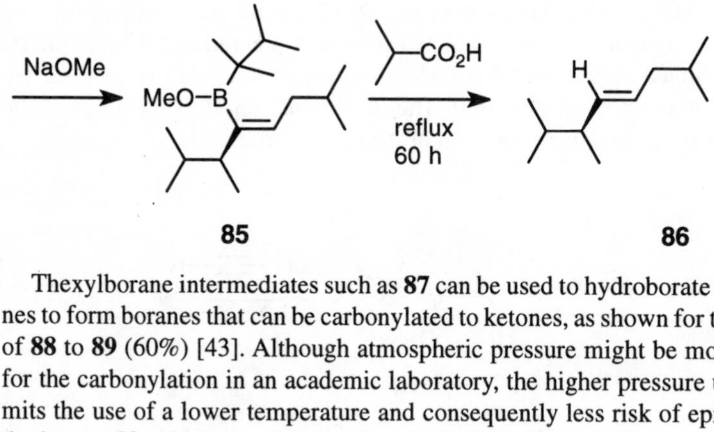

85   86

Thexylborane intermediates such as **87** can be used to hydroborate terminal alkenes to form boranes that can be carbonylated to ketones, as shown for the conversion of **88** to **89** (60%) [43]. Although atmospheric pressure might be more convenient for the carbonylation in an academic laboratory, the higher pressure used here permits the use of a lower temperature and consequently less risk of epimerization of the ketone **89**, which was found to be >99% diastereomerically pure.

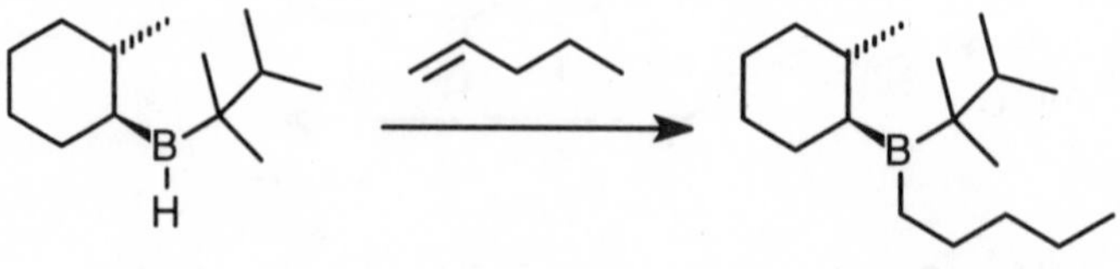

87   88

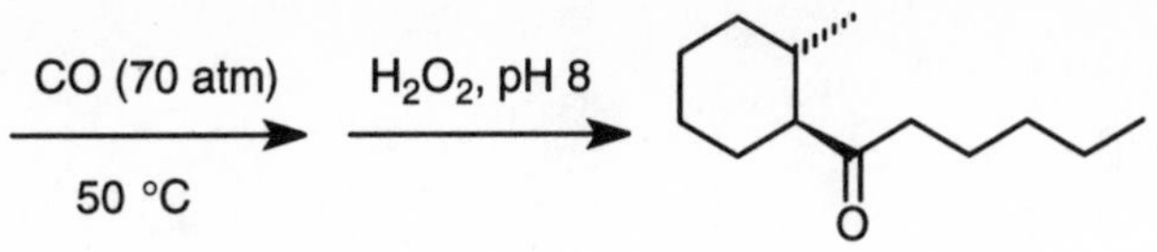

89

Pelter's sodium cyanide/trifluoroacetic anhydride reaction provides another means of carbonylation [57], and this was applied to the thexylborane **90** (derived from borohydride **49**) to yield the ketone **91** (75% based on **49**, >99% de) [43].

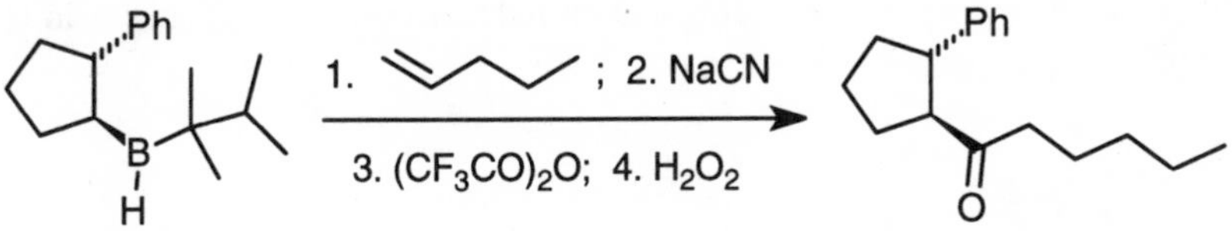

90   **91**

Those alkenes that can be hydroborated asymmetrically with monoisopino-campheylborane (**6**) can be converted to α-chiral (*E*)-alkenones easily [58, 59]. One of a number of illustrative examples is the conversion of methylcyclopentenone (**92**) to the chiral borane **93**, which hydroborates alkynes such as cyclopentylacetylene to form the (*E*)-alkenylborane intermediate **94**. The usual α-pinene cleavage with acetaldehyde leads to borinic ester **95**, which is converted to the α-chiral alkenyl ketone **96**. Alternatively, an enantiomerically pure diethyl alkylboronate, R*B(OEt)$_2$ (see Table 6-1), can be treated with an (*E*)-alkenyllithium to form a borinic ester analogous to **95**, and the synthesis finished in the same manner.

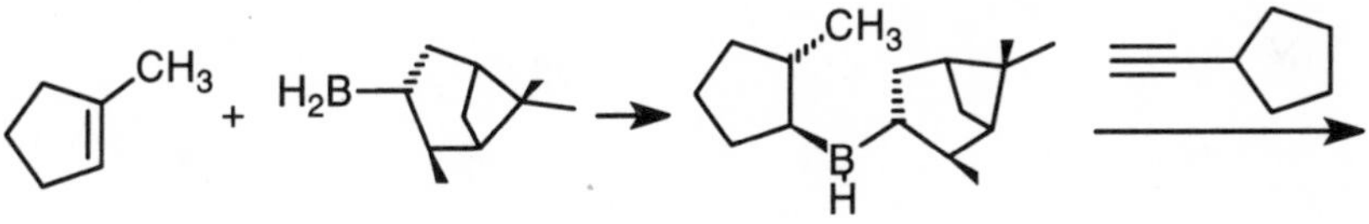

92   **6**   **93**

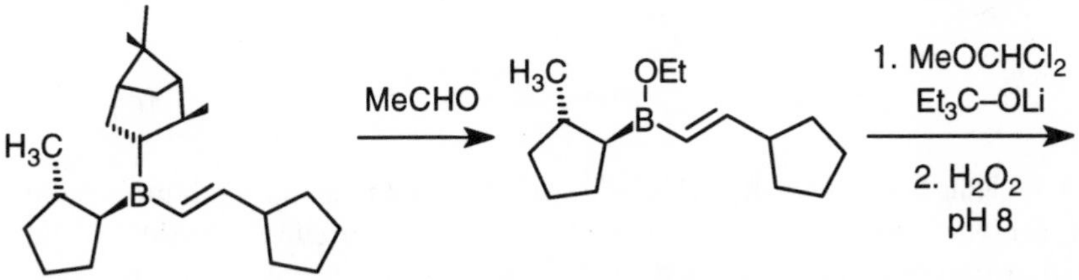

94   **95**

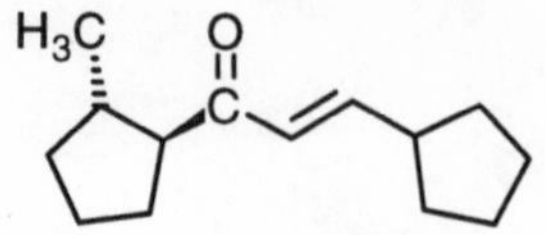

96

(α-Haloalkenyl)(alkyl)boranes undergo migration of the alkyl group from boron to the alkenyl carbon, with stereospecific retention of the configuration of the alkyl group and inversion of the alkenyl geometry [60]. An example is the hydroboration of 1-bromohexyne with borane **97** to form (α-bromoalkenyl)(dialkyl)borane **98**, cleavage of α-pinene with acetaldehyde to borinic ester **99**, base catalyzed rearrangement to alkenylboronic ester **100**, and protodeboronation with acetic acid to alkene **101**.

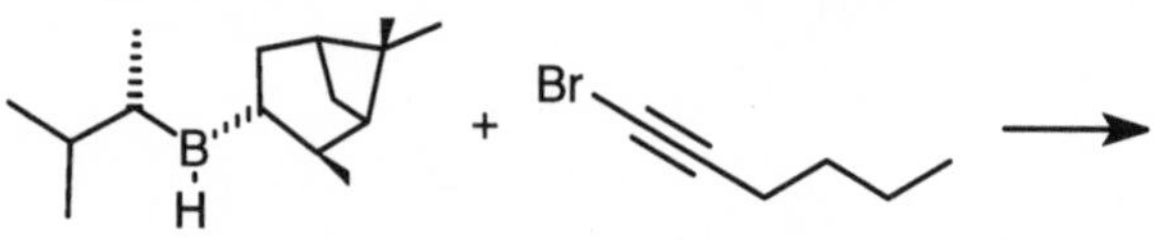

97

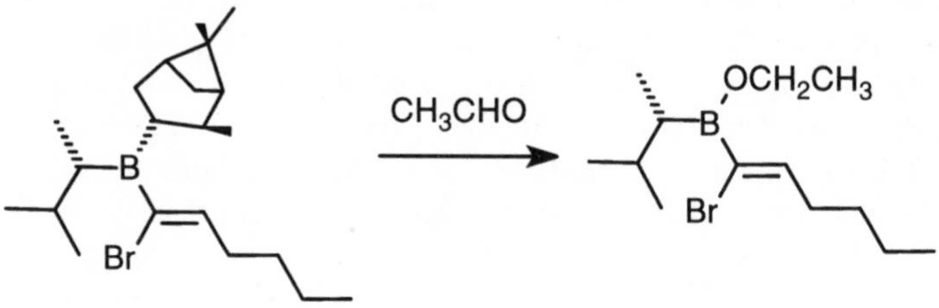

98   **99**

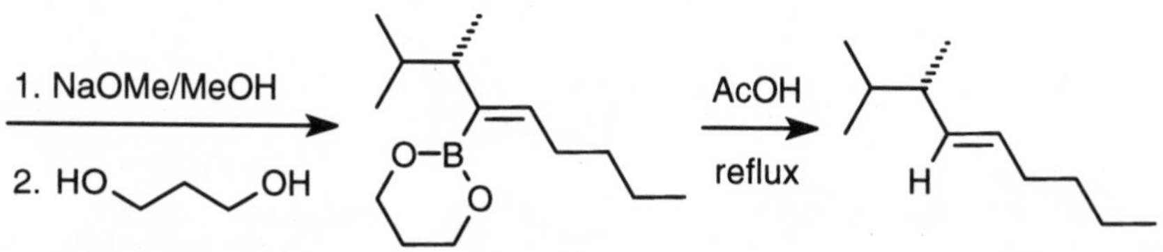

100   **101**

Acetylenic ketones result if a boronic ester is treated first with a lithium acetylide and then with dichloromethyl methyl ether and lithium triethylcarboxide, as illustrated by the conversion of **47** to **102** [61]. Several other examples were also chosen from the usual list of chiral substrates .

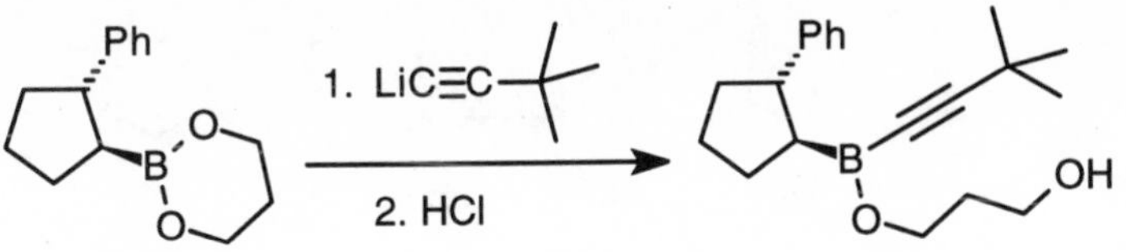

47

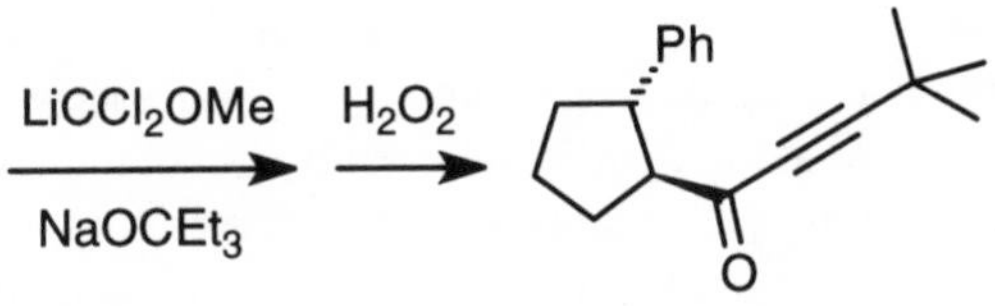

102

Direct oxidation of terminal organoboranes to carboxylic acids with various chromium(VI) reagents has been studied [62, 63]. Chromium trioxide in aqueous 90% acetic acid was found to be the most effective reagent for converting *n*-hexyl-boronic acid to hexanoic acid. Sodium dichromate in aqueous sulfuric acid or pyridinium dichromate in dimethylformamide also accomplish this conversion. The procedures were successful with a variety of borane structures. For earlier work on dichromate oxidations of boranes, and for oxidation of α-chloro boronic esters (or aldehydes) with sodium chlorite, see Section 3.2.2.6.

6.3.3 Carbon—Nitrogen Bond Formation

Conversion of enantiomerically pure boronic esters into alkylboron dichlorides followed by reaction with alkyl azides leads to enantiomerically pure secondary amines [64]. As usual, one of a half-dozen examples used to illustrate the reaction is the asymmetric 2-butyl group. In this case, the azide was also derived from a borane, (*R*)-2-butylborane **2**, via the derived alcohol **3**, which was converted to its tosylate and treated with sodium azide. The enantiomer of **2** was the precursor of the (*S*)-enantiomer of the alkylborohydride **103**, which was converted in the usual manner with hydrogen chloride to the alkylboron dichloride (**104**). The dichloroborane **104** with (*S*)-2-butyl azide (**105**) presumably forms a complex **106** which eliminates nitrogen to form the dichloroaminoborane **107**. (*S,S*)-Di(2-butyl)amine (**108**) was obtained in 65% yield, >99% ee [64].

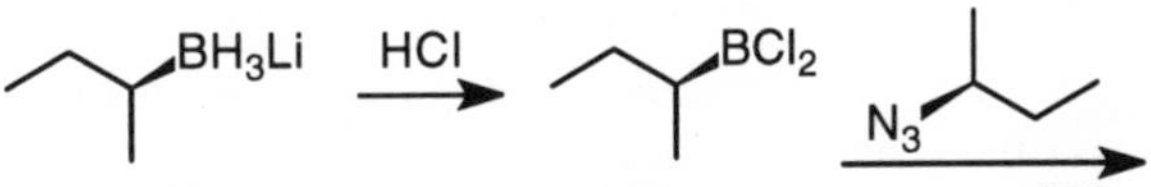

103   **104**   **105**

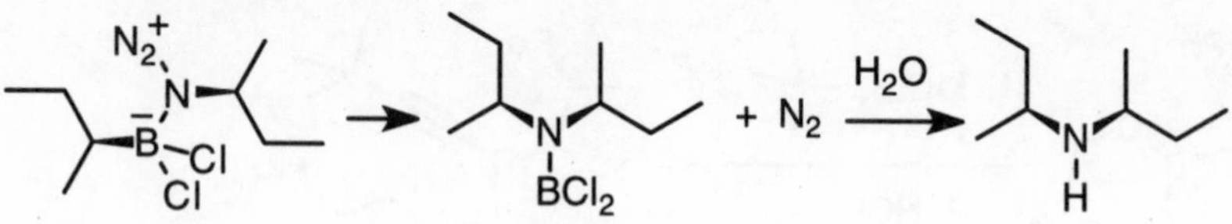

106   **107**   **108**

Though reported with racemates, the synthesis of cycloalkanopiperidines and cycloalkanopyrrolidines has the potential to be developed into an asymmetric synthesis [65]. The synthesis demonstrates the steric discrimination of hydroboration in the regioselective hydroboration of **109** with disiamylborane. After straightforward conversion to the ω-bromoalkylcyclohexene **110**, hydroboration with dichloroborane dimethyl sulfide/boron trichloride and hydrolysis to **111**, and conversion to the azide **112**, treatment with boron trichloride to form intermediate **113** results in ring closure with boron displacement to **114**, which is easily hydrolyzed to cyclohexanopiperidine **115**. Similar syntheses of piperidine fused *trans* with 5- and 7-membered rings and pyrrolidine fused with 6- and 7-membered rings were reported. Obviously, all of this is simply a test run for the planned asymmetric synthesis, which will be accomplished by using monoisopinocampheylborane (**6**) in the second hydroboration step and will involve no new principles.

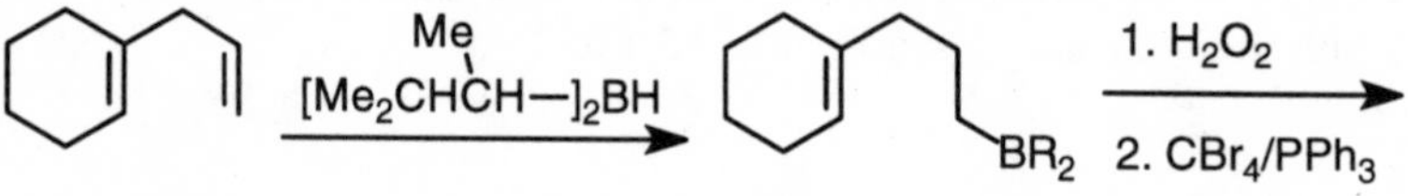

109

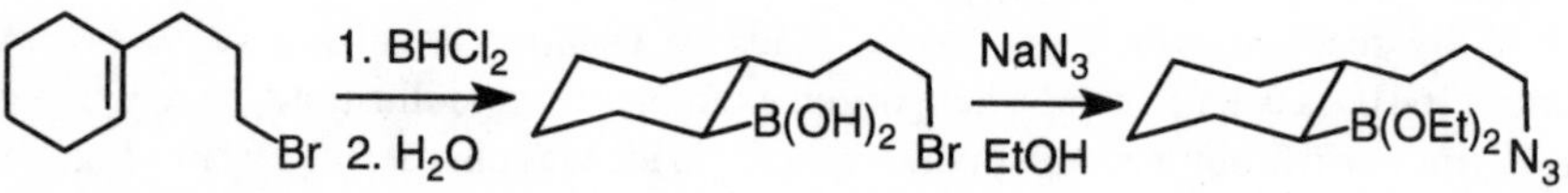

110   **111**   **112**

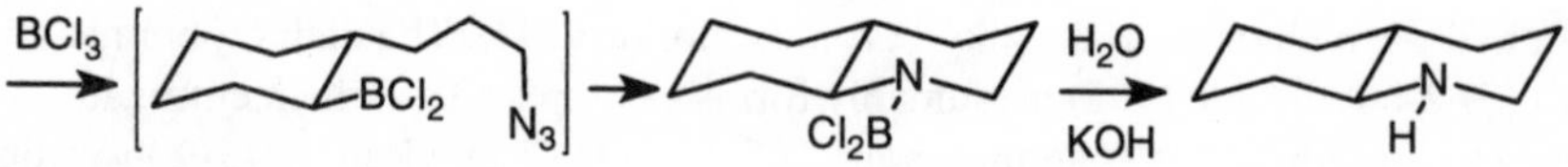

113   **114**   **115**

A particularly interesting example of borane–azide chemistry was preceded by a substrate directed hydroboration and is described in Section 6.4.1.2.

6.4 Substrate Directed Hydroboration

6.4.1 Noncatalytic methods

6.4.1.1 Introduction. No attempt has been made to search the literature systematically for examples of substrate induced diastereoselection in hydroboration, but enough examples are included here to provide some insight into the possibilities. In general, there is a tendency toward relatively low levels of diastereoselection because the boron atom goes to the least hindered site, which is often the most remote from the preexisting asymmetric centers. However, there are also situations in which reaction of a chiral substrate with a borane leads to a borane intermediate having a B—H bond strategically located for highly diastereoselective internal hydroboration.

6.4.1.2 Isopropenyl group hydroboration. It has been noted in Section 6.2.2 that an exomethylene unit flanked by a methyl group and a large chiral chain in structure **33** undergoes hydroboration by 9-BBN to produce a 2:1 diastereomer ratio [32]. Analogous substrate directed acyclic diastereoselection in the hydroboration of **116** with the more sterically demanding thexylborane to form **117** has yielded a modest but synthetically useful 85:15 diastereomeric ratio [66, 67].

116   **117 (85% this isomer)**

Allylic hydroxyl functions protected as boryloxy, silyloxy, alkoxy, or acyloxy can lead to considerable diastereoselection in hydroboration of the double bonds. A fundamental study by Still and Barrish yielded diastereomeric preferences of **119** to **120** in the 5:1 to >15:1 range in the hydroboration of various versions of protected allylic alcohol **118** with 9-BBN or other common hydroborating agents [68]. These results then enabled design of a route to the stereochemical array of ansa chain of rifamycin S (in racemic form) via meso intermediates **121** and **122**.

118   **119 (major)**   **120 (minor)**

R^1 = H, Bu; R^2_2 = (cyclohexyl)$_2$, 9-BBN, etc.; Y = BR^2_2, SiR_3, CPh_3, COMe, etc.

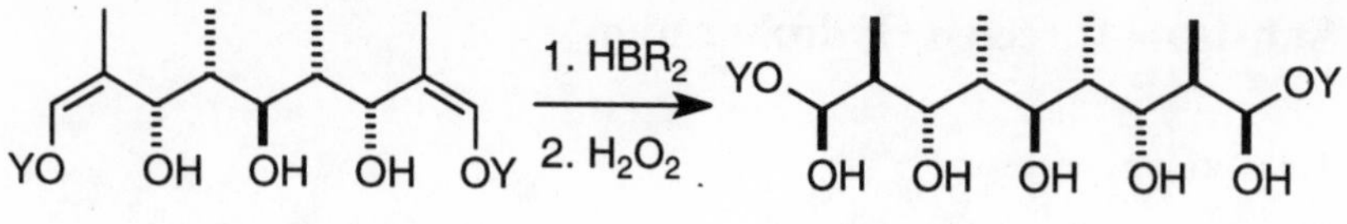

121 (Y = Ph₃C)   **122** (major product, 4:1)

Building on the foregoing results, Evans and Weber reported a felicitous one-pot sequence of substrate directed hydroboration of azido ester **123** that yielded 72% of >97% pure substituted proline ester **127**, an intermediate for the planned synthesis of the antifungal cyclic hexapeptide echinocandin D [69]. The stereodirection in the hydroboration to form postulated intermediate **124** follows the precedent seen with **119**. Subsequent cyclization to the azido borane complex **125** is then followed by alkyl migration/ring contraction to **126**, which is converted to **127** by aqueous work up. It was necessary to use dicyclohexylborane to achieve this result, as a 9-BBN group underwent ring expansion to the nitrogen instead of allowing the desired ring contraction of **125** to **126**.

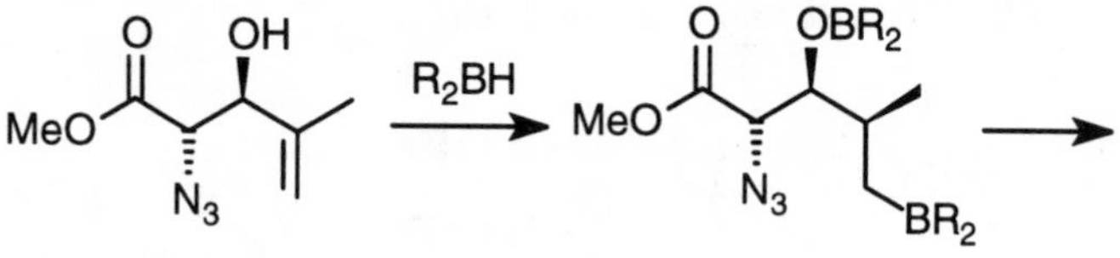

123   (R = cyclohexyl)   **124**

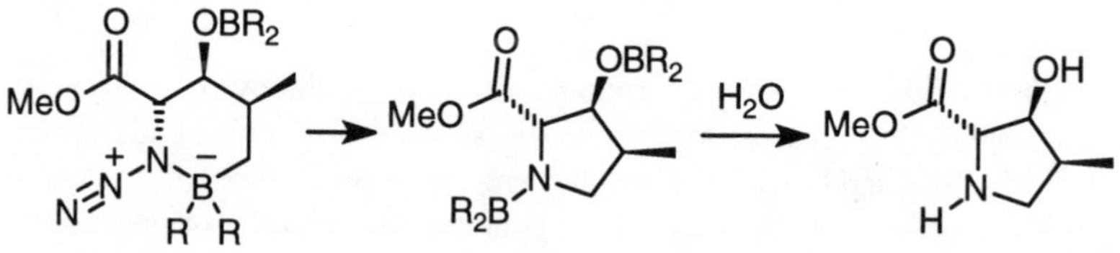

125   **126**   **127**

6.4.1.3 Cyclic borane intermediates. Stereocontrol through intramolecular cyclic interactions was observed by Brown and coworkers early in the history of hydroboration, for example, in the hydroboration—oxidation of limonene stereospecifically to diol **128** via a cyclic borane intermediate [70]. The excellent stereocontrol in this case presumably results from initial hydroboration of the terminal double bond of the limonene followed by intramolecular hydroboration of the only accessible side of the remaining double bond.

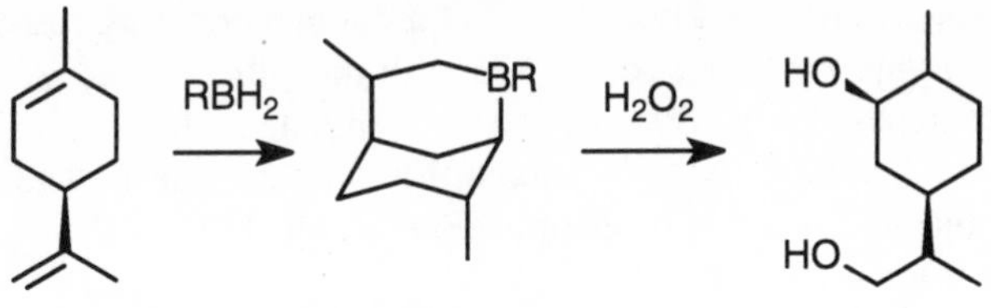

$(R = -CMe_2CHMe_2)$ **128**

One of the applications of diastereocontrolled hydroboration was a key step in the synthesis of (±)-estrone methyl ether by Bryson and Reichel [71]. Diene **129** was hydroborated with thexylborane, which presumably attacks the terminal vinyl group first and then closes to the trisubstituted double bond from the back side as illustrated. Cyanidation, treatment with trifluoroacetic anhydride (TFAA), and oxidation replaces the boron by a carbonyl group and completes the synthesis of intermediate **130** [71].

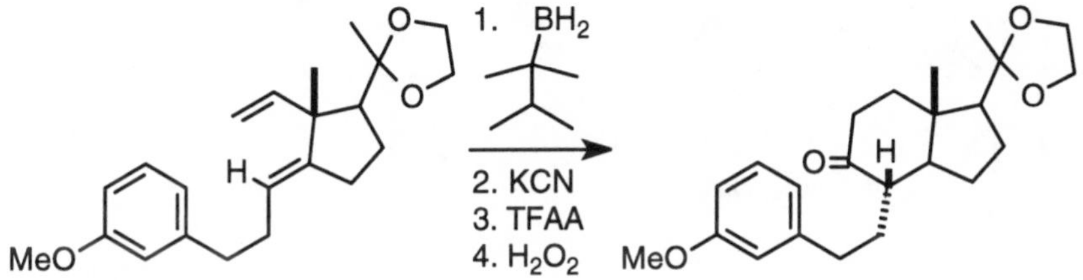

129   **130**

The foregoing is, of course, merely an extension of earlier work by Brown and Negishi, who made perhydroindanone from 1-vinylcyclohexene via hydroboration/ carbonylation [72], and Pelter's group, who developed cyanidation as an alternative to carbonylation (see Section 3.4.3.9). However, extension of chemistry based on model systems is frequently not trivial, and attempts to extend this type of closure to **131** were unsuccessful [73]. It was hoped that a single molecule of BH_3 might hydroborate all three double bonds to provide a polycyclic borane that could be carbonylated directly to a steroid analogue, but the best that was achieved was hydroboration by thexylborane to provide **132** in only 15% yield.

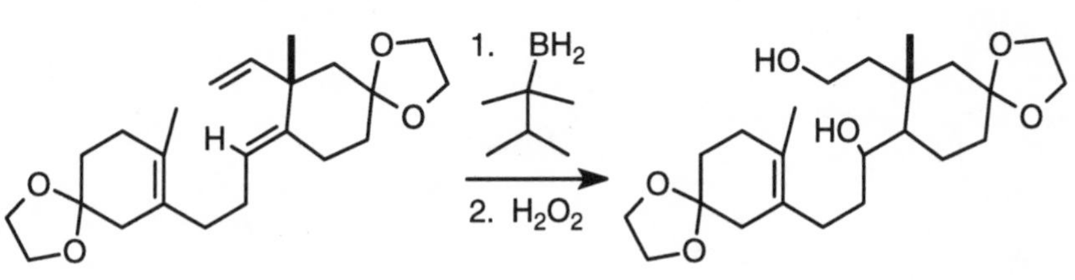

131   **132** (15% yield)

A more successful result was obtained in a closure of a 7-membered ring fused with a 5-membered ring. Hydroboration of diene **133** with thexylborane led to a diastereomeric mixture containing more of isomer **134**, which led to a formal total syntheses of the terpene (±)-confertin. Hydroboration of **133** with the less hindered reagent bromoborane led mainly to **135**, a precursor to (±)-helenalin [74, 75]. Yields and diastereomeric ratios were mediocre.

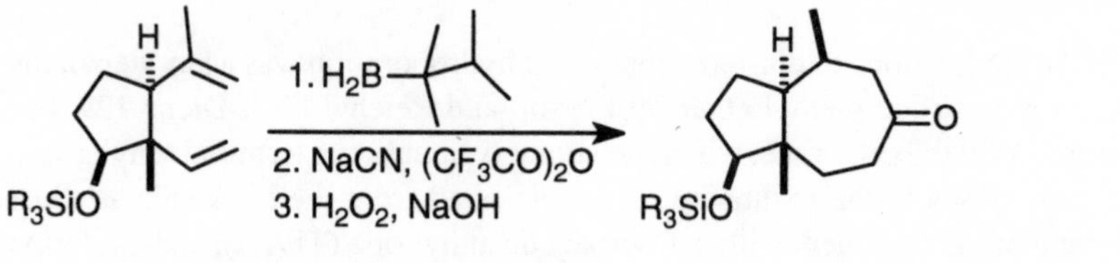

133   (R₃Si = *tert*-BuMe₂Si)   **134** (70:30 this isomer)

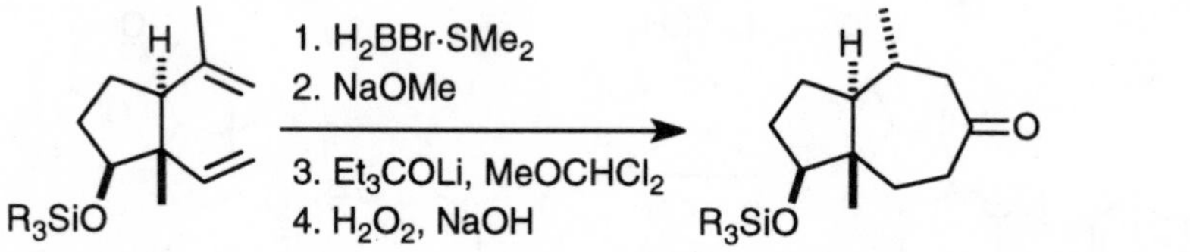

133   **135** (77:23 this isomer)

High diastereoselectivity was achieved in the hydroboration—oxidation of the allylic tin compound **136** to diol **137** [76]. In this case, the tin directs the initial hydroboration of the ring, boron—oxygen elimination follows, and the resulting exomethylene group is hydroborated rapidly from the less hindered side. It was not proved whether the final hydroboration product was cyclic, as illustrated, or acyclic.

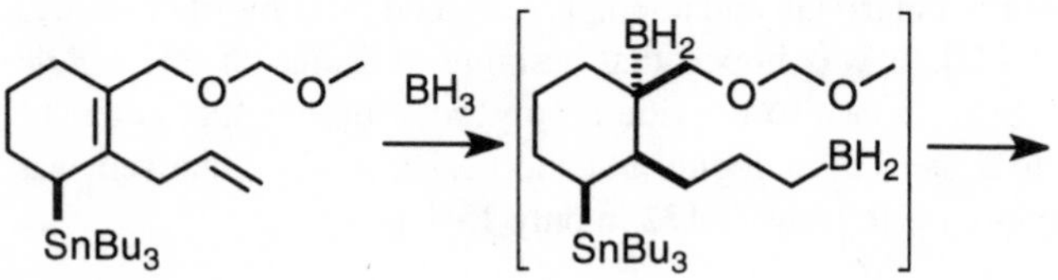

136

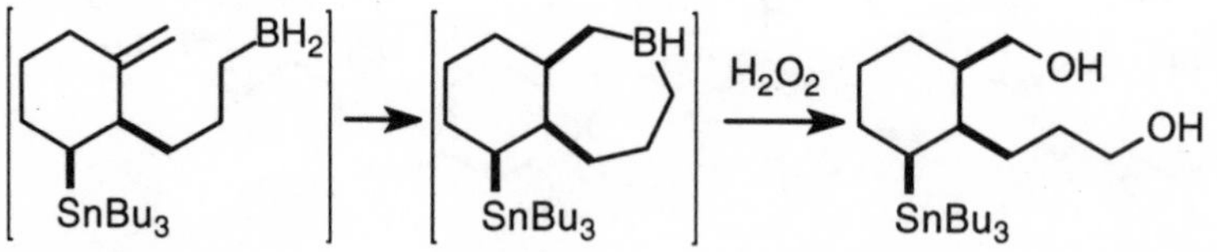

137

Borane dimethyl sulfide reduces α-alkoxy carboxylic esters, and the dialkoxy-borane formed can hydroborate a strategically placed double bond diastereo-selectively [77]. As is often the case with such internally controlled stereochemistry, the diastereoselection is modest and varies according to the substituent pattern, but the α-alkoxy group appears to exert the major influence. Thus, α-methoxy ester **138** is converted mainly to borolane **139**, and diastereomeric α-methoxy ester **140** yields mostly **141**.

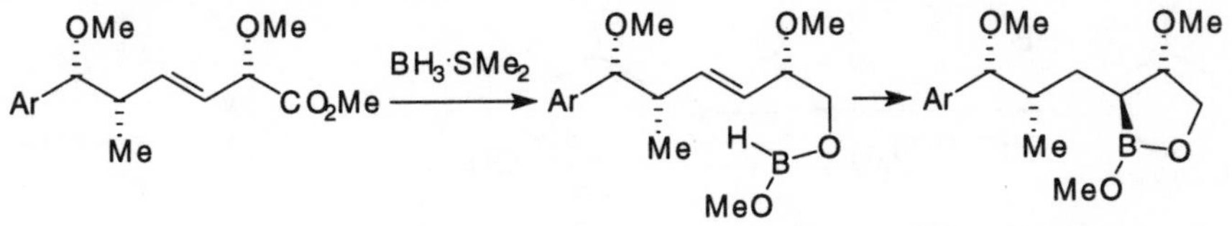

138   **139** 8:1 this isomer

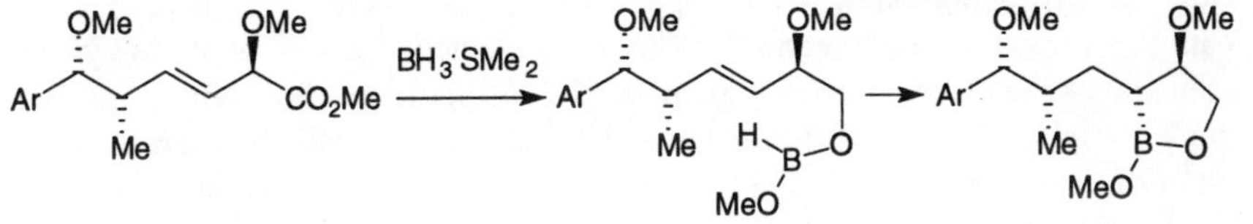

140   **141** 4:1 this isomer

6.5 Catalytic Asymmetric Hydroboration

6.5.1 Catalyst Effects in Substrate Directed Hydroboration

Hydroboration of 2-cyclohexen-1-ol or certain of its derivatives with catechol-borane in the presence of Wilkinson's catalyst [RhCl(PPh$_3$)$_3$] results in different regioselectivity from uncatalyzed hydroboration [78, 79] (see Section 2.3.4). Uncatalyzed hydroboration with 9-BBN followed by peroxidic oxidation yields 70-86% of a mixture containing 68-83% of the *trans*-1,2-cyclohexanediol derivative (**142**), but catalyzed hydroboration with catecholborane and Wilkinson's catalyst yields 79-87% of a mixture containing 72-86% of the *trans*-1,3-cyclohexanediol derivative (**143**). These are each accompanied by some of the other *trans*-regioisomer as well as the *cis*-1,3-isomer, with little of the *cis*-1,2-isomer. It might be noted that where the starting material is unmasked 2-cyclohexen-1-ol, the first step is borylation of the hydroxyl group, so that free hydroxyl is never involved in the hydroboration process itself.

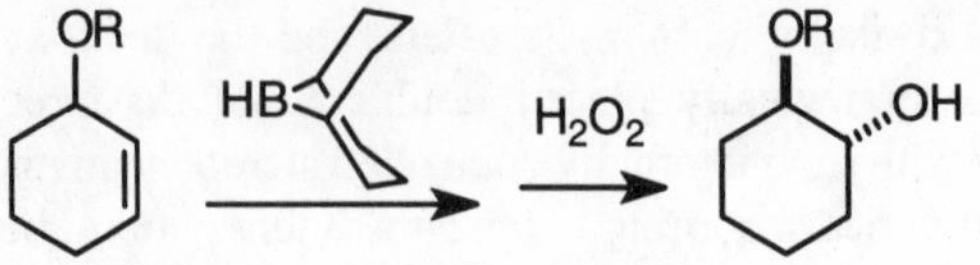

(R = H, CH$_2$Ph, SiMe$_2$-t-Bu)   **142**

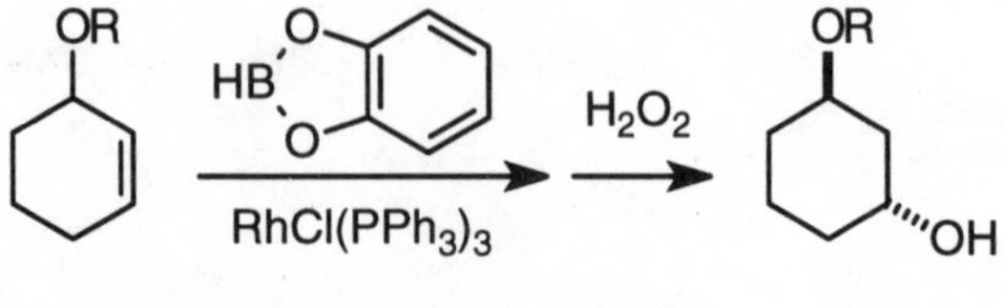

(R = H, CH$_2$Ph, SiMe$_2$-t-Bu)   **143**

A related preference for *syn* selectivity has been observed in the rhodium catalyzed hydroboration of several open chain allylic alcohol derivatives [78, 80]. With R = benzyl or isobutyl and Y = COCF$_3$, protected allylic alcohols **144** yielded a ratio of **145** to its *anti* diastereomer in the 9:1 to 14:1 range [80]. With R = *n*-butyl and Y = triphenylmethyl, **145** was favored by 18:1 [81], or with Y = *tert*-BuMe$_2$Si, 24:1 [78]. However, most **144** did not react so diastereoselectively, and some types of structures, especially R = C$_6$H$_5$ or C$_6$F$_5$, favored the *anti* isomer of **145**. An interpretation of the electronic factors affecting the stereoselectivity of the reaction has been presented [81, 82]. It appears that the favored route is the one that places the most electron-withdrawing substituent *anti* to the metal in the metal—alkene complex that is an intermediate in the reaction (see Section 2.3.4) [82].

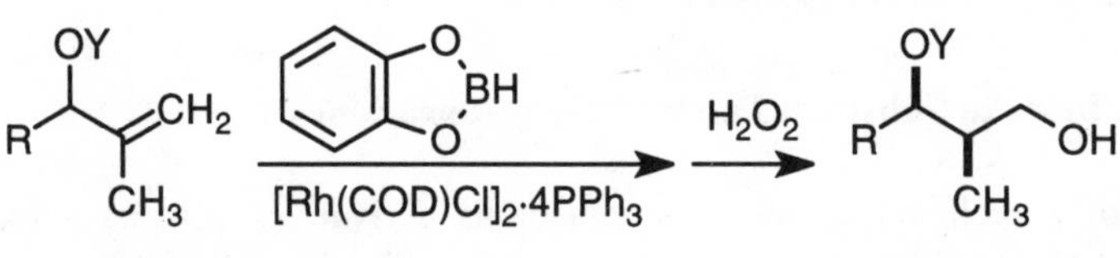

144   **145**

Good stereocontrol has been achieved in both catalyzed and uncatalyzed hydroborations of amine derivatives analogous to **144** [83, 84]. In this case it was shown that the tosylated allylic amines **146** could be derived from amino acids in optically active form, though test reactions were run on racemates. Catalyzed hydroboration with catecholborane yielded ratios of *syn* isomer **147** to *anti* isomer **148** in the range 6:1 to 18:1 (R = PhCH$_2$, Bu, Me$_2$CHCH$_2$). Noncatalytic hydroboration with 9-BBN yielded ratios in the range 7:1 to 25:1. However, hydroboration with excess BH_3·THF produced mainly the *anti* isomer **148**, with ratios in the range 17:1 to 21:1 [84].

146 **147** **148**

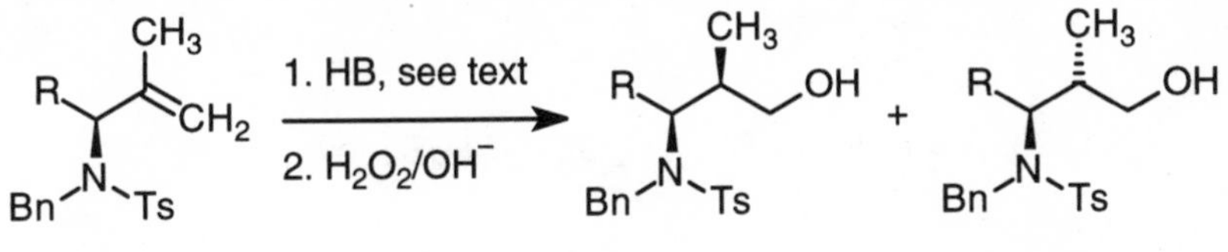

The use of the iridium(I) complex [Ir(cod)(PR_3)(py)]PF_6, R = cyclohexyl, to catalyze hydroboration of cyclohexenyl amides has provided a high degree of diastereoselection, with predominant *cis*-substitution in the ring. Examples include the conversion of **149** to **150** and **151** to **152** [85].

149 **150** (95% this isomer)

151 **152** (88% this isomer)

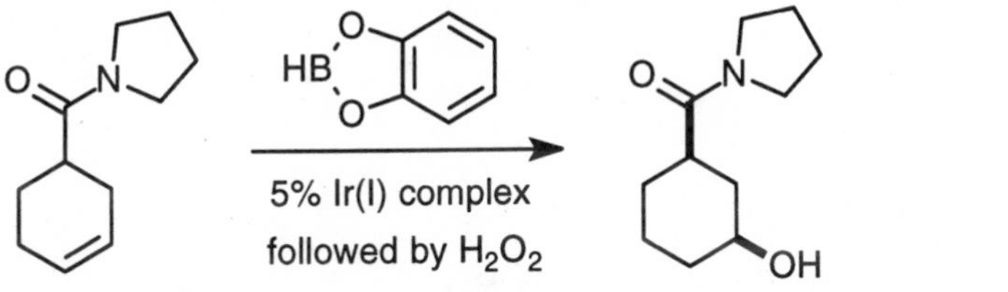

It has been reported that no selectivity could be achieved between the double bonds in the uncatalyzed hydroboration of **153** with 9-BBN or disiamylborane [86], but that catalyzed hydroboration of limonene occurs only at the less hindered site to produce **154** [79], and that catalyzed hydroboration of **155** selectively hydroxylates only the less hindered double bond to produce **156**, in contrast to the unselectivity of epoxidation or uncatalyzed hydroboration [87].

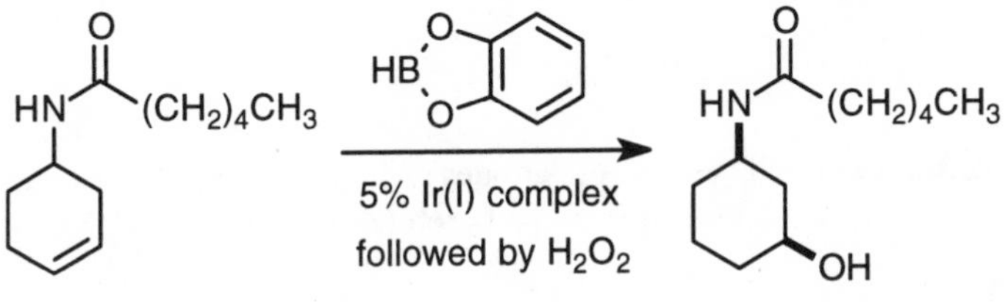

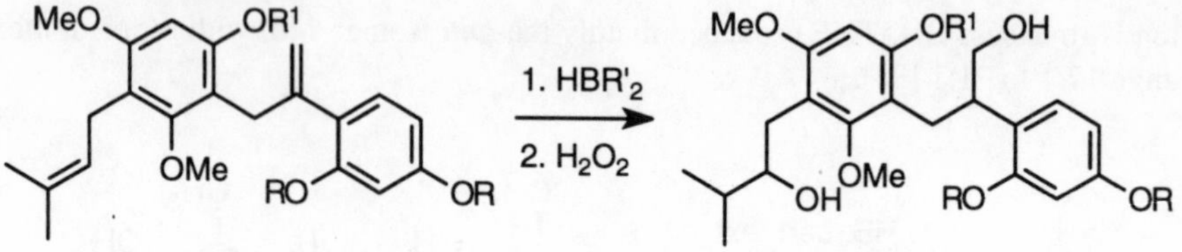

153 (R^1 = COCH$_3$, R = CH$_2$OCH$_2$CH$_2$SiMe$_3$)

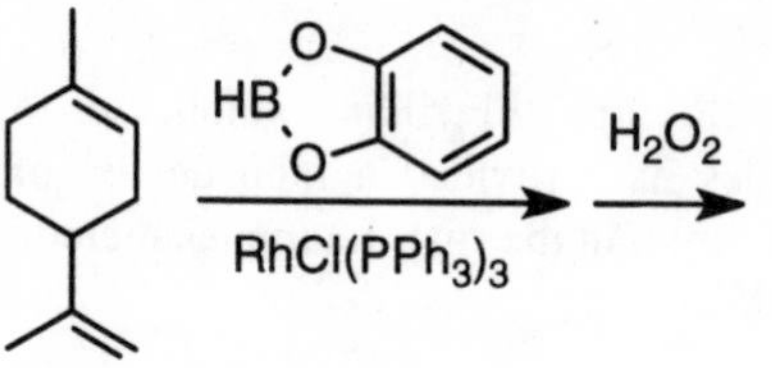

154

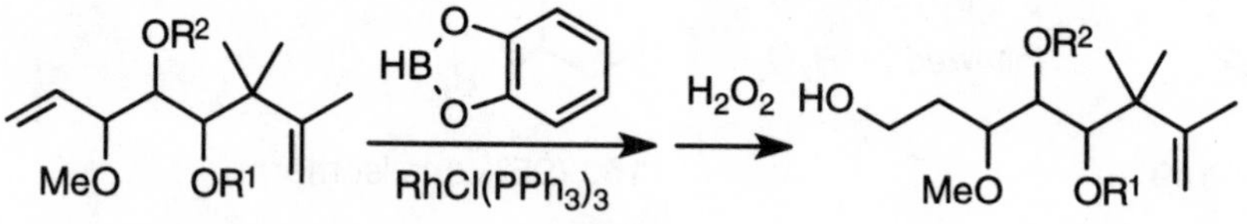

155 (R^1 = *p*-methoxybenzyl, R^2 = SiEt$_3$)   **156**

Treatment of α,β-unsaturated esters or amides with catecholborane and Wilkinson's catalyst results in 1,4-reduction to the boron (*Z*)-enolate, not in hydroboration [88].

6.5.2 Hydroboration with Asymmetric Catalysts

Asymmetric catalytic hydroboration is still in the early stages of development, and at this point there are more research opportunities than useful procedures. Asymmetric diphosphines that have been tested include **157-160**, which are illustrated with their common literature mnemonic abbreviations. The rhodium bis(cyclooctadiene) complex **161** is often used as the rhodium source.

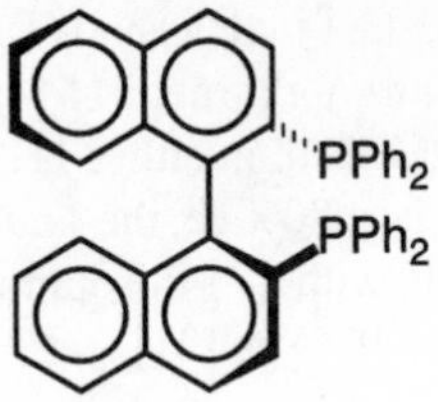

157
(S)-BINAP

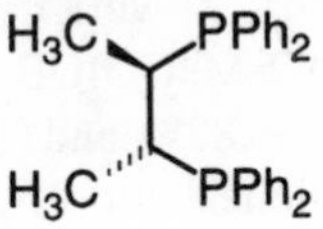

158
(R,R)-CHIRAPHOS

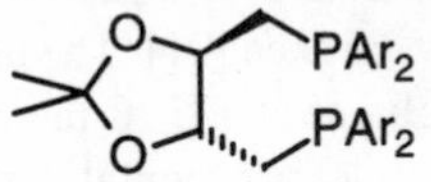

159
(R,R)-DIOP (Ar = Ph);
(R,R)-2-MeODIOP (Ar = 2-MeOC6H4)

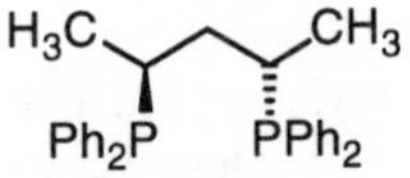

160
(S,S)-BDPP

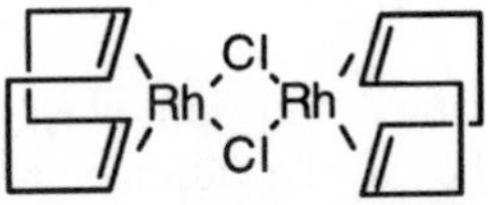

161
[RhCl(COD)]2

Catecholborane with catalyst prepared from [Rh(COD)₂]BF₄ and (R)-BINAP converted styrene to boronic ester which was oxidized to (R)-1-phenylethanol (**162**) in 91% yield, 96.2% ee, at –78 °C in dimethoxyethane, with no detectable 2-phenylethanol [89]. With catalyst prepared from (+)-(S,S)-DIOP and [ClRh(C₂H₄)₂]₂ the hydroboration-oxidation of indene to (–)-(R)-1-indanol (**163**) was accomplished in 74% ee [90].

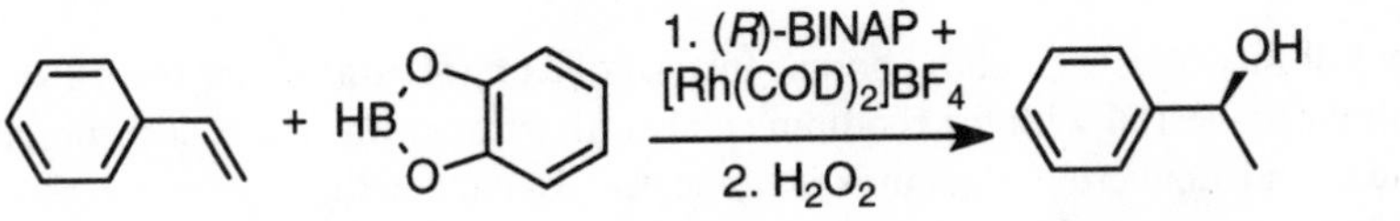

162 (96.2% ee)

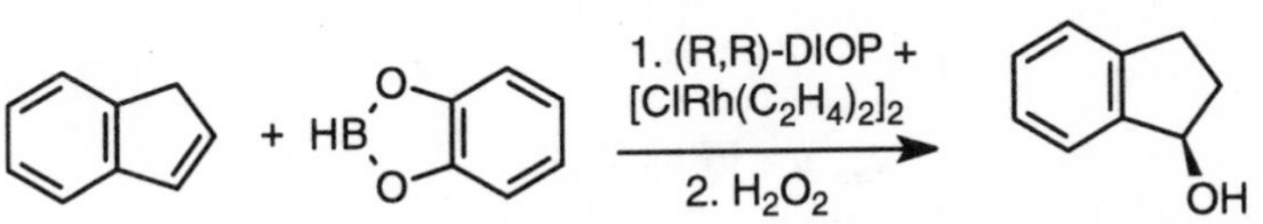

163 (74% ee)

Norbornene with catecholborane and a catalyst prepared from (+)-(S,S)-DIOP and [ClRh(C₂H₄)]₂ followed by in situ oxidation led to 2(S)-2-*exo*-norborneol in 59% ee, 81% yield [90]. The reaction of norbornene with catecholborane in THF at

−25 °C in the presence of a catalyst prepared from [RhCl(COD)]$_2$ (1 mol %) and (R)-BINAP (2 mol %) has led after oxidative work up to 2(R)-2-*exo*-norborneol **164** in up to 65% ee [91]. The use of (R,R)-2-MeODIOP with [RhCl(COD)]$_2$ in either THF or toluene at −25 °C increases the ee to 82%, and (S,S)-BDPP gave 80% ee, the same within experimental error [91]. Noncatalytic hydroboration with (+)-diisopino-campheylborane from (−)-α-pinene (Section 6.2.2) has yielded 97% ee [21].

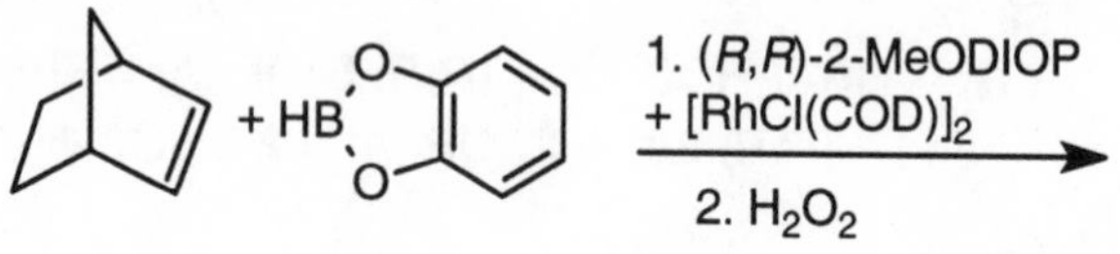

164 (82% ee)

Results from hydroboration-oxidation of 1,1-disubstituted ethylenes, where the induced asymmetry is at the site of addition of the hydrogen, have been less successful. The best result reported has been the conversion of 2,3,3-trimethyl-1-butene to (R)-2,3,3-trimethylbutanol (**165**) in 69% ee with catalyst prepared from (R,R)-DIOP and [RhCl(COD)]$_2$ [92, 91]. Analogous asymmetric hydroboration of α-methyl-styrene to (S)-2-phenyl-1-propanol has given 38% ee's with (+)-(S,S)-DIOP or (+)-(S)-BINAP.

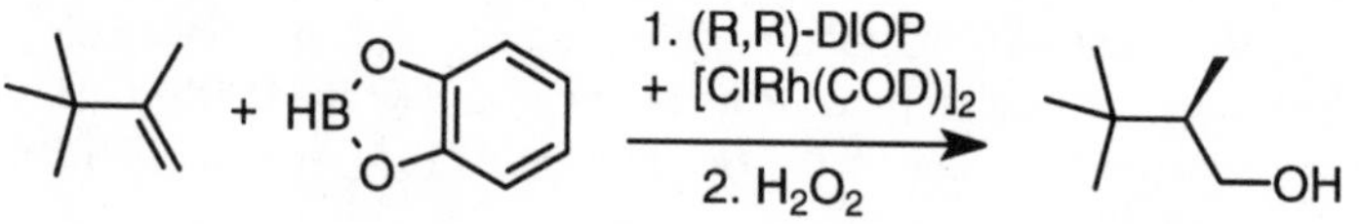

165 (69% ee)

The possibility of using a chiral borane instead of a chiral catalyst has been tested [93]. Chiral borane **166** with the rhodium catalyst illustrated was the best combination tested, and converted *p*-methoxystyrene to (S)-1-(p-methoxyphenyl)ethanol (**167**) in 76% ee, 82% yield, with 14% 2-arylethanol and 4% reduction. Attempts to use a chiral catalyst containing BINAP with this system resulted in low yields and low asymmetric induction.

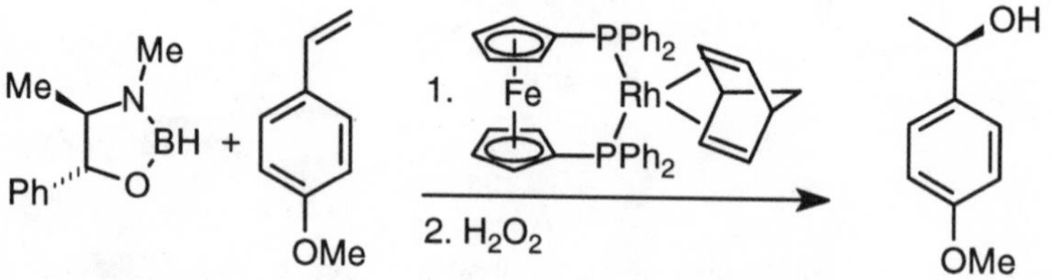

166   **167** (76% ee)

p-Methylstyrene with [RhCl(COD)]$_2$ (1 mol %) and (*S*)-(−)-BINAP (2 mol %) treated with catecholborane (22% excess) at −78 °C for three days followed by oxidation in situ with alkaline hydrogen peroxide has yielded 95% of (*S*)-(−)-1-(*p*-methylphenyl)ethanol in 96% ee [94]. However, the generality appears to be limited, as indene and vinylferrocene gave very little asymmetric induction under these conditions..

6.6 References

1. Brown HC, Zweifel G (1961) J. Am. Chem. Soc. 83:486
2. Zweifel G, Brown HC (1964) J. Am. Chem. Soc. 86:393
3. Brown HC, Singaram B (1984) J. Org. Chem. 49:945
4. Brown HC, Desai MC, Jadhav PK (1982) J. Org. Chem. 47:5065
5. Brown HC, Ayyanger NR, Zweifel G (1964) J. Am. Chem. Soc. 86:397
6. Brown HC, Singaram B (1988) Acc. Chem. Res. 21:287
7. Brown HC, Jadhav PK, Mandal AK (1981) Tetrahedron 37:3547
8. Brown HC, Jadhav PK (1983) "Asymmetric Synthesis", Vol. 2, Academic Press, New York, p. 1
9. Brown CA (1978) Synthesis 754
10. Brown HC, Yoon NM (1977) Isr. J. Chem. 15:12
11. Brown HC, Singaram B (1984) J. Org. Chem. 49:945
12. Brown HC, Schwier JR, Singaram B (1978) J. Org. Chem. 43:4395
13. Brown HC, Jadhav PK, Desai MC (1982) J. Org. Chem. 47:4583
14. Catalog of Aldrich Chemical Company, Milwaukee, Wisconsin (1993).
15. Brown HC, Jadhav PK, Desai MC (1982) J. Am. Chem. Soc. 104:4303
16. Brown HC, Randad RS, Bhat KS, Zaidlewicz M, Weissman SA, Jadhav PK, Perumal PT (1988) J. Org. Chem. 53:5513
17. Brown HC, Weissman SA, Perumal PT, Dhotke UP (1990) J. Org. Chem. 55:1217
18. Brown HC, Vara Prasad JVN, Zaidlewicz M (1988) J. Org. Chem. 53:2911
19. Masamune S, Kim BM, Petersen JS, Sato T, Veenstra SJ, Imai T (1985) J. Am. Chem. Soc. 107:4549
20. Brown HC, Vara Prasad JVN (1986) J. Am. Chem. Soc. 108:2049
21. Joshi NN, Pyun C, Mahindroo VK, Singaram B, Brown HC (1992) J. Org. Chem. 57:504
22. Letsinger RL, Skoog I (1955) J. Am. Chem. Soc. 77:2491
23. Matteson DS, Ray R, Rocks RR, Tsai DJS (1983) Organometallics 2:1536
24. Tripathy PB, Matteson DS (1990) Synthesis 200
25. Brown HC, Vara Prasad JVN (1986) J. Org. Chem. 51:4526
26. Brown HC, Gupta AK (1988) J. Organomet. Chem. 341:73
27. Partridge JJ, Chadha NK, Uskokovic MR (1973) J. Am. Chem. Soc. 95:532
28. Partridge JJ, Chadha NK, Uskokovic MR (1973) J. Am. Chem. Soc. 95:7171
29. Rüttiman A, Mayer H (1980) Helv. Chim. Acta 63:1456
30. Rüttiman A, Englert G, Mayer H, Moss GP, Weedon BCL (1983) Helv. Chim. Acta 66:1939
31. Brown HC, Bhat KS, Jadhav PK (1991) J. Chem. Soc. Perkin 1 2633
32. Masamune S, Lu LDL, Jackson WP, Kaiho T, Toyoda T (1982) J. Am. Chem. Soc. 104:5523
33. Brown HC, Vara Prasad JVN, Gupta AK, Bakshi RK (1987) J. Org. Chem. 52:310
34. Brown HC, Jadhav PK, Mandal AK (1982) J. Org. Chem. 47:5074
35. Brown HC, Singaram B (1984) J. Am. Chem. Soc. 106:1797

36. Brown HC, Bhat NG, Somayaji V (1983) Organometallics 2:1311
37. Brown HC, Gupta AK, Vara Prasad JVN (1988) Bull. Chem. Soc. Jpn. 61:93
38. Brown HC, Cole TE, Srebnik M (1985) Organometallics 4:1788
39. Brown HC, Srebnik M, Bakshi RK, Cole TE (1987) J. Am. Chem. Soc. 109:5420
40. Singaram B, Cole TE, Brown HC (1984) Organometallics 3:774
41. Srebnik M, Cole TE, Ramachandran PR, Brown HC (1989) J. Org. Chem. 54:6085
42. Brown HC, Joshi NN, Pyun C, Singaram B (1989) J. Am. Chem. Soc. 111:1754
43. Brown HC, Bakshi RK, Singaram B (1988) J. Am. Chem. Soc. 110:1529
44. Sadhu KM, Matteson DS (1985) Organometallics 4:1687
45. Masamune S, Kennedy RM, Petersen JS, Houk KN, Wu Y (1986) J. Am. Chem. Soc. 108:7404
46. Brown HC, Imai T, Perumal PT, Singaram B (1985) J. Org. Chem. 50:4032
47. Brown HC, Naik RG, Bakshi RK, Pyun C, Singaram B (1985) J. Org. Chem. 50:5586
48. Brown HC, Naik RG, Singaram B, Pyun C (1985) Organometallics 4:1925
49. Matteson DS, Majumdar D (1983) Organometallics 2:1529
50. Brown HC, Singh SM (1986) Organometallics 5:994
51. Brown HC, Singh SM (1986) Organometallics 5:998
52. Brown HC, Singh SM, Rangaishenvi MV (1986) J. Org. Chem. 51:3150
53. Matteson DS, Hurst GD (1986) Organometallics 5:1465
54. Brown HC, Gupta AK, Rangaishenvi MV, Vara Prasad JVN (1989) Heterocycles 28:283
55. Rangaishenvi MV, Singaram B, Brown HC (1991) J. Org. Chem. 56:3286
56. Brown HC, Mahindroo VK, Bhat NG, Singaram B (1991) J. Org. Chem. 56:1500
57. Pelter A, Smith K, Hutchings MG, Rowe K (1975) J. Chem. Soc., Perkin Trans. I 129
58. Brown HC, Mahindroo VK (1993) Tetrahedron: Asymmetry 4:59
59. Brown HC, Mahindroo VK (1992) Synlett 626
60. Brown HC, Iyer RR, Mahindroo VK, Bhat NG (1991) Tetrahedron: Asymmetry 2:277
61. Brown HC, Gupta AK, Vara Prasad JVN, Srebnik M (1988) J. Org. Chem. 53:1391
62. Brown HC, Kulkarni SV, Khanna VV, Patil VD, Racherla US (1992) J. Org. Chem. 57:6173
63. Racherla US, Khanna VV, Brown HC (1992) Tetrahedron Lett. 33:1037
64. Brown HC, Salunkhe AM, Singaram B (1991) J. Org. Chem. 56:1170
65. Brown HC, Salunkhe AM (1993) Tetrahedron Lett. 34:1265
66. Evans DA, Bartroli J, Godel T (1982) Tetrahedron Lett. 23:4577
67. Evans DA, Bartroli J (1982) Tetrahedron Lett. 23:807
68. Still WC, Barrish JC (1983) J. Am. Chem. Soc. 105:2487
69. Evans DA, Weber AE (1987) J. Am. Chem. Soc. 109:7151
70. Brown HC, Pfaffenberger CD (1975) Tetrahedron 31:925
71. Bryson TA, Reichel CJ (1980) Tetrahedron Lett. 21:2381
72. Brown HC, Negishi E (1968) J. Chem. Soc., Chem. Commun. 594
73. Fulmer TD, Bryson TA (1989) J. Org. Chem. 54:3496
74. Welch MC, Bryson TA (1988) Tetrahedron Lett. 29:521
75. Welch MC, Bryson TA (1989) Tetrahedron Lett. 30:523
76. Bryson TA, Akers JA, Ergle JD (1991) Synlett 499
77. Panek JS, Xu F (1992) J. Org. Chem. 57:5288
78. Evans DA, Fu GC, Hoveyda AH (1988) J. Am. Chem. Soc. 110:6917
79. Evans DA, Fu GC, Hoveyda AH (1992) J. Am. Chem. Soc. 114:6671
80. Burgess K, Ohlmeyer MJ (1989) Tetrahedron Lett. 30:5857
81. Burgess K, Cassidy J, Ohlmeyer MJ (1991) J. Org. Chem. 56:1020
82. Burgess K, van der Donk WA, Jarstfer MB, Ohlmeyer MJ (1991) J. Am. Chem. Soc. 113:6139
83. Burgess K, Ohlmeyer MJ (1989) Tetrahedron Lett. 30:395
84. Burgess K, Ohlmeyer MJ (1991) J. Org. Chem. 56:1027
85. Evans DA, Fu GC (1991) J. Am. Chem. Soc. 113:4042

86.   Shih TL, Wyvratt MJ, Mrozik H (1987) J. Org. Chem. 52:2029
87.   Evans DA, Gage JR (1992) J. Org. Chem. 57:1958
88.   Evans DA, Fu GC (1990) J. Org. Chem. 55:5678
89.   Hayashi T, Matsumoto Y, Ito Y Tetrahedron: Asymmetry 2:601
90.   Sato M, Miyaura N, Suzuki A (1990) Tetrahedron Lett. 31:231
91.   Burgess K, Ohlmeyer MJ (1991) Tetrahedron: Asymmetry 2:613
92.   Burgess K, Ohlmeyer MJ (1988) J. Am. Chem. Soc. 110:5178
93.   Brown JM, Lloyd-Jones GC (1990) Tetrahedron: Asymmetry 1:869
94.   Zhang J, Lou B, Guo G, Dai L (1991) J. Org. Chem. 56:1670

7 Allylboron and Boron Enolate Chemistry

7.1 Introduction

Reactions of allylic boranes with aldehydes have proved to be among the most efficient means of building a stereocontrolled set of several adjacent chiral centers. Boron enolates follow a similar mechanistic path and have also provided significant advances in asymmetric synthesis. The chemistry described in this chapter has been the key to a considerable number of major natural products syntheses, especially of macrolide or ionophore antibiotics.

The reaction of triallylborane with aldehydes and ketones to form boronic esters of homoallyl alcohols was discovered by Mikhailov and Bubnov [1], and observation of allylic inversion in the reaction of tricrotylborane with formaldehyde soon followed [2]. These and other historical developments, as well as a variety of interesting allylborane reactions that do not have any obvious applications in stereo-directed synthesis, have been described in Chapter 3.4.5.

The stereoselectivity of the reaction of (Z)-crotylboronic esters with aldehydes to produce *anti*-homoallylic alcohols was first reported by Hoffmann and Zeiss [3]. The mechanistically similar stereoselective aldol reaction of dialkylboron enolates with aldehydes had been noted earlier by Fenzl and Köster [4, 5]. Practical application to synthetic problems by the Masamune [6-9] and Evans [10] groups dates from 1979, the same year as Hoffmann's allylborations.

A useful review of stereoselective allylmetal chemistry, including a comprehensive summary of the reactions of allylic boron compounds with aldehydes and related substrates, has been published recently [11].

In this chapter, allylic boronic esters will be discussed first (Section 7.2), beginning with sources of allylboronic esters (Section 7.2.2) and their use in diastereocontrolled synthesis (Section 7.2.3), then moving on to enantiocontrolled reactions (Section 7.2.4) and their applications in natural product synthesis (Section 7.2.5). There are a few examples of the use of allenic boronic esters (Section 7.2.6) and allylic borinic esters (Section 7.2.7) Allyldialkylboranes, which are not so easily manipulated for synthetic purposes, follow (Section 7.3). Finally, the use of boron enolates in asymmetric synthesis is described (Section 7.4).

7.2 Allylic Boronic Esters

7.2.1 General Advantages

Allylic boronic esters, though more easily oxidized or hydrolyzed than simple alkyl or aryl boronic esters, are stable and isolable compounds and do not isomerize at room temperature. In this respect, they offer important synthetic advantages over the more reactive allylic trialkylboranes, which undergo rapid and reversible allylic rearrangement to form the equilibrium mixture of isomers.

For purposes of asymmetric synthesis, chiral diols are used as the chiral directing groups. Readily available tartrate esters are among the best of chiral directors. Simple ligand exchange with achiral allylic boronic esters readily generates the chiral reagent, and the (cheap) chiral director is regenerated in aqueous work up of the products and is readily recoverable.

7.2.2 Sources of Allylic Boronic Esters

7.2.2.1 Allylic metal compounds. Unsubstituted allylboronic esters can be made from allylmagnesium chloride and a borate ester in the usual way (see Chapter 2.2.1.1). Acyclic esters of allylboronic acid undergo protonolysis of the allyl group very easily, and losses during work up may make yields erratic. It often works better to use allylmagnesium halide directly with a cyclic borate ester. An example is the synthesis of pinanediol allylboronate [12]. Alternatively, triallylborane with diols can yield allylboronic esters efficiently, though with sacrifice of two *B*-allyl groups [13].

Pinacol (Z)-crotylboronate (**1**) was first obtained from (Z)-crotylpotassium and a suitable boron compound [14, 15]. Hoffmann and coworkers avoided isomerization by the use of ClB(NMe$_2$)$_2$ as the boron source, and also obtained (*E*)-isomer **2** from crotylmagnesium chloride and ClB(NMe$_2$)$_2$ [16].

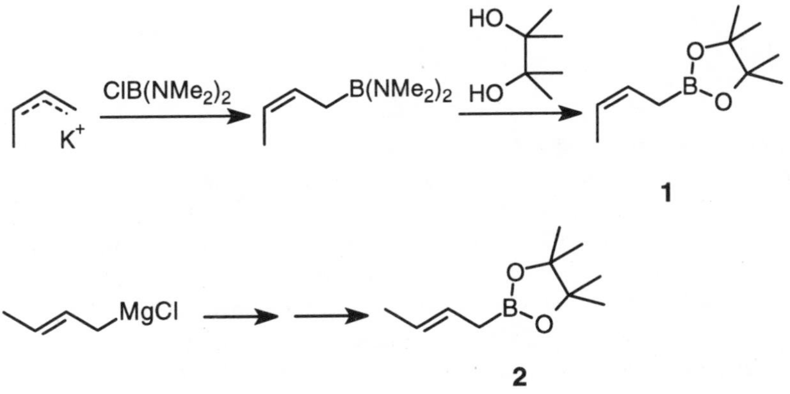

It was also shown that (E)- and (Z)-γ-methoxyallylboronates could be synthesized in a similar manner and yielded methoxy alcohols with good stereocontrol [17, 18]. The requisite (Z)-methoxyallyllithium was obtained by metalation of allyl methyl ether with butyllithium and TMEDA, the (E)-isomer by addition of thiophenol to methoxyallene and reductive cleavage of the resulting (E)-methoxyallyl phenyl sulfide with lithium naphthalenide [17]. The lithiated intermediates reacted with ClB(NMe$_2$)$_2$ as expected and treatment with pinacol yielded the boronic esters. The (Z)-isomers can be obtained from other allyl ethers [18, 19].

Improved preparations reported by Roush and coworkers are more efficient [20, 21]. These involve lithiation of cis- or trans-2-butene with butyllithium and potassium tert-butoxide, reaction with triisopropyl borate, and quenching with aqueous hydrochloric acid. Allylic boronic acids or their acyclic esters decompose readily. To avoid erratic yields, the boronic acid was extracted directly into an ethereal solution of diisopropyl tartrate to form the stable cyclic tartrate boronic esters of (Z)- and (E)-crotylboronic acids (3 and 4, respectively)

3

4

7.2.2.2 Retroracemization of a Grignard reagent. Racemic (1-methyl-2-butenyl)-boronates can be made from the corresponding Grignard reagent and various borate esters [22]. Reaction of (S,S)-DICHED isopropyl borate (5) with 2-pentenyl-4-magnesium chloride results in selective consumption of one of the two equilibrating enantiomers of the Grignard reagent, in effect driving racemization backwards so that the product is derived mainly from one enantiomer. (S,S)-DICHED (S)-(Z)-1-methyl-2-butenylboronate (6) was produced in 89% yield, 94% de [23]. Although this is not as high as the 99% de obtained via the α-chloroethyl boronic ester and

1-propenyllithium, Section 7.2.2.3, it is high enough to be useful, and the simplicity of the process is highly significant.

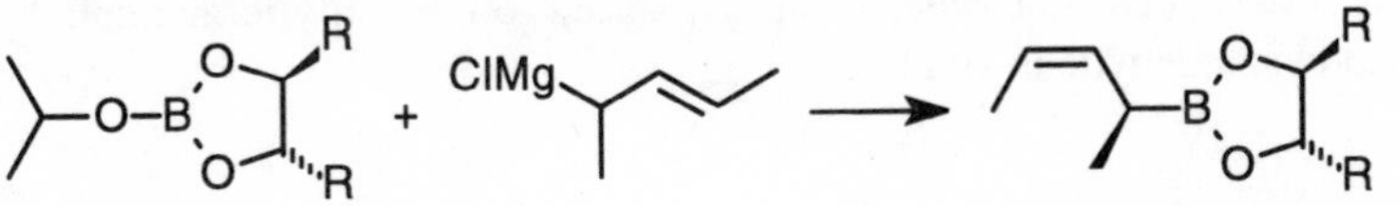

5 (R = cyclohexyl)   **6** (89% yield, 97% this isomer)

7.2.2.3 Via (alkenyl)(halomethyl)borate rearrangement. A more general approach to allylic boronic esters is to add an isomerically pure alkenylmetallic to a (halomethyl)boronic or (dichloromethyl)boronic ester [24, 25, 26] (see Sections 2.2.1.1, 5.2.3.6). An asymmetric example is the preparation of **6** from **7** via **8** (see Section 5.3.7.2 for details).

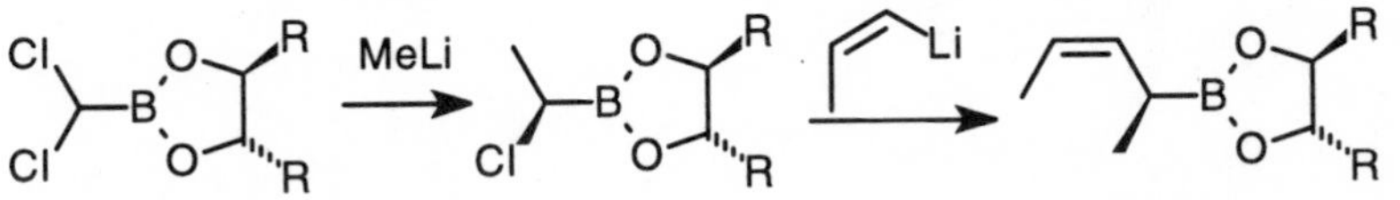

7 (R = cyclohexyl)   **8**   **6** (99.5% this isomer)

Alternatively, insertion of a methylene group into a stereodefined alkenylboronic ester (Section 4.2.2, 4.2.3) with (chloromethyl)lithium [27] accomplishes the same result [28]. Yet another alternative is coupling of pinacol (iodozincmethyl)boronate, $Me_4C_2O_2BCH_2ZnI$, with a 1-iodoalkene in the presence of $Pd(PPh_3)_4$ [29, 30, 31].

7.2.2.4 Catalyzed hydroboration. A different kind of source of (Z)-allylic boronic esters is provided by the $Pd(PPh_3)_4$ catalyzed hydroboration of dienes with catecholborane [32]. Unlike ordinary hydroborations, this one proceeds in a conjugate manner. For example, isoprene yields isomer **9** in high isomeric purity, which was not isolated but treated with benzaldehyde to yield the homoallylic alcohol **10**, the expected product (see Section 7.2.3.1).

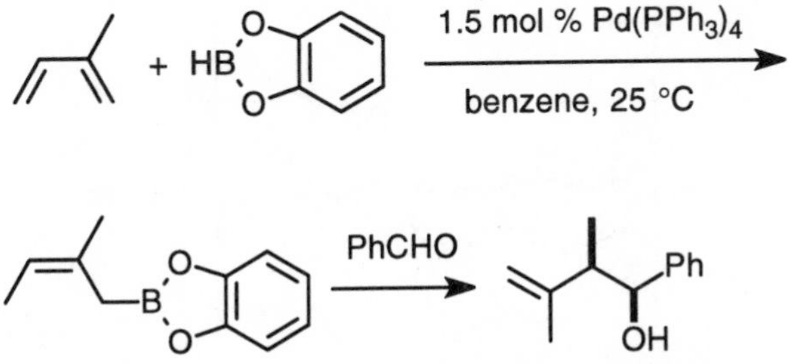

9   **10** (89%)

Hydroboration of isopropenylacetylene catalyzed by Pd(PPh$_3$)$_4$ yielded the allenic boronic ester **11** (see Section 7.2.6 on allenylboronic esters for related compounds). Cyclohexadiene required Rh$_4$(CO)$_{12}$ as catalyst to form **12**. In each case the allylic boronic esters were not isolated but converted to homoallylic alcohols by treatment with benzaldehyde in situ [32].

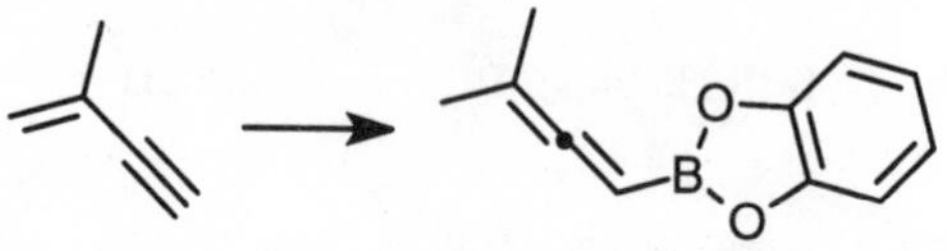

11

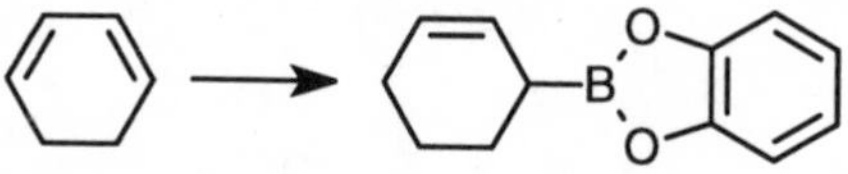

12

7.2.3 Diastereoselection with Racemates

7.2.3.1 Mechanism. The use of asymmetric boronic esters for the asymmetric synthesis of homoallyl alcohols was first developed by Hoffmann and coworkers, and Hoffmann has provided a review of the early work [33]. The mechanism of the process is illustrated with the reaction of pinacol (Z)-crotylboronate (**1**) with benzaldehyde, which is believed to proceed via a six-membered ring transition state **13** to transfer the boron from allylic carbon to the aldehyde oxygen to form intermediate boric acid ester **14**. Treatment with triethanolamine removes boric acid to provide the racemic *erythro* (or *syn*) homoallylic alcohol, (R*,S*)-2-methyl-1-phenyl-3-buten-1-ol (**15**), in 88% de (94:6 diastereomeric ratio) [3, 16].

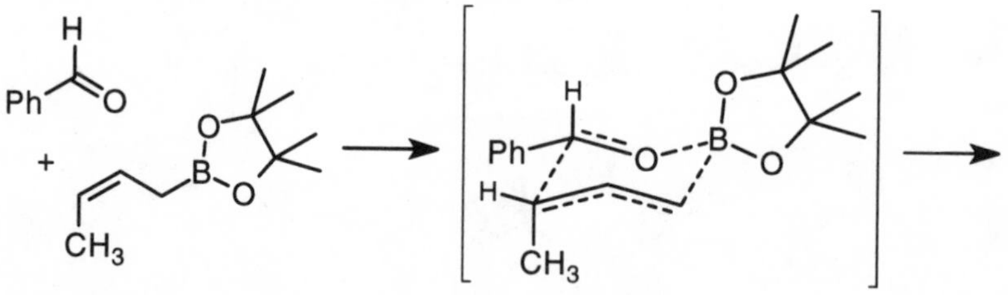

1   **13**

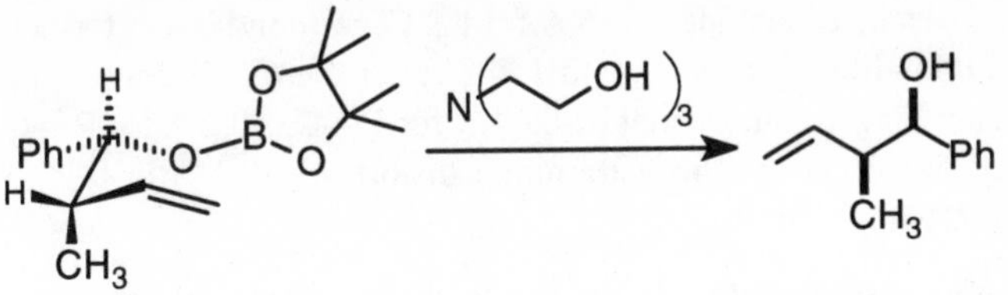

14   **15**

The geometry of the transition state **13** is postulated to have the cyclohexane chair-form geometry with the phenyl group of the benzaldehyde preferentially equatorial and the methyl group of the crotylboronic ester fixed as axial by the geometry of the C—C double bond. The use of an (*E*)-crotylboronate (**2**) produces the *threo* (or *anti*) homoallylic alcohol, (*R**,*R**)-2-methyl-1-phenyl-3-buten-1-ol (**18**), in 92% de (96:4 diastereomeric ratio) [3, 16]. Again, the transition state **16** has the phenyl group preferentially equatorial, and the methyl group is necessarily equatorial because of the olefinic geometry. Several other aldehydes with **1** or **2** yielded similar results.

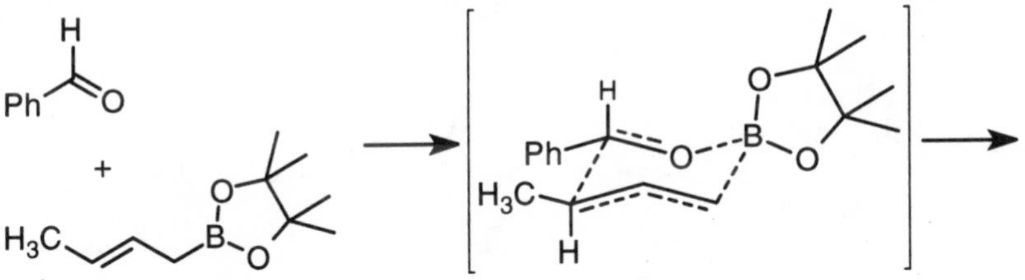

2   **16**

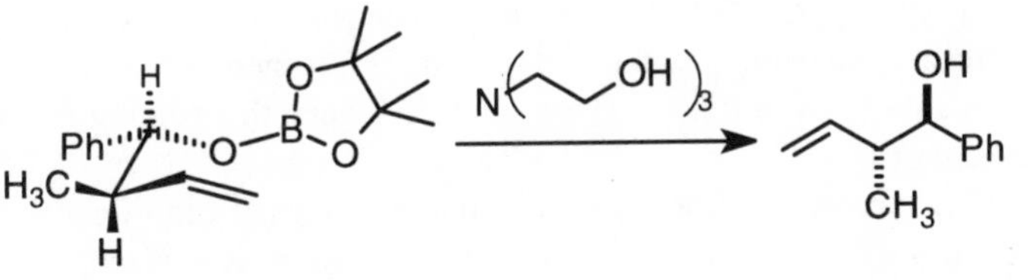

17   **18**

In subsequent illustrations, it will be taken for granted that intermediates analogous to **14** or **17** are formed and then cleaved, and the alcohols will be written directly as products.

This type of transition state is supported by ab initio calculations at the 3-21G level for the reaction of formaldehyde with allylboronic acid [34]. Li and Houk calculated the chair conformation to be 34 kJ mol^{-1} more stable than the twist-boat.

The most critical bond distances include 2.176 Å for the C—C bond being formed, 1.572 Å for the B—O bond being formed, and 1.708 Å for the C—B bond being broken. These values are longer than normal by 0.64 Å for C—C, ~0.2 Å for B—O, and ~0.1 Å for B—C. The other distances are minor distortions (0.1 Å or less) of those of the starting materials.

7.2.3.2 Range of useful substituents. In addition to the successful stereocontrol with crotylboronic esters, it was also shown that (*E*)- and (*Z*)-methoxyallylboronates yielded methoxy alcohols with good stereocontrol [17, 18]. The utility of this chemistry was illustrated with a synthesis of racemic *exo*-brevicomin (**22**), starting from the easily accessible ketoaldehyde **19** and the (*Z*)-alkoxyallylboronic ester **20**. Intermediate dehydrobrevicomin **21** was isolated in 77% yield, 88% de. One enantiomer of *exo*-brevicomin is a pine beetle pheromone (Section 5.4.3.2), and the opposite enantiomer does not inhibit the activity.

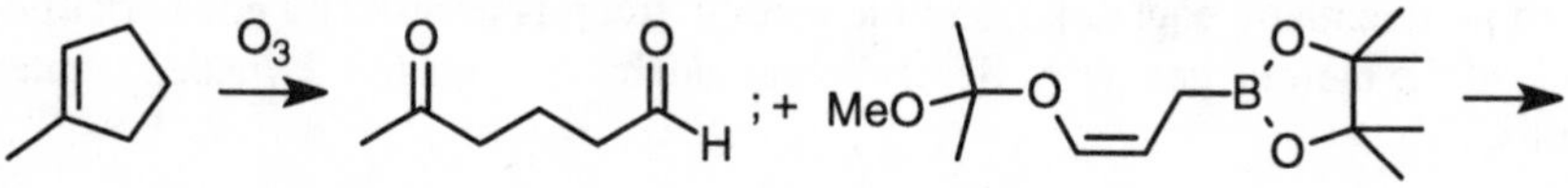

19   **20**

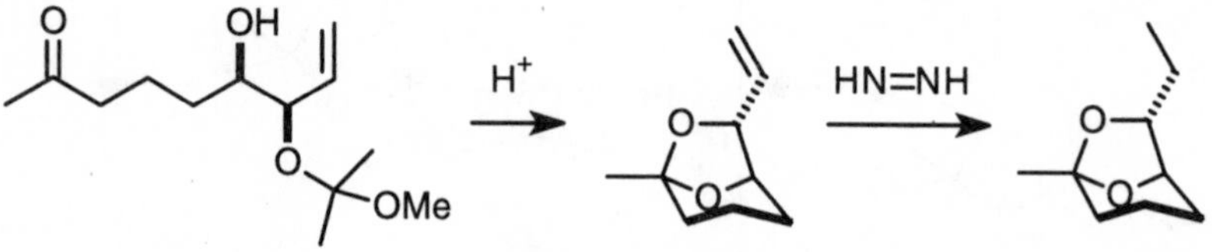

21   **22**

An analogue of **20** having 2,2-dimethyl-1,3-propanediol in place of pinacol in the boronic ester and tetrahydropyran (THP) as the enol ether blocking group showed evidence of strong diastereoselection by the THP group [35]. If there were a simple way to obtain pure enantiomers or diastereomers of THP ethers, this finding might have synthetic applications.

Pinacol (*E*)-γ-methoxycrotylboronate (**23**) with aldehydes gave high de's (80-96%) of **24** over **25**, though the reaction became very slow when R was isopropyl and proceeded well only at 8 kbar pressure [36]. However, the de's of **25** produced by the (*Z*)-isomer of **23** were only 40-46%.

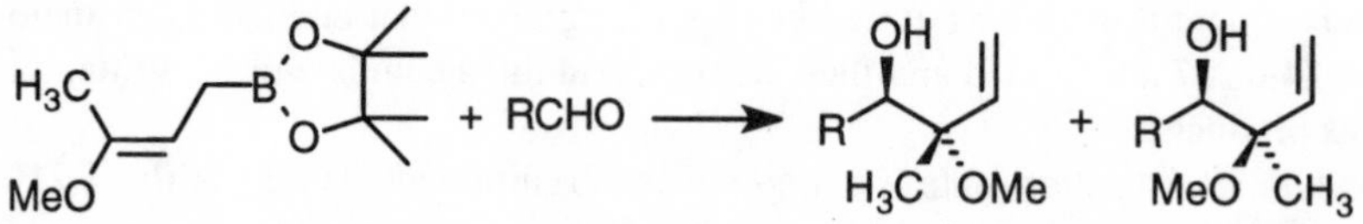

23   (R = Me, Et, Ph, i-Pr)   **24** (major)   **25** (minor)

Both (*Z*)- and (*E*)-γ-[(alkylthio)allyl]boronic esters have been prepared and found to undergo reactions similar to those of their oxygen analogues [37].

Racemates are perfectly satisfactory if the ultimate aim is merely to control olefinic geometry. An example is the highly stereoselective synthesis of (*E*)- or (*Z*)-1,3-dienes, starting from pinacol [γ-(trimethylsilyl)allyl]boronate (**26**) and an aldehyde [38]. The resulting homoallylic silanols (**27**) with sulfuric acid yielded (*E*)-1,3-dienes, isomeric purity >99%, or with potassium hydride yielded (*Z*)-1,3-dienes, isomeric purity 98%. Where R = AcO(CH$_2$)$_8$, the final dienes are the constituents of the red bollworm moth pheromone.

26   **27** (racemic)
[R = Ph, CH$_3$(CH$_2$)$_6$, AcO(CH$_2$)$_8$]

7.2.3.3 Imines. Schiff bases or oximes can be used in place of aldehydes with relatively unhindered allylic boronic esters, but the reactions are not very efficient. Dimethyl allylboronate was found to allylate several imines, R^{1}CH=NR^2, to form homoallylic secondary amines [39]. The reaction of ethylene glycol crotylboronates was very sluggish but proceeded at high pressures (3.6-9 kbar) to form hydroxylamines [40]. A chiral cyclic allylboronate with the oxime of a chiral aldehyde has produced a hydroxylamine product in good de [39]. Hydroxylamines produced by this process can also be converted to nitrones, the utility of which has been demonstrated in a racemic alkaloid synthesis [41]. Sulfenimines, RCH=N-SAr, react with pinacol or ethylene glycol allylboronate to form sulfenamines, which can be converted easily to primary amines [42]. Yields were promising, but stereochemistry was not tested.

7.2.3.4 Ketones. The reaction of racemic pinacol (*E*)- and (*Z*)-2-pentenylboronates with ketones has been examined and compared with similar reactions of other pentenylmetallics [43]. An example is the reaction of pinacol (*Z*)-(1-methyl-2-butenyl)boronate (**28**) with 3-methyl-2-butanone at 4-8 kbar pressure, which yields a mixture of (*E*)-homoallylic alcohols **29** and **30** with only 1.6% of (*Z*)-isomer (**32**). Pinacol (*E*)-(1-methyl-2-butenyl)boronate (**31**) was much less (*E*)/(*Z*) selective but much more diastereoselective, and produced a mixture containing only **29** and **32**. However, it should be noted that the enantiomer of **31** that produces **29** will produce the opposite absolute configurations in **32** (see transition states **109** and **111** in Section 7.2.4.4), and this is therefore not a promising process for asymmetric synthesis. The diastereomer of **32** was not observed in either case.

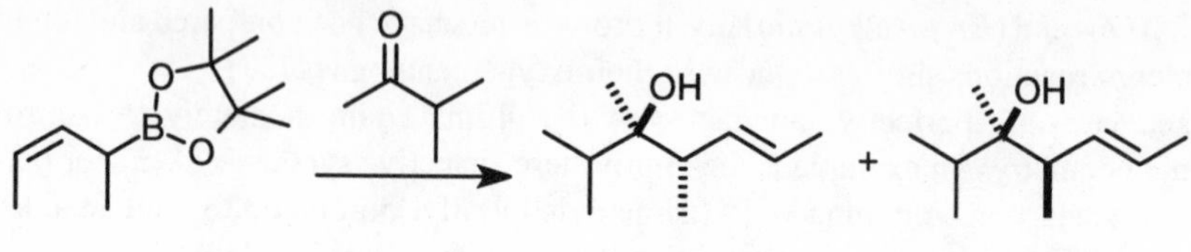

28   **29** (47.4%)   **30** (51.0%)

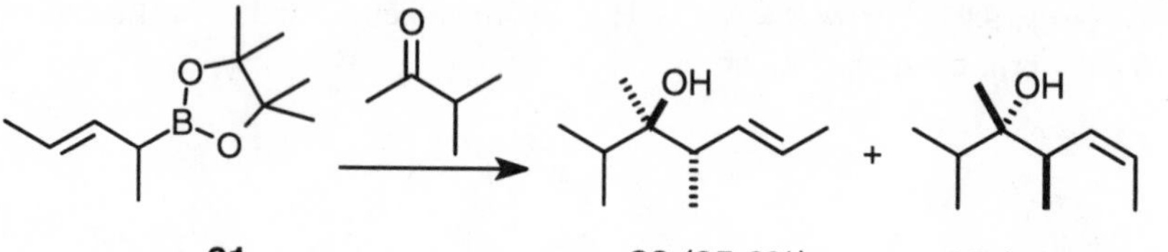

31   **29** (25.6%)   **32** (74.4%)

Reactions of diisopropyl allylboronates with α-hydroxy ketones [44] and α-keto acids [45, 46] evidently involve ligand exchange at boron and cyclic transition states capable of high diastereoselection. For example, diisopropyl (*E*)-crotylboronate (**33**) with benzoin yields diol **34**, and with α-ketobutyric acid the hydroxy acid **35**, both in ≥98% de. Similarly, diisopropyl (*Z*)-crotylboronate (**36**) yields highly pure diastereomers **37** and **38**.

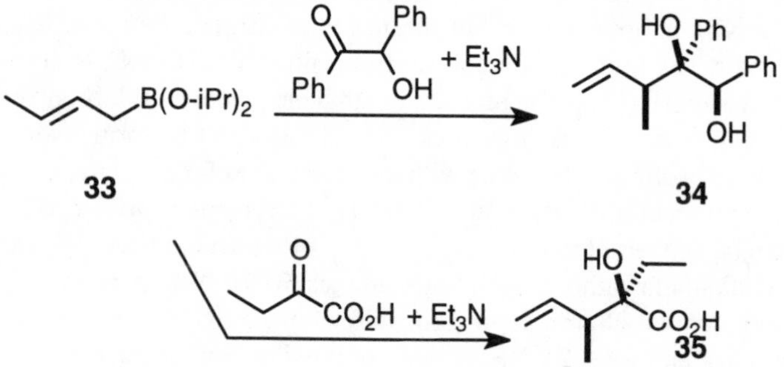

33   **34**

35

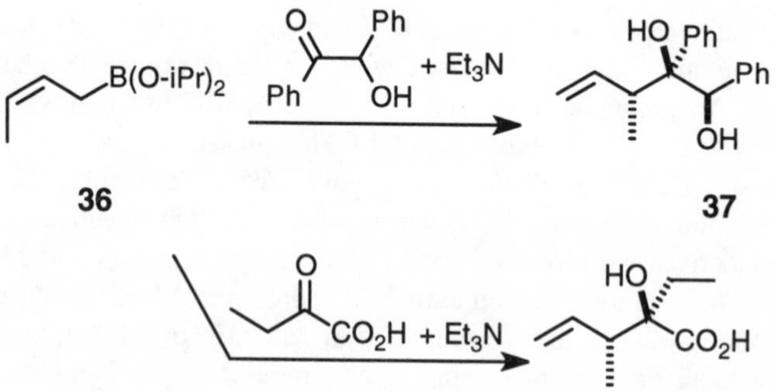

36   **37**

38

7.2.3.5 Cyclizations. The cyclizations reported to date have been done with achiral and racemic models, though some of the results make it obvious what is to be expected of chirally directed closures of enantiomerically pure compounds.

Achiral allylic boronates **40** and **42** also containing an aldehyde function were prepared by known chemistry [25] from the acetylenic acetal **39** [28]. These cyclized preferentially to *cis*- and *trans*-2-vinylcyclopentanols **41** and **43**, respectively. Racemic α-chloro boronic ester **44** cyclized to a mixture of (*Z*)- and (*E*)-*trans*-2-(chlorovinyl)cyclopentanols **45** and **46**, a disappointing result inasmuch as opposite absolute configurations result from the differing geometries (see transition states **109** and **111** in section 7.2.4.4) and chiral control with enantiomerically pure **44** is therefore expected to be poor [28].

39

40 **41**

42 **43**

44 (racemic) **45** (80%) **46** (20%)

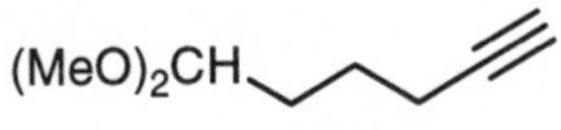

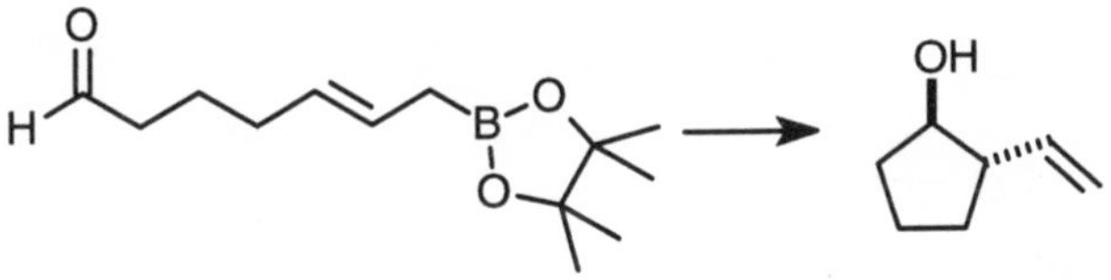

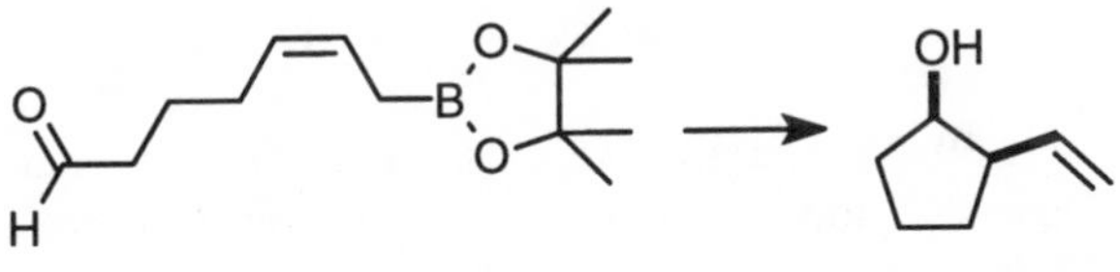

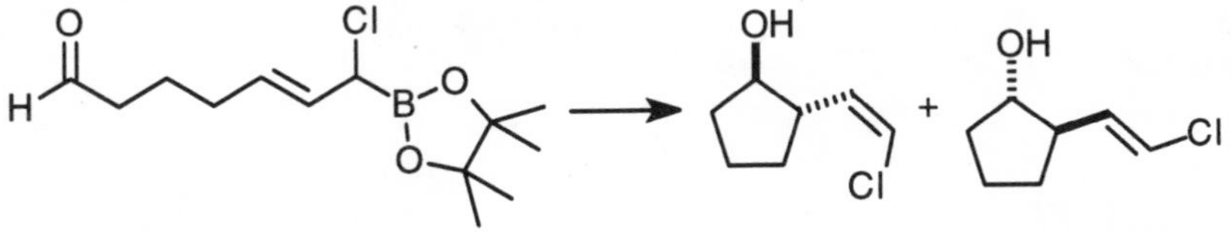

Six-membered ring formation has been investigated more extensively. Closures of unsubstituted 8-carbon (*E*)- and (*Z*)-allylic boronic esters to a freshly unblocked ω-aldehyde group were very highly diastereoselective [31]. Racemic methyl substituted (*E*)-allylic boronic esters **47**, **48**, and **49** cyclized with very high diastereoselectivity at the newly formed bond and good diastereoselection at the methyl substituent [47]. The transition state geometry has to resemble a *trans*-decalin. If the starting materials had been enantiomerically pure, each diastereomeric product would have been enantiomerically pure.

47 (85%) (15%)

48 (97%) (3%)

49 (95%) (5%)

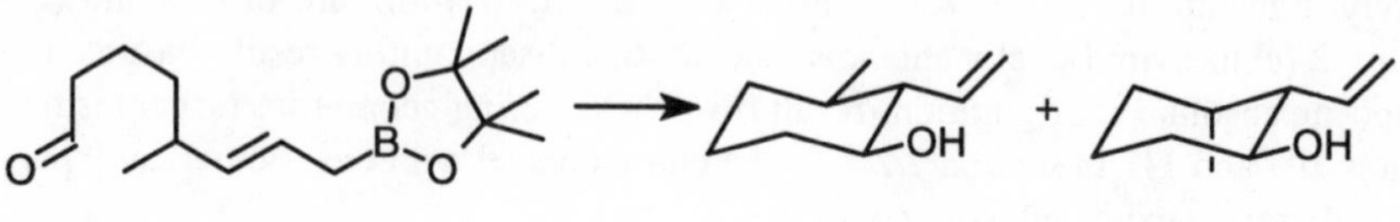

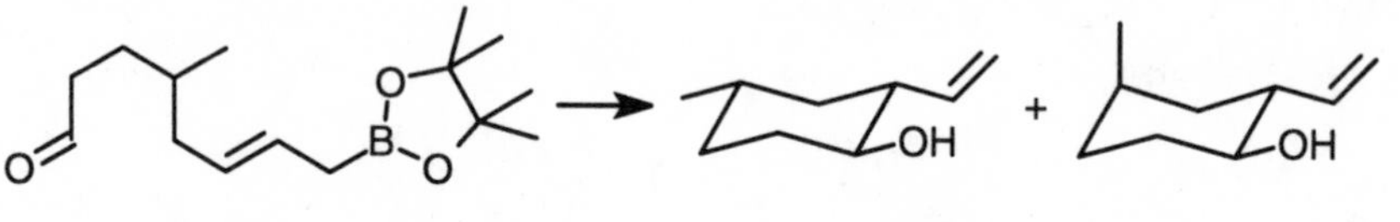

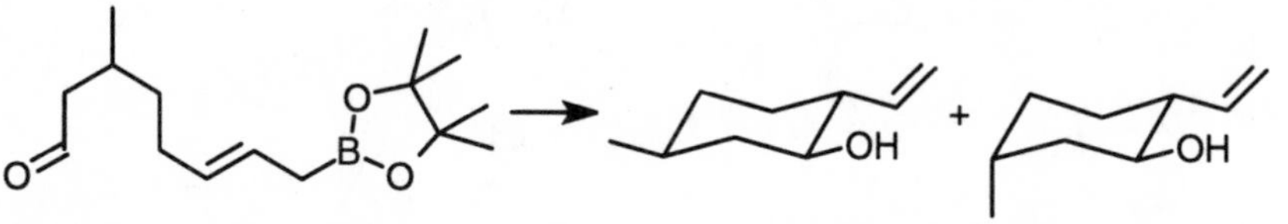

The (*Z*)-allylic boronic esters **50**, **51**, and **52** must close via a transition state having *cis*-decalin geometry, and the products illustrated are all formed in ≥98% diastereomeric purity [47]. Yields (from the aldehyde precursor) were generally ~45%.

50

51

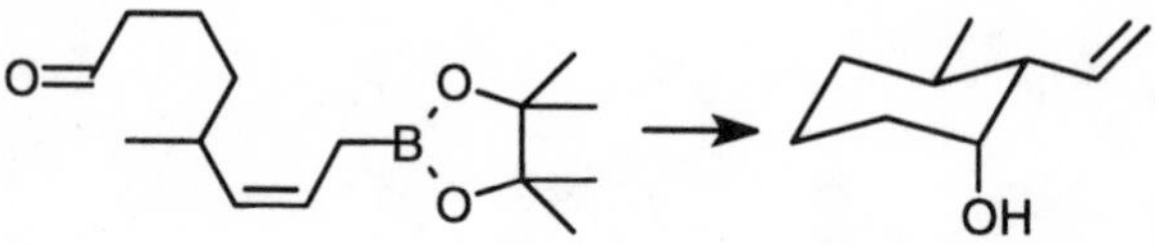

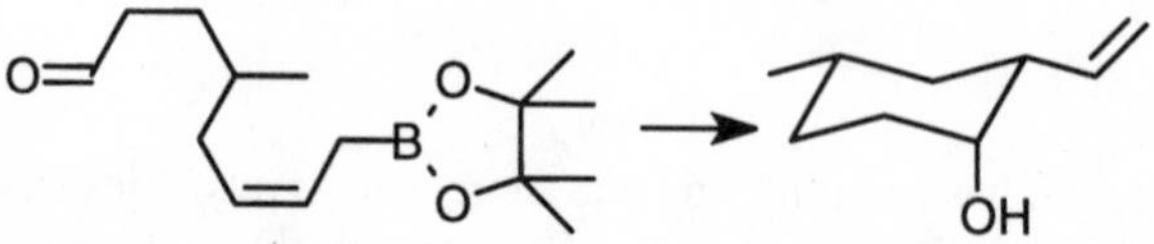

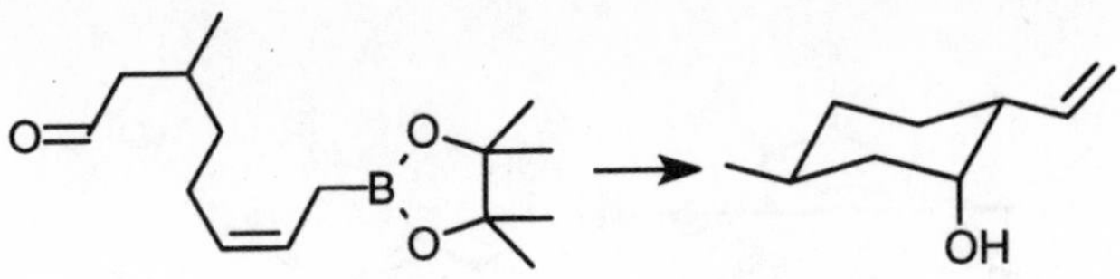

52

The benzyloxy substituted (Z)-allylic boronic esters **53** and **54** yielded only the cyclization products illustrated. The other four benzyloxy analogues of **47-49** and **51** proved unselective at the benzyloxy site, yielding diastereomeric ratios in the range 51:49 to 70:30 [48].

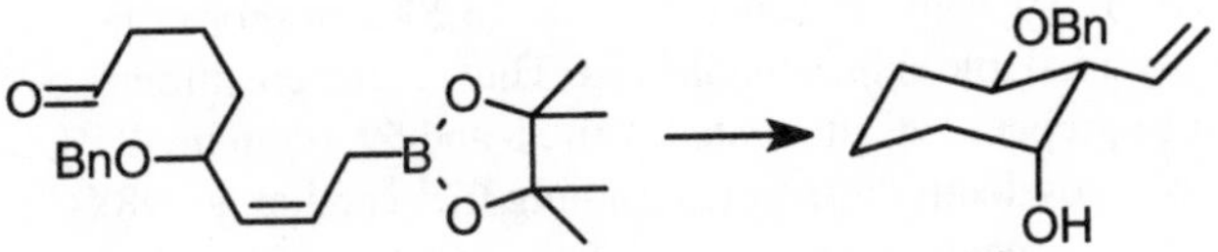

53

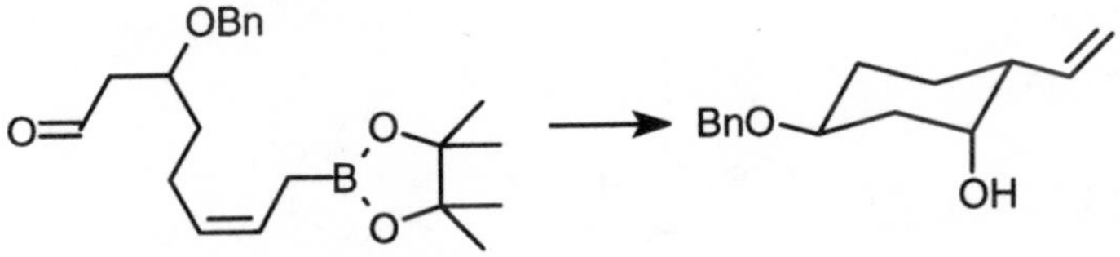

54

7.2.4 Asymmetric Reactions of Boronic Esters

7.2.4.1 Boronic esters of chiral diols with achiral aldehydes. Early attempts to direct the chirality of the product by the use of an asymmetric boronic ester met with limited success [13, 49]. Several asymmetric allylboronic esters were tested with acetaldehyde [49]. The pinanediol ester provided 8% asymmetric induction, and several diols derived from camphor resulted in 0-72% ee's in the screening tests. Phenylbornanediol allylboronate **55** proved most practical, and under optimized conditions (propane solvent, −90 °C) produced (R)-(−)-1-penten-4-ol (**56**) in 86% ee. From other aldehydes with **55**, ee's were often near 70% [13]. Important as it was to the historical development of the methodology, this chiral director is obsolete for most purposes, and further discussion of it will be minimized.

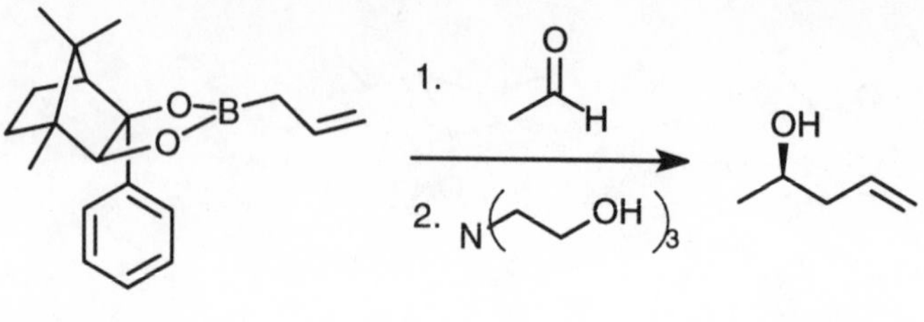

55   **56**

Roush and coworkers have found that tartrate ester allylboronates provide enantioselectivities that are as good as or better than those obtained by Hoffmann's group, as well as much higher reaction rates [50]. Both enantiomers of useful tartrate esters, for example, (*R,R*)-diisopropyl tartrate (**57**) and (*S,S*)-diisopropyl tartrate (**ent-57**), are commercially available at reasonable cost. Thus, either enantiomer of a homoallylic alcohol can be prepared, as illustrated with **58** and **59**, often in ~85% ee [51]. To achieve this diastereoselectivity requires running the reaction at –78 °C in toluene. A number of other solvents were tested and all gave substantially lower ee's. The reaction is also run in the presence of 4Å molecular sieves to remove traces of water, which reduces the ee's, presumably by hydrolyzing the tartrate boronic esters to achiral allylboronic acid (or its anhydride) [51].

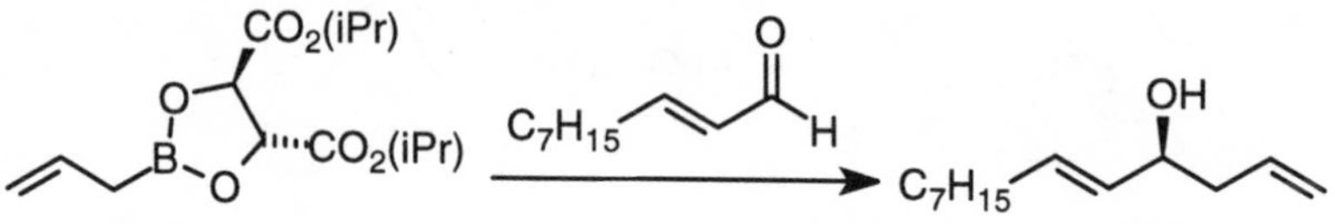

57   **58 ee 86%**

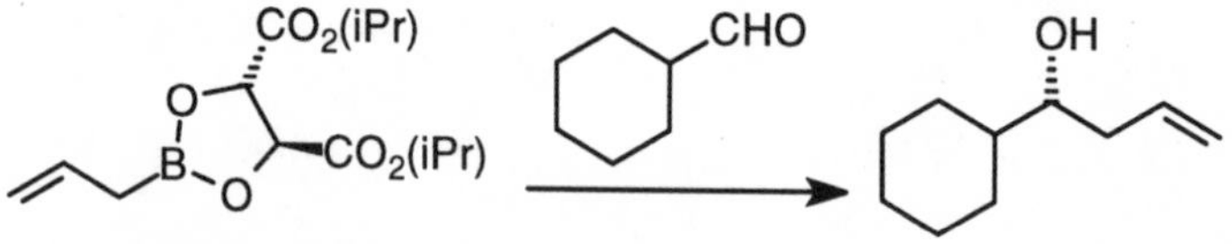

ent-57   **59 (ee 86%)**

(*R,R*)-Diisopropyl tartrate (*E*)-crotylboronate (**4**) and (*Z*)-crotylboronate (**3**) have been tested with a variety of aldehyde substrates [20, 21, 52]. Achiral aldehyde R groups tested with **4** and the ee's observed in the homoallylic alcohol products **60** included n-C_9H_{19}, 84-88%; $PhCH_2CH_2$, 84-85%; n-C_5H_{11}, 85%; cyclohexyl, 86-87%; t-$BuMe_2SiOCH_2CH_2$, 85%; t-Bu, 73%; n-$C_7H_{15}CH=CH$, 74%; Ph, 62-66%. Lower ee's were generally obtained with **3**. The ee's observed in the homoallylic alcohol products **61** included R = n-C_9H_{19}, 77-82%; cyclohexyl, 83%; t-

BuMe$_2$SiOCH$_2$CH$_2$, 72%; *t*-Bu, 70%; *n*-C$_7$H$_{15}$CH=CH, 62%; Ph, 28%. Yields were usually ~85-90%, the only notable exceptions being those with R = *t*-BuMe$_2$SiOCH$_2$CH$_2$ (68-71%) and *t*-Bu (very slow, 41-66%). In reactions of (*E*)-isomer **4**, diastereomeric ratios of **60** to **61** were generally within 1-2% of the isomer ratios of **4** to **3**, often ≥ 99:1. Slightly less diastereoselectivity was observed with the (*Z*)-isomer **3**, but **4** to **3** ratios were usually ≤ 3:97.

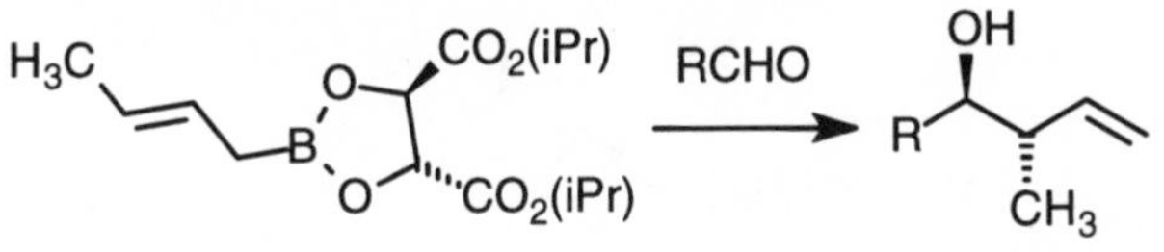

4   **60** (R: see text)

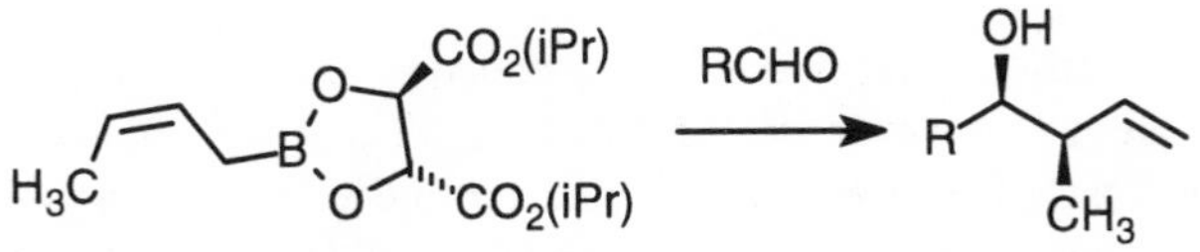

3   **61**

Aromatic and acetylenic aldehydes do not work as well as the others, but metal carbonyl complexes of these substrates give higher ee's [53]. The ee of **62** is substantially higher than that with benzaldehyde, and that of **64** is much improved over the 72% ee obtained with **57** and uncomplexed 2-decynal. (*R,R*)-Diisopropyl tartrate (*E*)-crotylboronate (**4**) and (*Z*)-crotylboronate (**3**) yielded the *anti*- and *syn*-homoallylic alcohols **65** and **66**, respectively. The diastereomeric purities of both were very good. Enantiomeric purities were in the order **65** > **64** > **66**.

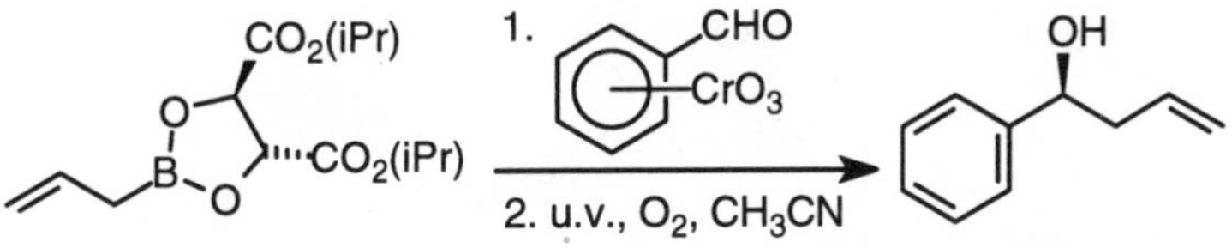

57   **62** (83% ee, 90% yield)

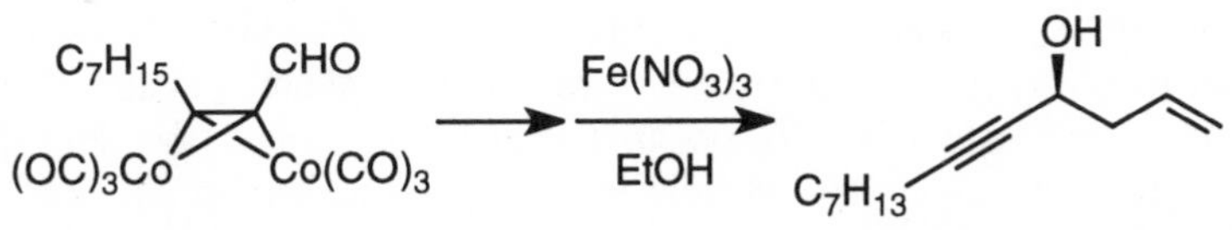

63   **64** (92% ee, 85-95% yield)

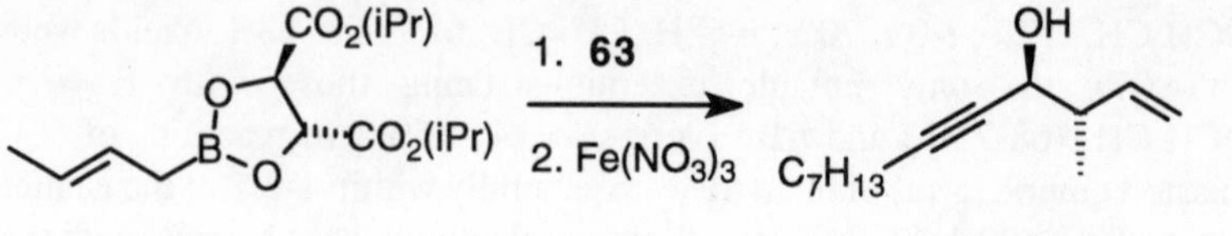

4

65 (96% ee, 94% de)

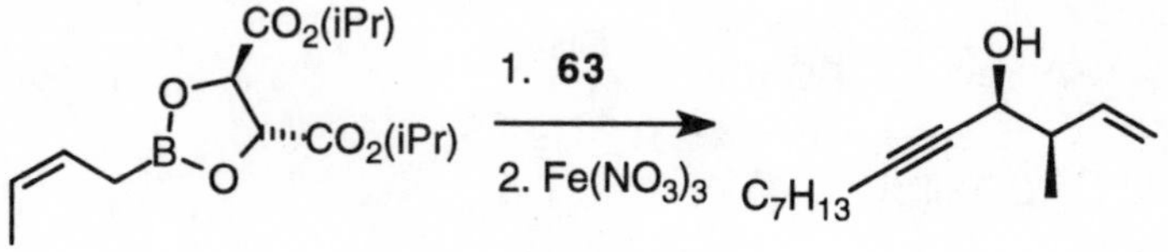

3

66 (86% ee, 94% de)

The enantioselectivity of tartrate esters has been interpreted in terms of lone pair repulsions between the carbonyl oxygens of the tartrate ester and the aldehyde substrate [50]. In this analysis it is assumed that the tartrate ester groups are pseudoaxial to the dioxaborolane ring in the transition state, which is in accord with X-ray and solution NMR data for the boronic esters [54]. It was suggested that the transition state **67** that leads to the favored enantiomer keeps the lone pairs of the aldehyde and tartrate oxygens farther apart than does the disfavored transition state **68** [54]. The carbonyl oxygen in **67** is pointed away from the nearest ester oxygens, and in **68** it is pointed toward them.

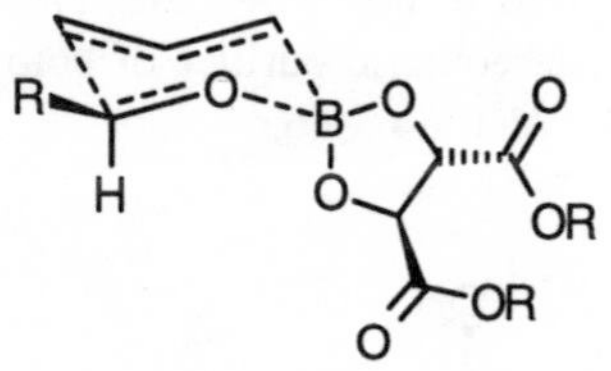

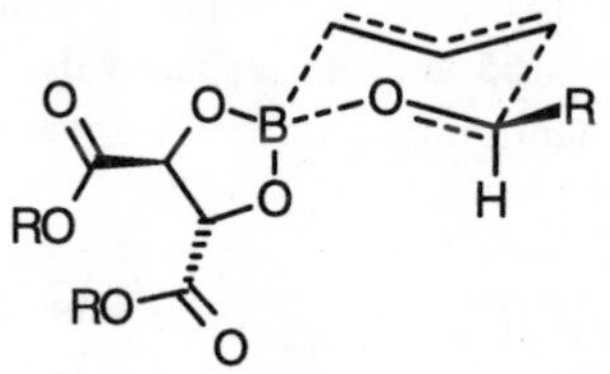

67   **68**

An attempt by the reviewer to model this stereochemistry with the MMX program failed to yield any indication of stereoselectivity. Although eclipsed orientation of the carboxylic ester groups is favored in the parent dioxaborolanes, in which the boron atom is trigonal, conversion toward a tetrahedral boron in any reasonable sort of transition state model favors pseudoequatorial orientation. Perhaps the modeling program does not take the relevant lone pair interactions into account adequately. Roush has pointed out that steric repulsions do not explain the stereoselection [50]. Interpretations regarding the cause of the stereoselectivity have to be considered tentative until better calculations are done, but Roush's suggestion appears reasonable in principle.

Whatever the exact nature of the interactions, there is good evidence that the polarity and orientation of the tartrate ester carbonyl groups are a strong factor in the stereoselectivity. This assumption led to design of the cyclic tartramide allylboronate **69**, which has been found to yield significantly higher ee's than simple tartrate esters or acyclic tartramides [55]. For example, (S)-1-benzyloxy-4-penten-2-ol (**70**) was obtained in 86% ee from **69** and benzyloxy acetaldehyde, but only in 60% ee from **57**. The reagent was designed to orient the carbonyl dipoles properly to achieve maximum diastereoselection. However, the low solubility of **69** in toluene makes it react slowly, conversions tend to be low, and **57** remains the more practical reagent for synthetic purposes.

69   **70** (86% ee)

Other alternative chiral directors to tartrate esters have been explored, mostly with negative results. Analogues of **57** in which the $CO_2(iPr)$ groups were replaced by amide, dimethyloxazoline, methyl, isopropyl, phenyl, or glycol acetonide groups reacted generally much more slowly than **57**, and the ee's were generally well below any useful range [56]. Pinacol allylboronate reacted even more slowly than the chiral boronic esters.

7.2.4.2 A chiral boron amide. Chiral directors based on arenesulfonyl derivatives of (R^*,R^*)-1,2-diphenylethane-1,2-diamine have been developed by Corey and coworkers. The diamine is easily prepared and resolved [57]. With the amine groups sulfonated, the 1,3,2-diazaborolidine derivatives behave much like boronic esters, but have one considerable advantage as chiral directors inasmuch as the arenesulfonyl groups are forced into conformations pointing away from the phenyls [58]. 2-Allyl-1,3,2-diazaborolidines **71** were prepared by reaction of the corresponding allyltributyltin and 2-bromo-1,3,2-diazaborolidine and found to react with aldehydes at −78 °C to provide homoallylic alcohols **72** in high yields and ee's. With Ar = p-tolyl, X = H, and R = Ph, (E)-PhCH=CH-, cyclohexyl, or n-pentyl, the ee's of the derived **72** were 95-98%, and with X = Cl or Br the ee's for the same series of R were 84-99% [58]. With Ar = 3,5-bis(trifluoromethyl)phenyl, ee's in the 88-92% range were observed [59]. With Ar = p-tolyl, X = CH_3, and R = n-pentyl, the ee of **72** was 88%, and with the (E)-crotyl analogue of **71**, ee's with the series of test aldehydes were 91-95% [58].

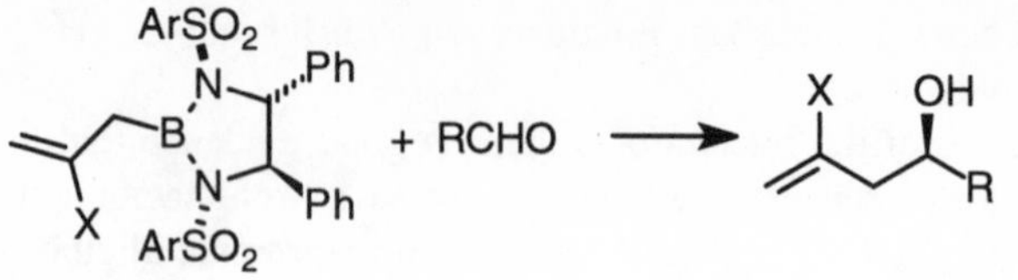

71   **72**

7.2.4.3 Chiral Aldehydes. The chirality of the aldehyde may assist or oppose the chiral induction of the boronic ester. Where the interaction is cooperative, the aldehyde and boronic ester are considered to be a "matched" pair, and where the influences are opposed, they are said to be "mismatched".

(*S*)-*O*-Benzyllactaldehyde (**73**) yields mixtures of separable diastereomers with achiral allylic boronic esters [60]. The (*Z*)-isomer **74** yielded 87% of an 82:18 mixture of **75** and **76**, and the (*E*)-isomer **77** yielded 97% of a 40:60 mixture of **78** and **79**.

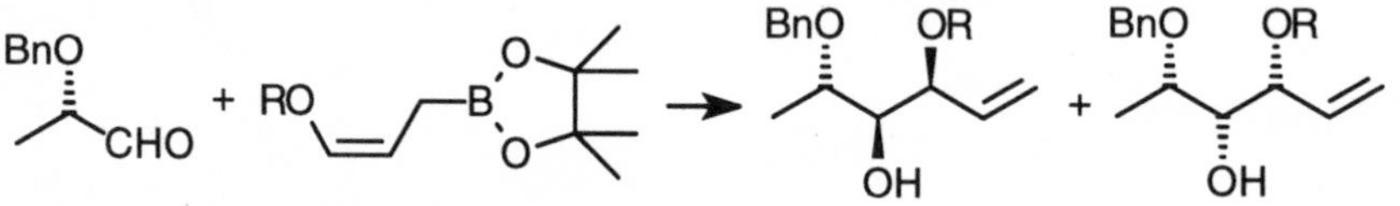

73   **74 (R = Me₃SiCH₂CH₂)75**   **76**

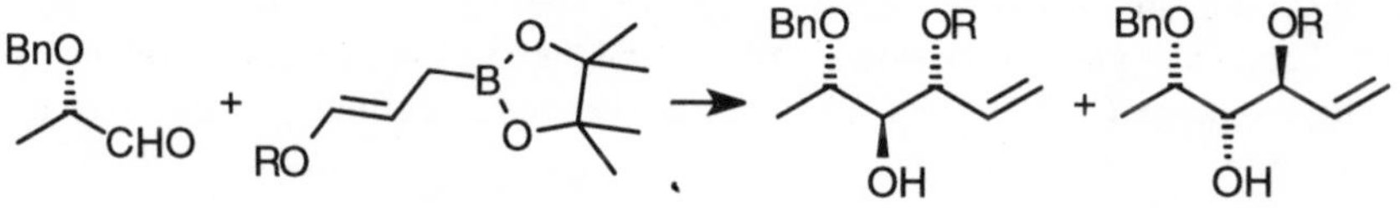

73   **77 (R = Me₃SiCH₂CH₂)78**   **79**

The core transition state geometry found by Li and Houk [34] was used by Hoffmann and coworkers to calculate relative transition state energies of isomeric pathways involving γ-substituted allylboron compounds [61]. First, it was found experimentally that the stereoselectivity of reaction of several γ-substituted allylboronic esters with 2-methylbutanal was independent of the size of the γ-substituent. The isomer ratios increased only from 72:28 to 77:23 for the series of (*E*)-γ-substituents (iPr)$_3$Si, Me$_3$Si, Et$_3$Si, *t*-Bu, Me, respectively. For the reaction of (*E*)-crotylborane with 2-methylbutanal, the four lowest energy transition states, **80**, **81**, **82**, and **83**, calculated with the MMX force field using the core geometry of Li and Houk are illustrated together with their relative energies in kJ mol^{-1}. Note that transition state **81** leads to the opposite diastereomer to the other three, that the energy differences between **80**, **81**, and **82** are negligible, and that a nearly statistical 2:1 product mixture is predicted, close to the 3:1 observed. However, if the real effect were totally entropic, the product ratio would be independent of temperature, but in fact the small temperature dependence showed that $\Delta\Delta H^{\ddagger}$ is about three times the very small $T\Delta\Delta S^{\ddagger}$ in the temperature range studied. It was concluded that the MMX force field is not adequate to deal with predicting stereochemical preferences [61]. However, it might be noted that the accuracy of the MMX calculations is within ~3 kJ mol^{-1}, which is predictive for any but the most subtle differences.

80 (0.0 kJ)   **81 (0.79 kJ)**   **82 (0.84 kJ)**   **83 (2.13 kJ)**

Reactions of the pinacol (*E*)- and (*Z*)-crotylboronates **2** and **1** with six different chiral aldehydes have been compared with MACROMODEL force field calculations [62]. The core geometry calculated for the reaction of allylborane with formaldehyde by Li and Houk [34] was used, and all of the significant most stable conformations of the substituents were used in a Boltzmann distribution to calculate the relative transition state energies. For RCH(CH$_3$)CHO with **2**, experimental and calculated free energies showed average agreement within about ±1.4 kJ mol^{-1}, with the largest deviation 2.7 kJ mol^{-1}. For RCH(CH$_3$)CHO with **1**, experimental and calculated free energies showed average agreement within about ±2.5 kJ mol^{-1}, with the largest deviation (R = *tert*-Bu) 5.1 kJ mol^{-1}. It was regarded as possible that the structural assignment was wrong in this worst case, where the predicted Cram/anti-Cram ratio was predicted to be 60:40 and found to be 16:84. The next worse case, R = Bn$_2$N experimentally, 49:51 found, R = Me$_2$N for calculations, 78:22 found, error 3.2 kJ mol^{-1}, was considered to indicate "unaccounted shortcomings" and "failure" of the theoretical model [62]. The reviewer, who does not expect miracles, would consider the general agreement between theory and experiment to be very good, since the energy differences involved are exceedingly small.

As part of the foregoing investigation, it was also found that pinacol (*E*)-γ-(trimethylsilyl)allylboronate yielded approximately the same diastereoselectivities with the same set of six aldehydes as did the (*E*)-crotylboronate **2**. The largest difference $\Delta\Delta G^{\ddagger}$ that could be measured accurately was for R = Ph, 2.7 kJ mol^{-1} [62], with the silyl boronic ester more selective.

Other examples of the stereocontrol achievable with chiral aldehydes and achiral (*Z*)- or (*E*)-allylic boronic esters have been reported and correlated with a self consistent set of $\Delta\Delta G^{\ddagger}$ values [63]. Several of the diastereoselectivities were as high as 80-90% de.

Double diastereodifferentiation was found to have a strong influence on the product ratios [64]. Reaction of (*S*)-2-methylbutanal (**85**) with (−)-phenylbornanediol

(E)-crotylboronate **84** yielded (3S,4R,5S)-3,5-dimethyl-1-hepten-4-ol (**86**) in 84% de over the (3S,4R,5S)-diastereomer (**87**), but **85** with the enantiomer of **84** yielded **86** in 20% de. (−)-Phenylbornanediol (Z)-crotylboronate (**88**) with **85** gave 10% de of **89**, and the racemate of **88** yielded diastereomer **90** in 40% de, suggesting that the matched pair would be highly diastereoselective for **90**.

84 (matched)   **85**   **86** (92%)   **87** (8%)

88 (mismatched)   **85**   **89** (55%)   **90** (45%)

α-Alkoxy aldehydes such as D-glyceraldehyde acetonide provide stronger chiral directing influence [65]. However, Hoffmann's chiral director being harder to make and usually not quite as selective as Roush's more recent tartrate boronates, the latter will be chosen to illustrate the general principles.

Roush and coworkers have studied the reaction of tartrate allylboronates with chiral alkoxyaldehydes [50, 66]. Typical examples include the reactions of enantiomers **57** and **ent-57** with aldehydes **91**, **94**, and **73** [66]. An (R,R)-tartrate ester with an (αR)-α-alkoxyaldehyde, or an (S,S)-tartrate with an (S)-aldehyde, constitutes a "matched" pair and produces anti-α-alkoxyalcohols (**92**, **95**, and **97**, respectively) in good to excellent de's. (This is an example where the customary nomenclature of aldol condensations, which would label these products "threo", is particularly confusing, inasmuch as the analogous sugar is clearly erythrose.) The opposite pairing of an (R,R)-tartrate with an (S)-α-alkoxyaldehyde or vice versa constitutes a "mismatched" pair and produces syn-α-alkoxyalcohols (**93**, **96**, and **98**, respectively) in fair to good de's.

57 (R,R) ("natural")   **ent-57** (S,S) ("unnatural")

| | from 57 (matched) | 92 98% | 93 2% |
| | from ent-57 (mismatched) | 7% | 93% |

91   92   93

94   95   96
from ent-57 (matched)   99.7%   0.3%
from 57 (mismatched)   27%   73%

73   97   98
from ent-57 (matched)   84%   16%
from 57 (mismatched)   28%   72%

These results were interpreted in terms of additive free energy effects according to Masamune's analysis [67]. The transition states leading from the matched pair (R,R)-tartrate 57 + (R)-α-alkoxyaldehyde 91 to (S)-homoallyl alcohol 92 + (R)-homoallyl alcohol 93 differ in free energy by an amount $\Delta\Delta G_M^{\ddagger}$ [66]. The transition states leading from the mismatched pair (R,R)-tartrate 57 + (S)-enantiomer of 91 to the enantiomeric pair of products differ by $\Delta\Delta G_{MM}^{\ddagger}$. If the intrinsic diastereofacial selectivity of the tartrate allylboronate reagent is defined as $\Delta\Delta G_R^{\ddagger}$ and the intrinsic diastereoselectivity of the α-alkoxy boronate substrate is $\Delta\Delta G_S^{\ddagger}$, then $\Delta\Delta G_M^{\ddagger} = (\Delta\Delta G_R^{\ddagger} + \Delta\Delta G_S^{\ddagger})$ and $\Delta\Delta G_{MM}^{\ddagger} = (\Delta\Delta G_R^{\ddagger} - \Delta\Delta G_S^{\ddagger})$. From these relationships, $(\Delta\Delta G_M^{\ddagger} + \Delta\Delta G_{MM}^{\ddagger}) = 2\Delta\Delta G_R^{\ddagger}$, or $\Delta\Delta G_R^{\ddagger}$ is the average of $\Delta\Delta G_M^{\ddagger}$ and $\Delta\Delta G_{MM}^{\ddagger}$. (It might be noted that $\Delta\Delta G_S^{\ddagger}$ can be similarly derived as the average between $\Delta\Delta G_M^{\ddagger}$ and $-\Delta\Delta G_{MM}^{\ddagger}$.) It is to be expected that $\Delta\Delta G_R^{\ddagger}$ will be approximately constant for a given chiral reagent with a series of chiral substrates, provided that the free energies are not altered too grossly by other interactions between the reactant and substrate $(\Delta G_{RS}^{\ddagger})$ and thus remain approximately additive.

This linear free energy model seems to hold reasonably well for a series of α-alkoxy aldehydes that have the alkoxy group constrained by a ring, such as **91** and **94**, but a series of open chain α-alkoxy aldehydes, including **73**, gave poor diastereoselection and almost no difference between matched and mismatched reactants. The proportions of (R,S)-homoallyl alcohol **97** from matched pair (S,S) **ent-57** and (S) **73** and of (S,S)-homoallyl alcohol **98** from mismatched pair (R,R) **57** and (S) **73** are very similar. This might suggest that $\Delta\Delta G_s^{\ddagger}$ is near zero, but the real source must be reactant–substrate interactions having different $\Delta G_{RS}^{\ddagger}$ and $\Delta G'_{RS}^{\ddagger}$ for the two types of pairing [66]. Interactions involving a lone pair of electrons on the tartrate ester group and a lone pair on the α-alkoxy group of the aldehyde are though to be the source of the discrepancy.

Tartrate crotylboronates incorporate (E)- and (Z)-geometry, which lead to an additional chiral center in the products and an additional factor in $\Delta\Delta G^{\ddagger}$. In general, the (E)-crotylboronates give better de's than the (Z)-crotylboronates [52, 68, 69]. However, with α-methyl aldehydes, the most important factor is that the product should have a 1,3-*anti* relationship of the methyl groups. Reactions of (S)-2-methyl-3-silyloxypropanals (**99**) with the four isomeric isopropyl tartrate crotonates illustrate this point. The 3-silyloxy substituents of **99** tend to lower the diastereoselectivities. The percentage of the indicated major diastereomers **100**, **101**, **102**, and **103** in the mixture of the four diastereomeric products with the type of silyl group optimized is indicated in each case [69].

99	**(R,R)-4** matched	**100** R = t-BuMe₂Si 97%
99 +	**(S,S)-ent-4** mismatched	**101** R = t-BuPh₂Si 90%
99 +	**(S,S)-ent-3** matched	**102** R = t-BuMe₂Si 95%

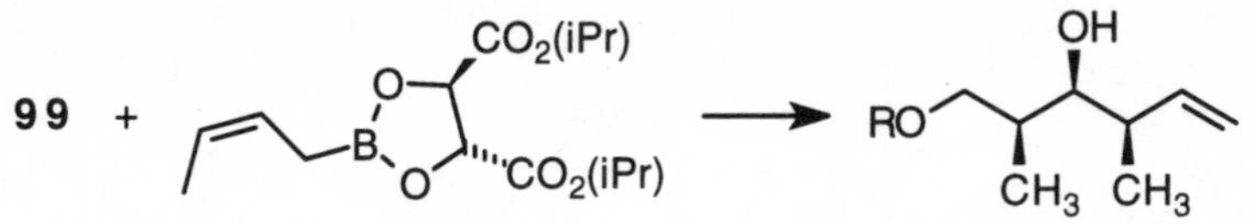

99 +

(R,R)-**3**
mismatched

103
R = t-BuPh₂Si   64%

If aldehydes **99** having ROCH₂ groups are replaced by aldehydes having larger substituents, the matched pairs become more stereoselective and the mismatched pairs less so as the size of the substituent is increased [69]. It was concluded that the synthesis of polypropionate structures should be planned so that (1) any necessary 1,3-*syn* methyl relationships should be introduced as early as possible and (2) that the strategy maximize the number of 1,3-*anti* relationships in the C–C bond connection steps. It was further noted that 1,2-*anti*-2,3-*anti* substructure **101** is the most significant problem, since the 1,2-*syn*-2,3-*syn* substructure **103** is accessible via aldol technology (see Section 7.4) or crotyltin chemistry (outside the scope of this review). The application of these principles to the synthesis of the C(19)-C(29) segment of rifamycin S is summarized in Section 7.2.5.2.

Roush has discussed theoretical reasons for the observed steric preferences in crotylboron and boron enolate reactions with aldehydes [70]. The standard cyclic transition state model was used. It was concluded that flat phenyl or vinyl groups behave as less bulky than methyl groups, and that some of the previously observed preferences for 1,3-*syn*-dimethyl products are actually preferences for placing the less sterically hindered pair of groups 1,3-*anti*. Transition state **104** is the most stable of several conformers that can be derived from an (*E*)-crotylboronate and a chiral aldehyde with an R substituent larger than methyl. The R group extends away from other substituents, and the methyl groups are aligned so that they do not run into each other. The product from **104** has the stereochemistry depicted in **105**. In transition state **106** derived from the (*Z*)-crotylboronate, the positions of the relevant methyl group and hydrogen atom are interchanged. These groups must be interchanged on the aldehyde reactant as well, which is illustrated here as the enantiomeric aldehyde leading via transition state **106** to product **107**, which has the same absolute configuration as **105** at the CHOBX₂ group and the opposite configuration at both of the other chiral centers. Although illustrated here with crotylboronates, these arguments were originally presented for generalized metal enolates [70].

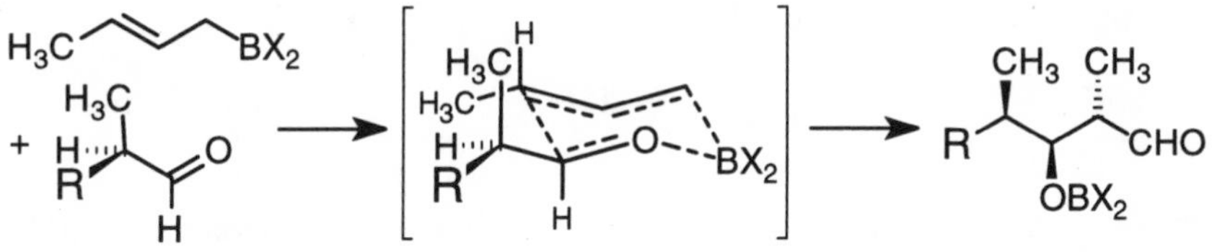

104   **105**

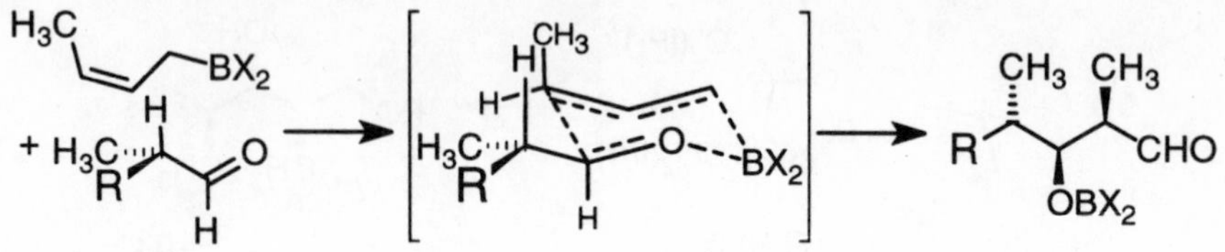

106   **107**

7.2.4.4 Chiral allylic groups.

With α-substituted allylboronic esters (**108**), the favored product is usually the (Z)-alkene (**110**), which requires that the substituent X be axial in the transition state (**109**) [71]. The axial substituent X in **109** encounters less steric repulsion from the boronic ester group than does the equatorial substituent X in alternative transition state **111**, which leads to the minor product **112**. It should be noted that **110** and **112** have opposite absolute configurations when both are formed from the same enantiomer of **108**. Various groups X tested included Cl, Br, OCH₃, and EtS. The percentage of (Z)-isomer was highest, >97%, with X = CH₃ and R = isopropyl, but 96% (Z)-isomer was obtained with X = Cl or Br and R = isopropyl, and the yields (~83%) were considerably higher [71].

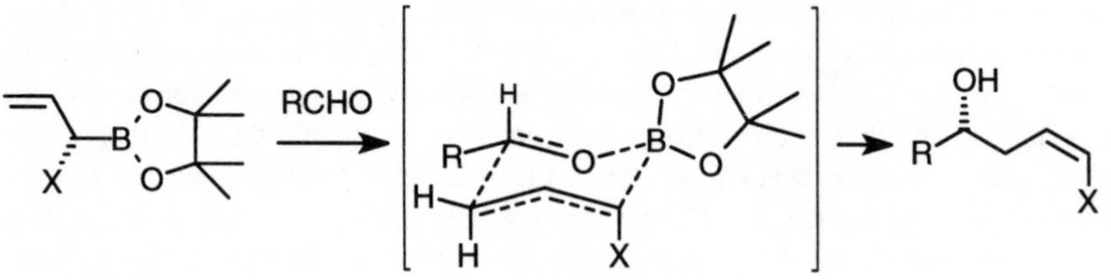

108   **109**   **110**   (major)

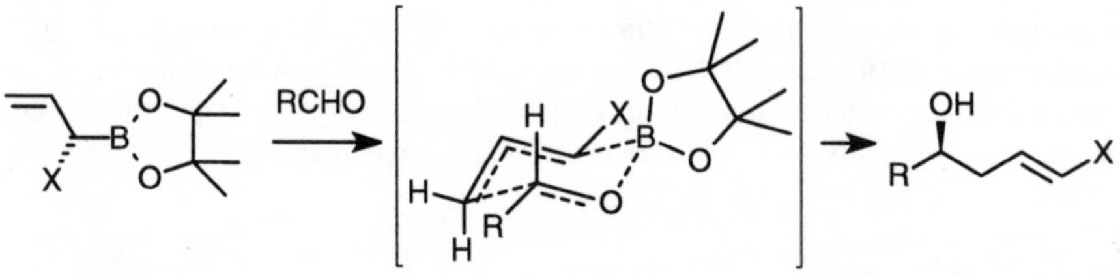

108   **111**   **112**   (minor)

The same reactions were then run with pinacol (S)-(α-chloroallyl)boronate (**108**, X = Cl) of 92% ee, which had been prepared via the reaction of butane-2,3-diol (dichloromethyl)boronate with vinylmagnesium chloride (see Section 5.3.1) and transesterified with pinacol [72]. (The butanediol ester did not yield good ratios of **110** to **112**.) With aldehydes having R = Me, Et, i-Pr, or Ph, alcohols **110** were obtained in 89-92% ee, with the ratio of separable isomers **110** to **112** in the range 94:6 to 96:4. It was also found that with Δ-glyceraldehyde acetonide (**91**), the enantiomer of **108** (X = Cl) gave a higher proportion of (Z)-product **110** (98.5%) than did **108** itself (86%).

An important recent development is the improvement of the foregoing reaction by the use of (R,R)-DICHED (αS)-(α-chloroallyl)boronate (**113**), which with a series of achiral aldehydes yielded 76-89% (Z)-chlorohomoallylic alcohols **114** in ≥99% isomeric purity [73]. For the mismatched pairs **113** with chiral aldehyde **91** or enantiomer **ent-113** with **73**, the 92:8 (Z)/(E) ratios of **115** to **116** and of **117** to **118**, respectively, were equal to or better than with the tartrates, but not quite as good as with the more reactive chiral allyl(dialkyl)boranes (see Section 7.3). With the matched pairs **115** with **91** or **113** with **73**, (Z)/(E) ratios were 96:4 to 97:3. These last results are anomalous. If the linear free energy relationships discussed in Section 7.2.4.3 and below in this section were valid, the matched pairs should provide even higher (Z)/(E) ratios than those with achiral aldehydes, and the (E)-isomers should be undetectable.

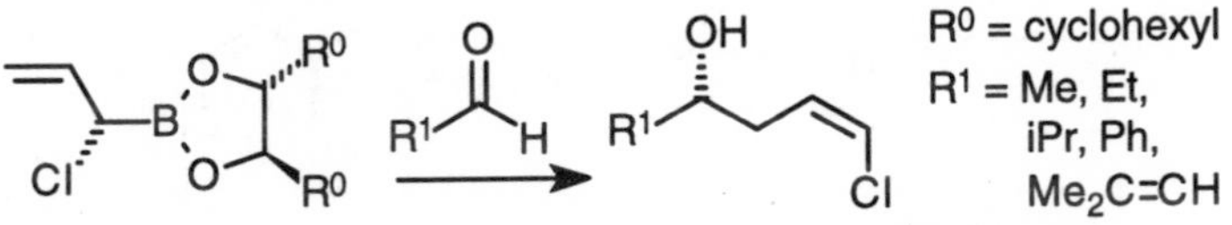

R⁰ = cyclohexyl
R¹ = Me, Et,
    iPr, Ph,
    Me₂C=CH

113   **114 (≥99%)**

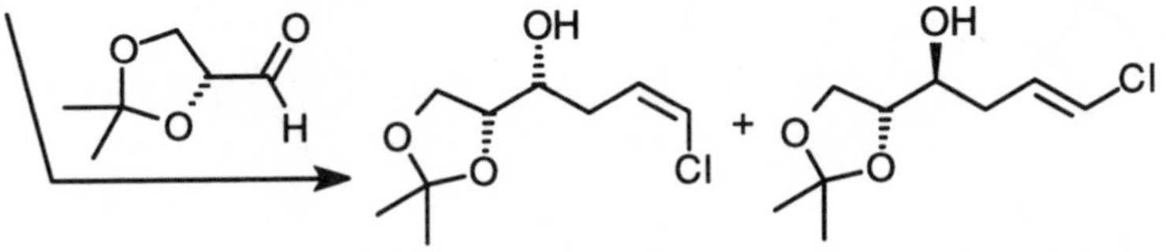

91   **115 (92%)**   **116 (8%)**

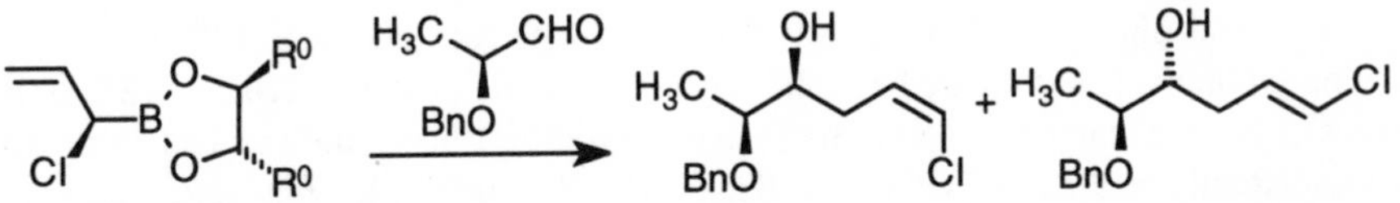

ent-113 (R⁰ = cyclohexyl) **73**   **117 (92%)**   **118 (8%)**

Reaction of functionally substituted alkynes with dicyclohexylborane followed by cleavage of the cyclohexyl groups with trimethylamine *N*-oxide has yielded 3-substituted (*E*)-1-alkenylboronic esters [74]. Pinacol (*E*)-3-trimethylsilyloxy-1-butenylboronate was converted by thionyl chloride and a limited amount of water to pinacol (*E*)-1-chlorocrotylboronate [74]. Pinacol (*Z*)-1-chlorocrotylboronate (**119**) was prepared from pinacol (*Z*)-1-propenylboronate and (dichloromethyl)lithium [75]. Reaction with aldehydes yielded mixtures containing 85-93% (*E*)-1-chloro homoallylic alcohols (**120**) with the remainder being the (*Z*)-isomers (**121**). Although racemates, the steric relationships were known from the chloroallyl boronic ester work to be those depicted. Optically active pinacol (*E*)-1-chlorocrotylboronate (**122**) [74] with aldehydes yielded mixtures containing 95% (*Z*)-1-chloro homoallylic alcohols (**123**) in high enantiomeric purity, from which the 5% of (*E*)-isomers **124** of opposite absolute configuration could be separated chromatographically [75]. Racemic functionally substituted analogues of **122** prepared via the (dichloromethyl)lithium reaction also showed high selectivity for (*Z*)-1-chloro homoallylic alcohol formation [75].

119   (R = Me, Et, i-Pr, Ph) **120**   **121**

122   (R = Me, Et, i-Pr, Ph, Bu) **123**   **124**

The stereoselectivity of (*E*)-(α-chlorocrotyl)boronate **122** has also been studied with chiral aldehydes [76]. It was found that the matched pair **122** with (*S*)-2-methylbutanal (**125**) yielded a >98:2 ratio of isomeric products **126** and **127**. However, the mismatched enantiomeric (α-chlorocrotyl)boronate *ent*-**122** with **125** yielded only an 90:10 mixture of **128** and **129**. Reaction of the racemate **122** + *ent*-**122** with racemic aldehyde **125** produced an 80:20 mixture of racemic **126** and racemic **128**, with negligible amounts of the (*E*)-isomers **127** and **129**.

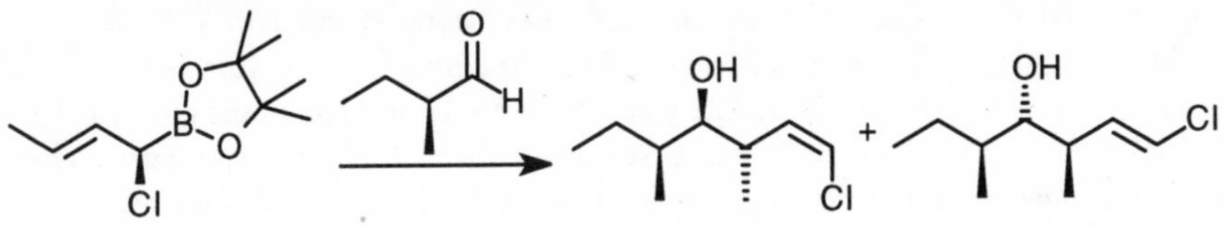

122   **125** (matched)   **126** (>98%)   **127** (not detected)

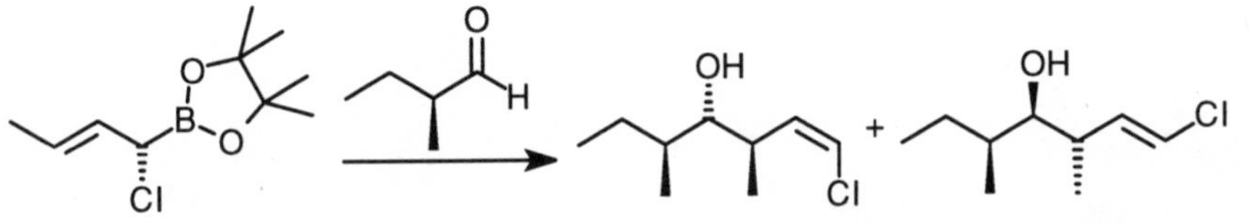

ent-122   **125** (mismatched)   **128** (90%)   **129** (10%)

These results are consistent with roughly additive $\Delta\Delta G^{\ddagger}$ effects for the chirality of the boronic ester and the aldehyde [76]. The typical 97:3 isomer ratio produced in reactions of **122** with achiral aldehydes corresponds to $\Delta\Delta G^{\ddagger} = 8.4$ kJ mol^{-1}, the 80:20 isomer ratio from racemic **122** and racemic **125** to $\Delta\Delta G^{\ddagger} = 3.4$ kJ mol^{-1}. Assuming that $\Delta\Delta G^{\ddagger}$ between the transition states leading to minor byproducts **128** and **129** is also 3.4 kJ mol^{-1}, the difference, $\Delta\Delta G^{\ddagger} = 5.0$ kJ mol^{-1}, should correspond closely to the 90:10 ratio of **128** to **129** obtained from mismatched **ent-122** + **125**, which it does. The predicted $\Delta\Delta G^{\ddagger}$ for the reaction of matched pair **122** + **125** is then 11.8 kJ mol^{-1}, which corresponds to a 500:1 ratio of **126** to **127** and is consistent with the experimental result that the ratio is >50:1.

The diastereoselectivity of **ent-122** is insufficient to overcome that of the alkoxy-substituted mismatched aldehyde **130**, which slowly produces the (E)-1-chloroalkene **131** as the only isolable product (56%).

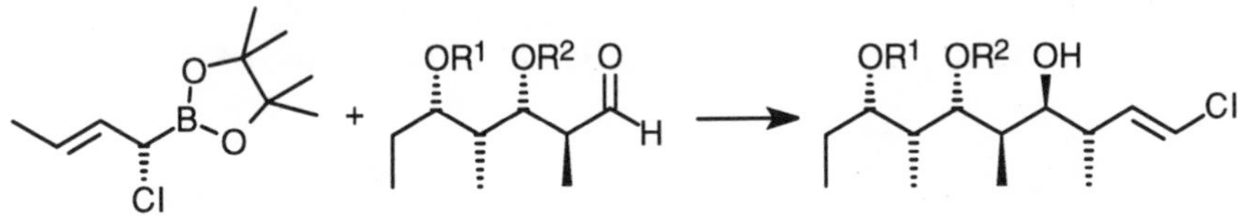

ent-122   **130**   **131**

($R^1 = t$-BuMe$_2$Si, $R^2 = p$-methoxybenzyl)

Although ozonolysis of the chlorovinyl compounds to aldehydes is the usual application, routes for replacement of the chlorine by alkyl, vinyl, or (trimethylsilyl)methyl groups have been reported [77].

Pinacol (α-methoxycrotyl)boronate (**132**) has been prepared in 90% ee from (α-chlorocrotyl)boronic ester **122** and lithium methoxide [78]. Reaction of **132** with mismatched chiral aldehyde **125** yielded (Z)-enol ether **133** as the only isolable product, indicating that **132** is more stereoselective than **122**. However, when the enantiomeric (α-methoxycrotyl)boronate *ent*-**132** was paired with more hindered mismatched aldehyde **134**, isomeric products **135** and **136** were obtained in a 9:1 ratio. The matched pair **132** + **134** produced the expected product **137** in high purity.

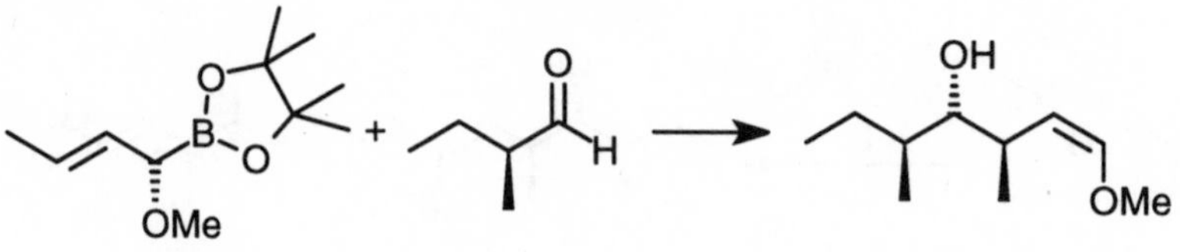

132   **125**   (mismatched)   **133**

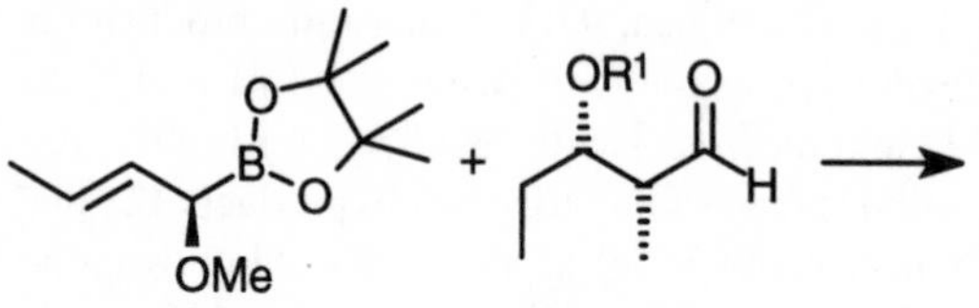

ent-**132**   **134** (R^1 = *t*-BuMe$_2$Si) (mismatched)

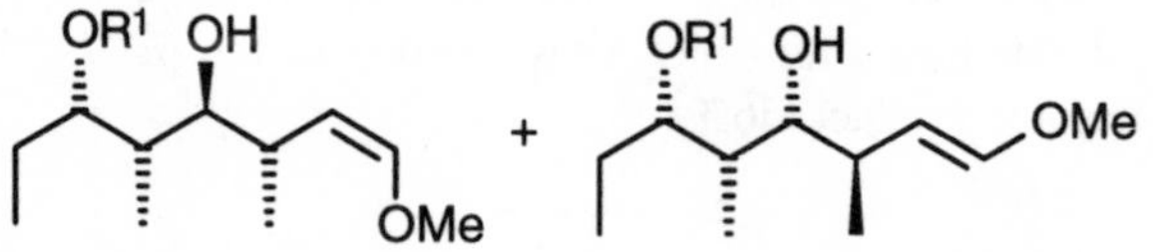

135 (90%)   **136** (10%) (total yield 66%)

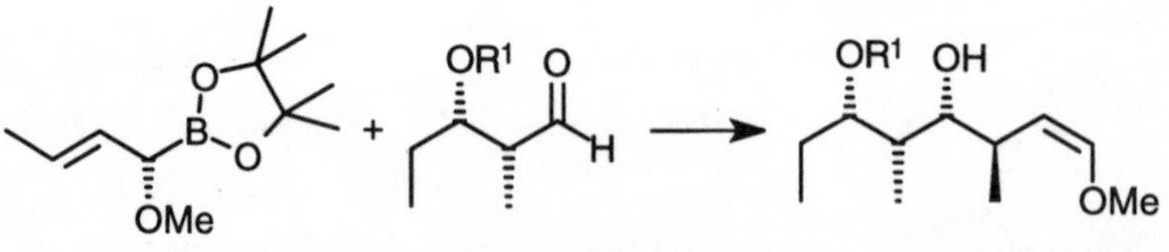

132   **134**   (matched)   **137** (only product)

Kinetic studies have been carried out with various α-substituted pinacol allyl-boronates [79]. A small but representative cross-section of the data is shown in Table 7-1. The relative rates and stereoselectivities appeared to represent a combination of electronic and steric effects.

Table 7-1. Second-Order Rate Constants (L mol^{-1} sec^{-1}) and Percent (Z)-Product for Reaction of α-Substituted 2-Allyl-1,3,2-dioxaborolanes with Propanal in CDCl$_3$.

Substituent	$10^4 k$, –10 °C	% (Z), –10 °C	$10^4 k$, 22.5 °C	% (Z), 22.5 °C
H	3.53	(50)	16.9	(50)
CH$_3$	7.25	67	26.5 ^{a}	64.4
Cl	3.97	95.4	13.3	95.9
Br	0.44	93.2	2.42	95.4
CH$_3$O	6.00	—	29.9	97.0
CH$_3$S	1.60-3.0	81.8	7.19	76.6

Chirality transfer has also been used to generate quaternary chiral centers. First, the diastereoselectivity of γ,γ-disubstituted pinacol allylboronates was verified with several aldehydes [80], and α-methyl-γ,γ-disubstituted pinacol allylboronates were also shown to give good diastereoselection [81]. The stereochemically defined allylic boronic esters **138** and **140** were then used with acetaldehyde to generate chiral homoallylic alcohols **139** and **141** [81]. The enantiomeric series was also prepared. The de's of the products were 96-97%, but the ee's were only 84-93%, implying that the **138** and **140** were not obtained diastereomerically pure, even though the same route that had led to very pure **6** was used for their preparation. The reasons for this discrepancy are not understood.

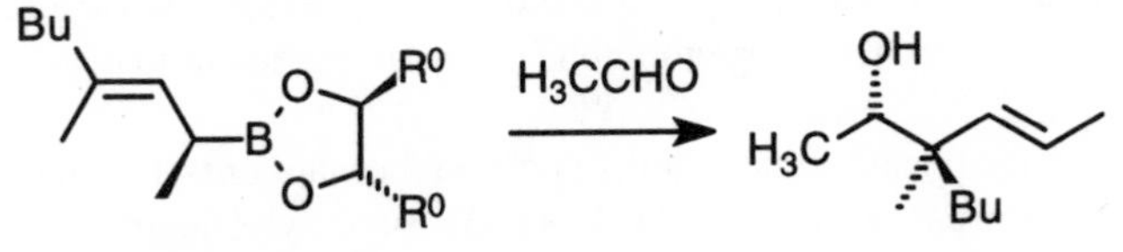

138 (R^0 = cyclohexyl)   **139**

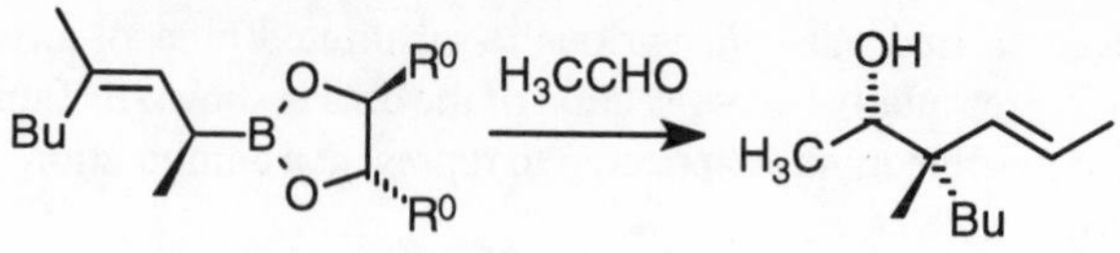

140 (R⁰ = cyclohexyl)   **141**

2-Allyl-1,3-ditosyl-4,5-diphenyl-1,3,2-diazaborolidine (**142**) was tested with the chiral aldehydes **143** and *ent*-**143** [58]. In this case, the matched pair with **143** yielded **144** in 96% diastereomeric excess, and the mismatched pair with *ent*-**143** yielded **145** in 92% de.

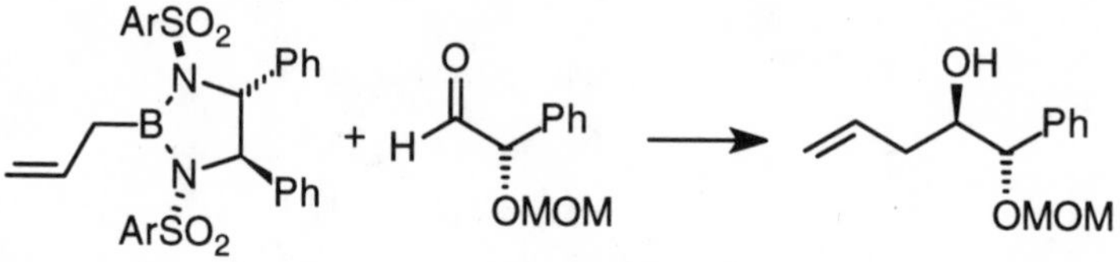

142   **143** (MOM = CH₂OCH₃) **144** (matched)

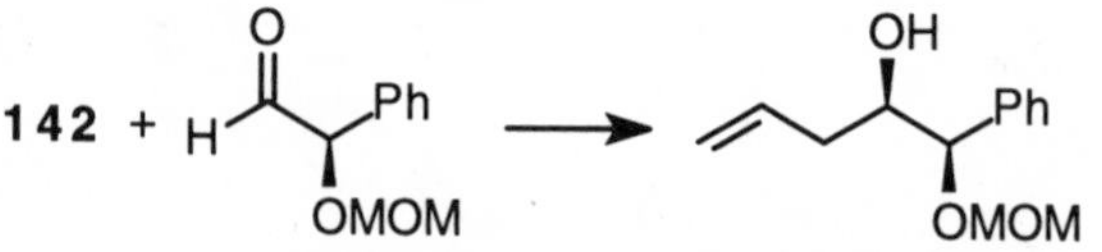

142 +

ent-**143**   **145** (mismatched)

7.2.5 Asymmetric Syntheses with Boronic Esters

7.2.5.1 Insect pheromones. An important early application of allylic boronic ester chemistry was in the first synthesis of stegobinone (**149**), the pheromone of the drugstore beetle *Stegobium paniceum*, as well as its enantiomer and several diastereomers [82, 83]. This synthesis established the stereochemistry and absolute configuration of the active pheromone, though the activity of the impure **149** obtained was only a small fraction of that of the natural product [82], evidently because small amounts of the repellent epimer epistegobinone (**150**) are sufficient to deactivate the attractant.

The synthesis began with phenylbornanediol (Z)-crotylboronate *ent*-**88**, which with acetaldehyde and the usual work up yielded (3S,4S)-3-methyl-1-penten-4-ol (**146**), 93%, de 89%, ee 68-75%. Suitable protection and ozonolysis steps led to intermediate ester **147**, which contains the correct chiral centers at C(2) and C(3) of stegobinone. Acylation of the appropriate diketone dianion and desilylation yielded triketone alcohol **148**, which retained the correct stereochemistry only at C(2). Inasmuch as the stereocontrol at C(3) was lost in any event, the production of **147** and its diastereomer via yeast reduction of ethyl 2-methylacetoacetate was then investigated and found to yield mostly the wrong but irrelevant absolute configuration at the carbon which becomes C(3), but the correct absolute configuration (>95% ee) at the future C(2), and the best sample of stegobinone (**149**) was obtained by this route, not via boron chemistry. Cyclization with sulfuric acid yielded a mixture of four isomers, stegobinone (**149**), epistegobinone (**150**), and the pair of epimers having the opposite configuration at C(3) (**151**) [83]. Chromatographic separation readily led to crystalline epistegobinone (**150**), on which an X-ray structure was determined. Frustratingly, stegobinone (**149**) crystallized only with difficulty, apparently retaining some oily phase. Spectroscopic data clearly identified **149** as the natural product, even though the synthetic **149** had limited attractant activity and was not stable to storage. For the more recent synthesis of pure **149**, see Section 5.4.3.5.

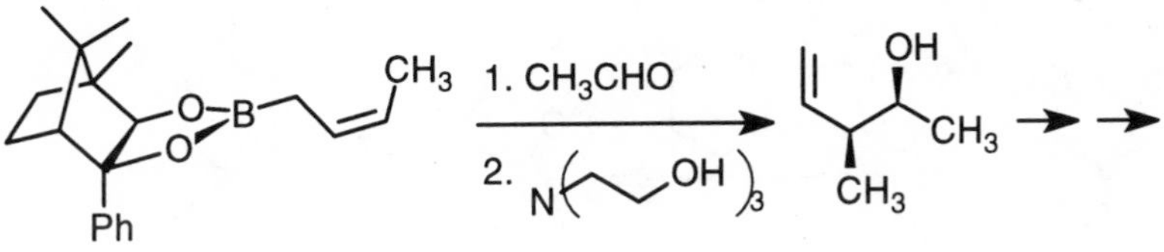

ent-88   **146**

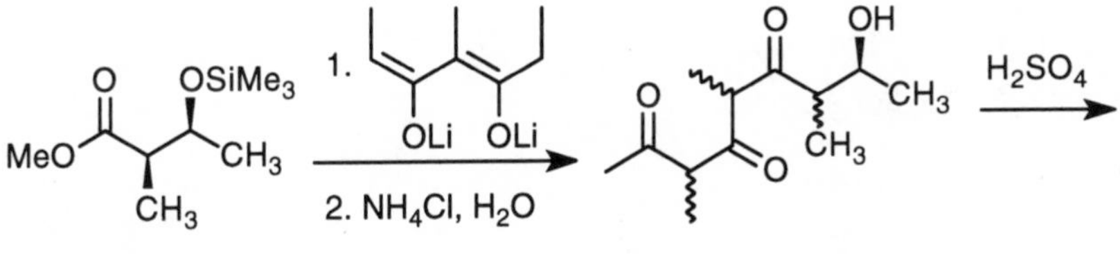

147   **148**
(alternatively derived via a yeast reduction)

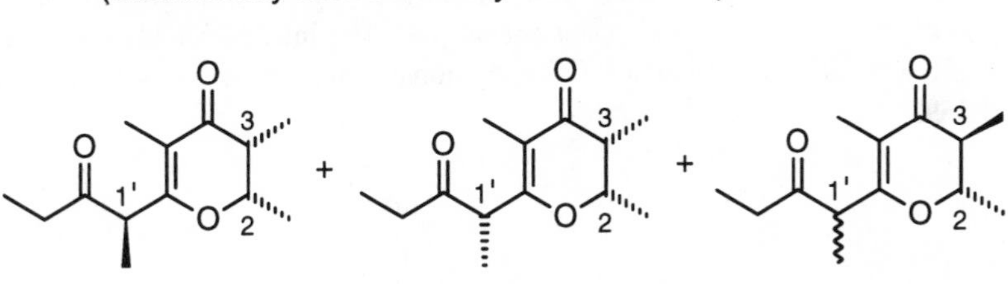

149   **150**   **151**

Phenylbornanediol (*E*)-crotylboronate **84** has been used to control the chirality in a synthesis of δ-multistriatin (**156**), the major component of the pheromone of the

European population of the elm bark beetle *Scolytus multistriatus* [84]. The homoallylic alcohol intermediate **152** was transformed by standard methods to the
tosylate **153**, which with the magnesioenamine **154** yielded intermediate **155** having
an indeterminate stereochemistry at the methyl group α to the imine. Hydrochloric
acid converted **155** to an isomeric mixture containing 70-85% δ-multistriatin, which
was separable by gas chromatography in 97% diastereomeric purity, 52% ee.

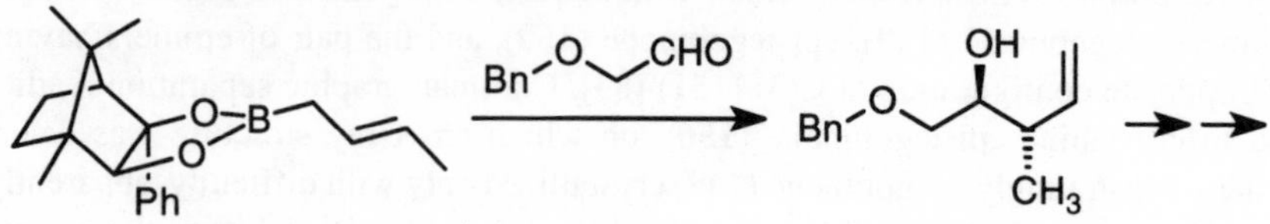

84   **152**

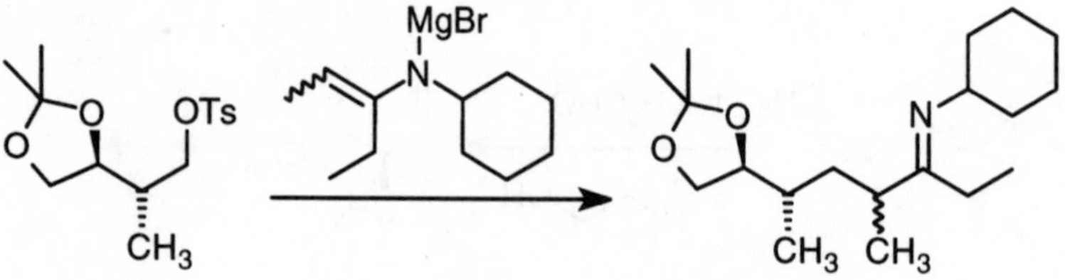

153   **154**   **155**

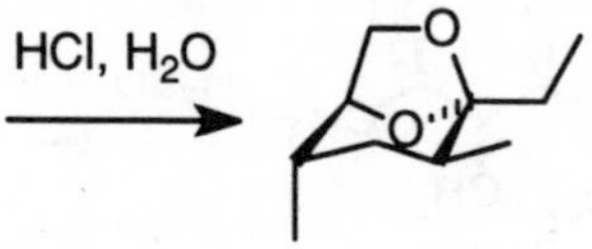

156

7.2.5.2 Macrolide and ionophore antibiotic syntheses. The Prelog-Djerassi
lactone (**159**) was synthesized from (2*S*,4*R*)-2,4-dimethyl-5-hexenal (**157**), obtained
via a resolution route, and boronic ester *ent*-**88** [64]. The intermediate homoallylic
alcohol **158** was formed in 70-80% de. Conventional protection and oxidation steps
yielded **159**, which crystallized readily.

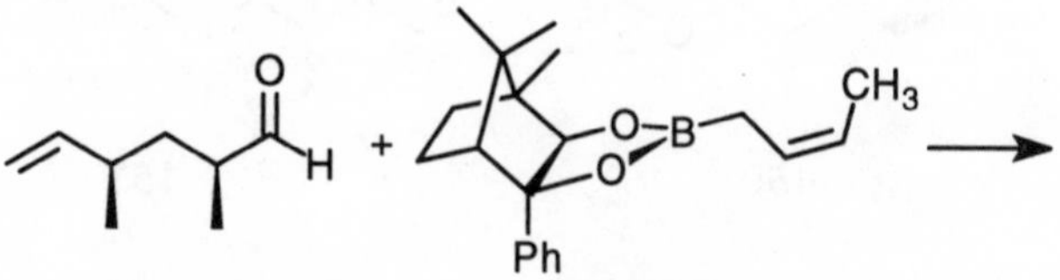

157   *ent*-**88**

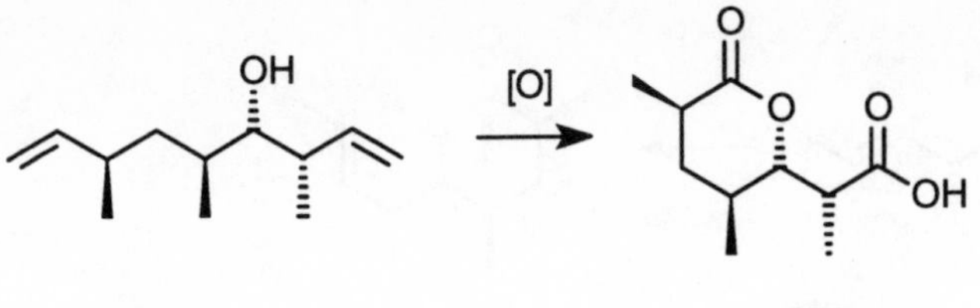

158 **159**

The more sophisticated reagent (*S,S*)-DICHED (*S,Z*)-(1-methyl-2-butenyl)-boronate (**6**) was used with matched asymmetric aldehyde **160** (from resolved 2,4-dimethylglutaric acid) to form the expected homoallylic alcohol **161** in high diastereomeric and enantiomeric purity [26]. Further conventional transformations provided an efficient synthesis of the natural product invictolide (**162**).

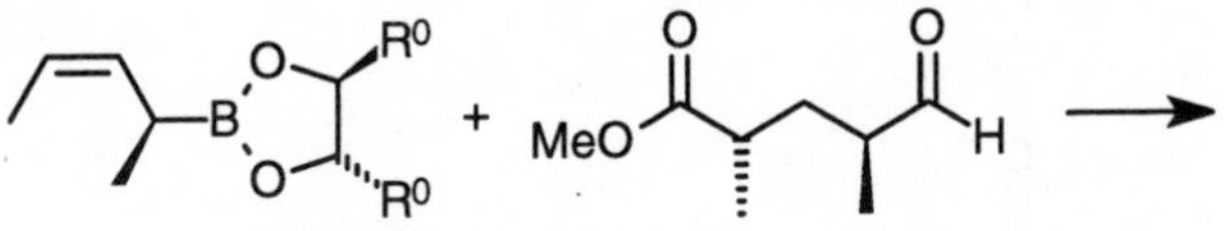

6 (R⁰ = cyclohexyl)   **160**

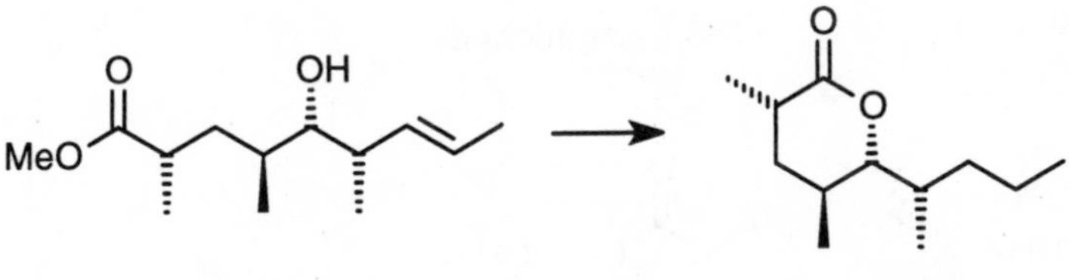

161 **162**

(*S,S*)-DICHED (*S,Z*)-(1-methyl-2-butenyl)boronate (**6**) was used with matched aldehyde **163** to provide intermediate **164** in >95% isomeric purity [85]. The derivative **165** was then used to provide the C(3)–C(9) fragment in the first synthesis of an macrolide antibiotic aglycon, mycinolide V [86, 87].

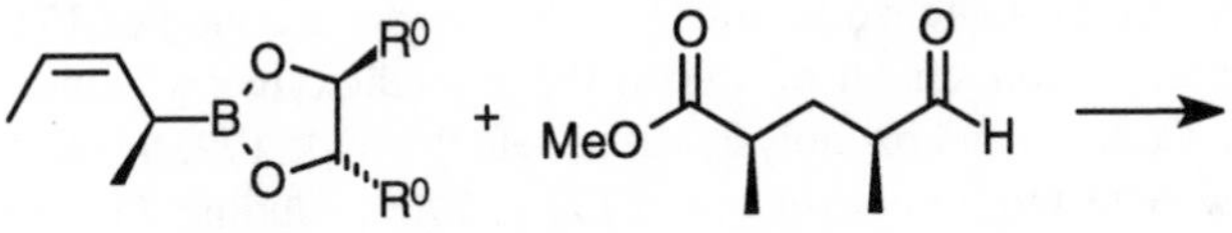

6 (R⁰ = cyclohexyl)   **163**

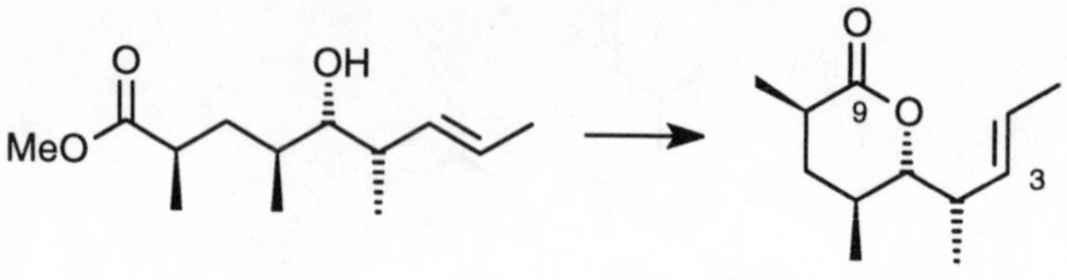

164   **165**

Intermediate **165** was also ozonized to the Prelog-Djerassi aldehyde **166** [85]. However, **6** is mismatched with **166** for further chain extension to the C(1)-C(9) building block **167** for 6-deoxyerythronolide, and considerable epimerization of **166** occurred during the course of the very slow reaction (35% in 12 h), producing **168** as 27% of the product mixture [85]. Since the acid that would be derived from ozonolysis of **167** has been synthesized efficiently from **166** via chiral enolate chemistry, this synthesis was not pursued further.

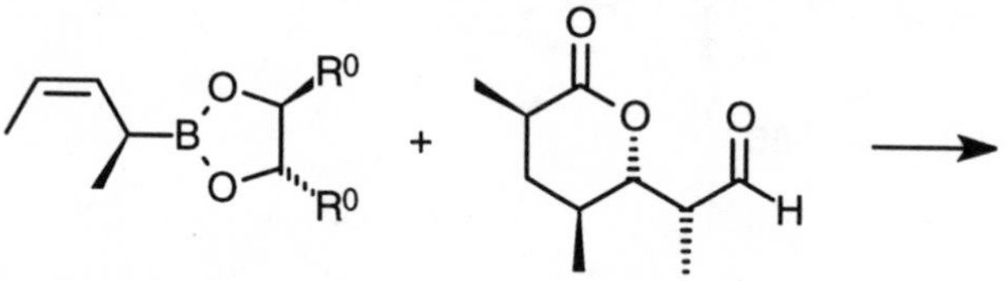

6 (R⁰ = cyclohexyl)   **166** (mismatched)

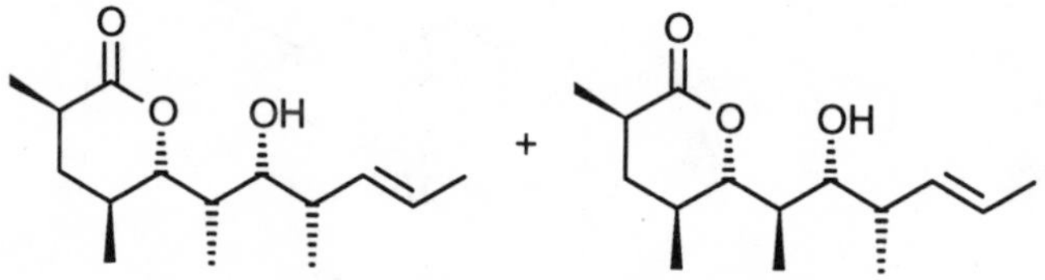

167 (63%)   **168** (27%)

The (Z)-(α-methylcrotyl)boronate **6** has also been used as the starting point for a synthesis of the denticulatins (**177**) [88]. Condensation of **6** with propanal provides the homoallylic alcohol **169**, which was silylated and then ozonized to the aldehyde **170**. Condensation of **170** with the matched achiral (E)-crotylboronate **2** yielded **171** in 95% diastereomeric purity. Conversion of **171** to the p-methoxybenzyl ether, desilylation, oxidation to a ketone, and ozonolysis led to ketoaldehyde **172**. Several crotylboronate reagents were tested for conversion of **172** to **173**, including **2** (ratio of desired (Z) to undesired (E) isomer with wrong configuration, 20:80), (E)-(crotyl)(dipinanyl)borane (ratio 55:45), the tartrate (E)-crotylboronate **ent-4** (ratio

68:32), and the chiral (*E*)-α-methoxycrotylboronate **132** (ratio 80:20). Ozonolysis of **173** to aldehyde **174** was followed by condensation with chiral boron enolate **175** to form intermediate **176**. Oxidation of the secondary alcohol function of **176** to a ketone followed by reductive cleavage of the *p*-methoxybenzyl group resulted in rearrangement of the cyclic hemiketal functionality to yield the denticulatins **177**, which consist of a mixture of two stereoisomers at C(10) because of the lability of the β-keto-hemiketal functionality [88].

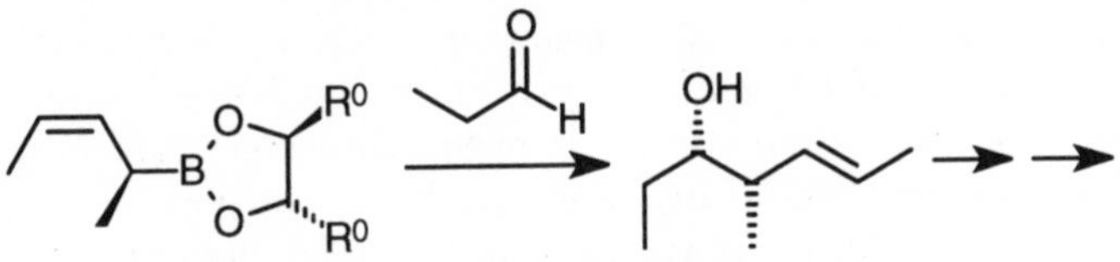

6 (R⁰ = cyclohexyl)   **169** (98.5% ee)

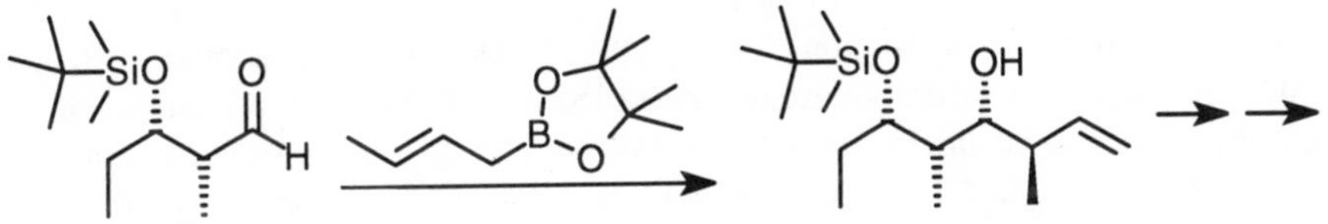

170   **2**   **171** (95% this isomer)

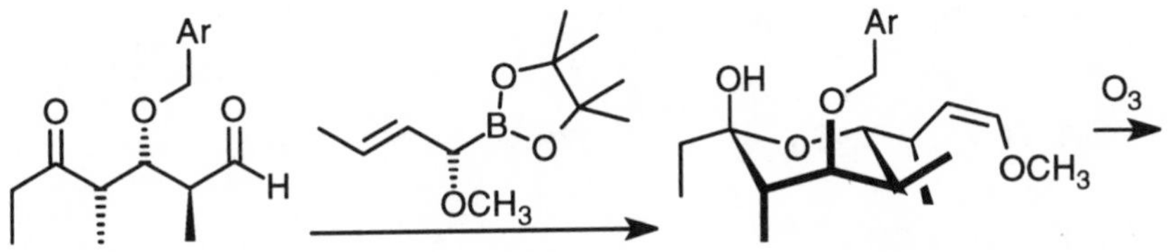

172 (Ar = *p*-MeOC₆H₄)   **132**   **173** (80% this isomer)

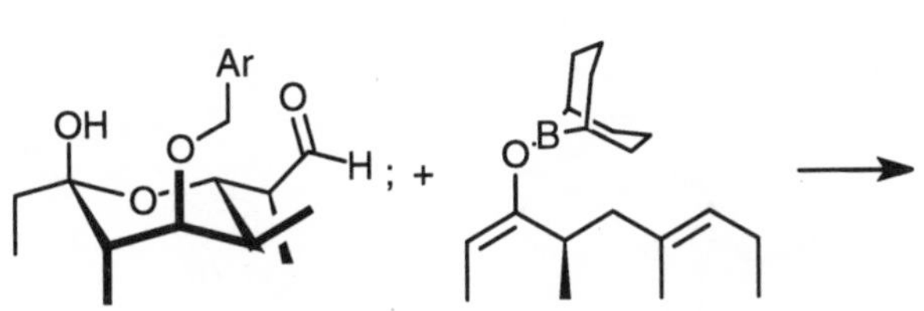

174   **175**

176 **177**

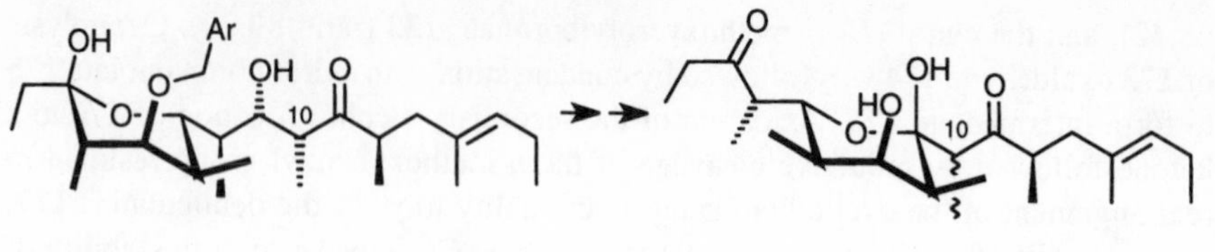

The C(19)–C(29) segment of rifamycin S has been constructed in the form of **184** by the use of (*S,S*)-isopropyl tartrate crotylboronate (***ent*-4**), the (*R,R*)-isomer (**4**), and the (*R,R*) allylboronate (**57**) [69, 89]. As discussed in Section 7.2.4.3, the relationship which causes the greatest problem is the mismatched *syn*-1,3-dimethyl substituents, which are best introduced early in the synthesis. The possibility of building **184** from one end, introducing the *syn*-1,3-dimethyl groups during the second crotylboronate–aldehyde connection, led to a 73:27 isomer ratio in the mismatched connection. It proved much more feasible to start with the preparation of **101** from aldehyde **99** and crotylboronate ***ent*-4**, which had been found to produce a mixture containing 90% of the desired diastereomer (see Section 7.2.4.3). Conversion to aldehyde **178** was followed by matched pair condensation with **4** to form **179**. Conventional manipulations led to aldehyde **180**, which in another matched pair condensation with crotylboronate ***ent*-4** led to **181**. Yet another conversion of the vinyl group to an aldehyde (**182**) was followed by the final C—C bond connection with allylboronate **57** to yield homoallylic alcohol **183**, which was methylated and hydrolyzed to aldehyde **184**.

101 (R = *t*-BuPh$_2$Si) **178** **4**

179 **180**

***ent*-4** **181**

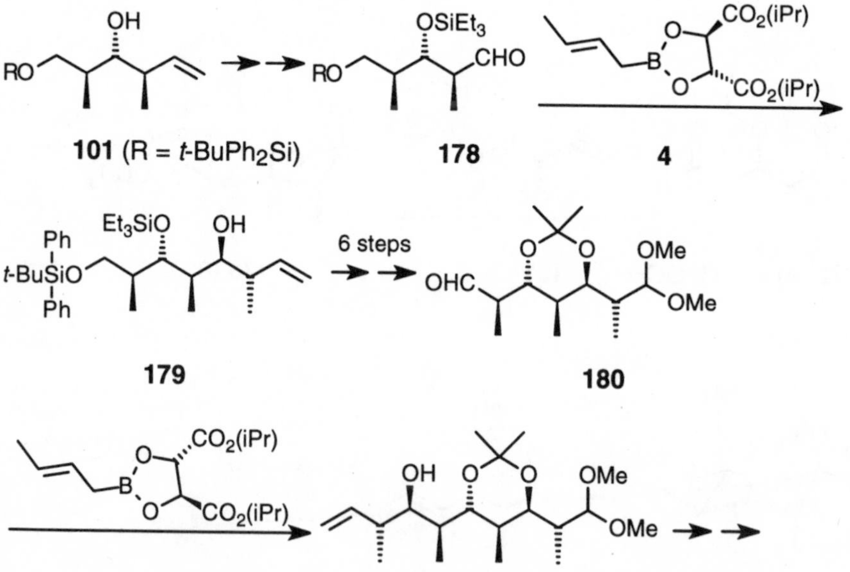

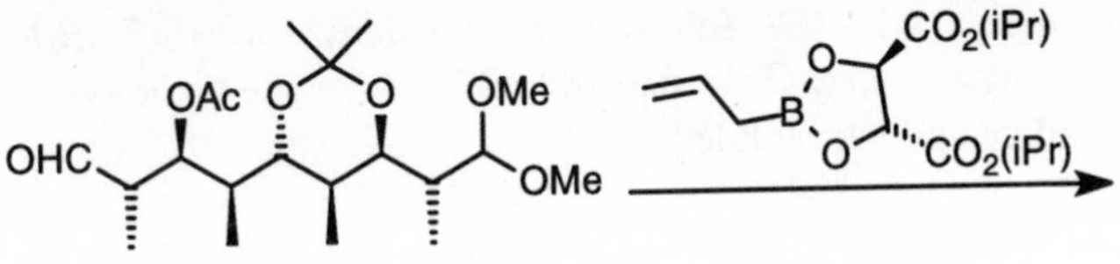

182   **57**

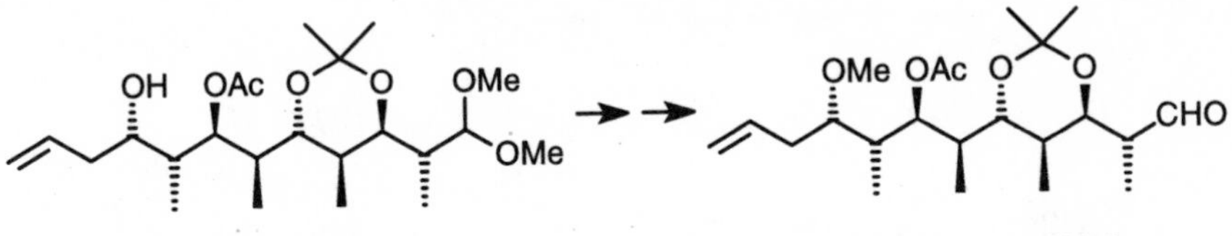

183   **184**

Intermediate **4** has been used for one of the intermediate steps in the synthesis of the C(1)–C(15) segment of streptovaricin D [90]. Aldehyde **185** forms a matched pair with **4**, and **186** is the only isolable product.

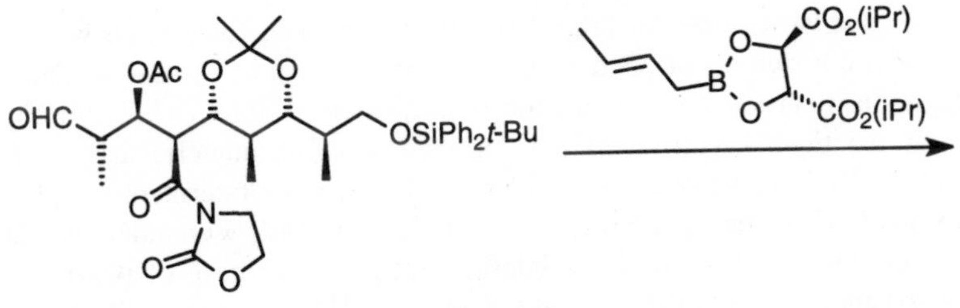

185   **4**

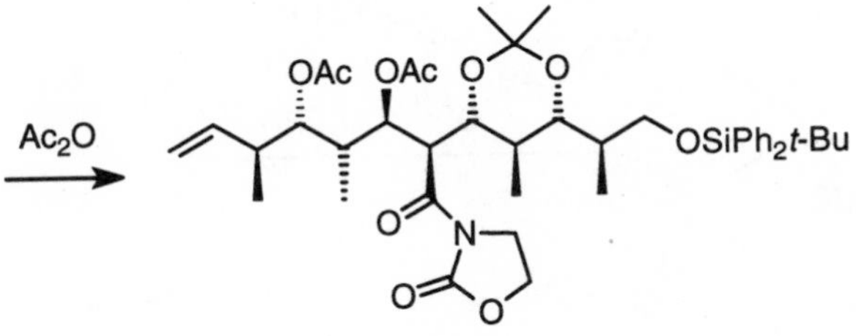

186

Glyceraldehyde acetonide (**91**) has been combined with tartrate esters of allylic boronic acids to provide starting materials for natural product syntheses. Product **92** from **91** with (R,R)-diisopropyl tartrate allylboronate (**57**) (Section 7.2.4.3) has been used as an intermediate in the synthesis of a segment of chlorothricolide [91]. With (R,R)-diisopropyl tartrate (E)-crotylboronate (**4**), intermediate **187** results, and further transformations to aldehyde **188** followed by condensation with (R,R)-diisopropyl tartrate (Z)-crotylboronate (**3**) lead to **189**, an intermediate in the synthesis of a subunit of kijanolide and tetronolide [92].

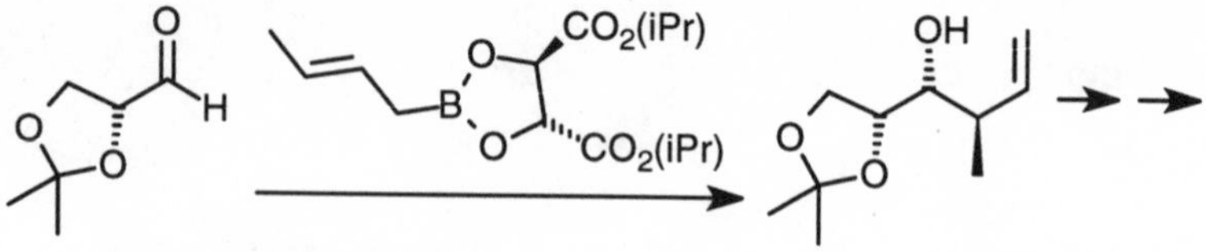

Condensation of **91** with tartrate crotylboronates and related compounds was explored as a possible stereocontrolled route to trienes that could be used in Diels-Alder cyclizations to produce precursors for the nargenicin antibiotics [93, 94]. The boronic ester reactions involved no new principles. Unfortunately, the stereochemistry of the Diels-Alder reactions did not yield the correct intermediates [94].

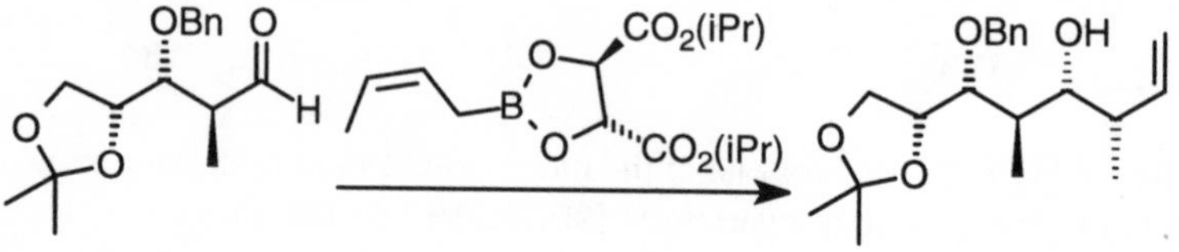

Reaction of (R,R)-diisopropyl tartrate (E)-crotylboronate (**4**) with aldehyde **190** yielded a 4:1 mixture of **191**, which is an intermediate in the synthesis of avermectin A_{1a}, and its diastereomer **192** [95]. The relatively low degree of stereoselection is typical of α,β-unsaturated aldehydes (see Section 7.2.4.1).

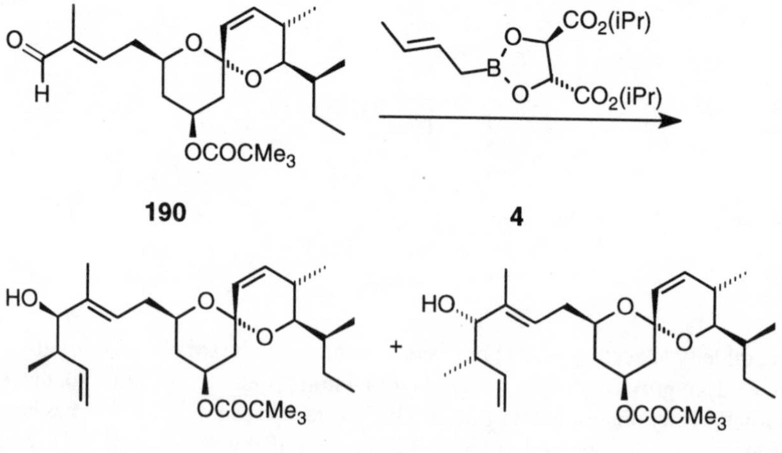

191 (80% of product)   **192** (20% of product)

Aldehyde **193**, derived from L-threonine, with matched (*S,S*)-diethyl tartrate allylboronate (**194**) efficiently yielded homoallylic alcohol **195** [96]. (Note the very similar reaction of **ent-57** and **94** to form **95** in 99.4% de in the Section 7.2.4.3.) Further conventional transformations led to deoxy sugar derivative **196** mixed with isomers that could be reequilibrated to form more **196**, which was then used to make both halves of the disaccharide unit of olivomycin A.

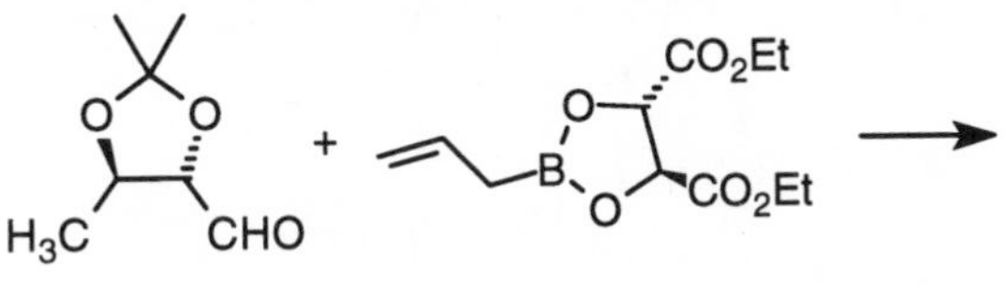

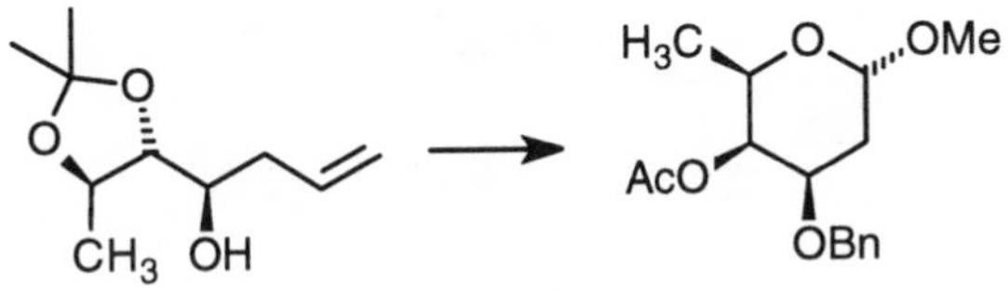

In related chemistry, intermediates derived from (*R,R*)- or (*S,S*)-diethyl tartrate allylboronate and the stereoisomers of 4-(*tert*-butyldiphenylsilyloxy)-2,3-epoxy-butanal have been used as intermediates for the synthesis of several deoxyhexoses [97].

Wuts and Bigelow used allylboronate chemistry to prepare a racemic intermediate **202**, which constitutes a formal synthesis of carbomycin B [25]. 3-(Benzyl-oxy)propanal with *meso*-butanediol (Z)-(γ-methoxyallyl)boronate (**197**) yielded *anti* methoxy alcohol **198**, which was converted to aldehyde **199**. A (Z)-γ-substi-tuted allylboronic ester with **199** yielded a gross mixture of isomers, a result consistent with others described above, and the synthetic strategy was modified accordingly. Reaction of **199** with (E)-γ-substituted allylboronic ester **200** yielded 85% of a mixture containing 75% **201** with lesser amounts of the three other separable diastereomers. The vinyl group of **201** was then elaborated to a methyl-substituted three-carbon side chain by conventional and unstereoselective means to provide the desired intermediate **202**.

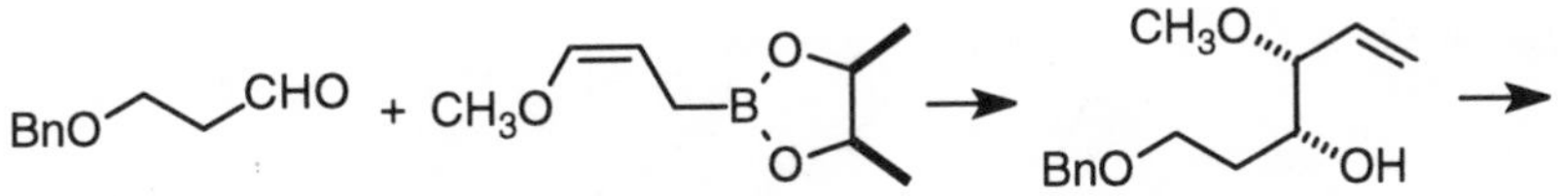

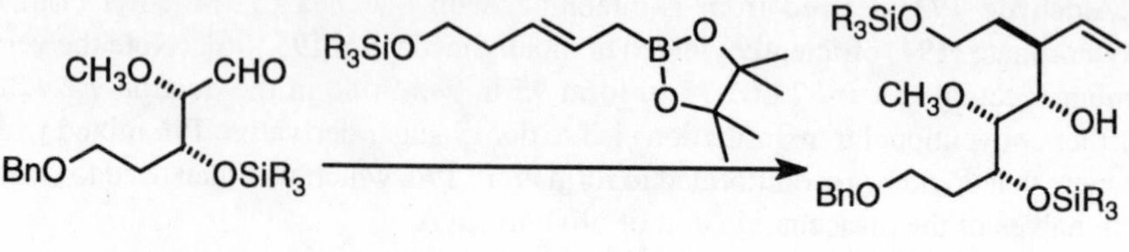

199 (SiR₃ = SiMe₂t-Bu)   **200**   **201**

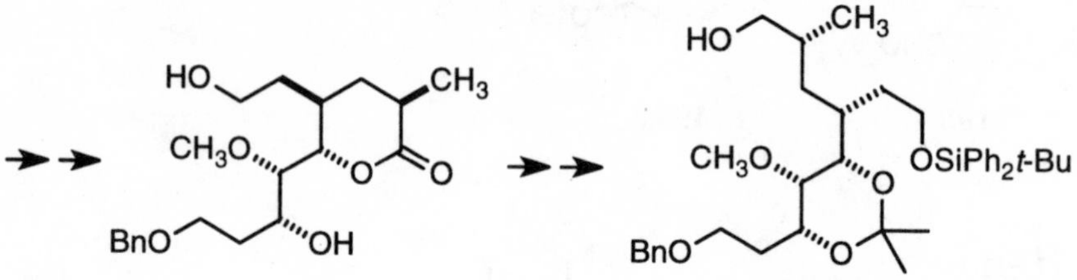

202 (racemic)

The 2-allyl-1,3,2-diazaborolidine **203** (= **71**, Section 7.2.4.2, with Ar = *p*-tolyl, X = acetoxy) has been condensed with aldehyde **204** to make **205**, which is an intermediate in the synthesis of the C(18)-C(35) segment of the immunosuppressant FK-506 [98]. Aldehyde **204** was made by protecting and reducing thiol ester **282** (see Section 7.4.3.2), which was made via an aldol condensation of an enolate analogous to **203**.

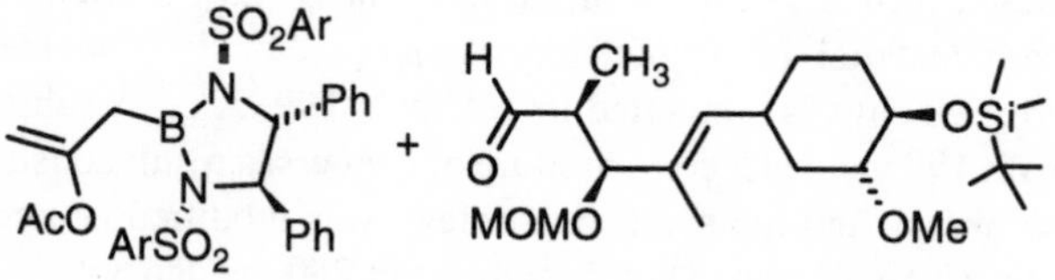

203   **204**

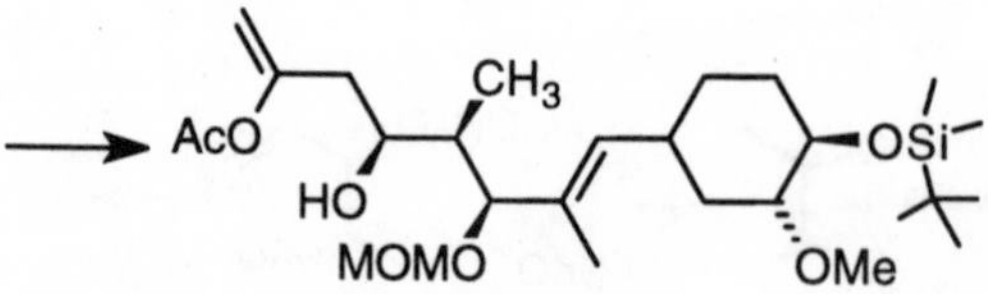

205

7.2.6 Allenyl and Propargyl Boronic Esters and 1,3,2-Diazaborolidines

Allenylboronic esters usually result when a propargylic organometallic such as a Grignard reagent is treated with a suitable borylating agent [99]. For an allenic boronic ester prepared by catalytic hydroboration of isopropenylacetylene [32], see the Section 7.2.2.4 on synthesis of allylic boronic esters.

Tartrate allenylboronic esters (206) react with aldehydes to form propargylic alcohols such as 207 in high ee's [99]. With 2,4-dimethylpentyl tartrate as chiral director, ee's were 94-97% with hexanal, 99% with cyclohexanecarboxaldehyde, and >99% with isovaleraldehyde. Cyclododecyl tartrate was equally effective, but ethyl and isopropyl tartrate had given lower ee's (62-95%) in earlier work [100]. Intermediate 207 was converted to the insect pheromone (S)-(−)-ipsenol (208).

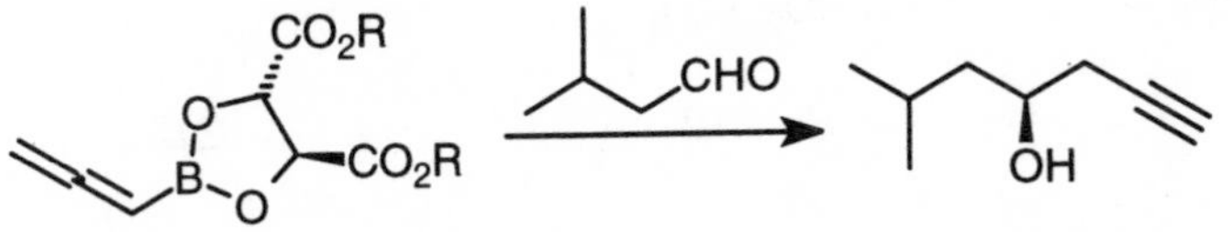

206

(R = 2,4-dimethylpentyl)

207 (99.6% ee)

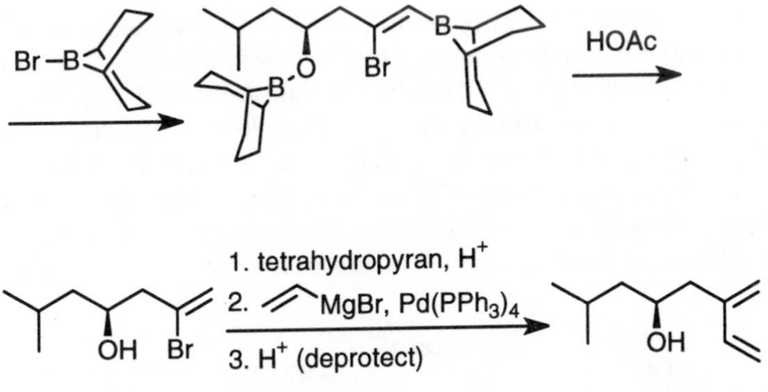

208

A 2-allenyl-1,3,2-diazaborolidine 210 has been made from bromo precursor 209 a propargyltin compound and with aldehydes yields homopropargylic alcohols 211 in 91-98% ee [R = CH$_3$(CH$_2$)$_4$, (CH$_3$)$_2$CH, (CH$_3$)$_3$C, cyclohexyl, Ph, PhCH=CH] [101].

Use of an allenyltin compound in place of the propargyltin yielded the 2-propargyl-1,3,2-diazaborolidine 212, which with the same set of aldehydes used above with 210 yielded 72-82% of the allenyl alcohols 213 with all ee's >99% [101]. Further transformations of 213 have been described [102].

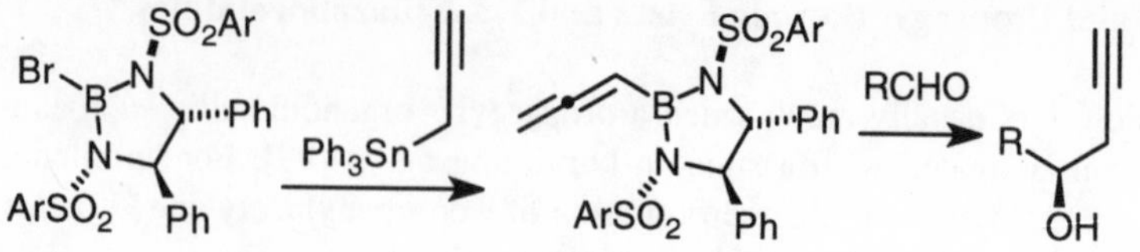

209 (Ar = *p*-tolyl)   **210**   **211**

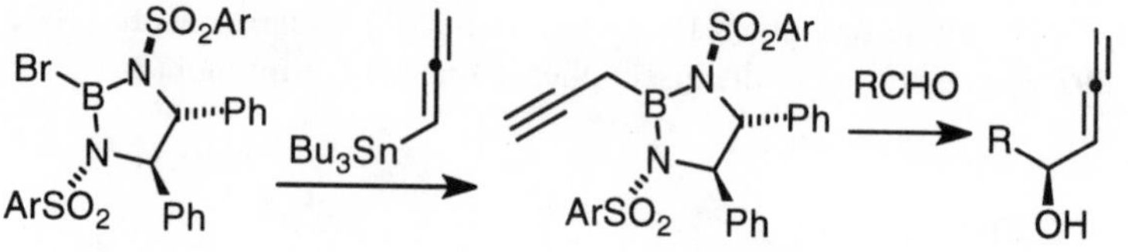

209   **212**   **213**

7.2.7 Borinic Acids

These little studied compounds have more in common with boronic esters than with trialkylboranes.

Propargylic alcohols derived from asymmetric reduction of acetylenic ketones have been converted to acetates, hydroborated to alkenyldialkylboranes **214**, and rearranged with base to provide asymmetric allylic borinic acids **215** [103]. These react with aldehydes to yield homoallylic alcohols **216** in 96-98% diastereomeric purity and 79-85% ee [104]. The ultimate utility of the products lies in conversion to β-hydroxy acids **217**, from which the group R^2 has been cleaved. Accordingly, R^2 is chosen as a disposable bulky group which goes to the equatorial position in the reaction of **215** with aldehydes and gives high *trans* selectivity.

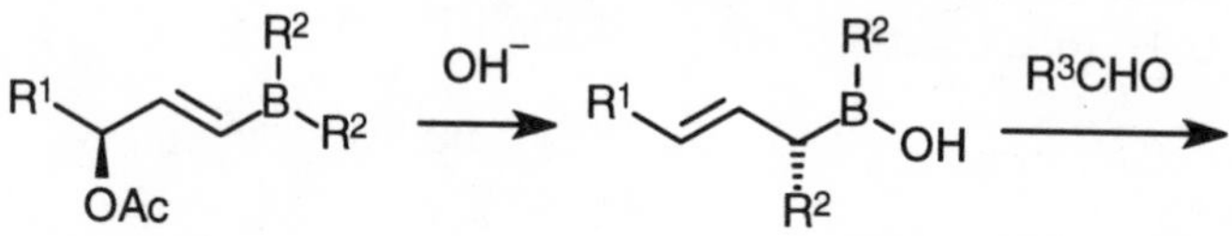

214   **215**

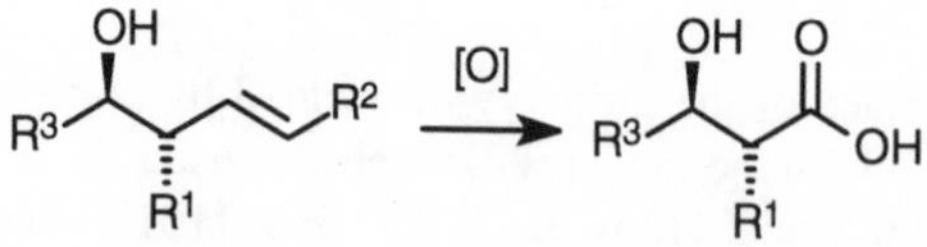

216   **217**

7.3 Allyldialkylboranes

7.3.1 From α-pinene and other terpenes

7.3.1.1 Unsubstituted allyl groups. *B*-(Allyl)diisopinocampheylborane (**218**), generated from allylmagnesium bromide and *B*-(methoxy)diisopinocampheylborane derived from (+)-α-pinene, allylates common aldehydes at low temperatures to form homoallylic alcohols **219** in 83-96% ee [105]. *B*-(Allyl)bis(4-isocaranyl)-borane (**220**) gives somewhat higher ee's (88-99%) [106]. *B*-(Allyl)bis(2-iso-caranyl)borane (**221**) yields the enantiomeric homoallylic alcohols *ent*-**219** in very high ee's [107].

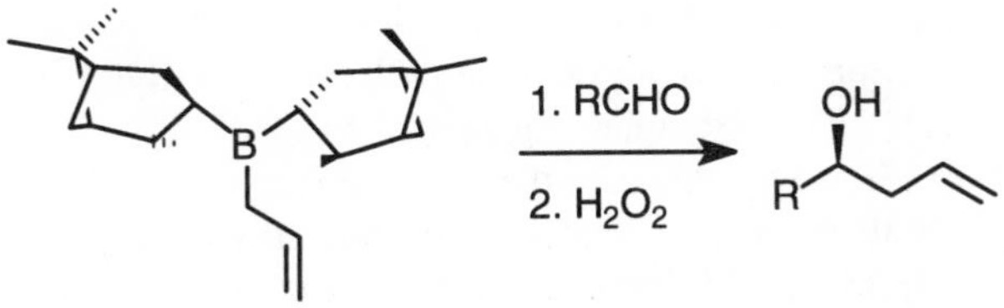

218   **219** (83-96% ee)
(R = Me, Et, Pr, iPr, *t*-Bu)

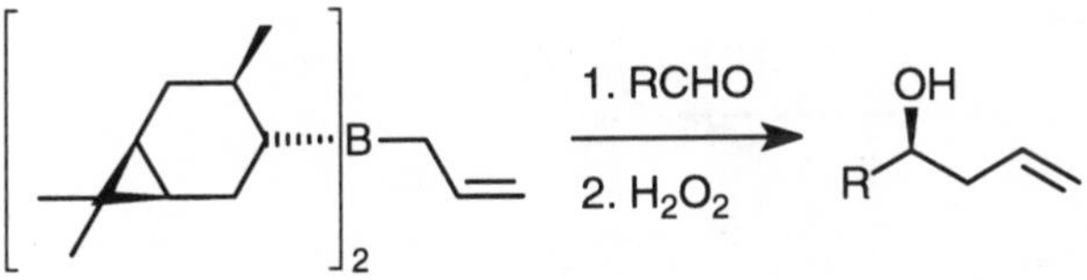

220   **219** (88-99% ee)
(R = Me, Et, Pr, iPr, *t*-Bu)

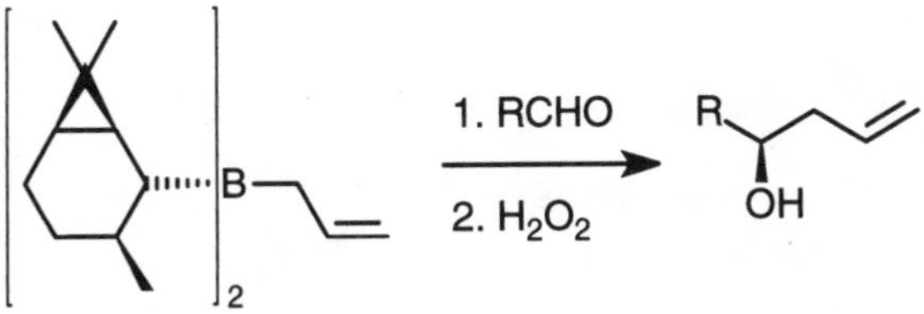

221   *ent*-**219** (94-99% ee)
(R = Me, Et, Pr, iPr, *t*-Bu, Ph, CH₂=CH-)

The rates of reaction of **218** and **220** with benzaldehyde are too fast to measure at −78 °C, and are much faster than rates measured for any of the boronic ester reagents in common use in these reactions [108].

Reactions of **ent-218** and **220** with bromoacetaldehyde have yielded **ent-219** and **219** (R = BrCH$_2$) in 89 and 92% ee's, respectively, and these have been converted to the corresponding epoxides [109]. The (R)-epoxide derived from **ent-218** was converted to (−)-(R)-γ-amino-β-hydroxybutyric acid.

The chiral auxiliary can be recovered by suitable cleavage of the borinate product [110]. Diisopinocampheylboranes are cleaved to α-pinene by acetaldehyde, and bis(2- or 4-isocaranyl)boranes can be cleaved to the corresponding carene with isobutyraldehyde. Alternatively, the borinate esters initially produced can be isolated as ethanolamine or 8-hydroxyquinoline derivatives of the bis(terpene)borinic esters.

The enantioselectivity of allylboration of aldehydes with B-(allyl)diisopinocampheylborane (**218**) and similar compounds is improved if the magnesium salts are removed from the borane before use [111]. Under these improved conditions, several heterocyclic aldehydes have been tested with B-(allyl)diisopinocampheylborane (**218**), B-(allyl)bis(4-isocaranyl)borane (**220**), and B-(allyl)bis(2-isocaranyl)borane (**222**) [111]. The examples illustrated yielded 78-91% of homoallylic alcohol, ee ≥99%. However, these are the best examples, and it should be noted that **218** with 3-furaldehyde only yielded 91% ee, or with 2- or 3-thiophenecarboxaldehyde, only 75-80% ee. The reasons for these uneven results are not readily apparent.

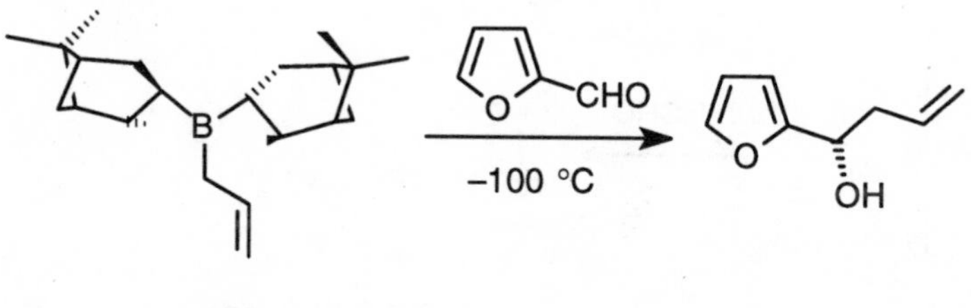

218

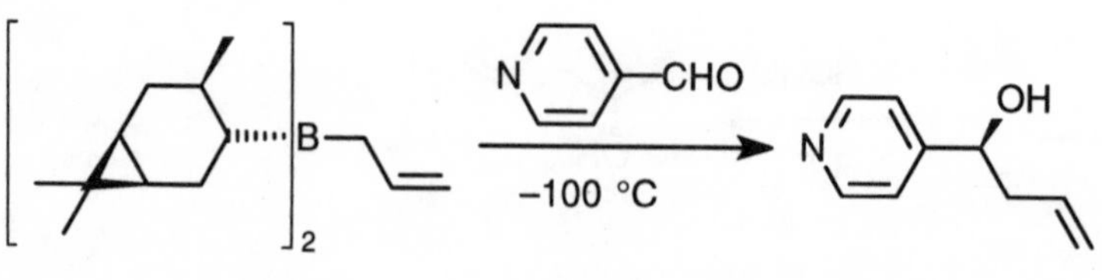

220

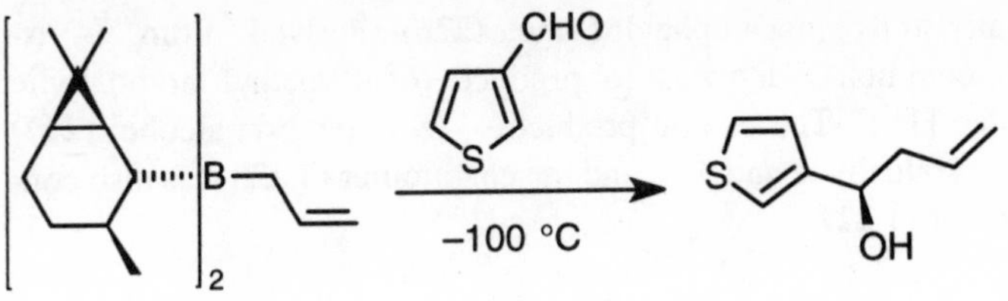

222

7.3.1.2 Methyl-substituted allyl groups. *B*-(Methallyl)diisopinocampheylborane (**223**) derived from (−)-α-pinene reacts with common aldehydes to yield the corresponding homoallylic alcohols in 90-96% ee.

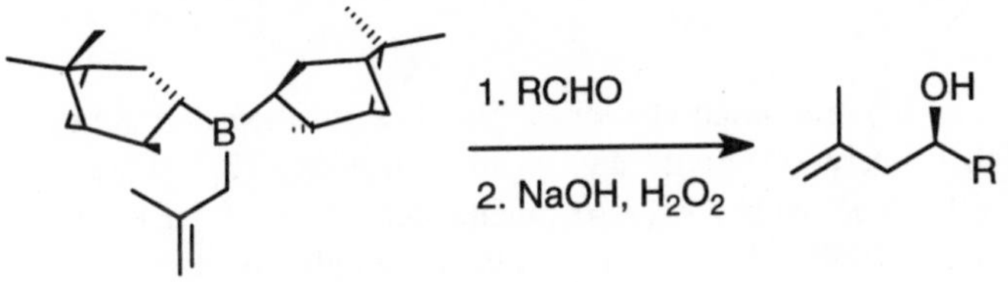

223

The reaction of 3-methyl-2-buten-1-al with *B*-(isoprenyl)diisopinocampheylborane (**224**) derived from (+)-α-pinene provides a simple route to the bark beetle pheromone (+)-(*S*)-ipsdienol (**225**) [112]. The enantiomer of **225** was similarly prepared from the enantiomer of **224**. Similar reactions with 3-methylbutanal led to both enantiomers of ipsenol (**208**, illustrated in Section 7.2.6). These syntheses are related to an earlier approach to the racemates in Russia, where the allylborane-aldehyde reaction was first discovered [113].

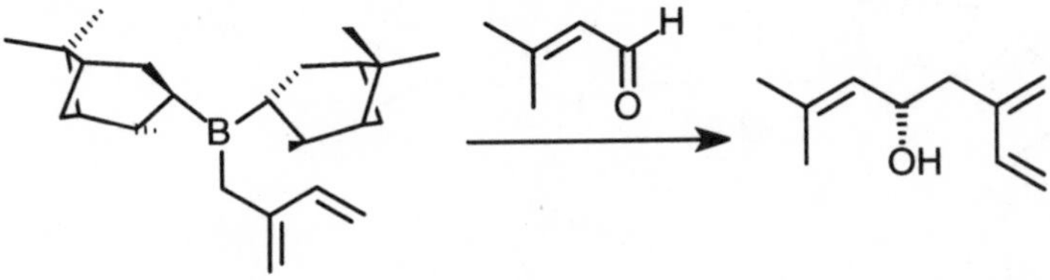

224   **225** (60% yield, 96% ee)

B-(3,3-Dimethylallyl)diisopinocampheylborane (**226**) derived from (–)-α-pinene reacts with common aldehydes to produce α,α-dimethyl homoallylic alcohols in 91-96% ee [114]. The natural product (–)-(*S*)-artemesia alcohol (**227**) was made from isobutyraldehyde and **226**, and the enantiomer of **226** was also converted to the enantiomer of **227**.

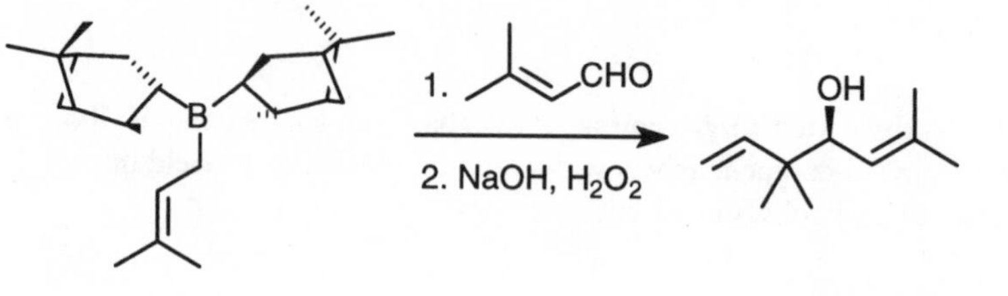

226   **227** (96% ee)

Reactions of (*E*)- and (*Z*)-crotyldialkylboranes, like those of their boronic ester analogues, show high diastereoselectivity and enantioselectivity [115, 116]. The general results are illustrated with the (+)-α-pinene derivatives *B*-[(*E*)-crotyl]-diisopinocampheylborane (**228**), which leads to *anti*-α-methylhomoallyl alcohols (**229**), and *B*-[(*Z*)-crotyl]diisopinocampheylborane (**230**) which leads to *syn*-α-methylhomoallyl alcohols (**231**). Both enantiomers of α-pinene as well as the carene isomers (see **220** and **222**) have been used as chiral directors [116, 117]. The required crotylpotassium reagents were made from the respective 2-butenes by the procedure of Fujita and Schlosser [118]. Diastereomeric impurities in the products **229** and **231** were not detected, and the ee's were 90(±2)% for all **229** and **231** prepared from **228** and **230**, 94% for **229** and **231** prepared from Δ³-carene as chiral director [116]. (Oddly, the table listing seven examples of **229** or their enantiomers and that listing similar examples of **231** show identical sets of figures for the ee's [116].)

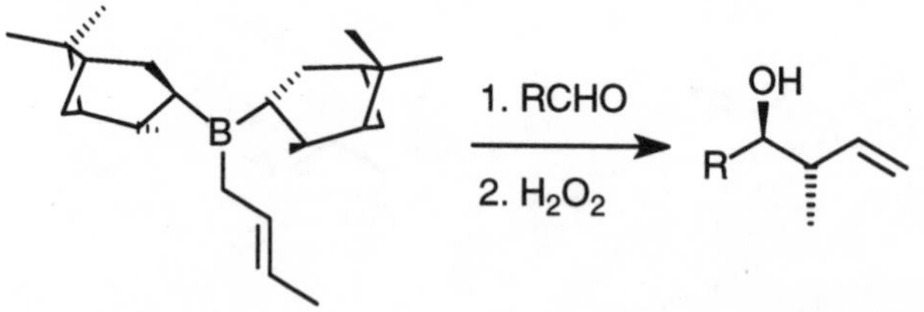

228   **229** (ee 88-92%)
(R = Me, Et, Ph, CH₂=CH-)

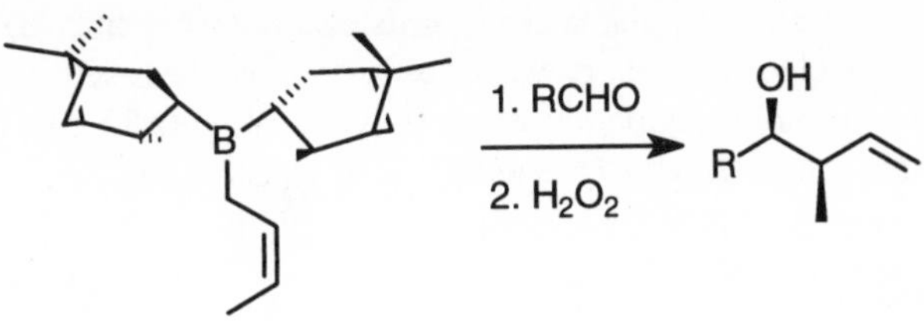

230   **231** (ee 88-92%)

(R = Me, Et, Ph, CH₂=CH-)

Barrett and Lebold have used (–)-(E)-crotyldiisopinocampheylborane (**228**) to prepare **232** (= **229**, R = p-pivaloyloxyphenyl) for use as a key intermediate in the synthesis of the antibiotic nikkomycin B [119, 120]. The p-pivaloyl group as the hydroxyl protection in **232** turned out to be essential, as a benzoyl group was partially cleaved in the course of the reaction with **228**, and *tert*-butyldimethylsilyl resulted in slow reaction and partial loss of stereocontrol. In further transformations, the vinyl group of **232** was ozonized to an aldehyde and converted to an additional chiral center in a reaction with ethoxyvinyllithium, which was guided by the already existing chirality.

228 +

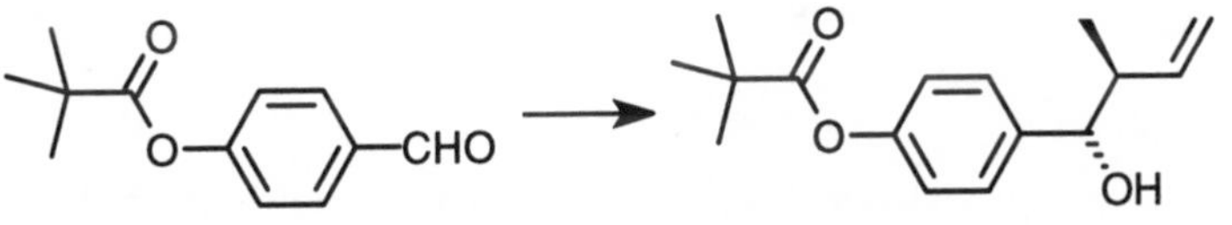

232 (ee ≥96%, diastereomeric ratio 96:4)

The available data indicate that the various allylic diisopinocampheylborane reagents **218**, **228**, and **230** and their enantiomers react more diastereoselectively than any of the boronic ester reagents with α-chiral aldehydes [121, 122, 123]. Thus, O-benzyllactaldehyde (**73**) yields excellent results with either **218** or its enantiomer, *ent*–**218** [121, 123], in contrast to the unsatisfactory ratios obtained from the tartrate allylboronates with **73** (see Section 7.2.4.3).

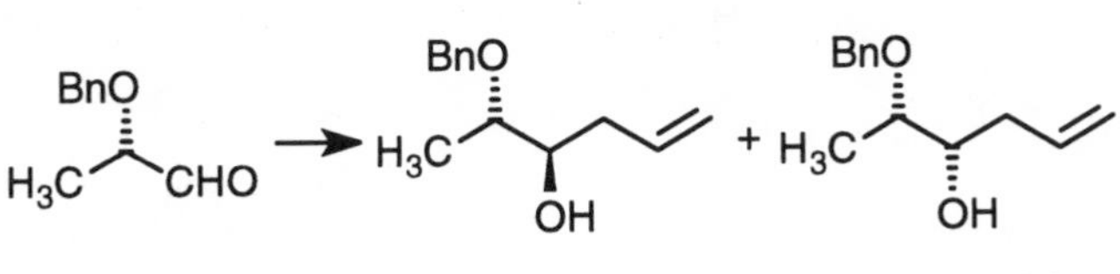

	97	**98**
73		
with **ent-218** (matched)	96%	4%
with **218** (mismatched)	6%	94%

The reaction of (E)-crotylborane **228** (mismatched) or its enantiomer (matched) with **73** results in excellent diastereoselection. However, as has proved the case with the analogous boronic esters, the (Z)-crotylborane **ent-230** (matched) with **73** gives excellent results, but **230** (mismatched) with **73** is unsatisfactory [122, 123].

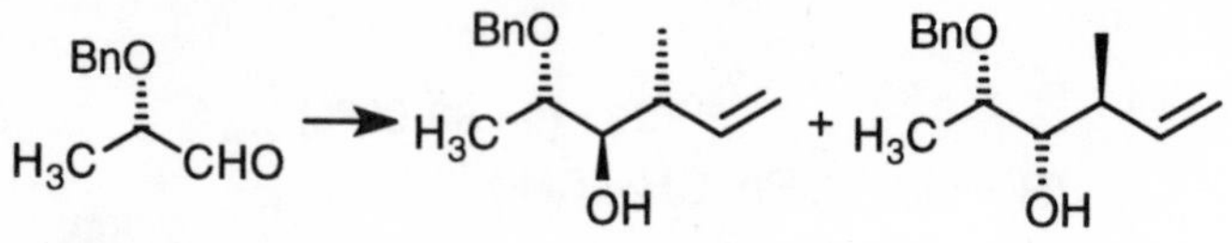

73
with **ent-228** (matched)	97%	3%
with **228** (mismatched)	5%	95%

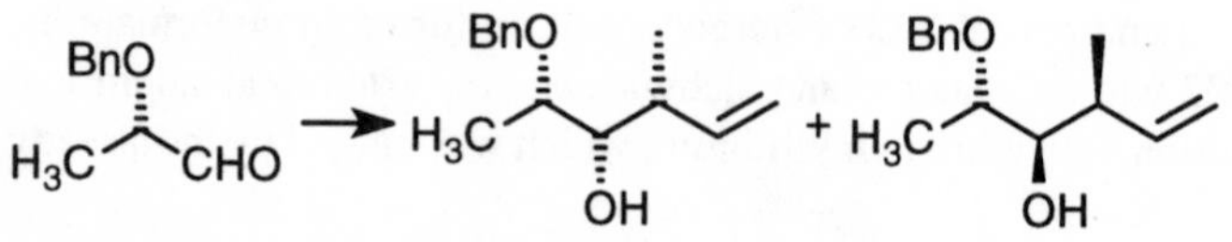

73
with **ent-230** (matched)	99%	1%
with **230** (mismatched)	27%	73%

Promising results were also obtained from (S)-2-methyl-3-(benzyloxy)propanal (**233**) and its enantiomer (not illustrated) with **228** and **230** and their enantiomers [123]. These are all distinctly better than what was found with tartrate crotyl-boronates with **233**, though Roush and coworkers solved the practical synthetic problem another way, by using silyl groups instead of benzyl [69] (see Section 7.2.4.3). The silyl analogues gave de's comparable to those from **233** with (E)-crotylboronate **228** or the matched pair with (Z)-crotylboronate **230**, but the mis-matched pair with **230** is far superior to the corresponding result with the boronic ester, a 64:36 diastereomeric ratio.

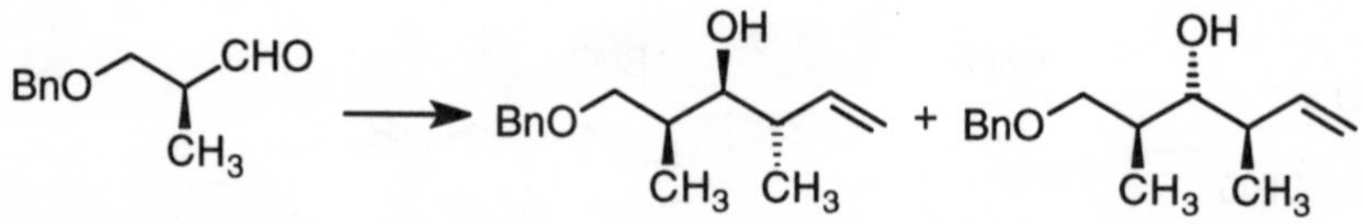

233
with **228** (matched)	98%	2%
with **ent-228** (mismatched)	5-6%	94-95%

233

with **ent-230** (matched)	94-95%	5-6%
with **230** (mismatched)	8-9%	91-92%

At this point it is difficult to say how useful the crotylboranes **228** and **230** might turn out to be in a major synthesis such as some of those with boronic esters described in Section 7.2.5. The marginally improved de's seen with matched pairs are not necessarily outside the range of experimental error, especially when comparisons have to be made between different research groups, and experimental convenience may become more important than the difference between 10:1 and 20:1 diastereomeric ratios in any event. Most of the synthetic routes have been designed to circumvent the problems of mismatched pairs, though in those situations where mismatching cannot be avoided, these reagents could prove useful. One unsolved problem is that as substituent groups become very large, mismatched pairs yield poor results. The mismatched pair **73** and **230** gives results that are not synthetically useful, and therefore the prospects for overcoming the problems inherent in larger mismatched substituents are not encouraging. The likelihood of functional group interference with the very rapid borane reactions does not seem great, but the acidity of the trialkylborane reagents could conceivably lead to unpleasant surprises. Until someone completes a major synthetic project with the aid of these reagents, their actual utility remains to be proved.

7.3.1.3 Other substituted allyl groups. Hydroboration of cyclohexadiene at –25 °C with diisopinocampheylborane derived from (+)-α-pinene yields mainly the allylic borane **234**, which reacts with aldehydes at –78 °C to form the diastereomerically pure 3-(1-hydroxyalkyl)cyclohexenes **235** in 58-71% yields, 90-94% ee [124].

234   **235** (ee 90-94%)
(R = Me, Et, iPr, t-Bu, CH₂=CH, Ph)

Another useful diisopinocampheylborane derivative is **235**, which has been reported by Barrett and Malecha to provide a good route to *anti* vicinal diols (**236**), which are accessible via peroxidic oxidation of the silicon function [125]. A limitation of the reagent appeared with the mismatched pair **ent-235** + **237**, which produced the two diastereomers **239** and **238** in a 1:2 ratio and 45% total yield.

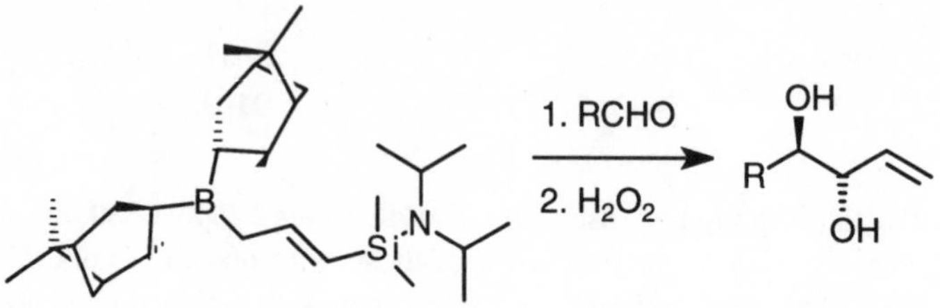

235

1. RCHO
2. H_2O_2

236 (ee >90%)
(R = Ph, thienyl, cyclohexyl, n-hexyl)

235 +

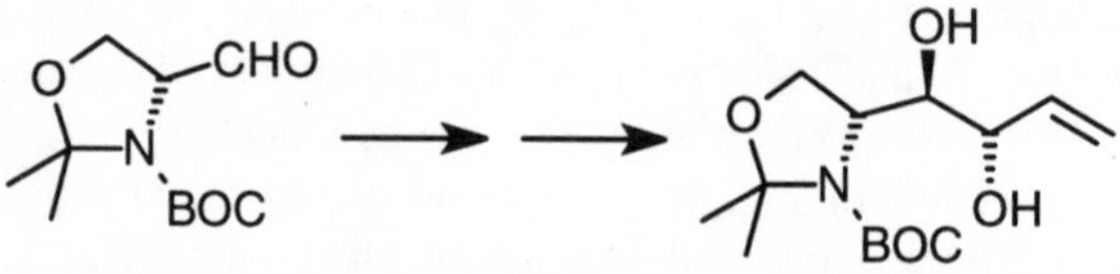

237

238 (98% this isomer)

ent-235 +

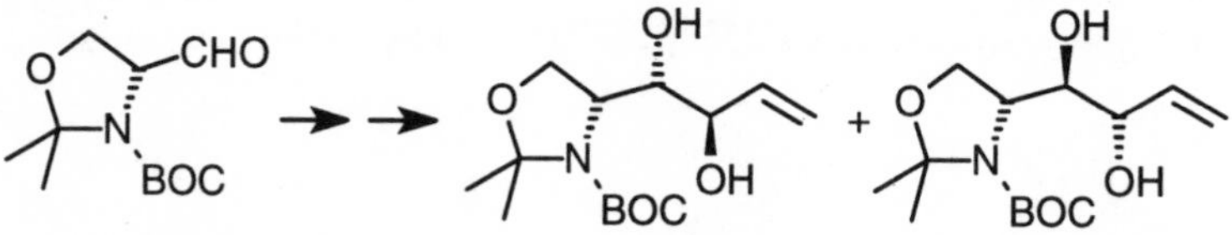

237   **239**   (ratio 33:67)   **238**

7.3.2 Borolanes as Chiral Directors

Although it is not as easy to prepare as the pinene derivatives, Masamune's chiral 2,5-dimethylborolane is an excellent chiral director. (R,R)-1-(E)-Crotyl-2,5-dimethylborolane (**240**) with aldehydes yields the *anti* (or *threo*) homoallylic alcohol series **241** in 86-92% de, 95-97% ee [126] (R = Et, Me$_2$CH, Me$_3$C). The chiral director can be recovered in the form of its aminoethanol derivative **242**. Similarly, (R,R)-1-(Z)-

crotyl-2,5-dimethylborolane (**243**) yields the *syn* (or *erythro*) products **244** with similarly high stereocontrol (except where R = Et, ee 86%).

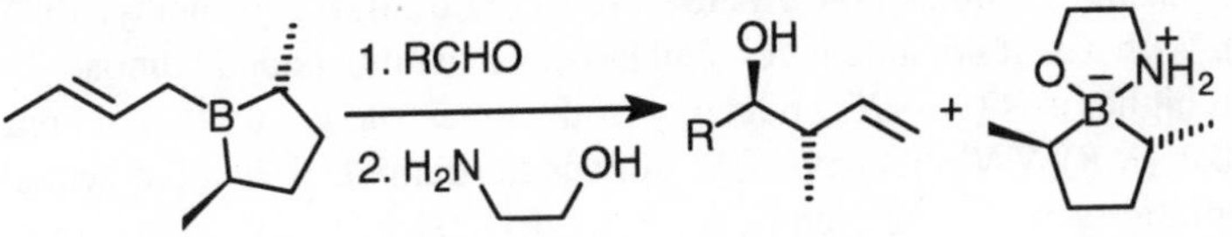

240   **241**   **242**

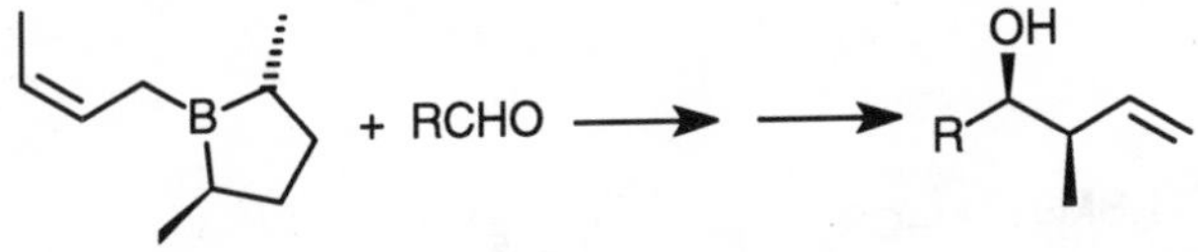

243   **244**

The reagents **240**, **243**, and their enantiomers also yield good results with glyceraldehyde acetonide (**91**). The percentage of the major product **245**, **246**, **187**, or **247** in the mixture of the four products is given for each of the isomeric reactants [126]. For comparison, Roush's tartrate (*E*)-crotylboronates (analogous to **206** or *ent*-**206**) yielded **187** in 96% isomeric purity, **245** in 87% isomeric purity [52]. This constitutes a reversal of which pair of isomers is matched or mismatched.

240, 243, *ent*-240, or *ent*-243

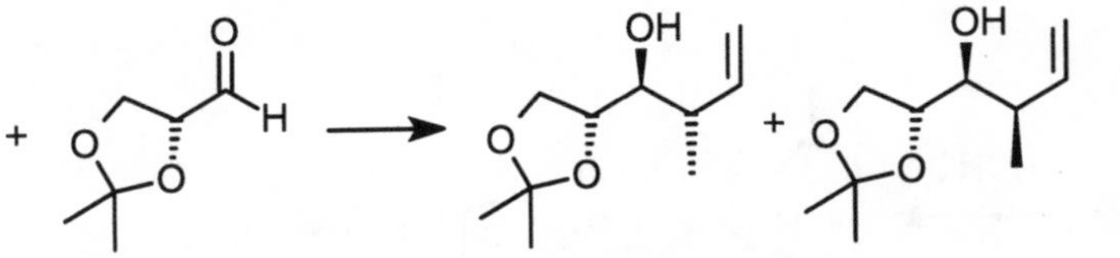

91   **245**   **246**
Matched pairs: from **240** 96.1%   from **243** 91.6%

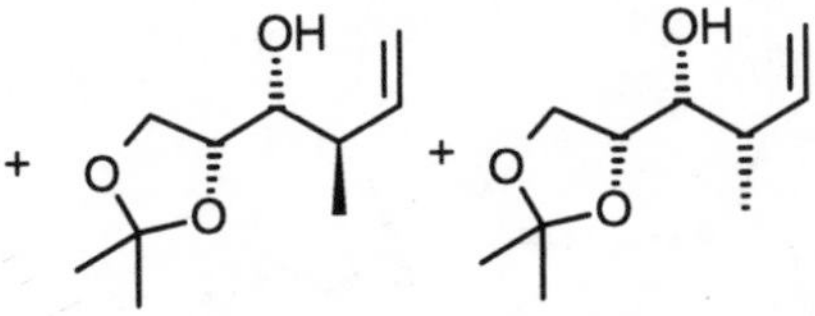

187   **247**
Mismatched pairs: from ***ent*-240** 85.6%   from ***ent*-243** 81.7%

2-(Trimethylsilyl)borolanes (**248**) are much easier to prepare and resolve than 2,5-dimethylborolanes [127]. The sterically favored (*S,S*)-(*N*)-methylpseudo-ephedrine derivative (*S*)-**249** crystallizes readily, leaving unchanged (*R*)-**248** in solution. Allylmagnesium bromide converts (*S*)-**249** to (*S*)-*B*-allyl-2-(trimethylsilyl)-borolane (**250**). With a series of aldehydes, **250** has consistently yielded homoallylic alcohols **252** in higher ee's (92-97%) than any of the other known allylborane reagents except the (*R,R*)-*N,N'*-dibenzyl-*N,N'*-ethylenetartramide derivative, which suffers from solubility problems (see Section 7.2.4.1). The postulated transition state **251** places the bulky trimethylsilyl group near the equatorial plane adjacent to the aldehyde carbonyl oxygen, avoiding the bulkier CH_2 of the allyl group. The six-membered ring is flexed away from the trimethylsilyl group, thus avoiding repulsive 1,3-interactions.

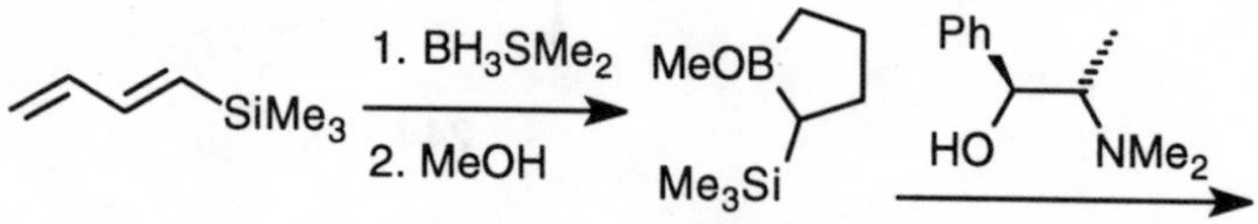

rac-**248**

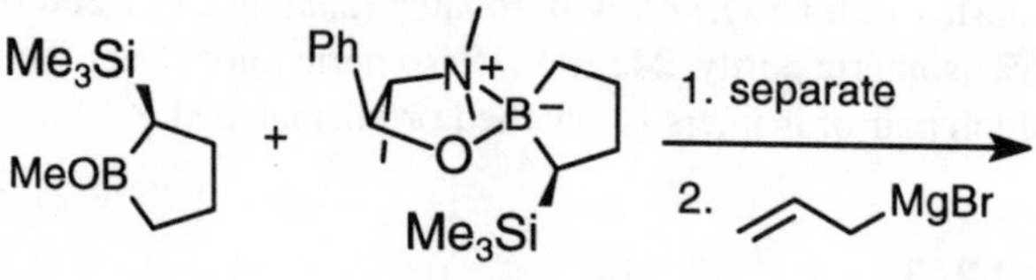

(*R*)-**248**   (*S*)-**249**

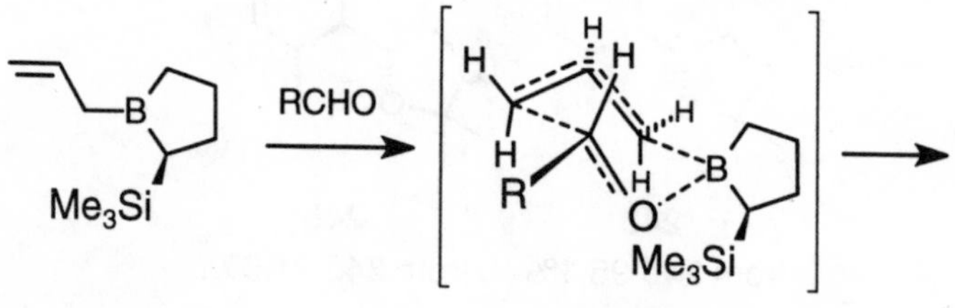

250   **251**

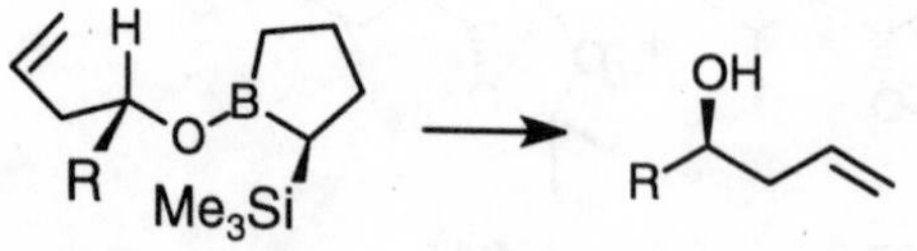

252

7.4 Boron Enolates

7.4.1 Introduction

Boron enolates have had a major impact on asymmetric synthesis. Their chemistry formally parallels that of allylboron compounds, but boron—carbon bonds are peripheral to the chemistry or not even present. In that sense this might not be considered to be "organoborane chemistry," and indeed, boron is but one of several elements that can serve as the connecting Lewis acid atom in stereodirected aldol condensations. Accordingly, this review is brief and confined to asymmetric syntheses plus new routes to boron enolates. Some historically important early papers have been cited briefly in Section 7.1.

7.4.2 Chiral Enolate Carbon Skeletons

7.4.2.1 Silylated hydroxy ketone enolates. The initially discovered diastereoselective aldol condensations were soon made enantioselective by the use of appropriate chiral directors [128-130]. Masamune and coworkers then used this chemistry to solve several major problems in the synthesis of macrolide antibiotics, including total syntheses of 6-deoxyerythronolide B [131, 132], narbonolide [133], and tylonolide [134], as well as the ansa chain of rifamycin S [135] and the stereocontrolled preparation of 3-hydroxy-2-vinylcarbonyl compounds [136, 137] used for rifamycin and also in the synthesis of amphotericin B [138].

Masamune's first efficient chiral director (253) was derived from commercially available (S)-mandelic acid in three steps, then converted to a boron enolate (254) [135]. The enantiomeric series from (R)-mandelic acid was equally readily available. With suitable choice of R^1, reaction of 254 with aldehydes yielded 255 with high diastereoselection. Desilylation of 255 and periodate cleavage led to the carboxylic acids 256 [135].

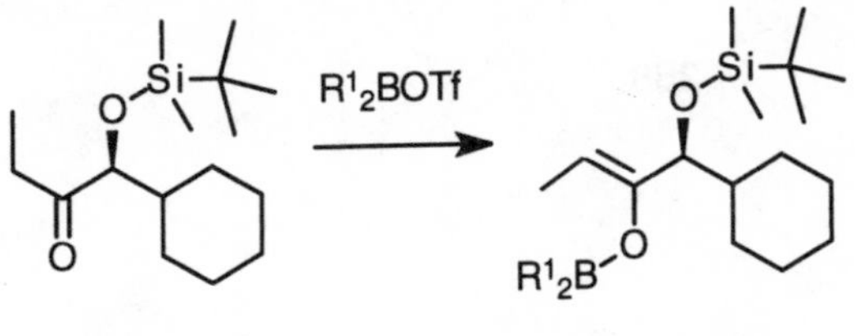

253

254

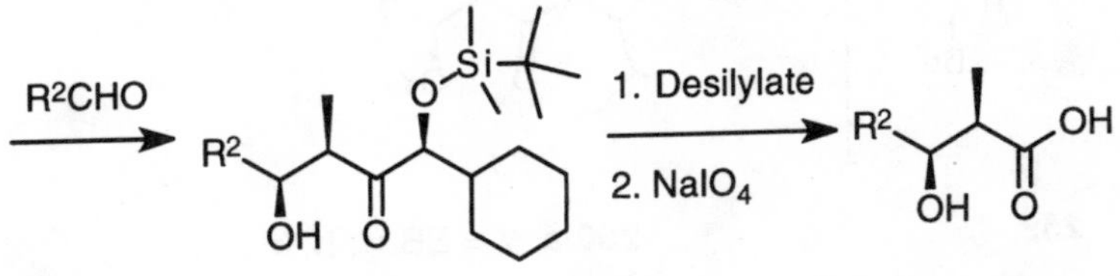

255

256

The only detectable diastereomer in **255** was the precursor of the enantiomer of **256**. The best choice of R^1 was cyclopentyl when $R^2 = Ph$, Et, or $BnO(CH_2)_2$, which yielded diastereomeric ratios 75:1, >100:1, and 100:1, respectively. When R^2 was isopropyl, $R^1 = $ cyclopentyl failed, but $R^1 = n$-butyl or $R^1_2B = $ 9-BBN yielded >100:1 diastereomer ratios for **255**.

The foregoing chemistry has been reviewed [139-141]. The particularly significant review by Masamune and coworkers in which the thermodynamics of double asymmetric induction are discussed has been noted in Section 7.2.4.3 [67].

7.4.2.2 Chiral amide enolates.

A practical limitation of the chemistry just described is that the chiral director cannot be recovered and recycled because it is destroyed in an essential cleavage step. Chiral oxazolidine **257** introduced by Evans and coworkers does not have this drawback [142]. It is easily prepared from readily available valinol and converted to the (Z)-enolate **258**. The postulated transition state **259** [143] is the same as those drawn for allylborane–aldehyde reactions, except that **259** has an O in place of the allylic CH_2. Formation of **260** is not as exothermic as the analogous allylborane rearrangements, but the equilibria are favorable enough that the yields can be high (75-95%), and the product usually consists of the single isomer illustrated in >99% purity. Methoxide cleavage of **260** yields the methyl ester **262**, and regenerates the chiral director in the form of oxazolidine **261**, which can be acylated to regenerate **257**.

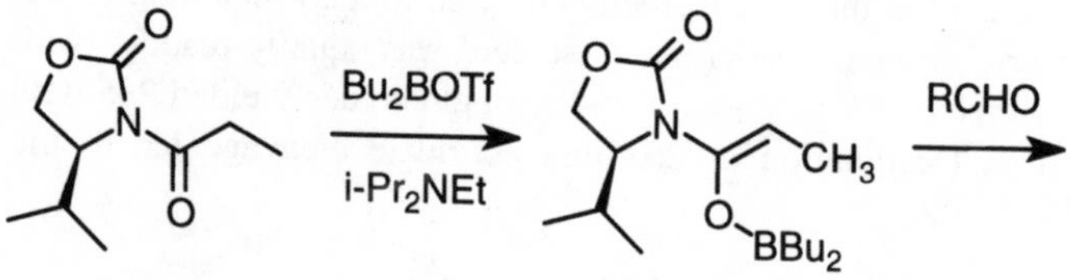

257   **258**

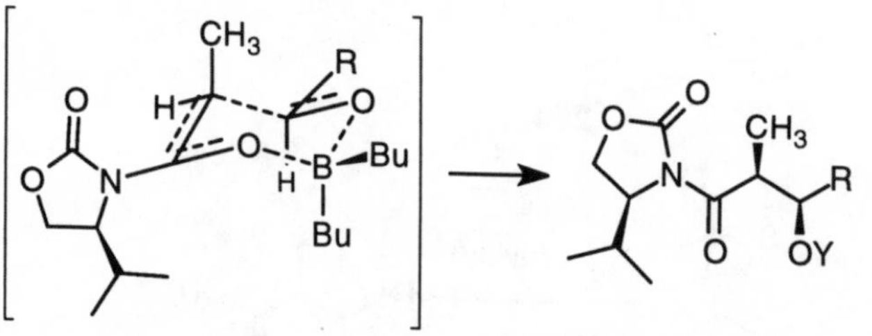

259   **260 a**, Y = BBu$_2$; **b**, Y = H

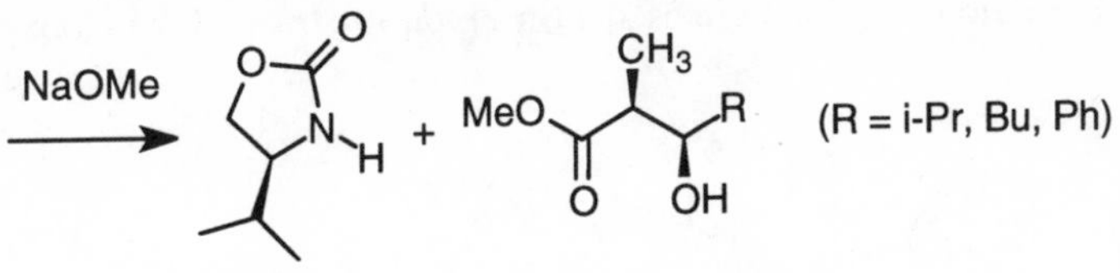

The enantiomer of **257** would have to come from a scarce D-amino acid, but **263** is easily accessible and serves the same function. Enolate **264** with aldehydes yields aldol products **265**, which can be cleaved to methyl esters **266**, enantiomeric to **262**.

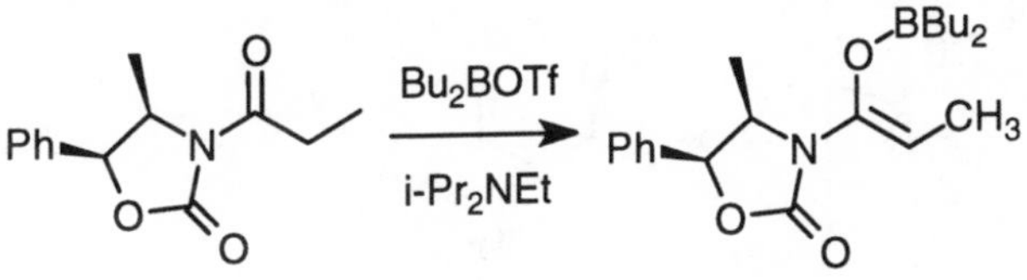

7.4.3 Chiral Boryl Groups

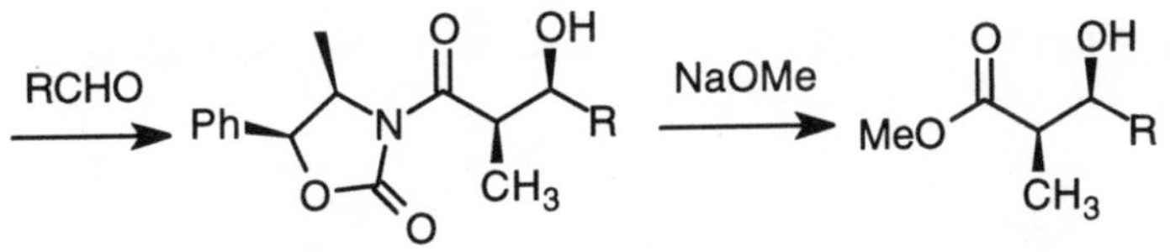

7.4.3.1 2,5-Dimethylborolanes. Masamune's chiral borolane **268** (see Section 6.2.1.3) can also serve as a chiral director in the aldol condensation, and in this reaction it is recyclable [144]. The use of (triethyl)methanethiol propionate (**267**) results in formation of the $E(O)$-enolate **269**, which with aldehydes (R = Pr, Me$_2$CH, Me$_3$C, cyclohexyl, Ph, PhCH$_2$OCH$_2$CH$_2$) leads to *anti* (or *threo*) α-methyl-β-hydroxy thiol esters **271b** in 94% de, 97-99.9% ee. The rationale for the chiral direction is evident from the steric interactions apparent in the drawing of the postulated transition state **270**, in which SCEt$_3$ avoids the borolane methyl groups. In contrast to the (triethyl)methanethiol ester **267**, 2-naphthalenethiol propionate forms $Z(O)$-

enolate in ~95% isomeric purity, and yields *syn* (or *erythro*) α-methyl-β-hydroxy thiol esters.

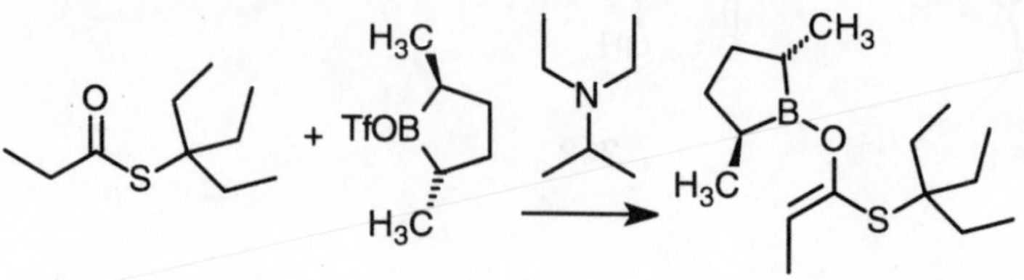

267   **268**   **269**

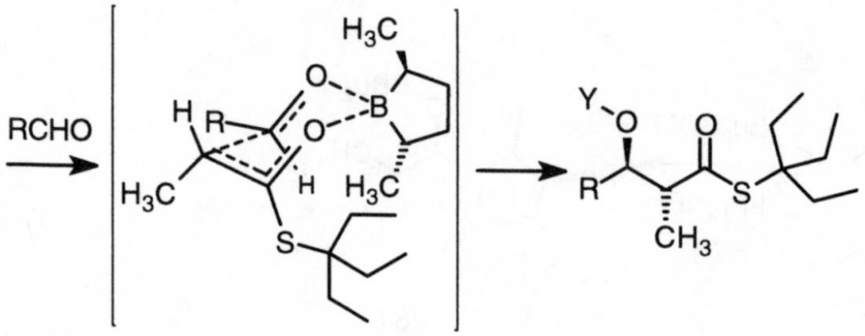

270   **271   a, Y = 2,5-di-Me-borolyl**
                                          b, Y = H

(Triethyl)methanethiol acetate (**272**) with the same set of aldehydes leads to (*R*)-β-hydroxy esters (**273**) in 89.4-98.4% ee [144].

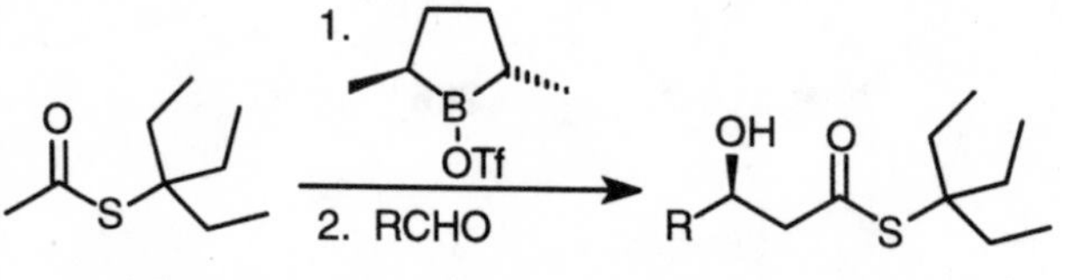

272   **273**

The 2,5-dimethylborolane chiral director was used in several key steps in the synthesis of the anticancer marine natural product bryostatin 7 [145, 146], and also in the synthesis of the antifungal antibiotic pimaricin, the structure of which was proved by the synthesis [147].

Masamune has reviewed the use of 2,5-dimethylborolane as a chiral director in aldol condensations as well as other contexts [148].

Reaction of **ent-267** with the matched chiral aldehyde **166** has led efficiently to **274**, which is the (triethyl)methanethiol ester of a diastereomer of the Prelog-Djerassi lactonic acid [149]. (Compare aldehyde **169**, Section 7.2.5). The mismatched pair *ent-267* with *ent-166* has yielded diastereomer 275 in slightly lower but still excellent diastereomeric excess.

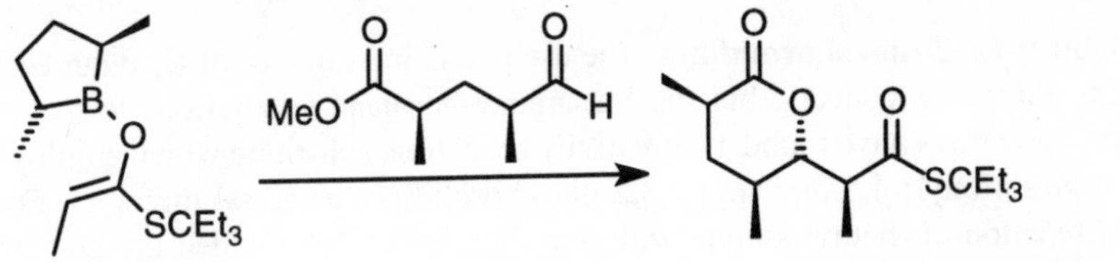

ent-267   166   (matched pair)   274 (de 99.0%)

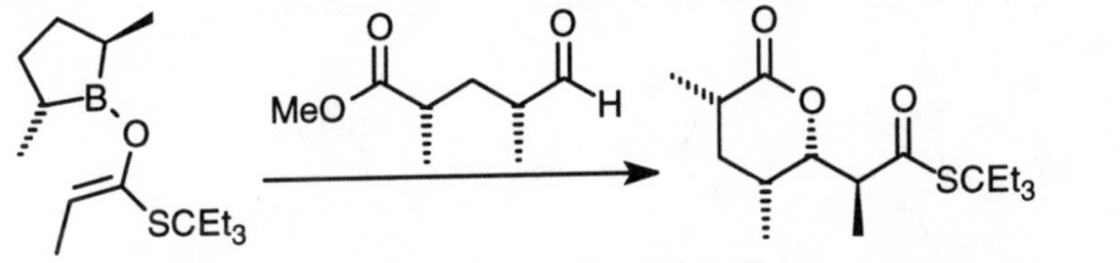

ent-267   ent-166   (mismatched pair) 275 (de 96.4%)

The 2,5-dimethylborolane chiral director can improve the stereoselectivity of matched pairs of chiral reagents. An example is the condensation of aldehyde 166 with the enolate derived from ketone 276, which with an achiral enolate favors the *anti* relationship between the methyl and newly formed hydroxyl groups of the resulting aldol 277 by a factor of 7:1. If the boron enolate is made from *ent-268*, the diastereomer ratio of the resulting 277 rises to 25:1 . With 268, the *anti/syn* ratio falls to 1:1 [150].

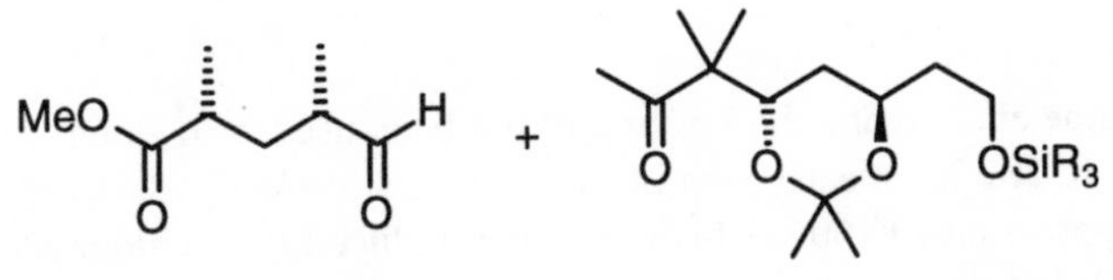

166   276

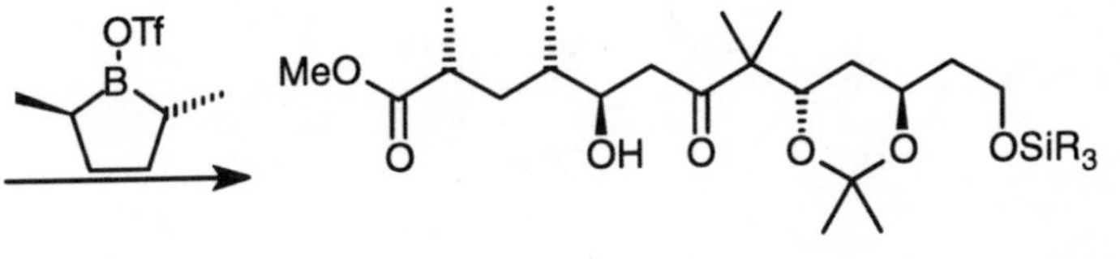

ent-268   277 (94% yield, 96% this isomer)

2-(Trimethylsilyl)borolanes (**248**), which are excellent chiral directors for allylborane reactions, yield ~1:1 mixtures of enantiomers in boron enolate chemistry because the mechanism of stereoselection involves discrimination between an O and a CH$_2$ [127].

7.4.3.2 Chiral 1,3,2-diazaborolidines. The use of a chiral director of C_2 symmetry is a highly important concept, but the Masamune borolane is laborious to make. Corey and coworkers have found a more easily accessible substitute which employs 1,3-bis(arenesulfonyl)-4,5-diphenyl-1,3,2-diazaborolidines as chiral directors. For example, reaction of diethyl ketone with diazaborolidine **209** yielded the enolate **278**, which with benzaldehyde, isobutyraldehyde, or propionaldehyde yielded the hydroxy ketone **279** in ≥95% ee [57]. Where R = Et, ee >98%, the compound is sitophilure, a grain weevil aggregation pheromone. It was suggested that this chiral director is superior to any other for the aldehyde-ketone reaction [57].

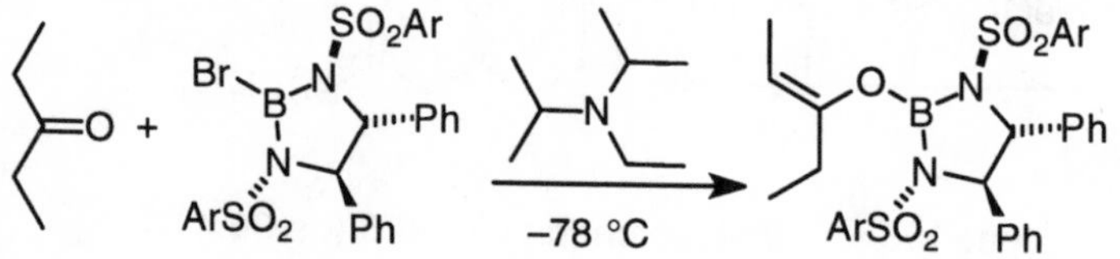

209 (Ar = *p*-tolyl)　　　　　　　　**278**

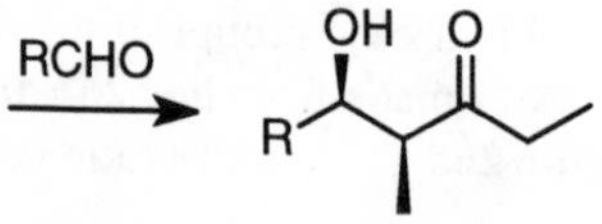

279

A 1,3,2-diazaborolidine enolate of a thiol ester **280** has been used with aldehyde **281** to make intermediate **282** in an early step of the synthesis of the C(18)–C(35) segment of immunosuppressant FK-506, which was then reduced to the aldehyde **204** (see Section 7.2.5.2) for reaction with a 2-allyl-1,3,2-diazaborolidine [98].

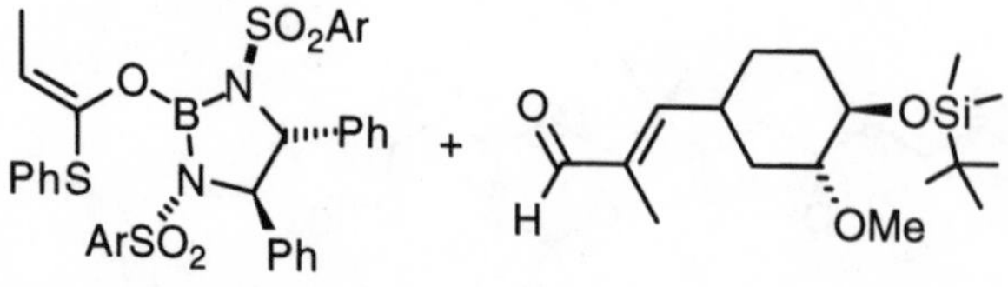

280 (Ar = *p*-O$_2$NC$_6$H$_4$)　　　　　**281**

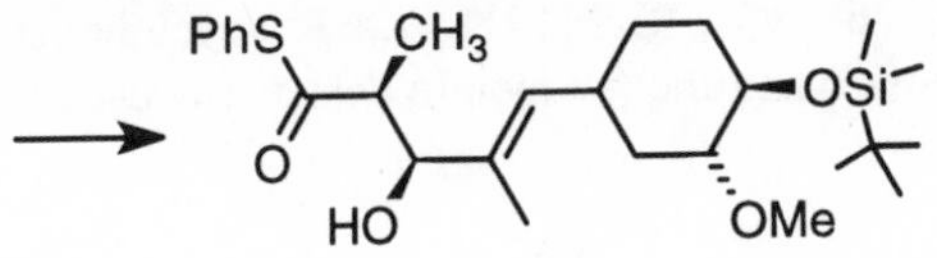

282

Acetate enolates usually fail to yield satisfactory ee's in asymmetric aldol condensations, but the phenyl thioacetate derivative **283** (Ar = *p*-tolyl) yielded **284** in 91% ee with benzaldehyde, 83% ee with isobutyraldehyde [57]. Note that **284** has the opposite absolute configuration from the usual result with propionate (*Z*)-enolates.

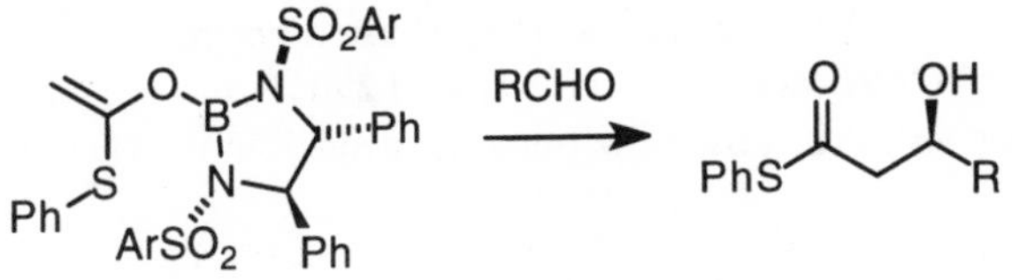

283   **284**

Propionate enolate **285** (Ar = *p*-O$_2$NC$_6$H$_4$) yields **286**, R = Ph, 95% ee, *syn/anti* 98.3:1.7 or R = (CH$_3$)$_2$CH, 97% ee, *syn/anti* 94.5:5.5 [57]. Although the usual type of chair-form transition state can be invoked to explain the results, this is evidently an oversimplified interpretation, as not only **283** (above) but also (*E*)-enolates yield the opposite enantiomer to that expected based on simple repulsive interactions (see below).

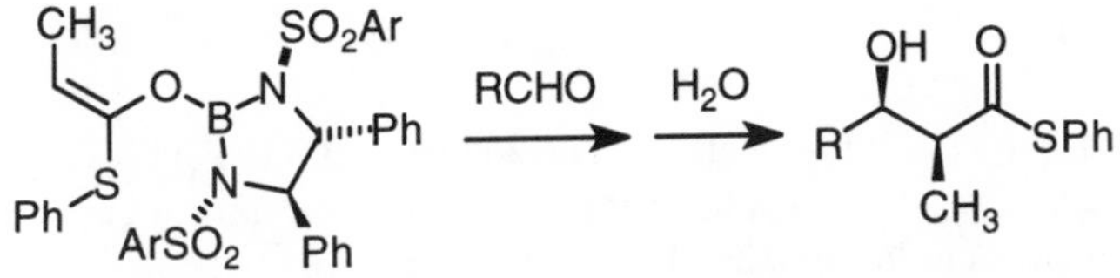

285   **286**

More recently, **209** [Ar = 3,5-(CF$_3$)$_2$C$_6$H$_3$] has been used successfully for preparing **286** via **285**, but more importantly, for making *anti* aldol products **288** from ester (*E(O)*)-enolates such as **287** [151]. The major enantiomer is the one that results from lining up the *tert*-butoxy group on the same side of the boron as the arenesulfonyl substituent. The difference has been noted and the mechanism discussed [152]. However, the possible interactions here are complex, and the usual simplistic steric arguments may be irrelevant. In any event, the stereocontrol is good, with ee's 94-

98% for R = Ph or PhCH=CH, *anti/syn* ratios 98:2-99:1 [151]. With R = cyclohexyl, the ee fell to 75%, but increased to 94% when the (+)-menthyl ester was used in place of *tert*-butyl.

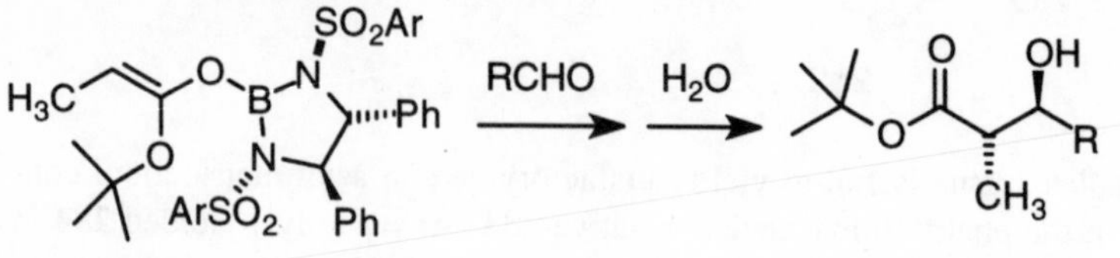

287   **288**

The ester enolate **289** has been made from *tert*-butyl bromoacetate and **209** [Ar = 3,5-$(CF_3)_2C_6H_3$], and the products **290** derived by condensation with aldehydes have been converted to epoxides **291** and other derivatives [153]. Diastereomeric ratios in favor of **290** were 95:5 to 99:1, ee's 91-98%. The conversion of **290** [R = $(CH_3)_2CH$] to (2*S*, 3*S*)-hydroxyleucine and also to its (2*R*, 3*S*)-isomer has been described [154].

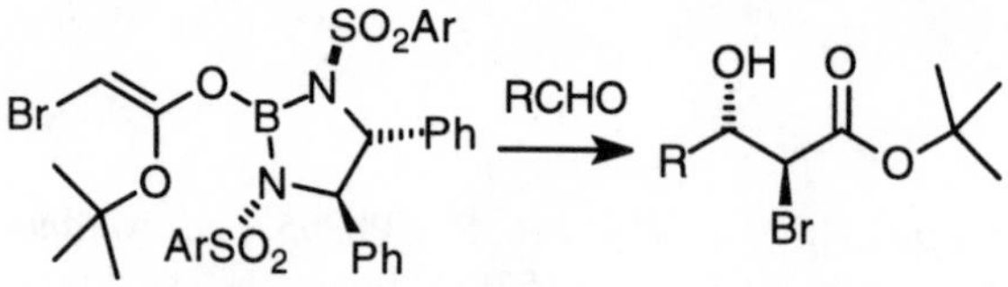

289   **290**

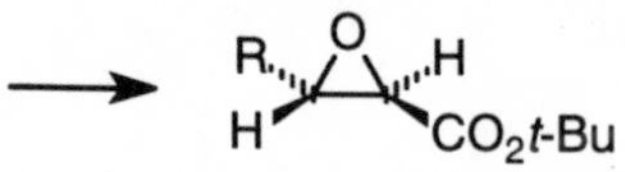

291

Enolate **292** from *tert*-butyl thiolpropionate and ***ent*-209** [Ar = 3,5-$(CF_3)_2C_6H_3$] in the presence of triethylamine reacts with aldimines to form *anti* β-amino thiol esters **293**, which are easily cyclized to lactams **294** [155]. The R^1 groups tested included phenyl, 1- or 2-naphthyl, cinnamyl, and 2-phenylethyl, with R^2 usually allyl but also phenyl or benzyl, and *anti/syn* ratios were ≥97:3, ee's 90 to >99%.

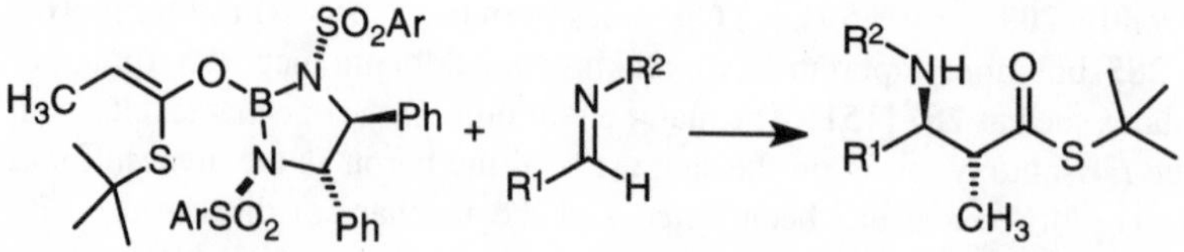

292   **293**

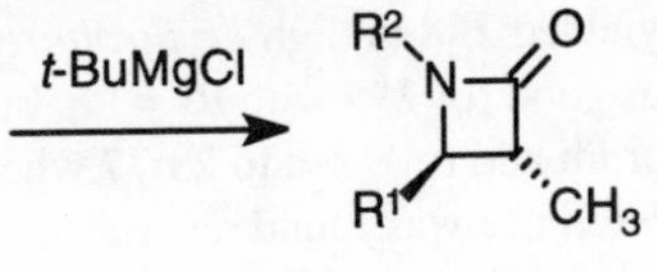

294

The Ireland-Claisen rearrangement is another type of enolate alkylation that can be controlled by the chiral diazaborolidine **ent-209** [Ar = 3,5-(CF$_3$)$_2$C$_6$H$_3$] [156]. The ability to prepare either the (E(O))-enolates **296** or (Z)-enolates **297** from the series of allylic esters **295** allowed full control of the predominant stereochemistry in the *erythro* products **298** and *threo* products **299**, respectively.

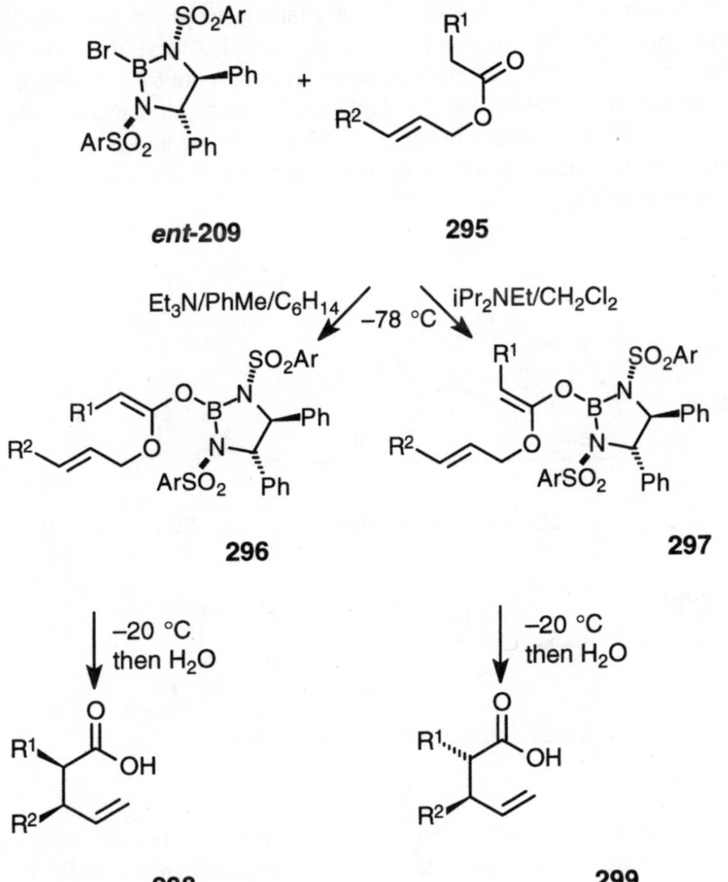

ent-209   **295**

Et$_3$N/PhMe/C$_6$H$_{14}$ / −78 °C \ iPr$_2$NEt/CH$_2$Cl$_2$

296   **297**

−20 °C then H$_2$O   −20 °C then H$_2$O

298   **299**

The best results were obtained with R^1 = R^2 = CH$_3$, which gave **298** in a 90:10 *erythro/threo* ratio and 96% ee or **299** in 99:1 *threo/erythro* ratio and >97% ee. Combinations having R^1 = Me, Et, or Ph and R^2 = Ph yielded **298** in high *erythro/threo* ratios, ≥95:5, and >97% ee. The ee's were nearly as good for **299** with R^2 = Ph, but the *threo/erythro* ratios fell to 91:9 with R^1 = Me or Ph and reversed to 23:77 when R^1 = Ph. For R^1 = SPh, R^2 = Me, similar *erythro* preference was found regardless of the enolate geometry, and where R^1 = ArCH$_2$ and R^2 = H the ee's fell to the 77-84% range.

7.4.4 Catalytic Boron Enolate Reactions

7.4.4.1 Silyl enol ethers. The catalyst **300** derived from tryptophan, which is the same as that used for Diels-Alder reactions (see Section 8.4.3), is effective for the reaction of aldehydes with silyl enol ethers, for example to produce simple β-hydroxy ketones **301** in 96-93% ee (R^1 = Ph, *n*-Pr, cyclohexyl, 2-furyl; R^2 = Ph, Bu) [157]. With R^2 = -CH=CH-OMe, ee's with the same set of R^1 were 67-82%, and the resulting **301** were cyclized with acid to chiral dihydropyrans. Cyclopentanone silyl enol ether and benzaldehyde were converted to **302** (92% ee) and **303** in 96:4 diastereomeric ratio. As is usual with these catalytic reactions, conditions were very mild, −78 °C in propionitrile.

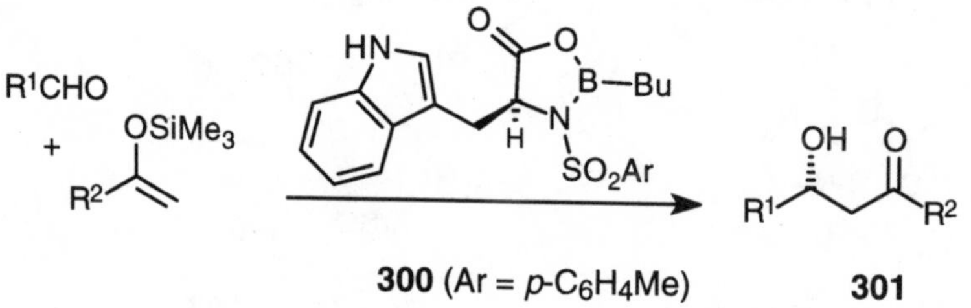

300 (Ar = *p*-C₆H₄Me)   **301**

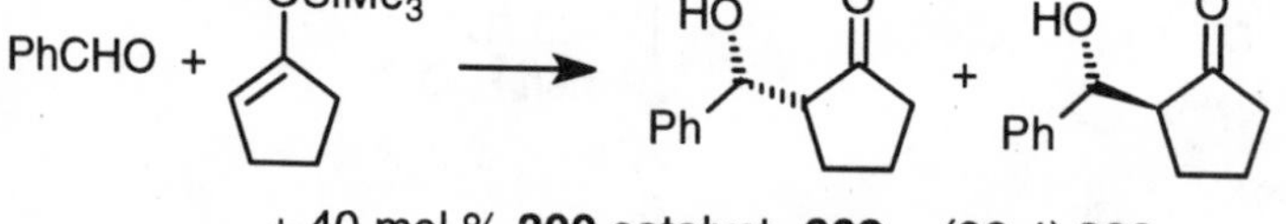

+ 40 mol % **300** catalyst **302**   (96:4) **303**

Certain *N*-tosyl amino acids such as **304** or **305** serve as efficient chiral directors in a catalytic aldol process involving a ketene silyl acetal and an aldehyde. The catalytically active species formed on treatment with borane-THF is believed to be an oxazaborolane **306**. The aldehyde was added slowly to a mixture of about 20 mol % catalyst and ketene silyl acetal for optimum results. The initial silylated products **307** hydrolyzed to β-hydroxy esters in 84-99% ee's [158].

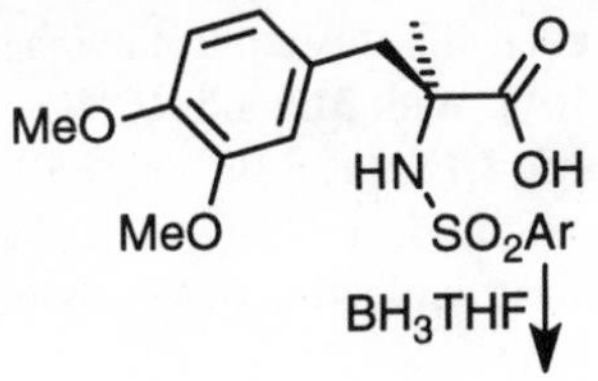

304 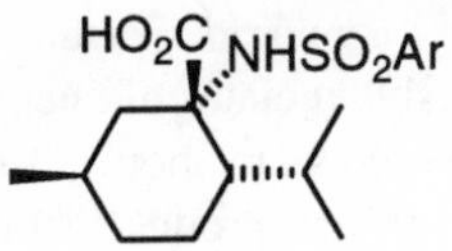   **305**

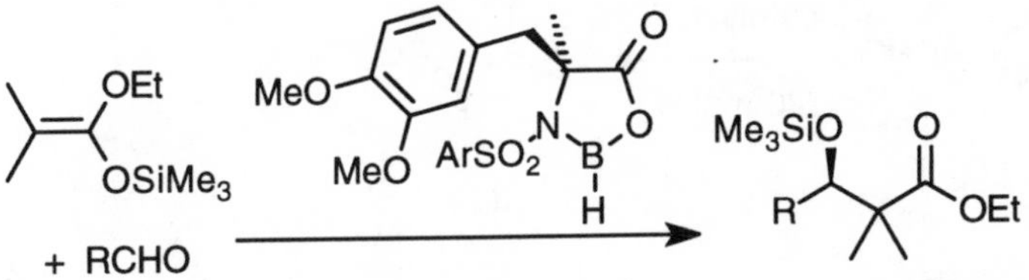

306   (Ar = *p*-tolyl)   **307**

The reaction of mono(2,6-dialkoxybenzoate) esters of (*R,R*)-tartaric acid with BH₃·THF or boronic acid derivatives yields acyloxy borane intermediates (**310**). These were first developed as catalysts for Diels-Alder reactions (see Section 8.4.1). The reaction of ketone silyl enol ethers such as **308** or **309** with aldehydes is catalyzed by **310a** to produce aldol condensation products in 80-96% ee's [159]. This reaction favors *erythro* products regardless of the geometry of the enolate, and either **308** or **309** yields 96-97% of material that is 93-94% *erythro* aldol product **311** in 94-96% ee [159]. Similar reactions were tested with the catalyst **310b** and in some cases the ee's were improved [160].

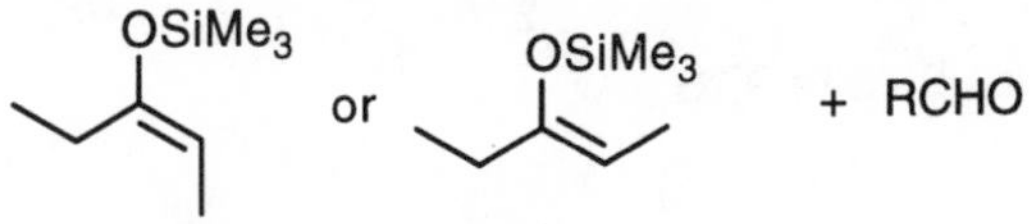

308   **309**

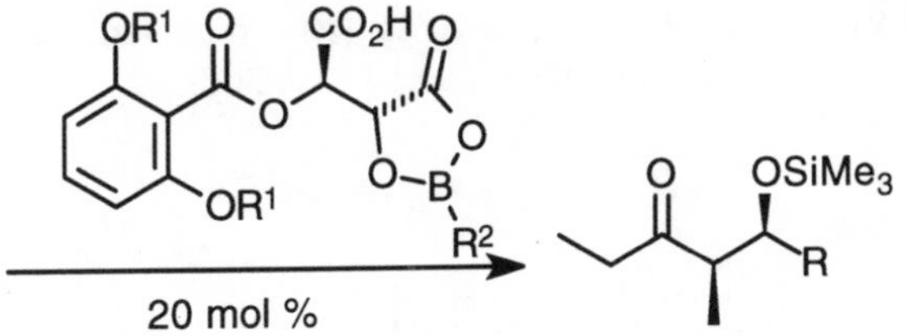

20 mol %

310   **311**
a, R¹ = CHMe₂, R² = H
b, R¹ = CHMe₂, R² = *o*-C₆H₅OC₆H₄
c, R¹ = Me, R² = H

The chiral acyloxy borane reagent also gave good chirality control in reactions of aldehydes with silyl enolates of phenyl esters [161]. With **312** and RCHO, *erythro/threo* ratios observed were about 2:1 for R = alkyl, 4:1 for R = Ph, and 25:1 for R = alkenyl, and the *erythro* products showed ee's ranging from 79-88% (R = alkyl) to 92% (R = Ph) to 94-97% (R = alkenyl). (*R*)-Tartaric acid led to (α*R*) products **313**.

catalyzed by **310a**

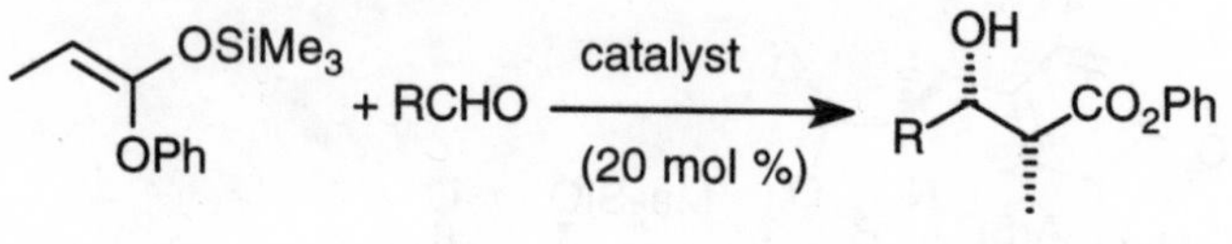

312   **313**

7.4.4.2 Allylsilanes. The analogous allylsilation reactions of aldehydes with allylic silanes such as a mixture of **314** and **315** (R^1 = Me, Et) yielded *erythro* homoallylic alcohols **316** in high ee's, with only minor amounts of *threo* isomers, though the reported yields varied from good (R^2 = Ph) to poor (R^2 = Pr, Bu, MeCH=CH) [161, 162].

310a catalysis

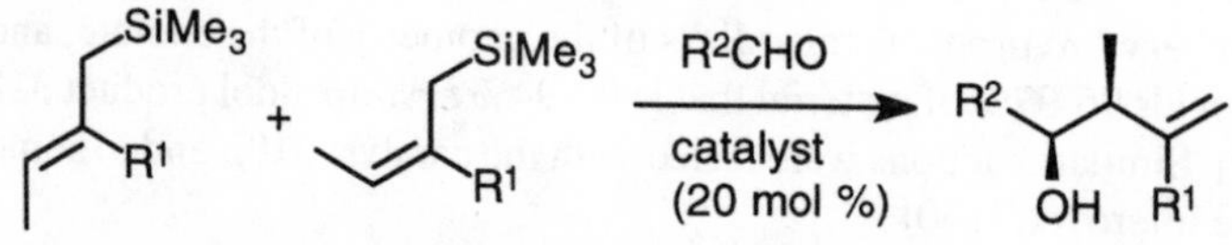

314 (61-65%)   **315 (39-35%)**   **316**
ee's 85-97%, *erythro/threo* ratios 95:5 to 97:3

Marshall and Tang tested tin reagents **317** in place of the allylsilanes [163]. A full equivalent of catalyst was required, together with trifluoroacetic anhydride to stop inhibition of the reaction by product. The series of homoallylic alcohols **318** included R = 1-hexynyl (yield 91%, *syn/anti* 71:29, ee 70%), (*E*)-1-propenyl (85%, *syn/anti* 92:8, ee 95%), n-propyl (61%, *syn/anti* 97:3, ee 81%), and others.

catalyzed by **310c**

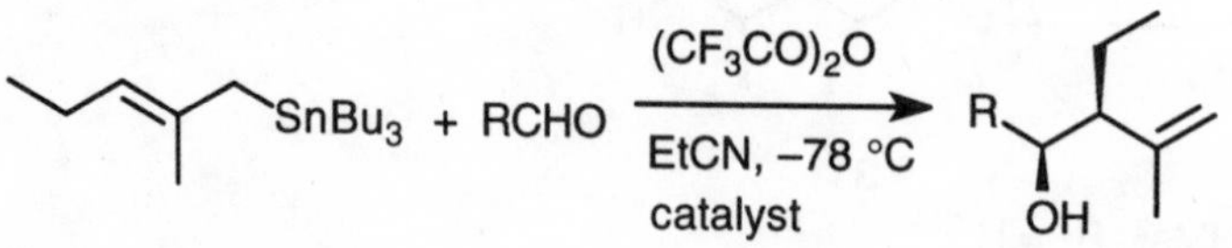

317   **318**

7.4.5 New Routes to Boron Enolates

7.4.5.1 Dialkylboron halides. In contrast to dialkylboron triflates, dialkylboron chlorides with ketones usually yield (*E*)-enolates [164, 165, 166, 167]. An example is the reaction of chlorodicyclohexylborane (**319**) and triethylamine with ethyl isopropyl ketone to form the (*E*)-enolate **320**, producing a >97:3 (*E*)/(*Z*) ratio [167]. The yield estimated by NMR was 97%. Reaction with benzaldehyde yielded the *anti* β-hydroxy ketone (**321**), which provided the basis for the estimate of the (*E*)/(*Z*) ratio.

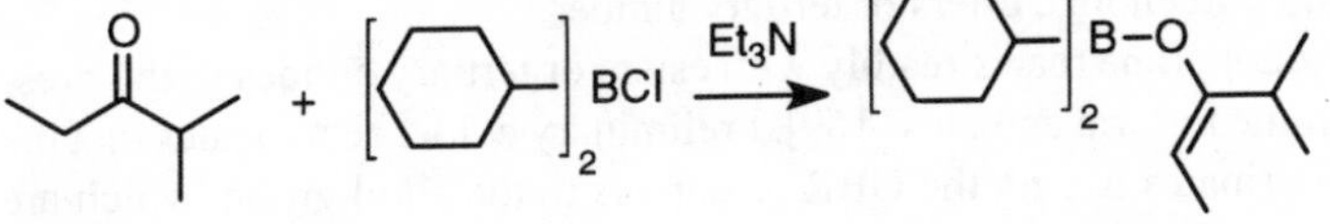

319   **320** [>97% (*E*)]

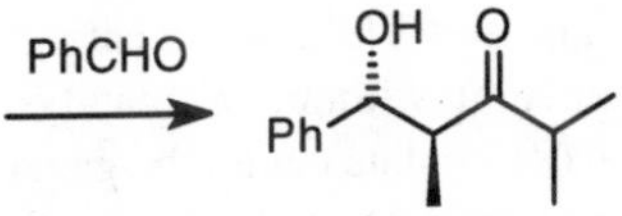

321

With propiophenone or ethyl *tert*-butyl ketone, **319** produced >97:3 (*E*)/(*Z*) ratios, but with diethyl ketone the (*E*)/(*Z*) ratio was only 79:21 [167]. The use of bromodicyclohexylborane in place of **319** yielded greater (*Z*)-isomer content. The iodo analogue usually yielded more (*Z*)- than (*E*)-isomer, and produced >97:3 (*E*)/(*Z*) ratios with propiophenone or ethyl *tert*-butyl ketone. 9-Iodo-9-BBN (**322**) consistently yielded predominantly (*Z*)-enolates such as **323**, (*Z*)/(*E*) ratio >97:3. 9-BBN triflate or mesylate also yielded (*Z*)-enolates, though not as consistently as **322** did.

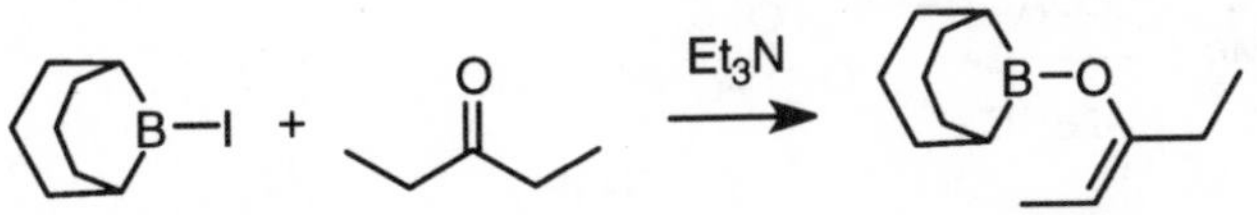

322   **323** [>97% (*Z*)]

Bis(bicyclooctyl)boron chloride (**324**) produces higher (*E*)/(*Z*) ratios than **319** with several ketones [168]. Diethyl ketone, which is especially prone to yield (*Z*)-enolate, yields (*E*)-enolate **325**.

324   **325 [97% (E)]**

Dicyclohexylchloroborane (**319**) with triethylamine also produces aldehyde enol-borinates, though apparently not with useful stereoselection, and with carboxylic acids yields the bis(dicyclohexylboryl) ester enolates, $RCH=C[OB(C_6H_{11})_2]_2$ [165]. However, **319** failed to enolize esters or tertiary amides.

Dicyclohexyliodoborane reacts readily with esters or tertiary amides in the presence of triethylamine to form enolates [169]. Preliminary evidence indicates that the esters form enolborinates having the OBR_2 group *cis* to the alkyl group, which are usually referred to as [Z(O)-enolates by analogy to the ketone series, even though systematic nomenclature would designate them as (E). The amides form (E)-enolates.

7.4.5.2 Other routes. Boron (Z)-enolates can be generated efficiently by treatment of α,β-unsaturated esters or amides with catecholborane and Wilkinson's catalyst [170]. α,β-Unsaturated ketones undergo similar reduction without the catalyst. Where the unsaturation is confined in a ring so that the (Z)-enolate cannot be generated, 1,2-reduction to alcohol borate occurs instead of conjugate reduction to the enolate.

A β-dioxyboryl substituent has been found to have little effect on the geometry of enolization of various carbonyl compounds, but where a (Z)-enolate is formed, the boron bonds to the enolate oxygen, as in **327**, and improved *syn* selectivity compared to a simple lithium enolate results [171]. The amide **326** is a model for possible asymmetric applications. The ratio of racemic diastereomers **328** to **329** was 20:1. β-Boryl ester and thiol ester enolates, which are normally [E(O)]-enolates, showed no effect due to the presence of the boron. The synthesis of **327** and related compounds is described in Section 3.4.3.6.

326   **327**

1. PhCHO

2. Me₃SiCl

328 (95%)   **329** (5%)

7.5 References

1. Mikhailov BM, Bubnov YuN (1964) Izv. Akad. Nauk SSSR, Ser. Khim. 1874
2. Mikhailov BM, Pozdnev VF (1967) Izv. Akad. Nauk SSSR, Ser. Khim. 1477
3. Hoffmann RW, Zeiss HJ (1979) Angew. Chem. 91:329; Angew. Chem. Int. Ed. 18:306
4. Fenzl W, Köster R (1975) Justus Liebigs Ann. Chem. 1322
5. Fenzl W, Köster R, Zimmermann HJ (1975) Justus Liebigs Ann. Chem. 2201
6. Masamune S, Mori S, Van Horn D, Brooks DW (1979) Tetrahedron Lett. 1665
7. Hirama M, Masamune S (1979) Tetrahedron Lett. 2225
8. Van Horn DE, Masamune S (1979) Tetrahedron Lett. 2229
9. Hirama M, Garvey DS, Lu LDL, Masamune S (1979) Tetrahedron Lett. 3937
10. Evans DA, Vogel E, Nelson JV (1979) J. Am. Chem. Soc. 101:6120
11. Yamamoto Y, Asao N (1993) Chem. Rev. 93:2207
12. Matteson DS, Campbell JD (1990) Heteroatom Chemistry 1:109
13. Hoffmann RW, Herold T (1981) Chem. Ber. 114:375
14. Schlosser M, Rauchschwalbe G (1978) J. Am. Chem. Soc. 100:3258
15. Schlosser M, Stähle M (1980) Angew. Chem. 92:497; Angew. Chem. Int. Ed. 19:487
16. Hoffmann RW, Zeiss HJ (1981)J. Org. Chem. 46:1309
17. Hoffmann RW, Kemper B (1981) Tetrahedron Lett. 22:5263
18. Hoffmann RW, Kemper B (1982) Tetrahedron Lett. 23:845
19. Hoffmann RW, Kemper B, Metternich R, Lehmeier T (1985) Justus Liebigs Ann. Chem. 2246
20. Roush WR, Ando K, Powers DB, Halterman RL, Palkowitz AD (1988) Tetrahedron Lett. 29:5579
21. Roush WR, Ando K, Powers DB, Palkowitz AD, Halterman RL (1990) J. Am. Chem. Soc. 112:6339
22. Andersen MW, Hildebrandt B, Köster G, Hoffmann RW (1989) Chem. Ber. 122:1777
23. Stürmer R (1990) Angew. Chem. 102:62; Angew. Chem. Int. Ed. 29:59
24. Brown HC, De Lue NR, Yamamoto Y, Maruyama K, Kasahara T, Murahashi S, Sonoda A (1977) J. Org. Chem. 42:4088
25. Wuts PGM, Bigelow SS (1988) J. Org. Chem. 53:5023
26. Hoffmann RW, Ditrich K, Köster G, Stürmer R (1989) Chem. Ber. 122:1783
27. Sadhu KM, Matteson DS (1985) Organometallics 4:1687
28. Hoffmann RW, Niel G (1991) Justus Liebigs Ann. Chem. 1195
29. Knochel P (1990) J. Am. Chem. Soc. 112:7431
30. Watanabe T, Miyaura N, Suzuki A (1993) J. Organomet. Chem. 444:C1
31. Hoffmann RW, Sander T, Hense A (1993) Justus Liebigs Ann. Chem. 771
32. Satoh M, Nomoto Y, Miyaura N, Suzuki A (1989) Tetrahedron Lett. 30:3789
33. Hoffmann RW (1982) Angew. Chem. 94:569; Angew. Chem. Int. Ed. 21:555
34. Li Y, Houk KN (1989) J. Am. Chem. Soc. 111:1236

35. Metternich R, Hoffmann RW (1984) Tetrahedron Lett. 25:4095
36. Hoffmann RW, Metternich R (1985) Justus Liebigs Ann. Chem. 2390
37. Hoffmann RW, Kemper B (1984) Tetrahedron 40:2219
38. Tsai JS, Matteson DS (1981) Tetrahedron Lett. 22:2751
39. Hoffmann RW, Eichler G, Endesfelder A (1983) Justus Liebigs Ann. Chem. 2000
40. Hoffmann RW, Endesfelder A (1987) Justus Liebigs Ann. Chem. 215
41. Hoffmann RW, Endesfelder A (1986) Justus Liebigs Ann. Chem. 1823
42. Wuts PGM, Jung YW (1986) Tetrahedron Lett. 27:2079
43. Hoffmann RW, Sander T (1990) Chem. Ber. 123:145
44. Wang Z, Meng XJ, Kabalka GW (1991) Tetrahedron Lett. 32:1945
45. Wang Z, Meng XJ, Kabalka GW (1991) Tetrahedron Lett. 32:4619
46. Wang Z, Meng XJ, Kabalka GW (1991) Tetrahedron Lett. 32:5677
47. Hoffmann RW, Sander T (1993) Justus Liebigs Ann. Chem. 1185
48. Hoffmann RW, Sander T (1993) Justus Liebigs Ann. Chem. 1193
49. Herold T, Schrott U, Hoffmann RW, Schnelle G, Ladner W, Steinbach K (1981) Chem. Ber. 114:359
50. Roush WR, Walts AE, Hoong LK (1985) J. Am. Chem. Soc. 107:8186
51. Roush WR, Hoong LK, Palmer MAJ, Park JC (1990) J. Org. Chem. 55:4109
52. Roush WR, Halterman RL (1986) J. Am. Chem. Soc. 108:294
53. Roush WR, Park JC (1990) J. Org. Chem. 55:1143
54. Roush WR, Ratz AM, Jablonski JA (1992) J. Org. Chem. 57:2047
55. Roush WR, Banfi L (1988) J. Am. Chem. Soc. 110:3979
56. Roush WR, Banfi L, Park JC, Hoong LK (1989) Tetrahedron Lett. 30:6457
57. Corey EJ, Imwinkelried R, Pikul S, Xiang YB (1989) J. Am. Chem. Soc. 111:5493
58. Corey EJ, Yu CM, Kim SS (1989) J. Am. Chem. Soc. 111:5495
59. Corey EJ, Kim SS (1990) Tetrahedron Lett. 31:3715
60. Hoffmann RW, Metternich R, Lanz JW (1987) Justus Liebigs Ann. Chem. 881
61. Hoffmann RW, Brinkmann H, Frenking G (1990) Chem. Ber. 123:2387
62. Brinkmann H, Hoffmann RW (1990) Chem. Ber. 123:2395
63. Hoffmann RW, Weidmann U (1985) Chem. Ber. 118:3966
64. Hoffmann RW, Zeiss HJ, Ladner W, Tabche S (1982) Chem. Ber. 115:2357
65. Hoffmann RW, Endesfelder A, Zeiss HJ (1983) Carbohydrate Research 123:320
66. Roush WR, Hoong LK, Palmer MAJ, Straub JA, Palkowitz AD (1990) J. Org. Chem. 55:4117
67. Masamune S, Choy W., Petersen JS, Sita LR (1985) Angew. Chem. Int. Ed. 24:1.
68. Roush WR, Palkowitz AD, Palmer MAJ (1987) J. Org. Chem. 52:316
69. Roush WR, Palkowitz AD, Ando K (1990) J. Am. Chem. Soc. 112:6348
70. Roush WR (1991) J. Org. Chem. 56:4151
71. Hoffmann RW, Landmann B (1986) Chem. Ber. 119:1039
72. Hoffmann RW, Landmann B (1986) Chem. Ber. 119:2013
73. Stürmer R, Hoffmann RW (1990) Synlett 759
74. Hoffmann RW, Dresely S (1988) Synthesis 103
75. Hoffmann RW, Dresely S, Lanz JW (1988) Chem. Ber. 121:1501
76. Hoffmann RW, Dresely S, Hildebrandt B (1988) Chem. Ber. 121:2225
77. Hoffmann RW, Giesen V, Fuest M (1993) Justus Liebigs Ann. Chem. 629
78. Hoffmann RW, Dresely S (1989) Chem. Ber. 122:903
79. Hoffmann RW, Wolff JJ (1991) Chem. Ber. 124:563
80. Hoffmann RW, Schlapbach A (1990) Justus Liebigs Ann. Chem. 1243
81. Hoffmann RW, Schlapbach A (1991) Justus Liebigs Ann. Chem. 1203
82. Hoffmann RW, Ladner W (1979) Tetrahedron Lett. 4653
83. Hoffmann RW, Ladner W, Steinbach K, Massa W, Schmidt R, Snatzke G (1981) Chem. Ber. 114:2786
84. Hoffmann RW, Helbig W (1981) Chem. Ber. 114:2802
85. Hoffmann RW, Ladner W, Ditrich K (1989) Justus Liebigs Ann. Chem. 883

86. Ditrich K, Bube T, Stürmer R, Hoffmann RW (1986) Angew. Chem. 98:1016; Angew. Chem. Int. Ed. 25:1028
87. Hoffmann RW, Ditrich K (1990) Justus Liebigs Ann. Chem. 23
88. Andersen MW, Hildebrandt B, Dahmann G, Hoffmann RW (1991) Chem. Ber. 124:2127
89. Roush WR, Palkowitz AD (1987) J. Am. Chem. Soc. 109:953
90. Roush WR, Palkowitz AD (1989) J. Org. Chem. 54:3009
91. Roush WR, Kageyama M (1985) Tetrahedron Lett. 26:4327
92. Roush WR, Brown BB, Drozda SE (1988) Tetrahedron Lett. 29:3541
93. Roush WR, Coe JW (1987) Tetrahedron Lett. 28:931
94. Coe JW, Roush WR (1989) J. Org. Chem. 54:915
95. Danishefsky SJ, Armistead DM, Wincott FE, Selnick HG, Hungate R (1987) J. Am. Chem. Soc. 109:8117
96. Roush WR, Straub JA (1986) Tetrahedron Lett. 27:3349
97. Roush WR, Straub JA, VanNiewenhze MS (1991) J. Org. Chem. 56:1636
98. Corey EJ, Huang HC (1989) Tetrahedron Lett. 30:5235
99. Ikeda N, Arai I, Yamamoto H (1986) J. Am. Chem. Soc. 108:483
100. Haruta R, Ishiguro M, Ikeda N, Yamamoto H (1982) J. Am. Chem. Soc. 104:7667
101. Corey EJ, Yu CM, Lee DH (1990) J. Am. Chem. Soc. 112:878
102. Corey EJ, Jones GB (1991) Tetrahedron Lett. 32:5713
103. Midland MM, Preston SB (1980) J. Org. Chem. 45:747
104. Midland MM, Preston SB (1982) J. Am. Chem. Soc. 104:2330
105. Brown HC, Jadhav PK (1983) J. Am. Chem. Soc. 105:2092
106. Brown HC, Jadhav PK (1984) J. Org. Chem. 49:4089
107. Brown HC, Randad RS, Bhat KS, Zaidlewicz M, Racherla US (1990) J. Am. Chem. Soc. 112:2389
108. Brown HC, Racherla US, Pellechia PJ (1990) J. Org. Chem. 55:1868
109. Bubnov YuN, Lavrinovich LI, Zykov AYu, Ignatenko AV (1992) Mendeleev Commun. 86
110. Brown HC, Racherla US, Liao Y, Khanna VV (1992) J. Org. Chem. 57:6608
111. Racherla US, Liao Y, Brown HC (1992) J. Org. Chem. 57:6614
112. Brown HC, Randad RS (1990) Tetrahedron 46:4463
113. Bubnov YuN, Etinger MYu (1985) Tetrahedron Lett. 26:2797
114. Brown HC, Jadhav PK (1984) Tetrahedron Lett. 25:1215
115. Brown HC, Bhat KS (1986) J. Am. Chem. Soc. 108:293
116. Brown HC, Bhat KS (1986) J. Am. Chem. Soc. 108:5919
117. Brown HC, Randad RS (1990) Tetrahedron 46:4457
118. Fujita K, Schlosser M (1982) Helv. Chim. Acta 65:1258
119. Barrett AGM, Lebold SA (1990) J. Org. Chem. 55:5818
120. Barrett AGM, Lebold SA (1991) J. Org. Chem. 56:4875
121. Brown HC, Bhat KS, Randad RS (1987) J. Org. Chem. 52:319
122. Brown HC, Bhat KS, Randad RS (1987) J. Org. Chem. 52:3701
123. Brown HC, Bhat KS, Randad RS (1989) J. Org. Chem. 54:1570
124. Brown HC, Bhat KS, Jadhav PK (1991) J. Chem. Soc. Perkin 1 2633
125. Barrett AGM, Malecha JW (1991) J. Org. Chem. 56:5243
126. Garcia J, Kim BM, Masamune S (1987) J. Org. Chem. 52:4831
127. Short RP, Masamune S (1989) J. Am. Chem. Soc. 111:1892
128. Evans DA, Taber TR (1980) Tetrahedron Lett. 21:4675
129. Choy W, Ma P, Masamune S (1981) Tetrahedron Lett. 22:3555
130. Masamune S, Choy W, Kerdesky FAJ, Imperiali B (1981) J. Am. Chem. Soc. 103:1566
131. Masamune S, Hirama M, Mori S, Ali SA, Garvey DS (1981) J. Am. Chem. Soc. 103:1568
132. Masamune S (1981) "Organic Synthesis Today and Tomorrow", Trost BM, Hutchinson CR, eds, Pergamon Press, Oxford, p. 197

133. Kaiho T, Masamune S, Toyoda T (1982) J. Org. Chem. 47:1612
134. Masamune S, Lu LDL, Jackson WP, Kaiho T, Toyoda T (1982) J. Am. Chem. Soc. 104:5523
135. Masamune S, Imperiali B, Garvey DS (1982) J. Am. Chem. Soc. 104:5528
136. Masamune S, Kaiho T, Garvey DS (1982) J. Am. Chem. Soc. 104:5521
137. Boschelli D, Ellingboe JW, Masamune S (1984) Tetrahedron Lett. 25:3395
138. Masamune S (1988) Annals of the New York Academy of Sciences 544:168
139. Masamune S, Choy W (1982) Aldrichimica Acta 15 No. 3:47
140. Masamune S, McCarthy PA (1984) "Macrolide Antibiotics", Omura S, ed, Academic Press, New York, P. 127
141. Masamune S (1984) Heterocycles 21:107
142. Evans DA, Bartroli J, Shih TL (1981) J. Am. Chem. Soc. 103:2127
143. Evans DA, Nelson JV, Vogel E, Taber TR (1981) J. Am. Chem. Soc. 103:3099
144. Masamune S, Sato T, Kim BM, Wollmann TA (1986) J. Am. Chem. Soc. 108:8279
145. Masamune S (1988) Pure Appl. Chem. 60:1587
146. Kageyama M, Tamura T, Nantz MH, Roberts JC, Somfai P, Whritenour DC, Masamune S (1990) J. Am. Chem. Soc. 112:7408
147. Duplantier AJ, Masamune S (1990) J. Am. Chem. Soc. 112:7079
148. Masamune S (1987) "Stereochemistry of Organic and Bioorganic Transformations", Bartmann W, Sharpless KB, eds, VCH Publishers, Weinheim, Germany, p. 49
149. Short RP, Masamune S (1987) Tetrahedron Lett. 28:2841
150. Duplantier AJ, Nantz MH, Roberts JC, Short RP, Somfai P, Masamune S (1989) Tetrahedron Lett. 30:7357
151. Corey EJ, Kim SS (1990) J. Am. Chem. Soc. 112:4976
152. Corey EJ, Lee DH (1993) Tetrahedron Lett. 34:1737
153. Corey EJ, Choi S (1991) Tetrahedron Lett. 32:2857
154. Corey EJ, Lee DH, Choi S (1992) Tetrahedron Lett. 33:6735
155. Corey EJ, Decicco CP, Newbold RC. (1991) Tetrahedron Lett. 32:5287
156. Corey EJ, Lee DH (1991) J. Am. Chem. Soc. 113:4026
157. Corey EJ, Cywin CL, Roper TD (1992) Tetrahedron Lett. 33:6907
158. Parmee ER, Tempkin O, Masamune S, Abiko A (1991) J. Am. Chem. Soc. 113:9365
159. Furuta K, Maruyama T, Yamamoto H (1991) J. Am. Chem. Soc. 113:1041
160. Furuta K, Maruyama T, Yamamoto H (1991) Synlett 439
161. Furuta K, Mouri M, Yamamoto H (1991) Synlett 561
162. Ishihara K, Mouri M, Bao Q, Maruyama T, Furuta K, Yamamoto H (1993) J. Am. Chem. Soc. 115:11490
163. Marshall JA, Tang Y (1992) Synlett 653
164. Brown HC, Dhar RK, Bakshi RK, Pandiarajan PK, Singaram B (1989) J. Am. Chem. Soc. 111:3441
165. Brown HC, Dhar RK, Ganesan K, Singaram B (1992) J. Org. Chem. 57:499
166. Brown HC, Dhar RK, Ganesan K, Singaram B (1992) J. Org. Chem. 57:2716
167. Brown HC, Ganesan K, Dhar RK (1993) J. Org. Chem. 58:147
168. Brown HC, Ganesan K, Dhar RK (1992) J. Org. Chem. 57:3767
169. Brown HC, Ganesan K (1992) Tetrahedron Lett. 33:3421
170. Evans DA, Fu GC (1990) J. Org. Chem. 55:5678
171. Curtis ADM, Whiting A (1991) Tetrahedron Lett. 32:1507

8 Diels-Alder Reactions

8.1 Introduction

Boron in Diels-Alder reactions falls into three categories: simple alkenylboranes as dienophiles, dienylboranes as dienes, and asymmetric boranes as acid catalysts that are chiral templates.

The first reports of vinyl- and ethynylboronic esters acting as dienophiles by Matteson and coworkers [1, 2, 3] and other early work in the field have been reviewed in Section 3.5.5.3. These studies established the fact that vinylboranes can be efficient dienophiles, but did not provide any reason to believe that they could be particularly useful for stereoselective synthesis. More recent work has provided examples of good regio- and diastereocontrol, though enantiocontrol is still a hope and not a reality. Dienylboranes have been developed more recently, and much the same can be said for them. Because this is a currently active research area with apparent potential for future developments, it is covered here in almost as much detail as enantioselective reactions.

Asymmetric boron compounds that act as chiral acid catalysts for Diels-Alder reactions provide some of the most efficient routes known to complex cyclic structures. These catalysts may or may not contain boron—carbon bonds, and might be considered marginal to the central topic of this book, but are covered here because of their great promise and the current interest in them.

8.2 Alkenylboranes

8.2.1 Dialkylalkenylboranes

Boronic esters are sluggish dienophiles [2, 3], and it might be expected that compounds lacking back bonding from the ligands to the vacant p-orbital of boron would be more reactive.

Singleton and Martinez have prepared 9-vinyl-9-BBN (**1**) from vinyltributyltin and 9-bromo-9-BBN. It was found that **1** is a much more reactive dienophile than vinylboronic esters, and reacts 200 times faster than methyl acrylate with butadiene at 25 °C [4]. In the reaction with isoprene the regioselectivity for 1-methyl-4-bo-

rylcyclohexene **2** over 1-methyl-5-borylcyclohexene **3** is high, and with 1,3-pen-
tadiene the product of *endo* addition **4** is strongly favored over the *endo* isomer **5**.
However, with cyclopentadiene the *endo/exo* ratio was only 2:1 [4]. The reactivity
of **1** is surprisingly insensitive to the structure of the diene, and does not correlate
with the electronegativity of substituents [5]. Accordingly, **1** may be considered an
"omniphilic dienophile." For example, 2-chlorobutadiene reacts a little faster than
butadiene or isoprene with **1**, and 13 times faster with **1** than with maleic anhydride,
or 650 times faster than tetracyanoethylene with **1**, to yield the 1-chloro-4-boryl
product **6** with 90:10 regioselectivity. However, **1** reacted only 0.13 times as fast
with 2-ethoxybutadiene as with 2-chlorobutadiene, or only 0.017 times the rate for
maleic anhydride with 2-methoxybutadiene, and the ratio of 1-ethoxy-4-boryl- to 1-
ethoxy-5-borylcyclohexene was only 69:31.

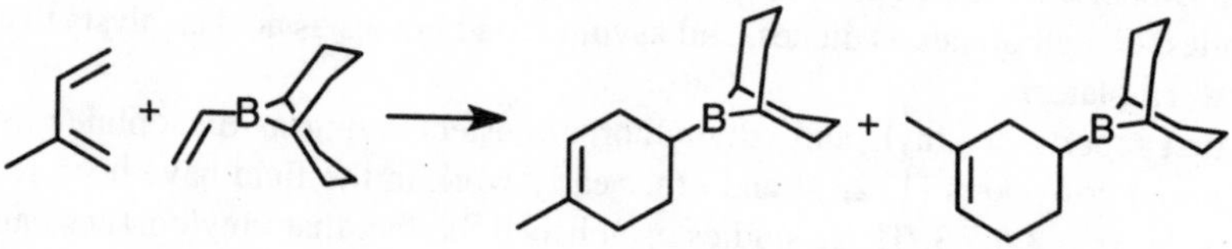

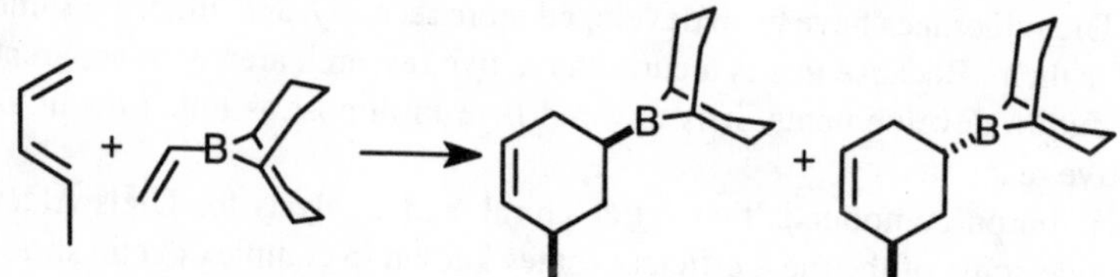

9-Vinyl-9-BBN (**1**) dimerizes slowly to 2-butenyl-1,4-bis(9-BBN), and this side
reaction decreases yields with relatively unreactive dienophiles. 1-Vinyl-3,6-
dimethylborepin (**7**) has been found to be more stable [6]. It appears that **7** is somewhat less regioselective than **1** in a number of cases, but it gives much better *endo*
selectivity in the formation of **8** with cyclopentadiene. Dimethylvinylborane and

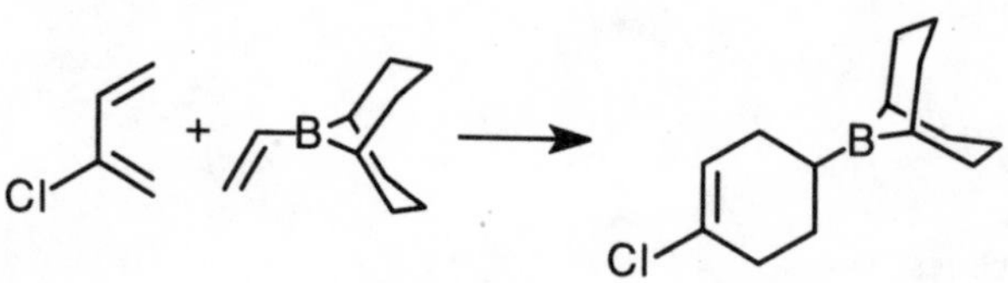

trivinylborane were also tested successfully but showed no special advantages. A procedure for generating the vinyldialkylboranes from vinyltributyltin and halodialkylboranes in situ has been developed, and avoids some of the hazards of handling and transferring spontaneously flammable boranes [7].

7   **8** (endo/exo 87:13)

Singleton has reported quantum mechanical calculations at the 6-31G* level for the transition state for Diels-Alder reaction of vinylborane with butadiene (**9**) and at the 6-21G level for the analogous transition state for vinyldimethylborane (**10**). Unexpectedly, these indicate that the boron atom bonds significantly to the terminal carbon atom of the diene in the *endo* transition state, and that the distances from the terminal carbon of butadiene to the boron atom are less than those to the α-carbon of the vinylboranes [8]. In the *exo* transition states, the B—C distances are longer. The relative transition state energies model the experimental results well. With ethynylborane and butadiene (**11**) at the 3-21G level, the corresponding C—B distance is even shorter [9].

9   **10**   **11**
(bond distances in Å)

8.2.2 Dichloroalkenylboranes

Vinyldichloroborane and (*E*)-1-hexenyldichloroborane have been found to be an active dienophiles toward cyclopentadiene, isoprene, and 2,3-dimethylbutadiene [10]. However, they are neither regioselective nor diastereoselective except for the diastereoselectivity enforced by the olefinic geometry of the hexenylborane. By treatment of the products with benzyl azide, the BCl_2 group was replaced with $NHCH_2Ph$. (*E*)-(5-Chloro-1-pentenyl)dichloroborane (**12**) undergoes Diels-Alder reactions that furnish products suitable for conversion to bicyclic amines **13** via the benzyl azide reaction [11].

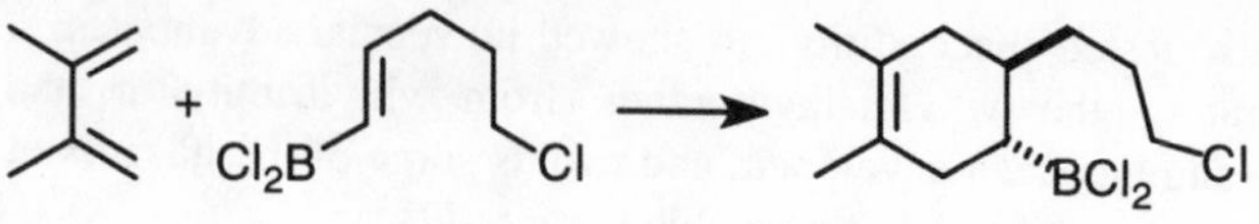

12

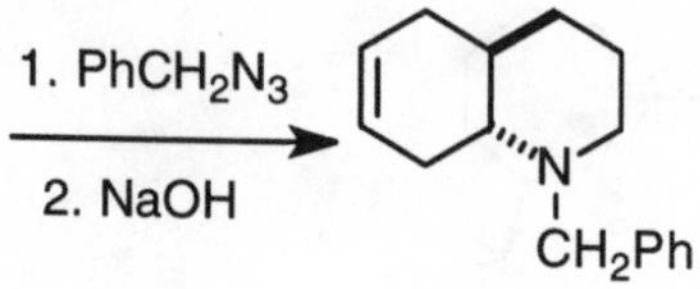

13

8.2.3 Electronegatively Substituted Alkenylboronic Esters

The synthesis of alkenylboranes bearing a second electron-withdrawing group has been accomplished by Vaultier's group [12, 13]. Hydroboration of methyl propiolate, known to direct boron to the β-carbon if substituents are bulky [14], was carried out with diisopinocampheylborane and the product was converted to the boronic ester **14** with acetaldehyde (see Section 6.2.1.2) followed by pinacol. Hydroboration of (phenylthio)acetylene followed by selective oxidation with *m*-chloroperbenzoic acid yielded the sulfoxide (**15**). The sulfone (**16**) was obtained by further oxidation, though an easier route utilized addition of toluenesulfonyl iodide to pinacol vinylboronate followed by dehydroiodination with diazabicycloundecane (DBU). The toluenesulfonyl iodide route has also been used to prepare α- or β-alkyl substituted derivatives of **16** [15].

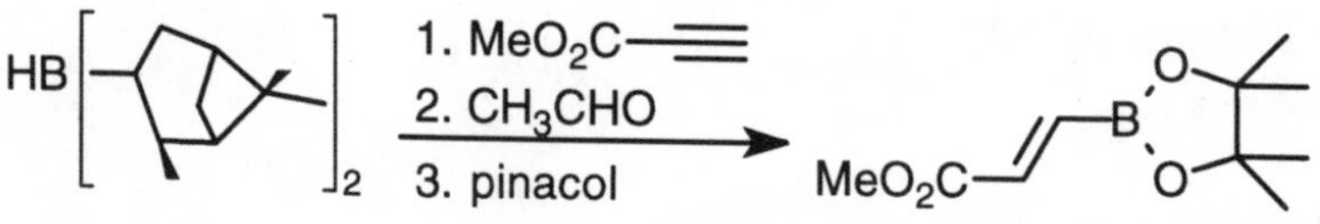

14

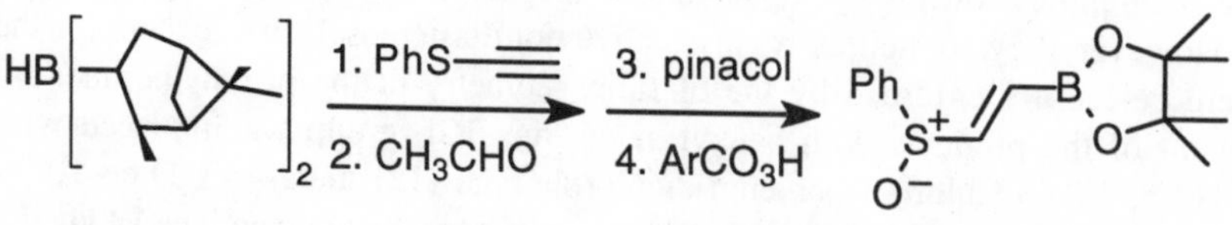

15

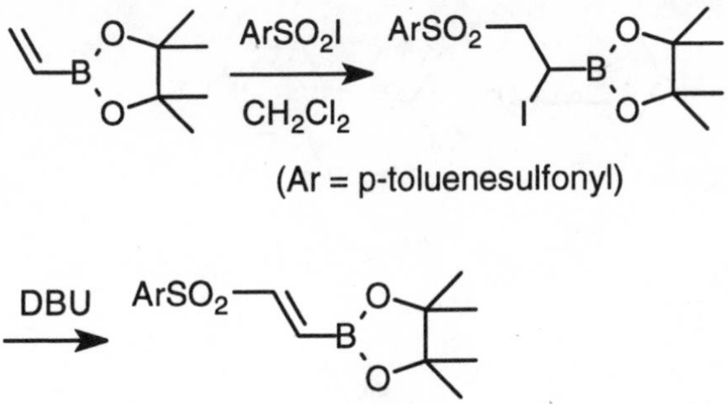

(Ar = p-toluenesulfonyl)

16 [100% (*E*)]

Boronic esters **14**, **15**, and **16** are good dienophiles [12, 16]. Examples include the reaction of **16** with cyclopentadiene to form **17** and **18**, and of **14** with butadiene to form **19**. The boronic ester can serve as a masked hydroxyl group, among other possibilities.

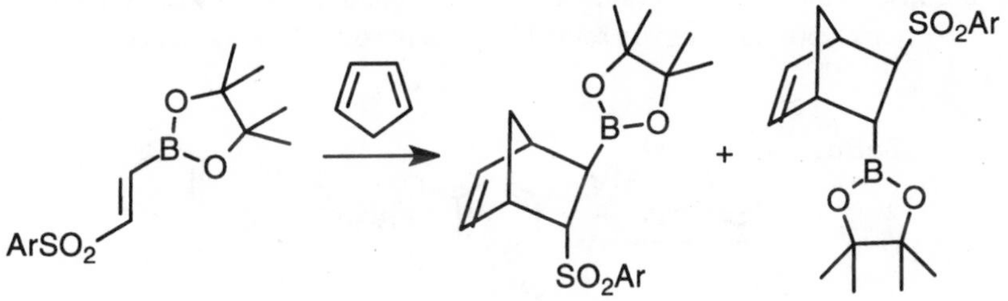

16   **17**   (86:14)   **18**

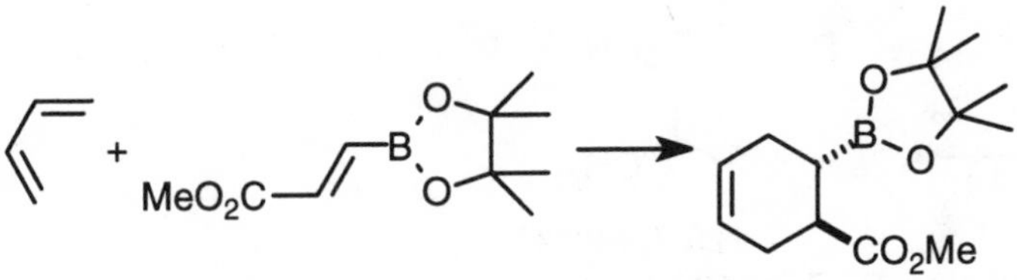

14   **19**

8.2.4 (Silylalkenyl)boranes and Related Compounds

9-(Trimethylsilylethynyl)-9-BBN (**20**) was found to be a useful dienophile. In the reaction with isoprene the majority product (76:24) was unexpectedly the 5-methyl-1-boryl-2-silyl-1,4-cyclohexadiene **21** [9]. (*E*)-9-(Trimethylsilylvinyl)-9-BBN (**22**) is also a useful dienophile, and with isoprene gives the expected product **23** in an 85:15 ratio to its regioisomer [17].

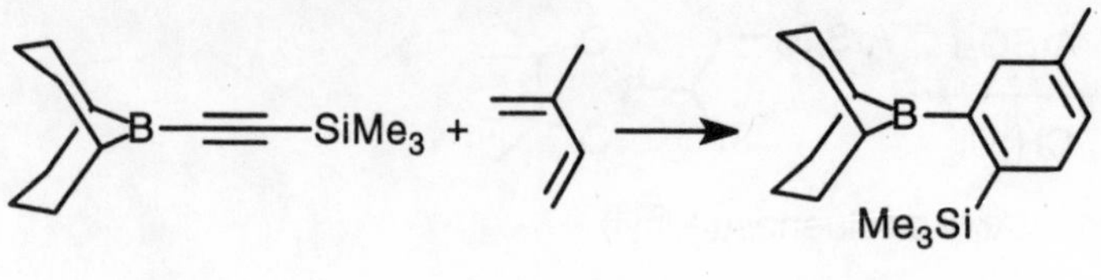

20   **21**

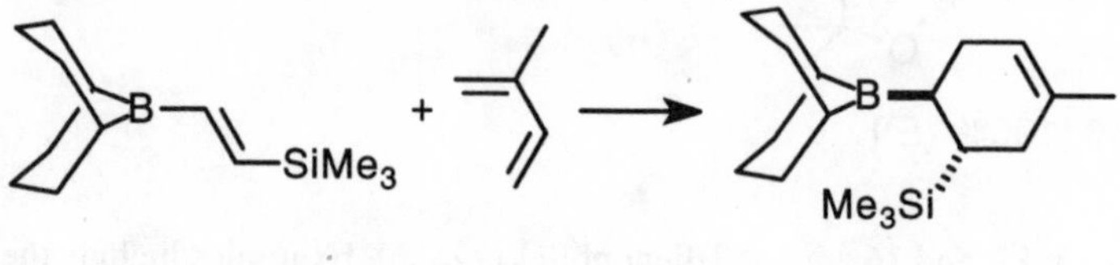

22   **23**

1,2-Diborylethenes can be prepared from 1,2-bis(tributylstannyl)ethene (**24**) and have proved to be good dienophiles [18]. The dialkylboryl derivatives have limited utility because of a tendency toward oxidative elimination during work up, but the diboronic ester **25** provides a good route to cyclic *trans*-1,2-diols such as **26**.

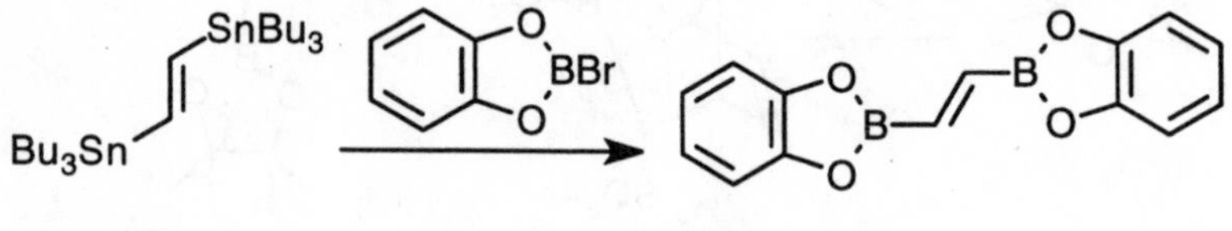

24   **25**

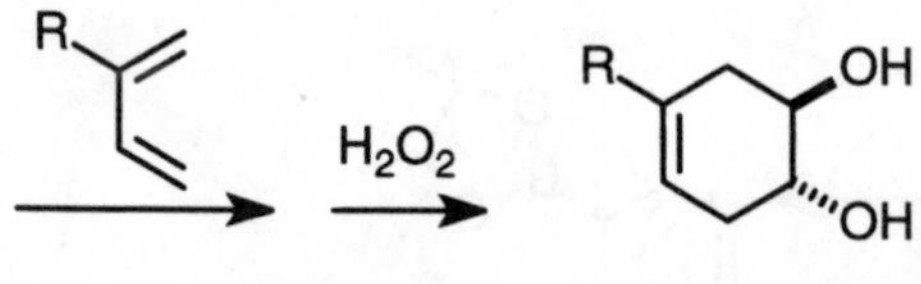

26 (69-88%)

(R = Me, t-Bu, Ph, Me$_2$C=CCH$_2$CH$_2$)

An analogous dienophile **27** having a boryl and a silyl substituent was also reported, and led to products such as *trans* silyl alcohol **28** [18].

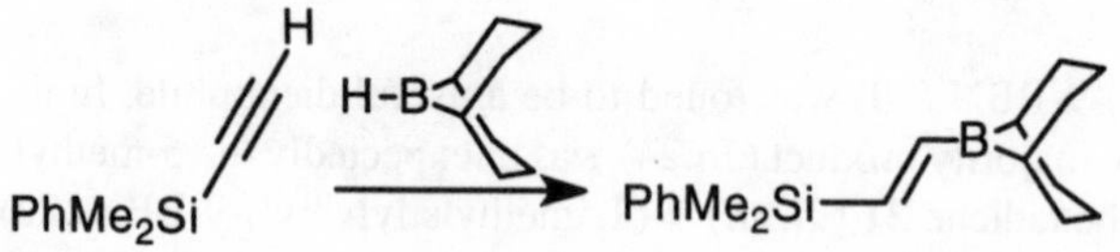

27

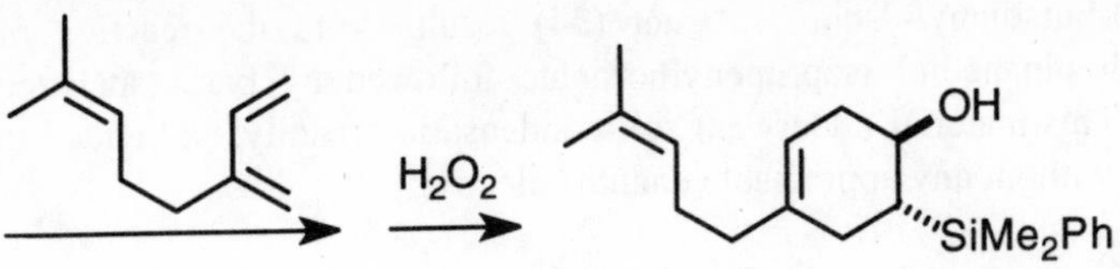

28 (77%; 85% this regioisomer)

Regioselectivity has been found to be strongly affected by the type of substituents on boron [19]. For example, isopropenyldibromoborane (**29**) tends to yield the unexpected 2,4,4-trisubstituted cyclohexenes **30**, while isopropenyl-9-BBN (**31**) strongly favors the 1,4,4-pattern **32**.

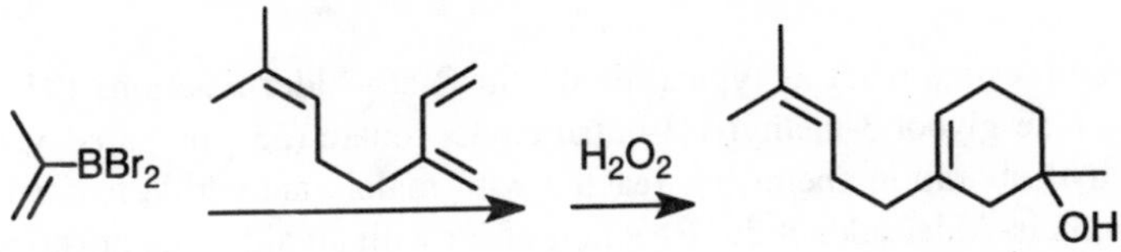

29

30 (77%; 100% this isomer)

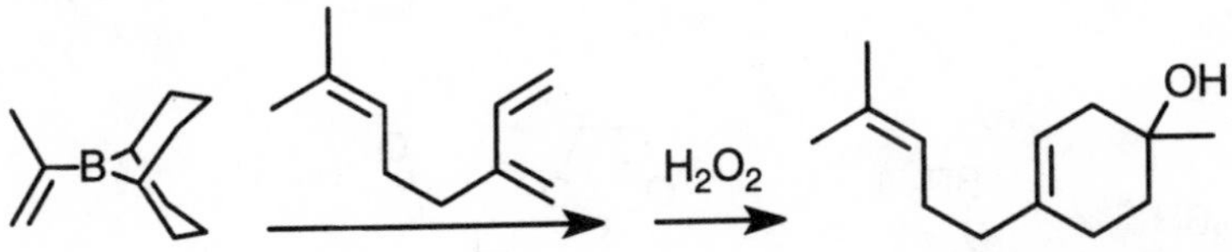

31

32 (75%; 90% this isomer)

8.3 Butadienylboranes

A general synthesis of 1,3-butadienyl-1-boronic esters (**33**) is provided by the Negishi coupling of 2-bromovinylboronic esters with alkenylzinc halides [20].

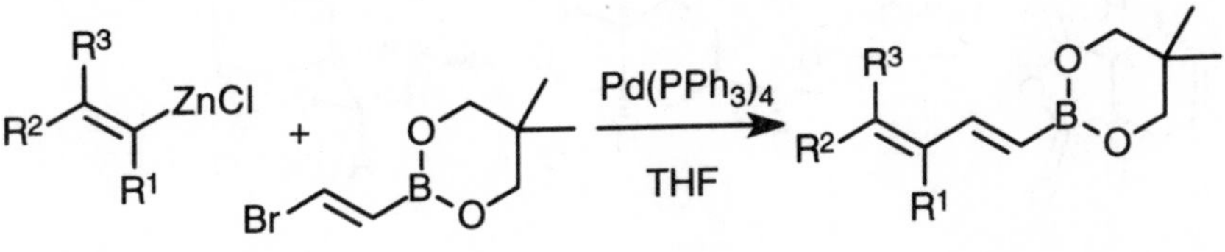

33

A pinanediol 1,3-butadienyl-2-boronic ester (**34**) results from the reaction of BrCH₂SO₂Br with pinanediol isopropenylboronate followed by base catalyzed elimination [15]. This material underwent self-condensation readily, and added to phenylmaleimide without any appreciable enantioselectivity

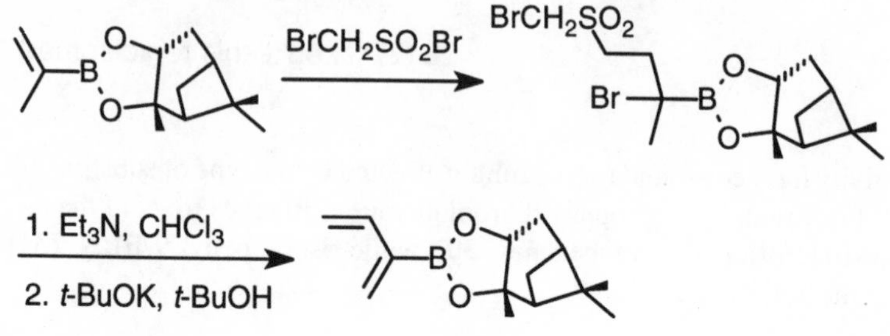

34

Butadienylboronic esters react as typical dienes in Diels-Alder reactions [21]. For example, ethylene glycol 3-methyl-1,3-butadienylboronate (**35**), prepared via well established hydroboration chemistry, reacted with maleic anhydride to form exclusively *endo* Diels-Alder adduct **36**. Reaction of **36** with an aldehyde at room temperature led to **37**, which underwent intramolecular self-acylation to form **38**. *N*-Phenylmaleimide underwent similar reactions up to the point of forming the phenylimido analogue of **37**. Diels-Alder reactions of pinacol 3-methyl-1,3-butadienylboronate with methyl acrylate or acrylonitrile gave diastereomer mixtures.

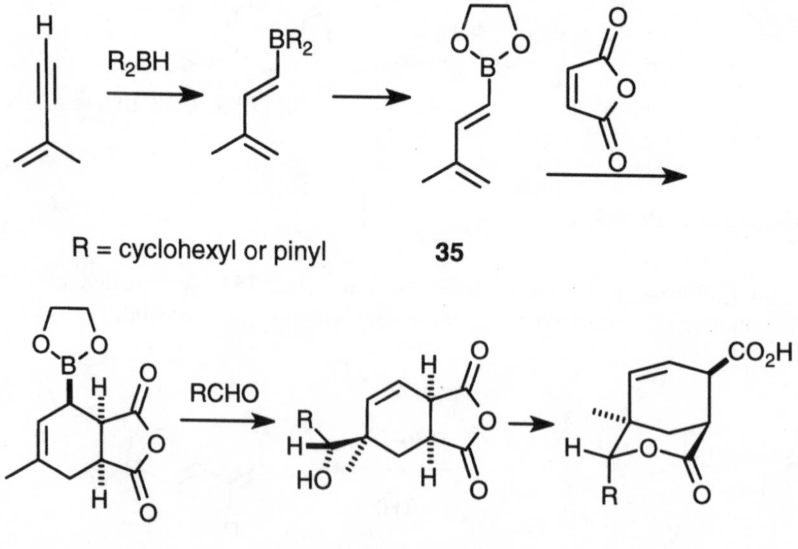

R = cyclohexyl or pinyl   **35**

36   R = Et, i-Pr, Ph   **37** (racemic)   **38**

A 2-butadienylboronic ester (**40**) is easily prepared via hydroboration of 1,4-dichloro-2-butyne to the boronic ester **39** followed by dechlorination with zinc, and has proved to be a highly reactive diene in Diels-Alder reactions [22]. An example is the reaction with dimethyl fumarate to produce adduct **41** (70%).

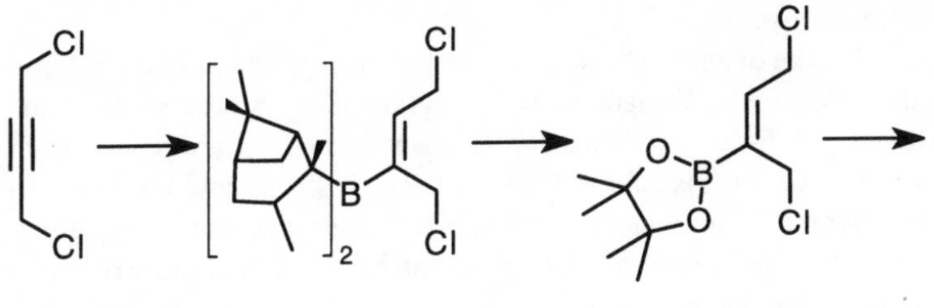

39

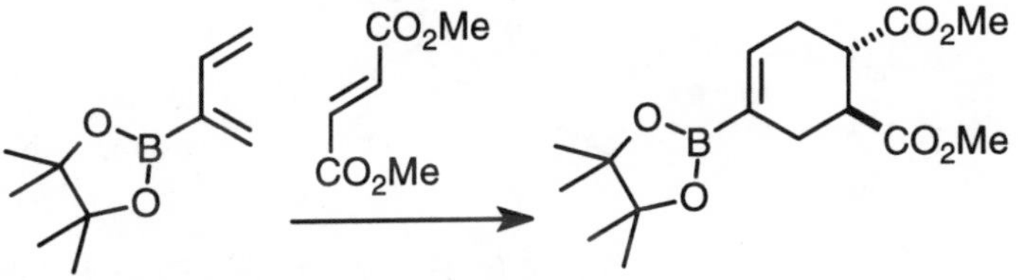

40   **41**

The boronic ester **42**, which is easily made from *o*-iodobenzaldehyde and (iodozincmethyl)boronic ester, can serve as a source of quinodimethane for Diels-Alder reactions [23]. The reactions are promoted either by light or by heat. The hydroxyl and carbonyl groups in typical products such as **44** are *cis*, and it was suggested that the intermediate quinodimethane **43** is therefore the (*E*)-isomer, which should undergo typical Diels-Alder reaction with *endo* geometry.

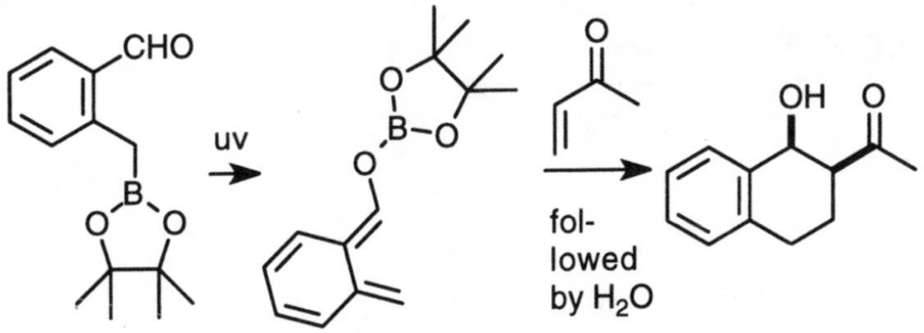

42   **43**   **44**

8.4 Catalyzed Diels-Alder Reactions

8.4.1 Catalysts Derived from Tartaric Acid

The use of chiral borane catalysts for Diels-Alder reactions has provided exciting results of major significance. The details are complex, and there are numerous possibilities for future discoveries.

A highly effective type of chiral Diels-Alder catalyst easily prepared from tartaric acid has been described by H. Yamamoto and coworkers [24]. A monoester of tartaric acid, preferably a 2,6-dialkoxybenzoate (**45**), was treated with borane-THF to yield the active catalyst, which is believed to have the structure **46** and is referred to as "chiral acyloxyborane," abbreviated "CAB" [25]. In this work, the alkoxy group was methoxy (**46a**), but the isopropoxy derivative **46b** has been used more recently, and a further modification having an aryl group in place of hydrogen on boron has proved useful in both kinds of reactions (see below).

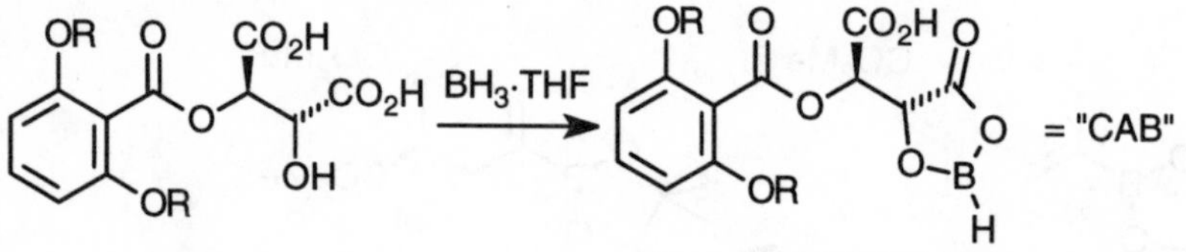

45   **46 (a, R = Me; b, R = CH(CH₃)₂)**

The CAB-catalyzed reaction of acrylic acid with cyclopentadiene has yielded 93% of a mixture in which **47** predominates, *endo/exo* ratio 96:4, ee 78% [24]. The same reagent catalyzes reactions of aldehydes with dienes, in several cases with excellent results [25]. The illustrated reaction yielded 85% **48**, *exo/endo* ratio 89:11, ee 96%. Other ee's with α-methylacrolein and isoprene, 2,3-dimethylbutadiene, or cyclohexadiene were 82-97% as determined by formation of chiral acetals, but absolute configurations were not determined. Acrolein yielded 80-84% ee's and (*E*)-α-methylcrotonaldehyde with cyclopentadiene gave 90% ee, but (*E*)-crotonaldehyde and cyclopentadiene furnished only 2% ee in the major *endo* isomer [25].

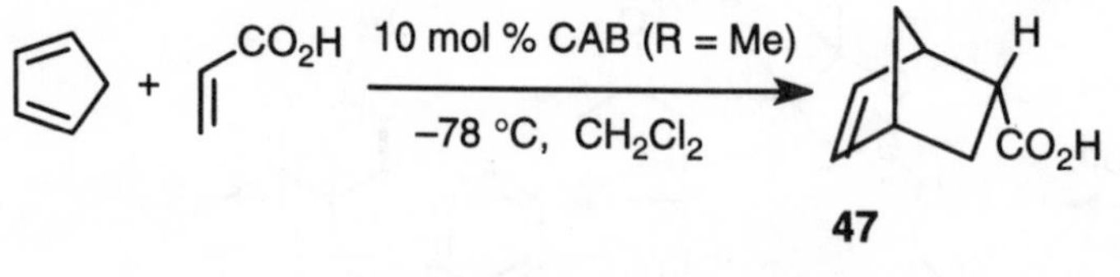

47

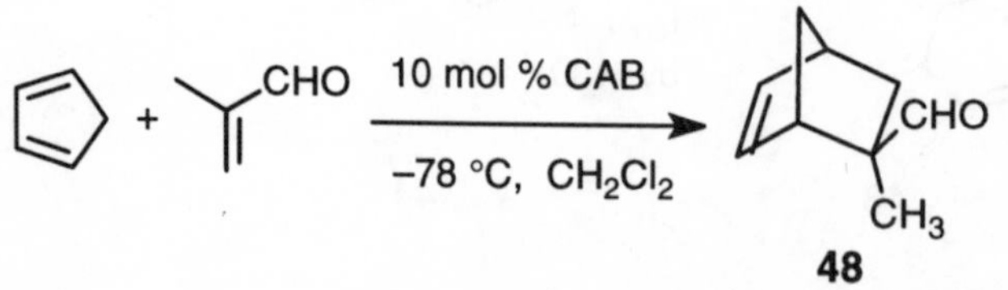

48

Further improvements in the stereoselectivity of these reactions have been made by using a modified "CAB" catalyst having an aryl or alkyl group in place of hydrogen on boron [26]. The best found had an *o*-phenoxyphenyl or *o*-naphthyloxyphenyl group. With **49a** as catalyst, the ee of **48** was increased to 93%, and with **49b** the reaction of crotonaldehyde became useful, yielding **50** in 77% ee.

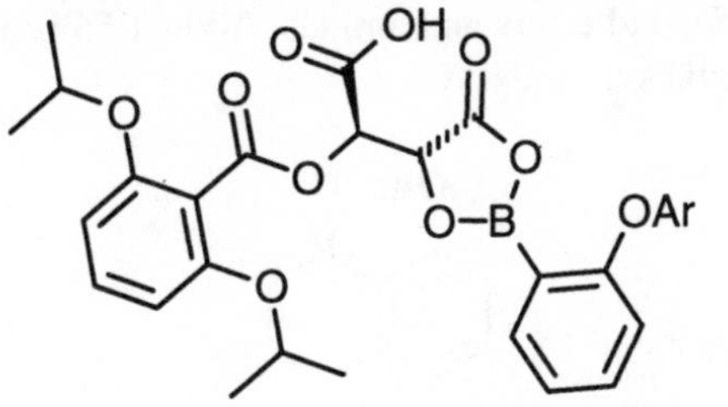

49 (a, Ar = Ph; **b**, Ar = naphthyl)

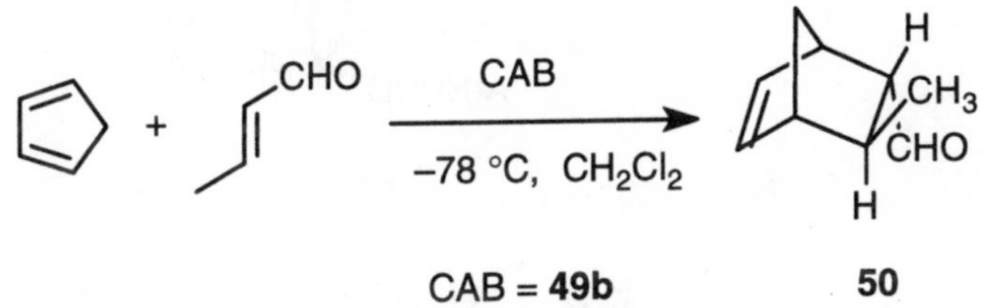

CAB = **49b**   **50**

Methacrolein with cyclopentadiene yields predominantly **48** with the (2*R*)-configuration regardless of the size of the substituent on boron. From NOE experiments it was concluded that methacrolein always prefers the *anti* conformation in its complexes with **46**, and that the methacrolein is positioned preferentially across the face of the 2,6-diisopropoxyphenyl group, as illustrated by **51** [26]. The major Diels-Alder product **48** is what would result if **51** is attacked from the top by cyclopentadiene.

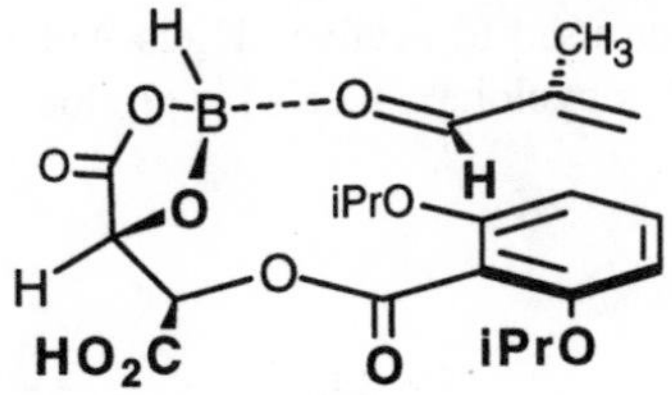

51

The situation is more complicated with acrolein and crotonaldehyde, which show NOE evidence of favoring a conformation similar to **51** with less hindered catalysts such as **46**. However, stereocontrol by **46** is weak, and with the more hindered **49** the enantioselection reversed to favor the (2*S*)-product as illustrated by **50**, as if the reacting species has the C=C group flipped over to the *syn* conformation [26]. Calculations by Birney and Houk have indicated that acrolein adopts the *syn* conformation in the transition state for the uncatalyzed reaction with a diene [27].

The 5-hydroxynaphthoquinone 52 with tartaric acid amide 53 in the presence of methyl borate yields chiral borate ester intermediate 54, which reacts with silyl enol ether 55 to yield 93% of the Diels-Alder adduct 56 , 88% ee [28]. These results were achieved after testing a series of 14 tartaric acid esters and amides. Adduct 56 is a useful intermediate for tetracycline antibiotic synthesis.

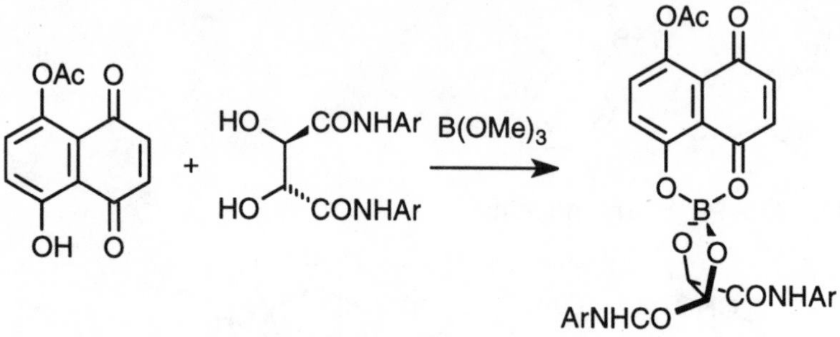

52   53 (Ar = m-tolyl)   54

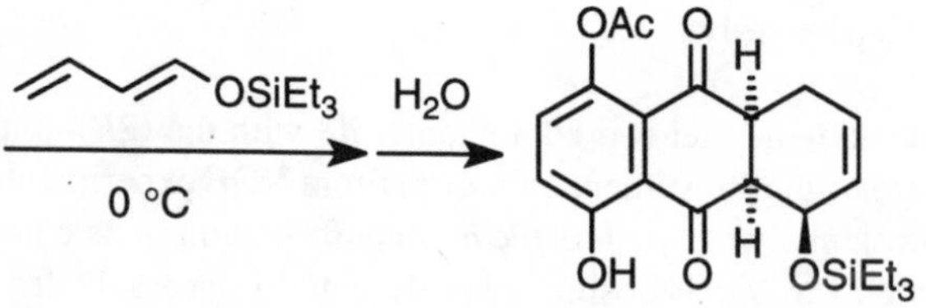

8.4.2 Catalysts Containing Naphthyl Groups

Catalysts based on binaphthol borates 57 have been used to catalyze Diels-Alder reactions of imines. For R^1 = H or Me and R^2 = aryl or cyclohexyl, ee's for products 58 were in the 72-87% range [29].

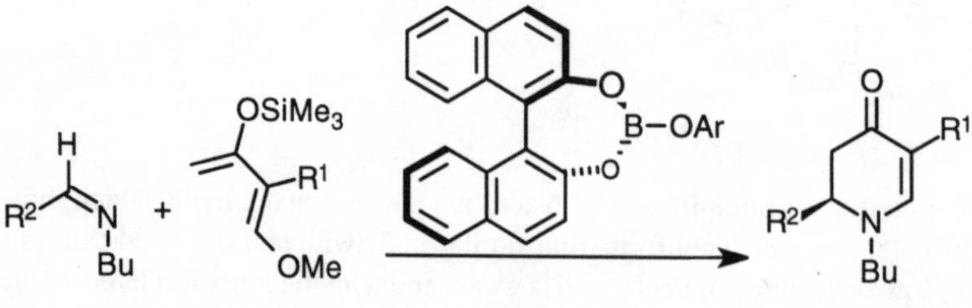

57   58

Although Yamamoto and coworkers appear not to have speculated publicly on the mechanism of action of their catalysts, it appears likely that the best ones incorporate the major binding features of the more explicitly designed catalysts discussed below. Hawkins and Loren designed a catalyst **59** having an acidic dichloroboryl group held to one side of a naphthalene ring by steric repulsions [30]. Methyl crotonate and **59** formed a stable crystalline complex **60**, which was found by X-ray crystallography to have approximately the conformation illustrated. NMR studies indicated that **60** maintains the illustrated conformation in solution. Note that only one face of the complexed dienophile is accessible to attack by diene.

Reaction of methyl crotonate with excess cyclopentadiene in the presence of 10 mol % of **59** produced **62** in 92% yield, 90-92% ee, presumably via transition state **61** [30, 31]. Only the endo isomer was reported. With methyl acrylate and cyclopentadiene, the ee was as high as 99.5%, yield 95%. Dimethyl fumarate and cyclopentadiene gave 90% ee, and methyl acrylate with cyclohexadiene 86% ee. A series of analogues of **59** having phenyl or 3,5-dichloro-, -dibromo-, or -dimethyl-phenyl in place of naphthyl also yielded crystalline complexes with methyl crotonate [31]. With decreasing polarizability of the aromatic ring, the carbonyl carbon shifts its position away from the face of the ring, toward the right as **60** is illustrated, and the dienophile becomes more exposed to attack on the opposite face. For example, with phenyl in place of naphthyl, the ee of **62** from methyl crotonate and cyclopentadiene falls to 70%. Evidence was presented that the attraction of the dienophile to the ring was a dipole and induced dipole interaction, not charge transfer.

59

60   **61**   **62**

(R = H, Me, CO$_2$Me)

Although it is possible that **60** might undergo internal rotation to some other structure before reacting with cyclopentadiene, for example a *syn* O=C-C=C orientation flipped over to the left in front of the ring (see **64** below), the double flip leaving the same side of the C=C bond exposed, there seems no reason to expect this. It was noted that **60** is in equilibrium with the starting materials and has a lower energy, the energy of the transition state is lowered even further with respect to the starting materials, and it seems unlikely that **60** would have to reorganize significantly in order to reach a transition state that is relatively close in energy [31]. Though highly significant for mechanistic understanding, the Hawkins catalyst is inconvenient for synthetic purposes, as it has to be resolved via a complex with menthone in a rather involved procedure [30].

8.4.3 Catalysts Derived from Amino Acids

An easily prepared series of asymmetric Diels-Alder catalysts **63** has been obtained from amino acid arenesulfonamides and borane THF [32]. Excellent regioselection and very good *endo/exo* ratios were achieved, but ee's were mediocre (51-74%).

63

Among the most useful catalysts for enantioselective Diels-Alder condensations are **64** reported by Corey and Loh [33]. In the example illustrated, α-bromoacrolein and cyclopentadiene with 5 mol % of catalyst **64a** at –78 °C for 1 h yielded 95% **66**, *exo/endo* ratio 96:4, ee 99%; catalyst **64c** yielded 98% **66**, *exo/endo* ratio 97:3, ee 96% [33]. These results are clearly superior to those obtained with a Diels-Alder catalyst of C_2 symmetry, 2-isobutyl-1,3-ditriflyl-4,5-diphenyl-1,3,2-diazaalumino-lidine [34], as well as any of the other catalysts described above. A titanium complex of *cis-N*-sulfonyl-2-amino-1-indanol has given **66** in a 67:1 *exo/endo* ratio but only 93% ee [35].

64   **a**, R^1 = H, R^2 = Bu   **65**
          b, R^1 = Me, R^2 = Bu
          c, R^1 = R^2 = H
          (Ar = p-C_6H_4Me)

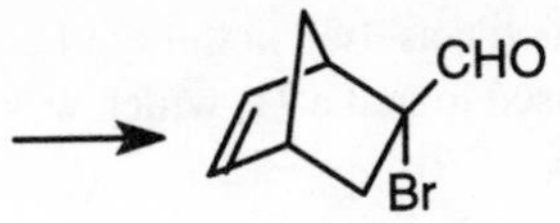

66

The mechanism appears to involve complexing of the dienophile to the indole ring of **64** in such a manner that the diene can approach easily only from one side and with only one orientation in the transition state (**65**) [36]. Although it was postulated that the catalytic system favors the *syn* arrangement of the double bond and the carbonyl group, the X-ray structure of the 2-methylacrolein-BF$_3$ complex shows an *anti* relationship in the crystal [37], consistent with the structures observed by Hawkins and coworkers [31]. The Hawkins transition state model here would predict the same major product, as it differs by a double flip of the C=C unit that leaves its same face exposed. However, the tether in **65** is longer than in **60/61**, perhaps just enough longer to make the system behave more like Yamamoto's **46/51** [26], where there is strong evidence that conformations centering the dienophile over the electron donating aromatic ring are favored. Thus, it is quite possible that each of the published mechanistic interpretations is correct for the particular system studied.

Other reactions tested successfully included preparations of **67** and **68** [33]. One important use for **68** is as a source of the prostaglandin intermediate **69**. This was accomplished via conversion to the oxime and the cyanohydrin followed by reversal of cyanohydrin formation to provide the ketone, after the procedure was first tested on **66**. The product **66** is also a versatile starting material for the preparation of a wide variety of derivatives [33, 38]. The 7-oxa analogue, prepared from furan and 2-bromoacrolein with the aid of catalyst **64b** in 92% ee, 99:1 *exo*, is also a broadly useful starting material [39].

64a catalysis:

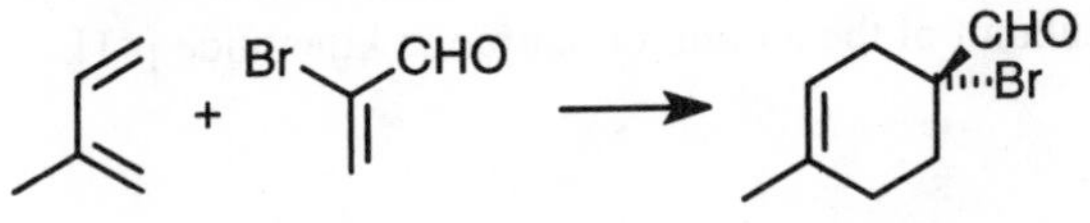

67

***ent*-64a catalysis:**

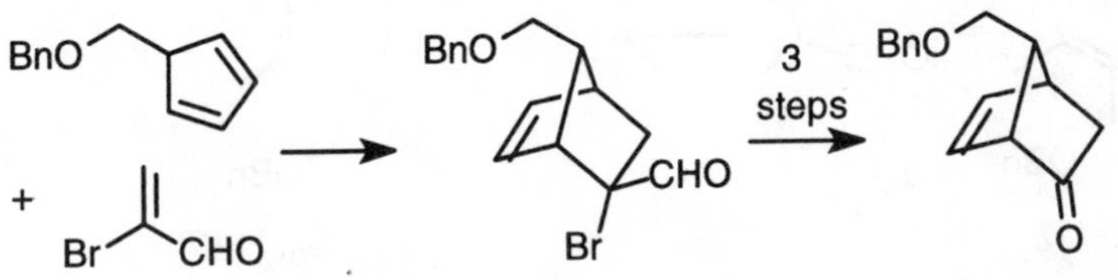

68   **69**

Catalyst **64b** has been used for the synthesis of **70**, which was used in the synthesis of antiulcer agent cassiol, and catalyst **64a** has been used to make **71**, which was used in the synthesis of gibberelic acid 40.

64b catalysis:

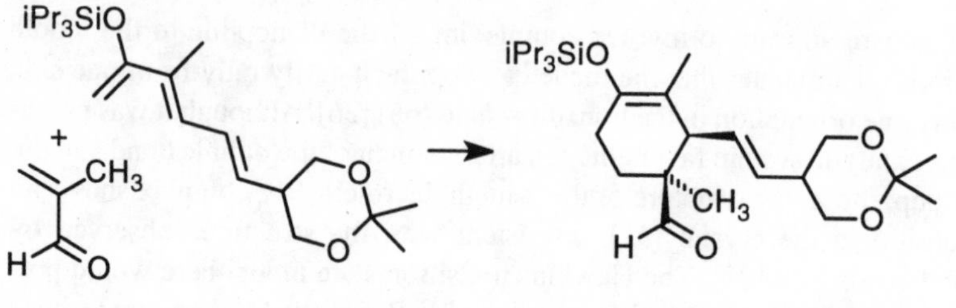

70 (97% ee, no detectable isomers)

64a catalysis:

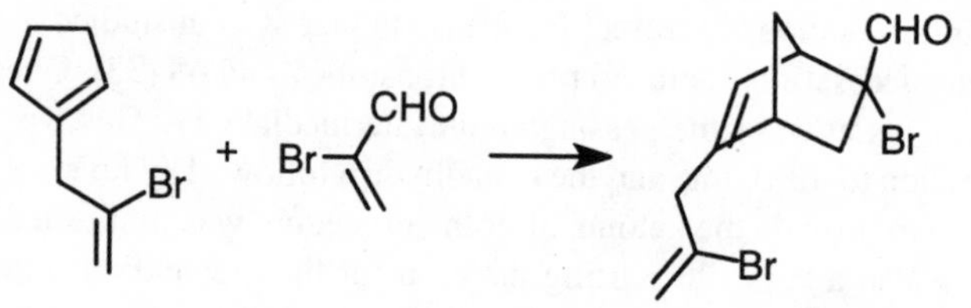

71 (99% ee, exo/endo 99:1)

Marshall and Xie used catalyst **64c** with 2-bromoacrolein to convert triene **72** to exclusively *endo* adduct **73** in 72% ee, which was used as a key intermediate in the synthesis of the spirotetronate subunit of the antitumor antibiotic kijanolide [41].

64c catalysis:

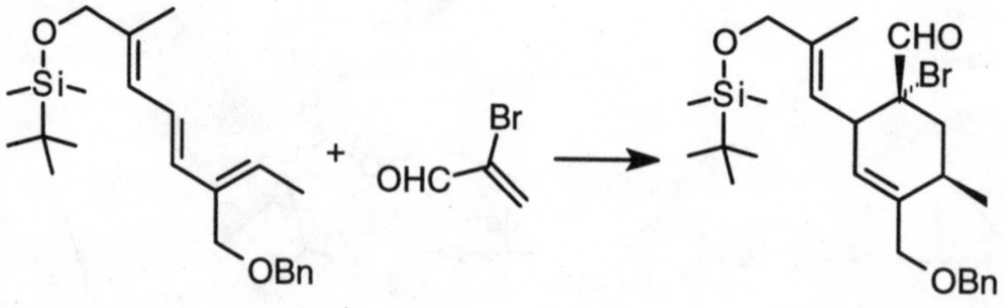

72   **73** ee 72%

8.5 References

1. Matteson DS, Peacock K (1960) J. Am. Chem. Soc. 82:5759
2. Matteson DS, Waldbillig JO (1963) J. Org. Chem. 28:366
3. Matteson DS, Talbot M (1967) J. Am. Chem. Soc. 89:1123
4. Singleton DA, Martinez JP (1990) J. Am. Chem. Soc. 112:7423
5. Singleton DA, Martinez JP, Watson JV (1992) Tetrahedron Lett. 33:1017
6. Singleton DA, Martinez JP, Watson JV, Ndip GM (1992) Tetrahedron 48:5831
7. Singleton DA, Martinez JP, Ndip GM (1992) J. Org. Chem. 57:5768
8. Singleton DA (1992) J. Am. Chem. Soc. 114:6563
9. Singleton DL, Leung SW (1992) J. Org. Chem. 57:4796
10. Noiret N, Youssofi A, Carboni B, Vaultier M (1992) J. Chem. Soc., Chem. Commun. 1105
11. Jego JM, Carboni B, Youssofi A, Vaultier M (1993) Synlett 595
12. Martinez-Fresneda P, Vaultier M (1989) Tetrahedron Lett. 30:2929
13. Rasset-Deloge C, Martinez-Fresneda P, Vaultier M (1992) Bull. Soc. Chim. Fr. 129:285
14. Negishi E, Yoshida T (1973) J. Am. Chem. Soc. 95:6837
15. Guennouni N, Rasset-Deloge C, Carboni B, Vaultier M (1992) Synlett 581
16. Rasset C, Vaultier M (1994) Tetrahedron 50:3397
17. Singleton DA, Martinez JP (1991)Tetrahedron Lett. 32, 7365
18. Singleton DA, Redman AM (1994) Tetrahedron Lett. 35:509
19. Singleton DA, Kim K, Martinez JP (1993) Tetrahedron Lett. 34:3071
20. Mazal C, Vaultier M (1994) Tetrahedron Lett. 35:3089
21. Vaultier M, Truchet F, Carboni B, Hoffmann RW, Denne I (1987) Tetrahedron Lett. 28:4169
22. Kamabuchi A, Miyaura N, Suzuki A (1993) Tetrahedron Lett. 34:4827
23. Kanai G, Miyaura N, Suzuki A (1993) Chem. Lett. (Jpn.) 845
24. Furuta K, Miwa Y, Iwanaga K, Yamamoto H (1988) J. Am. Chem. Soc. 110:6254
25. Furuta K, Shimizu S, Miwa Y, Yamamoto H (1989) J. Org. Chem. 54:1481
26 Ishihara K, Gao Q, Yamamoto H (1993) J. Am. Chem. Soc. 115:10412
27. Birney DM, Houk KN (1990) J. Am. Chem. Soc. 112:4127
28. Maruoka K, Sakurai M, Fujiwara J, Yamamoto H (1986) Tetrahedron Lett. 27:4895
29. Hattori K, Yamamoto H (1992) J. Org. Chem. 57:3264
30. Hawkins JM, Loren S (1991) J. Am. Chem. Soc. 113:7794
31. Hawkins JM, Loren S, Nambu M (1994) J. Am. Chem. Soc. 116:1657
32. Takasu M, Yamamoto H (1990) Synlett 194
33. Corey EJ, Loh TP (1991) J. Am. Chem. Soc. 113:8966
34. Corey EJ, Imwinkelried R, Pikul S, Xiang YB (1989) J. Am. Chem. Soc. 111:5493
35. Corey EJ, Roper TD, Ishihara K, Sarakinos G (1993) Tetrahedron Lett. 34:8399
36. Corey EJ, Loh TP, Roper TD, Azimioara MD, Noe MC (1992) J. Am. Chem. Soc. 114:8290
37. Corey EJ, Loh TP, Sarshar S, Azimioara M (1992) Tetrahedron Lett. 33:6945
38. Corey EJ, Cywin CL (1992) J. Org. Chem. 57:7372
39. Corey EJ, Loh TP (1993) Tetrahedron Lett. 34:3979
40. Corey EJ, Guzman-Perez A, Loh TP (1994) J. Am. Chem. Soc. 116:3611
41. Marshall JA, Xie S (1992) J. Org. Chem. 57:2987

9 Asymmetric Reductions and Miscellaneous Reactions

9.1 Introduction

The use of boranes as reducing agents is widespread, and only highly enantioselective reactions will be reviewed here, with a few minor exceptions. The two major categories of enantioselective reductions include those that donate hydride by β-elimination of boron from alkylboranes (Section 9.2), and those that donate hydride directly by breaking a B—H bond (Section 9.3). Some of the latter have been made catalytic (Section 9.3.3), and even though the amount of asymmetric catalyst required is often fairly high, these have generated major recent interest. Asymmetric reductions with boranes have been reviewed by Midland [1], and asymmetric boron catalyzed reactions by Deloux and Srebnik [2].

Boranes have been used as chiral directors for several other categories of reactions represented by few examples, which have been collected at the end of this chapter in Section 9.4.

9.2 Reductions with Alkylboranes

9.2.1 Introduction

The reaction of trialkylboranes, $(RCH_2CH_2)_3B$, with aldehydes, R'CHO, to form alkenes, $RCH=CH_2$, and alkoxyboranes, $(RCH_2CH_2)_2B-OCH_2R'$, was first reported by Mikhailov's group [3]. The temperatures required were high (~150 °C), and it was not until more reactive boranes were discovered some years later that this became a useful method for reducing carbonyl compounds. The special facility of pinylborane reductions of aldehydes and the use of the reaction with acetaldehyde to cleave α-pinene selectively from trialkylboranes has been noted in Section 6.2.1.2.

9.2.2 Pinylboranes

Hydroboration of (+)-α-pinene with 9-BBN yields isopinocampheyl-9-BBN (1), a valuable asymmetric reducing agent [1]. Midland and coworkers first discovered that 1 reduces aldehydes highly enantioselectively, so that if the aldehyde or the 1 is deuterated an asymmetrically labeled primary alcohol results [4, 5]. However, Midland's reagent is of much more interest for its highly stereoselective reduction of propargylic ketones. Both **1** and its enantiomer derived from (–)-α-pinene are easily prepared, and are sold as "Alpine-Borane" by the Aldrich Chemical Company.

The mechanism of the reduction evidently involves a cyclic transition state (**2**) [6, 7]. The model that predicts the asymmetric induction consistent with experiment has a boat conformation. The acetylenic substituent is smaller than any other alkyl, alkenyl, or aryl substituent R^1, and R^1 goes to an equatorial position in the transition state. The product from (+)-α-pinene after hydrolysis is always the (R)-propargylic alcohol **3** (assuming the acetylenic substituent has a higher priority than R^1 in the Cahn-Ingold-Prelog system) [1]. With 0.5 M **1** in THF, ee's in the range 73-100% were measured, after correction for the ee of the α-pinene used [8, 9]. Compounds in which R^1 was methyl yielded ee's in the lower part of the range.

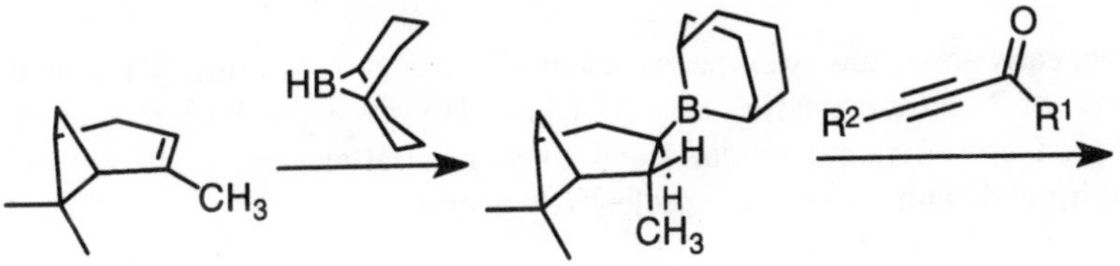

1

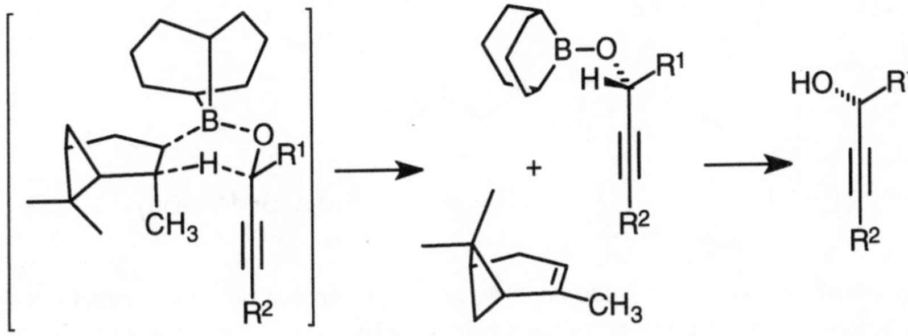

2   **3**

An early successful application of this reagent was the synthesis of the Japanese beetle pheromone **6** from the alkynyl keto ester **4** via propargylic alcohol ester **5** [10]. The ee was originally reported as 85-97%. Ester hydrolysis then provided an acid that was recrystallized as its cyclohexylamine salt to high enantiomeric purity, and the synthesis was completed by lactonization and catalytic hydrogenation to the (Z)-alkene. High purity is essential, as even a small proportion of the opposite enantiomer deactivates the pheromone.

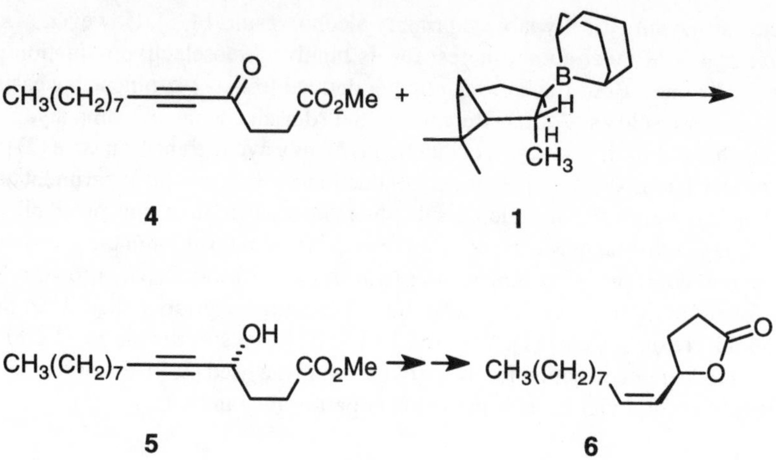

$$CH_3(CH_2)_7 \text{—}\equiv\text{—} \overset{O}{\text{C}}\text{—} CO_2Me \quad + \quad \mathbf{1} \quad \longrightarrow$$

4   **1**

$$CH_3(CH_2)_7 \text{—}\equiv\text{—} \overset{OH}{\underset{\vdots}{C}}\text{—} CO_2Me \quad \longrightarrow\longrightarrow \quad CH_3(CH_2)_7 \cdots \text{—} \mathbf{6}$$

5   **6**

Another early application was the reduction of steroidal side chains by **1**, as in the preparation of **7**, diastereomeric ratio 125:1 [11]. However, the diastereomer of **7** was not produced efficiently by *ent*-**1** and a symmetrical reagent, LiHB(*sec*-Bu)$_3$, was used instead, with 1:11 diastereoselectivity achieved.

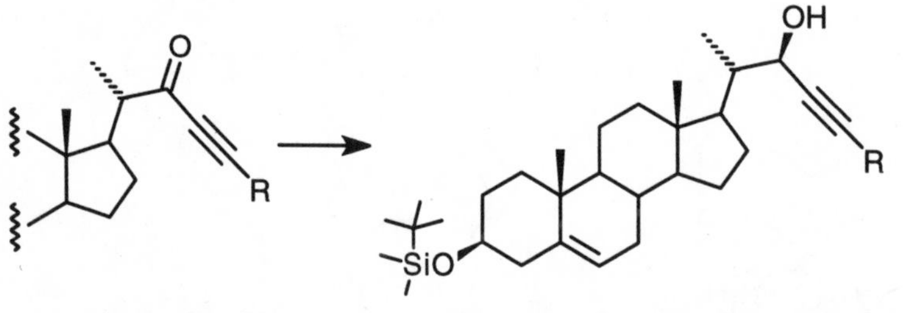

reduction with **1**   **7** (R = Me or CMe$_2$OSiMe$_2$*t*-Bu)

A problem with **1** is that it can undergo reverse hydroboration to free, achiral 9-BBN, which is a much faster reducing agent for ketones than **1** and leads to lowered ee's [12]. Heating makes the problem worse, and the reactions may take several days at room temperature if R^1 is secondary.

Since dissociation of **1** is first-order in **1** and reaction with an acetylenic ketone is second-order over all, the first measure to improve the results was to run the reaction without a solvent [13, 14]. Only acetylenic ketones were reduced under the original conditions, but neat **1** reduced aryl halomethyl ketones and yielded high ee's [14]. Further suppression of dissociation was achieved by running the reactions under high pressure (6000 atm.), which can be achieved in fairly simple apparatus with liquid samples [6, 7]. These conditions permitted reduction of acetophenone to 1-phenylethanol having nearly 100% ee, and also gave good results with acetylenic ketones [7]. However, ee's for dialkyl ketones or alkyl alkenyl ketones were not generally in the useful range.

The cyano group mimics the steric behavior of the acetylenic group, and acyl cyanides are reduced by neat **1** in 84-98% ee [15]. However, the initially produced cyanohydrin borinates decompose slowly back to aldehyde and cyanide. The reaction works efficiently if it is monitored by NMR and the initial product **8** is treated immediately with sodium borohydride and cobaltous chloride in methanol to reduce it to the aminomethylcarbinol **9**.

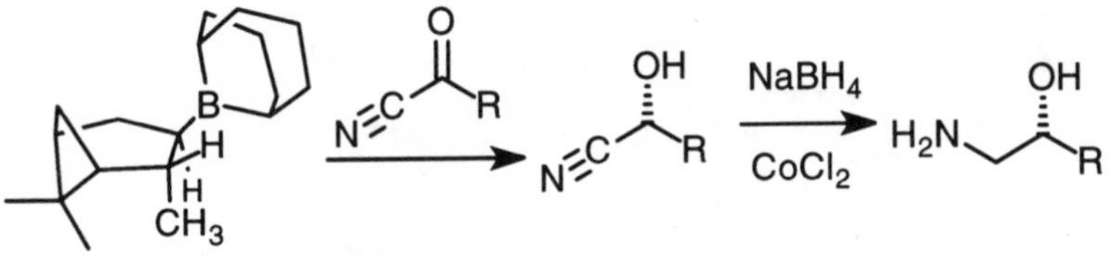

Reduction of acetylenic ketone **10** with isopinocampheyl-9-BBN from (−)-α-pinene (*ent*-**1**) has yielded (*S*)-alcohol **11** (89-94% ee), which served as an intermediate early in the construction of the carbon skeleton of the aglycone of the antibiotic chlorothricolide, and provided an essential element of chirality in the synthesis [16].

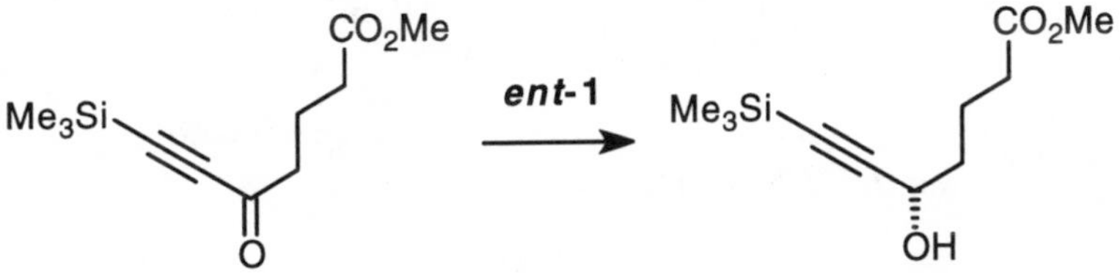

An alternative to the pinyl-9-BBN reagent **1** is the nopol benzyl ether analog **12** [17], which is sold by the Aldrich Chemical Company as NB-Enantrane®. The enantiomer illustrated is a derivative of natural (−)-β-pinene. Because of slightly higher steric requirements, **12** yields slightly higher ee's than **1** with some acetylenic ketones, especially with methyl ketones, but it also tends to react more slowly.

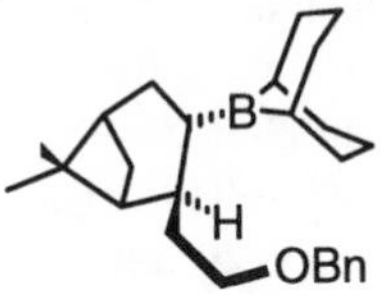

Another useful reagent is diisopinocampheylboron chloride (13), which yields
>90% ee's in reduction of several aryl alkyl ketones to (S)-1-phenylalkanols (14)
[18, 19]. A major virtue of 13 is its high reactivity, which allows reduction of *tert*-
alkyl ketones in high ee's, often >90% [20, 21]. Acylsilanes are reduced to (R)-1-
silyl alcohols (15) in 96-98% ee by 13 [22]. The use of 13 as a reducing agent and
chiral director for other reactions has been reviewed recently [23, 24]. Several other
terpene derivatives have been investigated, but these have generally given low or
mediocre ee's [1].

13   14 (92-98% ee)

13   15 (96-98% ee)

(R^1 = Me, Et; R^2 = Me, *n*-Bu, Me$_2$CH)

Brown and coworkers carried out a critical examination of the most stereoselective reducing agents for ketones available up to 1987 [25]. No one reagent
worked well for more than a few classes of ketones, and not all classes yielded
satisfactory ee's with any reagent. The pinylborane reagents gave high ee's with as
many substrates as any of the rest. The best alternative reagent to 1 or 13 in three
cases was the relatively expensive BINAP (chiral binaphthol) derivative of lithium
aluminum hydride introduced by Noyori.

An interesting recent variant of the pinylborane reductions is intramolecular reduction after hydroboration of allyl ketones (16) by (–)-diisopinocampheylborane
(17) [26]. The intermediate boryl ketones 18 evidently undergo intramolecular reduction via a transition state 19, generating (+)-α-pinene and boroxin 20 in 86-98%
ee and 85-98% ee {R = pentyl, cyclohexyl, Ph, Cl(CH$_2$)$_3$, [(CH$_2$)$_3$O$_2$CH](CH$_2$)$_4$,
MeO$_2$C(CH$_2$)$_4$} with one anomaly, R = NC(CH$_2$)$_{10}$, 75% ee. The products 20 were all
characterized as their 1,4-diol derivatives after oxidation with alkaline sodium
perborate.

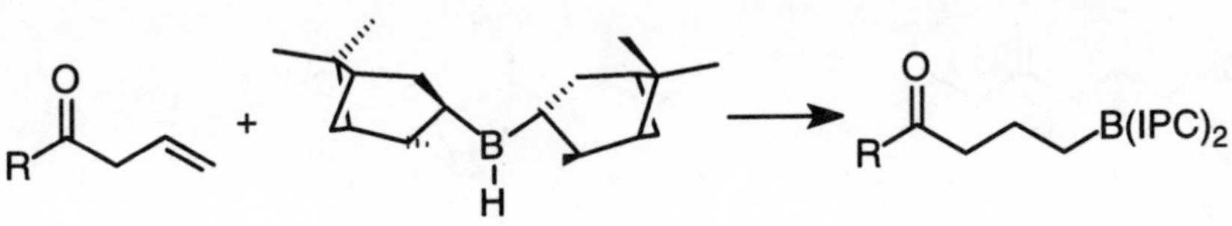

16   **17 = ("IPC)₂BH"**   **18**

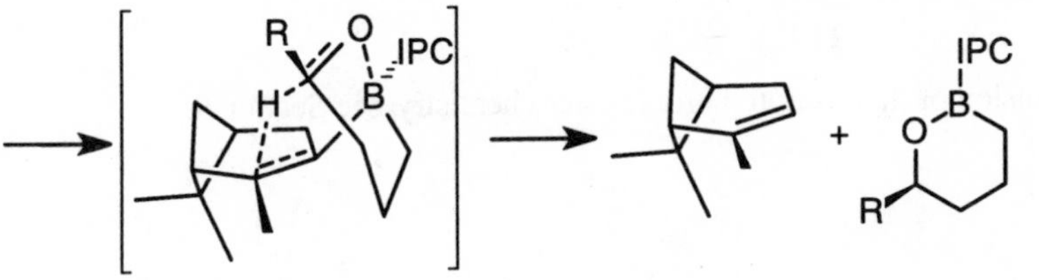

19   **20**

It was noted that once the ketone coordinates with the boron and makes it tetrahedral, the two pinyl groups become diastereotopic, one to the right and one to the left of the ring in the orientation shown. Postulated transition state **19** (or its boat form equivalent), the result of reaching for the pinanyl group on the left, appear to be relatively unhindered and lead to the observed product. If the boroxin ring flexes toward the pinyl group on the right, which would reduce the carbonyl group from the opposite side, the stable conformer **21** cannot react, and it appears likely that transition state **22** leading to the opposite enantiomer has unfavorable steric interactions compared with **19**.

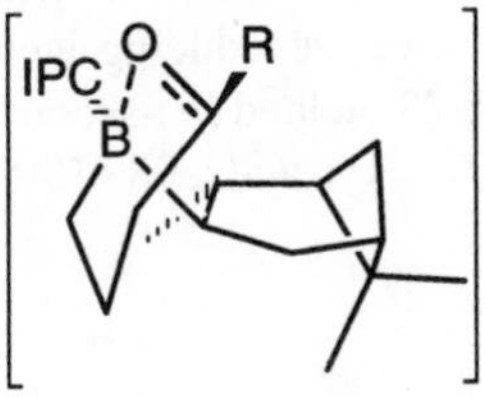

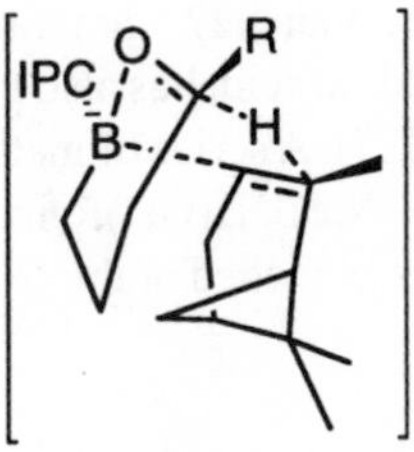

21   **22**

Where R = Cl(CH₂)₃, the intermediate **23** reacted with base under the basic deboronation conditions used to produce the substituted tetrahydrofuran **24**, 98% ee [26].

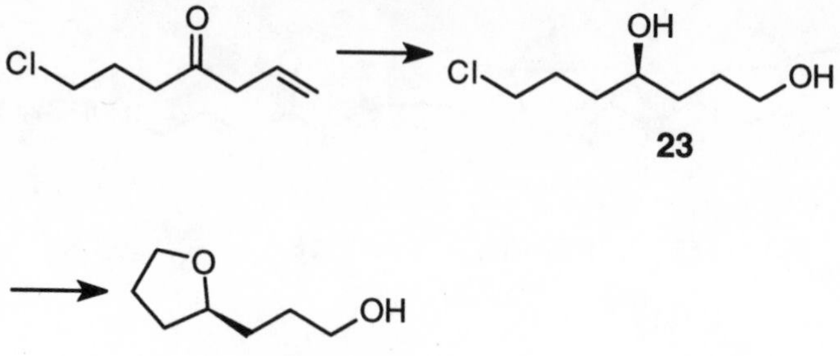

23

24

For extension of this work to boronic ester chemistry, see Section 9.3.4.

9.3 Reductions with Asymmetric B-H Compounds

9.3.1 Introduction

This section will be confined to a review of those asymmetric alkylboranes and borohydrides that have given the best results. No attempt will be made to review the very extensive literature of regioselective and diastereoselective reductions by achiral boranes and borohydrides.

9.3.2 Dimethylborolane

The utility of the (R^*,R^*)-2,5-dimethylborolane group as a chiral director in allylborane reactions has been described in Section 7.3.2, and in aldol reactions in Section 7.4.3.1. (R,R)-Dimethylborolane (**26**) containing ~20 mol % of the corresponding mesylate **27**, generated from the corresponding borohydride **25** with methanesulfonic acid, has also proved to be a highly enantioselective reducing agent for ketones [27]. A series of methyl ketones MeCOR with **26/27** yielded (R)-secondary alcohols MeC*H(OH)R in ~80% ee's where R was N-alkyl and ≥98.6% ee where R was branched at the α- or β-carbon.

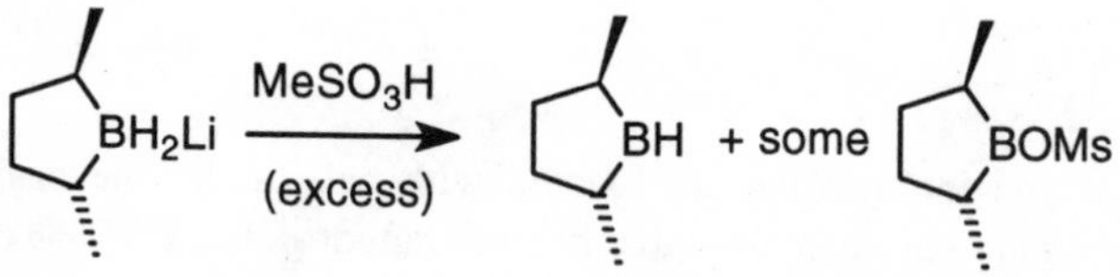

25   **26** (dimerizes)   **27**

The rate law indicated reversible formation of an intermediate **28** formed from the ketone and **27**, followed by reduction of **28** by the monomeric **26** present in equilibrium with its dimer. Ab initio calculations at the MP2/6-31G*/3-21G levels were used to estimate the optimum transition state geometry for reduction of a formaldehyde—borane complex by borane. Modified MM2 calculations for reduction of methyl ethyl ketone—dimethylborolane complex by dimethylborolane then indicated that the two major competing transition states should resemble structures **29** and **30**, with the latter favored by 5.02 kJ mol^{-1} (82% ee) if there is H in place of mesylate [28]. The observed ee of **31** was 80.4%, in better agreement with calculation than there was any reason to expect.

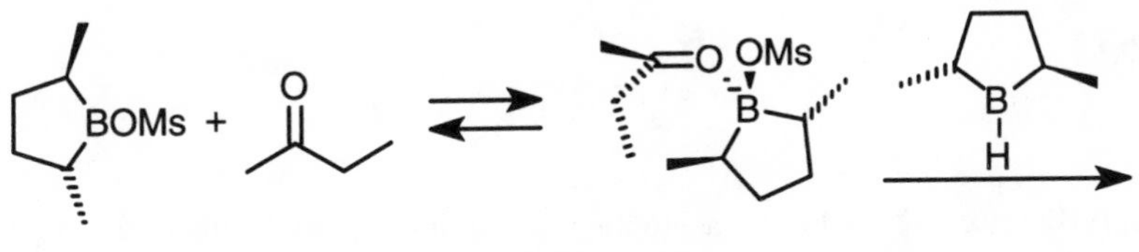

27   **28**   **26**

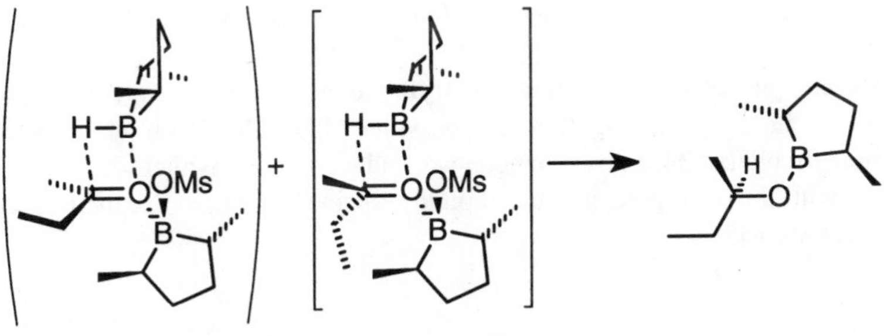

29 (disfavored)   **30** (favored)   **31** (80.4% ee)

As with the enolate reactions, the dimethylborolane chiral director is very valuable for the mechanistic insights it provides, but not competitive in cost with other chiral directors. The development of less costly chiral borane reducing agents for ketones is discussed below in Section 9.3.3.

9.3.3 1,3,2-Oxazaborolidine Catalyzed Asymmetric Reductions

The first report by Hirao, Itsuno and coworkers of modest ee's in the reduction of acetophenone by BH$_3$·THF in the presence of amino alcohols derived from amino acids [29] is of only historical interest. Soon afterward, Itsuno and coworkers

showed that 1,1-diphenylvalinol (32) with 2 moles of BH₃·THF reduces acetophenone, propiophenone, or butyrophenone to (R)-1-phenylalkanols (33) in 94-96%
ee [30, 31]. It was also stated that BuCOPh gave 100% ee, but this was based on
optical rotation, and old literature values are often low. Unsymmetrical dialkyl
ketones yielded ee's in the 55-78% range [31, 32].

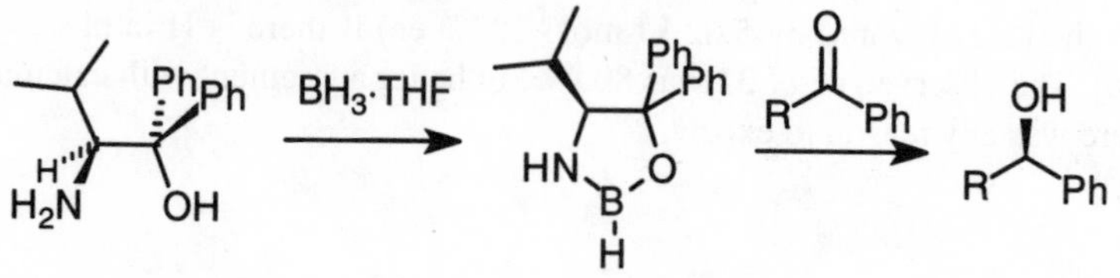

32   **33**

Chiral director **32** reduced acetophenone oxime methyl ether to (S)-1-
phenylethylamine in "100% ee," again based on optical rotation [33]. The absolute
configuration of this reduction product is opposite that of the alcohols produced by
ketone reduction, and the N-methoxy group must be directly involved in the
stereodirection. It is notable that the reagent was used catalytically, and 90% ee was
achieved in the presence of 25 mol % of **32**. Other oxime ethers did not yield such
high ee's.

Itsuno's group put considerable effort into trying to improve the recoverability of
their chiral agents by attaching them to polymers [34, 35]. With monomeric
phenylalanine derivative **34**, butyrophenone was reduced to (R)-1-phenylbutanol in
82% ee, and with the analogous polymer-bound tyrosine derivative **35**, the best ee
achieved was 88% [35].

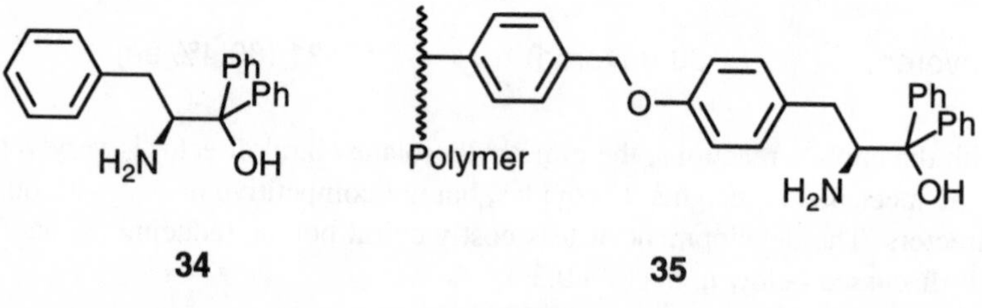

34   **35**

The performance of Itsuno's reagents was greatly improved by simple modifications
introduced by Corey. The first of these was the use of the proline derivative (S)-(–)-
2-(diphenylmethoxy)pyrrolidine (or its enantiomer) in place of the valine derivative
[36]. The resulting catalyst (**36a**) could be used at the 5-10 mol % level together with
0.6 mol of BH₃·THF and yielded 89-97% ee's in a series of reductions of ketones to
alcohols **37** having R^S = alkyl and R^L = aryl.

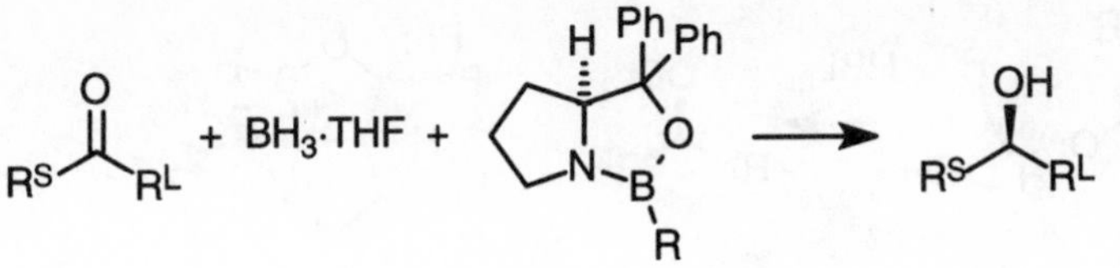

36

(36a, R = H; b, R = Me; c, R = Bu)

37

The second improvement was the use of **36b** having B-CH₃ in place of the BH in the ring [37]. This improved catalyst (**36b**) proved to be stable enough to weigh in air and store in sealed bottles, and usually gave better ee's than **36a**. Reaction times for ketones with 0.6 mol of BH₃·THF and 0.1 mol of **36b** were very short, often only 1-2 min, and yields were usually nearly quantitative. Typical examples of **37** included R^S = Me, R^L = Ph, ee 96.5%; R^S = Me, R^L = *tert*-Bu, 97.3; R^S = Me, R^L = cyclohexyl, 84; R^S = ClCH₂, R^L = Ph, 95.3; (*R*)-1-hydroxy-1,2,3,4-te-trahydronaphthalene from α-tetralone, 86% [37]. Catalysts having β-naphthyl groups in place of the phenyl groups of **36a** or **36b** yielded essentially the same results [38].

The mechanism of the reduction was assumed to involve an amine borane complex (**38**) [37], and the X-ray structure of **38** (L = H, R = CH₃) has recently been determined [39]. In the postulated transition state **39**, updated to incorporate more recent results and interpretation [40], the carbonyl oxygen coordinates with the boron atom in the oxazaborolidine ring, hydride is transferred from the borane unit, and the resulting borate **40** reacts with more borane to regenerate **38**. The smaller group R^S is in an axial orientation parallel to the B-alkyl group R, and the larger group R^L occupies an equatorial position farthest from other groups. The groups L may be H, as in the original work, or the oxygen atoms in catecholborane [41], a third useful improvement introduced by Corey and coworkers (see below).

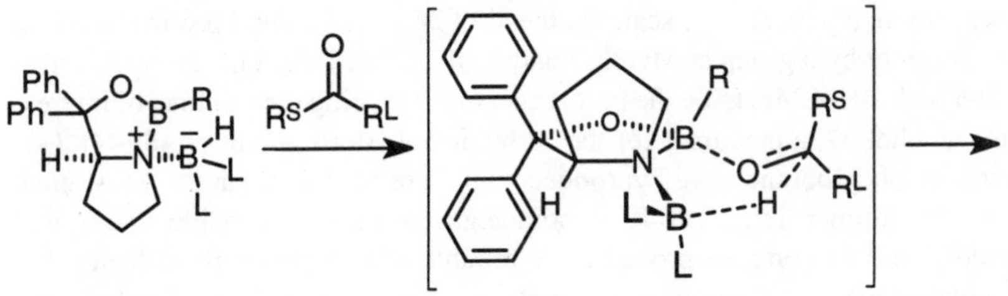

38

39

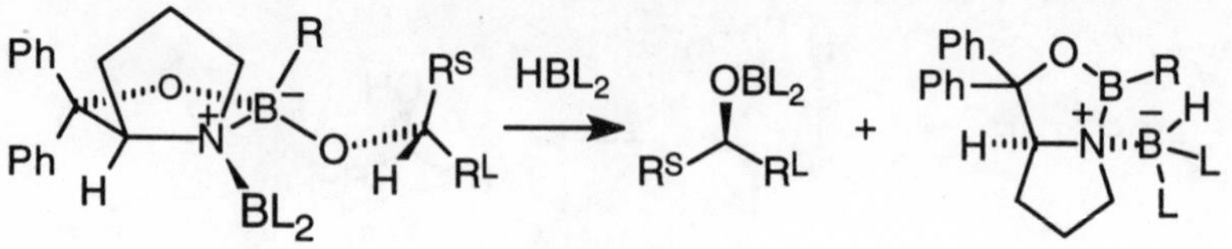

40   **38**

Exploratory syntheses of compounds chosen for potential utility in natural products synthesis included bromocyclohexenol **41**, allylic alcohol **42** and its epimer **43** as models for prostaglandin synthesis (diastereomeric ratios listed), and conversion of β-keto ester **44** to lactone **45** [37].

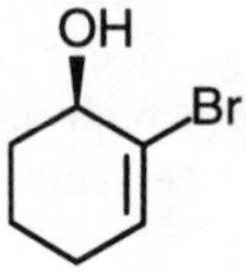

41 (ee 91%) 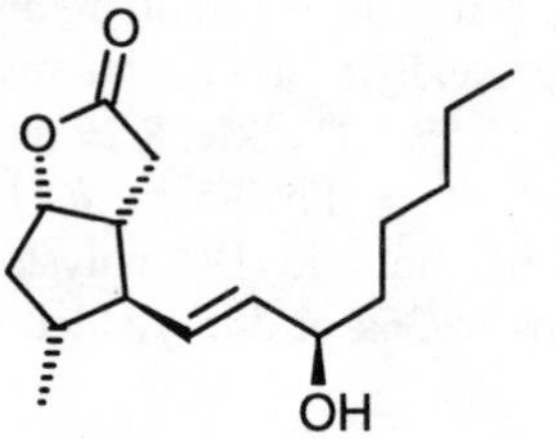   **42** (91:9) 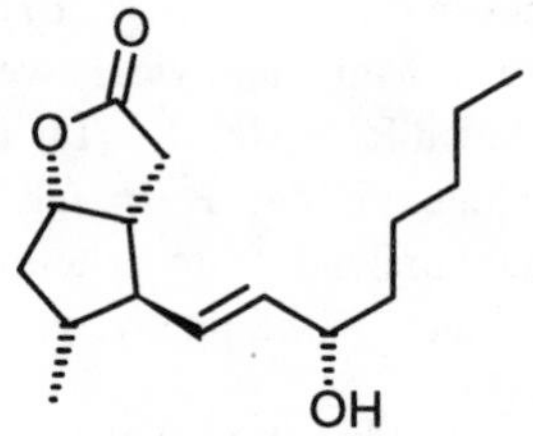   **43** (from *ent-36b*, 90:10)

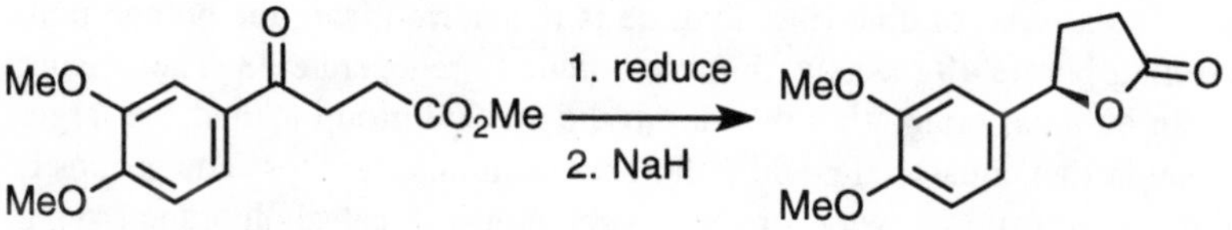

44   **45** (ee 95%)

An early note on the asymmetric synthesis of styrene oxide via 2-chloro-1-phenylethanol includes detailed directions for making the borane catalyst **36b** [42]. However, a more practical large scale synthesis of **36b** from natural proline has been reported recently by a group at Merck, Sharp and Dohme [43]. This group has also found **36b** with $BH_3 \cdot SMe_2$ to be the best reagent for reducing their ketone of interest **46** to the alcohol **47**, a precursor for the carbonic anhydrase inhibitor **48** [44]. (−)-Diisopinocampheylborane gave byproducts that were hard to separate. Yeast gave the wrong enantiomer. Tetrahydrofuran borane gave somewhat variable results, and the dimethyl sulfide complex proved more reliable. One highly critical factor that was discovered was that traces of water drastically lower the ee, 0.1% of water in the ketone being enough to reduce the ee from 95% down to 50%. Molecular sieves improved the ee's to the 95-99% range. The effect of varying the catalyst **36** by putting other groups in place of phenyl was also examined, but none of the replacements tested were quite as good as the original **36** with a set of six aryl alkyl ketones tested.

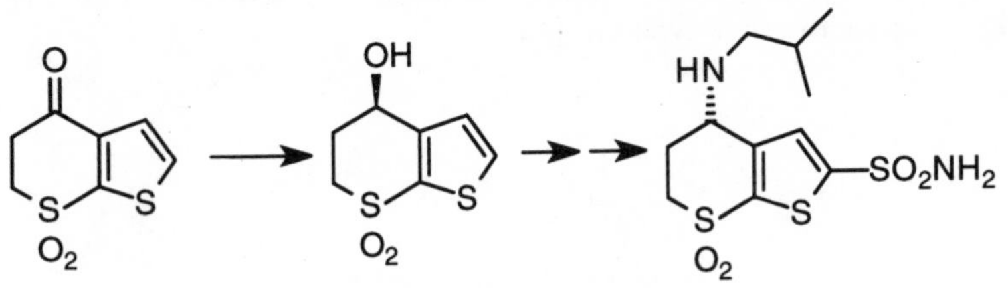

46   **47**   **48**

The third significant innovation introduced by Corey was the use of catecholborane as the reducing agent in place of BH$_3$·THF, which is not necessary in simple systems but can be very helpful in avoiding competing reduction of other functional groups [41]. This system was shown to work well with **36c** or **36b** as catalysts for the preparation of allylic alcohols **49, 50, 51, 52,** and **53**.

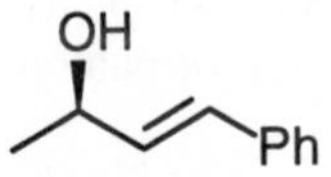

49 ee 92% 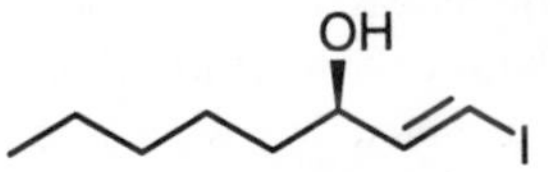   **50 ee 86%**

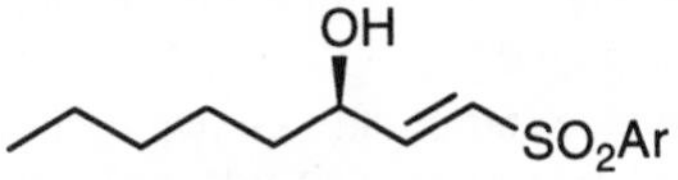

51 ee 91%. 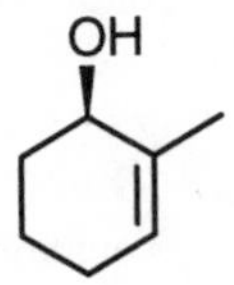   **52 ee 93%** 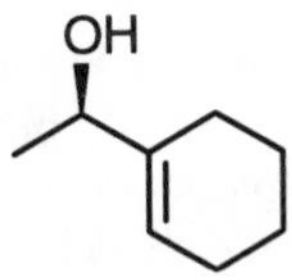   **53 ee 81%**

The preparation of trifluoroethyl boronic esters, RB(OCH$_2$CF$_3$)$_2$ (R = Et, Bu), and their convenient and rapid reaction with amino alcohols to form **36** has been reported [45]. Alternatively, bis(diisopropylamino)boranes, RB[N(i-Pr)$_2$]$_2$, are useful precursors to **36** and related compounds [46].

A more rigid bicyclic catalyst **54** was prepared (in both enantiomeric forms) and its methylboronic acid derivatives was found to give even higher ee's than obtained previously [47]. For example, the ee of (R)-bromocyclohexenol **41** (see above) was increased from 91% to 97.5%, (R)-1-cyclohexylethanol from 84% to 91%, and the reduction of α-tetralone improved from 86% to 97% ee. However, ee's for 1-phenylethanol, 1-*tert*-butylethanol, and **50** were improved by only 1-2%.

Bromocyclopentenol **58** was prepared, and it was shown that this material could be used as starting material in a multistep synthesis of the catalyst **54** used in its own preparation. The term "chemzymes" was introduced, with the implication that these catalysts are designed to mimic enzymes in the way they interact with substrates and the selectivity they show [47]. More recently, **55** has been prepared with **36b** as catalyst and used in a ginkgolide synthesis [48].

54   **55 ee 95%**

The new asymmetric catalyst system was used to solve a wide variety of synthetic problems during the course of its development. Corey's work is focused on reaching synthetic targets, and this review covering only the aspects involving boron chemistry cannot do justice to the significant concepts of synthetic strategy involved. What follows is a fairly comprehensive survey of the types of secondary alcohols that have been made, but with only brief mention of the type of natural product or pharmaceutical synthesis in which they were used.

In a very early application, after ordinary borohydride reduction had yielded only the wrong diastereomer, the older catalyst **36a** was used to reduce enantiomerically pure ketone **56** to alcohol **57**, diastereomeric ratio 91:9, 45% yield, for conversion to bilobalide, a type of ginkgolide [49]. Alcohol **58** was obtained with methylboronic acid derived catalyst **36b** in 93% ee and used in another synthesis of a ginkgolide, which is of interest as a platelet activating factor antagonist [50]. Catalyst **36b** was also used for preparation of a precursor **59** to fluoxetine (**60**), better known by its trade name, Prozac®, which was obtained enantiomerically pure in 82-88% over all yield [51].

56 [RO = (+)-menthyl-O]   **57**   **58**

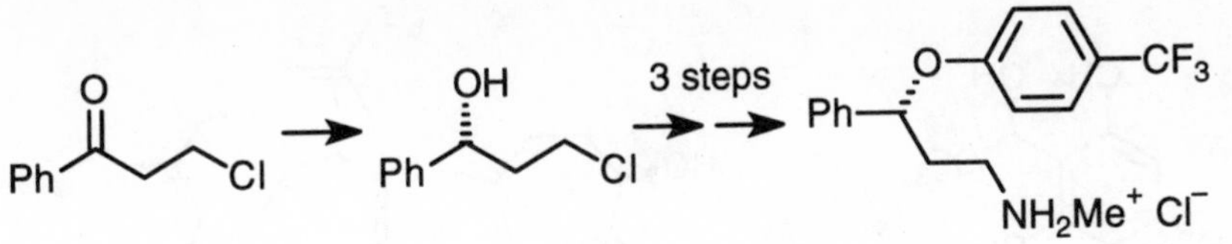

59 (94% ee)

60

(recryst. to 100% ee, 84% yield)

The catalyst **ent-36b** was used to prepare intermediate **61** and related compounds for use in the synthesis of forskolin, an activator of adenylate cyclase of pharmaceutical interest [52, 53]. An aryl chloromethyl ketone reduction catalyzed by **ent-CGb** provided the intermediate **62** for enantioselective synthesis of isoproterenol [54]. Intermediate **63** for the synthesis of antheridinic acid was made with **36b** [55]. The conversion of 1-phenylethanol to benzylic thiols, sulfinic esters, and sulfonic acids of high enantiomeric purity has been described [56].

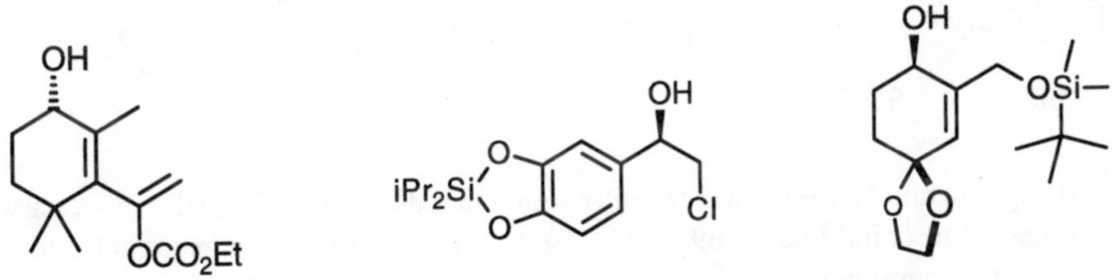

61 (93% ee)

62 (97% ee)

63 (92% ee, recryst. to >99%)

The trifluoromethyl group behaves as larger than mesityl, though the effect may be polar in nature, in the reduction of mesityl trifluoromethyl ketone catalyzed by **36c**, and the (R)-alcohol **64** is efficiently produced with no detectable enantiomer [57]. Acrylate esters of **64** give highly enantioselective acid catalyzed Diels-Alder reactions [58]. The R = isobutyl analogue of **ent-CGc** has been used with catecholborane to make the alcohol **65**, which can be converted to (3S)-2,3-oxidosqualene [59]. A closely related fluoro alcohol has been prepared and used as a starting material for the synthesis of oleanolic acid, erythrodiol, β-amyrin, and other pentacyclic terpenes [60].

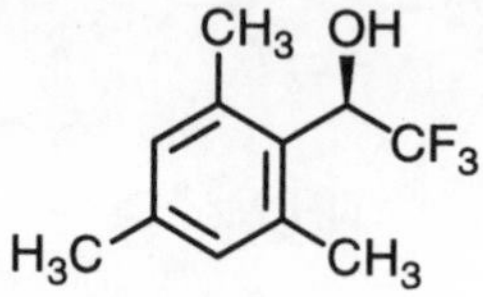

64

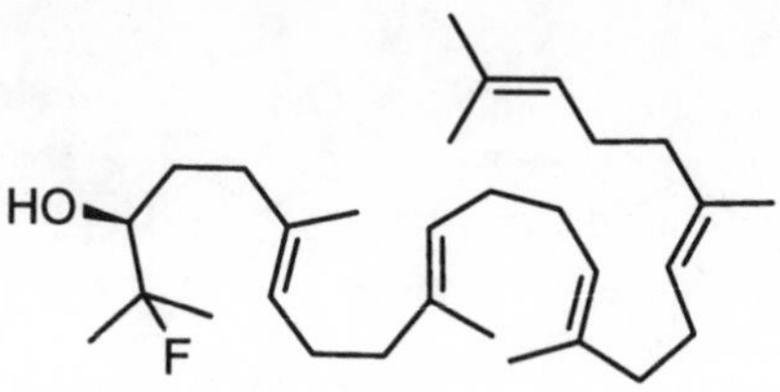

65

Reduction of trichloromethyl ketones with catecholborane catalyzed by **36c** yields alcohols **66** in 92-98% ee. Treatment of **66** with base and sodium azide proceeds via epoxides (**67**) to form azido acids, which can be reduced to amino acids [61]. The trichloromethyl alcohols **66** are also useful for making other derivatives such as hydroxy acids [62, 63], and general methods for making the trichloromethyl ketone precursors have been reported [64].

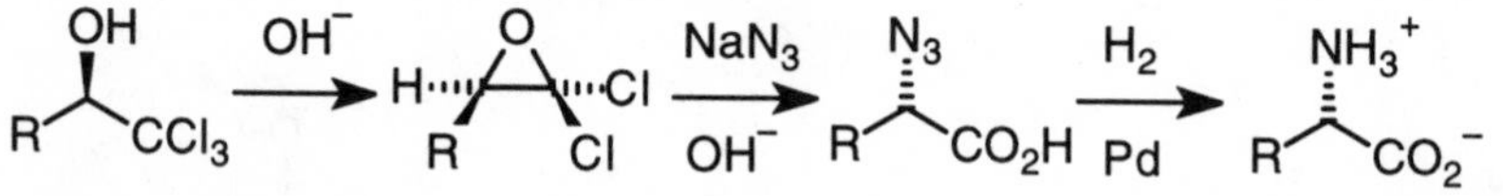

66   **67**

Bringman and Hartung have recently found that **36b** with $BH_3 \cdot THF$ reduces aryl lactones **68** to chiral biaryls **69** in 94-97% ee, and that **69** can be recrystallized to high enantiomeric purity [65].

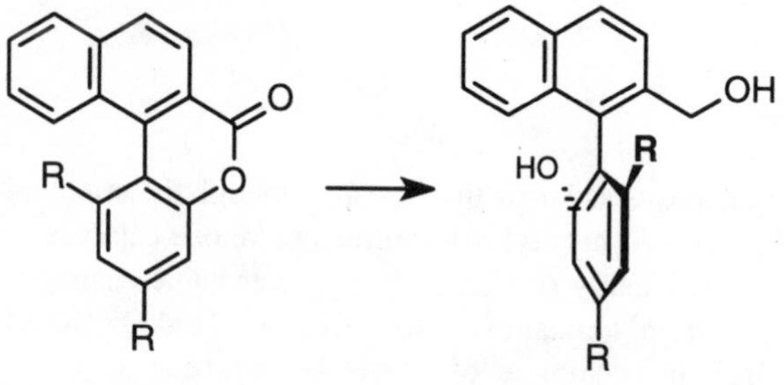

68   **69** (R = Me, OMe)

Although the Corey catalyst system has been optimized for many reductions, prolinol formed from in situ reduction of proline (**70**) at the 10 mol % level with excess $BH_3 \cdot THF$ has yielded 81 to >95% ee's in the reduction of several ketones to (R)-alcohols in refluxing toluene [66]. Brunel, Maffei, and Buono interpreted their results as implying that the relatively high temperature favored monomeric prolinol—borane complex, the most active and selective reducing agent, rather than the dimer that forms at lower temperatures. They also noted that their catalyst is an inexpensive, off the shelf reagent, and this procedure could be preferred in any situation where it produces ee's high enough for the intended purpose of the product.

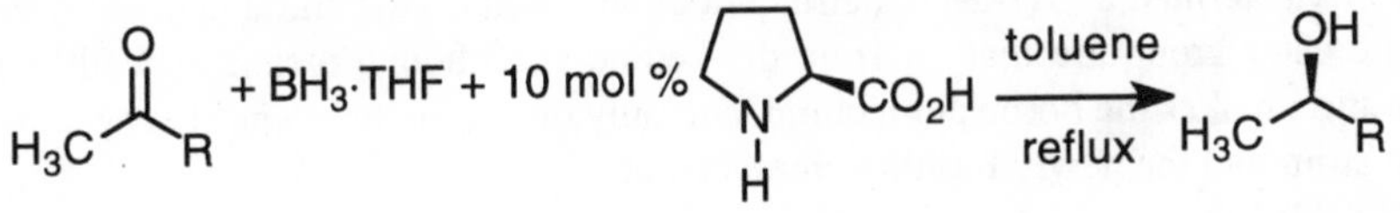

70

R = Ph, PhCH₂, tert-Bu, 2-furyl, isopropyl

9.3.4 Remote Asymmetric Induction in γ-Keto Boronic Esters

2-Alkyl-4-ketoalkylboronic esters **71**, which are synthesized easily in racemic form from organozinc reagents [67], react with borane-THF to yield the 2,4-*anti* diastereomers **73** in high de's [68]. A cyclic steric interaction in the postulated transition state **72** accounts for the results.

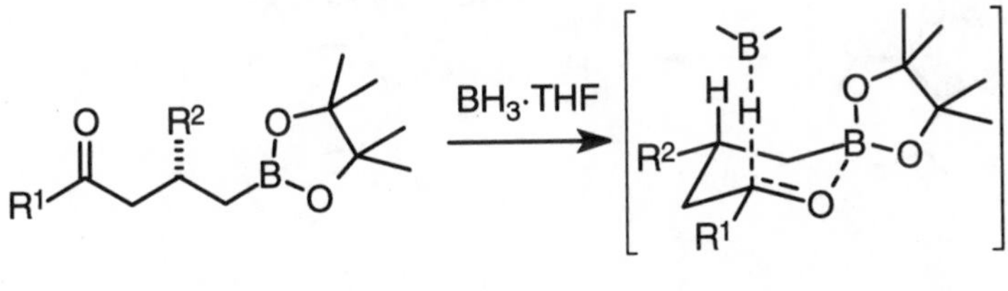

71   **72**

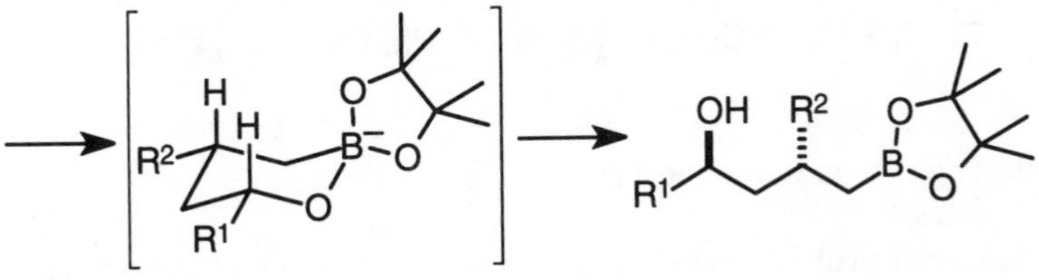

73

R¹ = Me, R² = Me, Et, iPr, Ph, de's 91-96%; R¹ = Et, Pr, iPr, Ph, R² = Me, de's 90-96%.

A variant of the foregoing chemistry permits asymmetric reduction of boronic esters of diols having C_2 symmetry (**74**) [69]. Transition state **75** would appear disfavored because the forward isopropyl group projects into the space required by the attacking boron hydride, but attack from the opposite side, transition state **76**, suffers no such interference, and the ultimate products **77** obtained after peroxidic deboronation had ee's in the range 92 to >98% except where R = CH3, 85%. The process was termed a "1,7-asymmetric induction" because the chiral centers of the boronic ester group are that far from the ketone site, though mechanistically the cyclization makes the boron atom chiral, and only one atom intervenes between the boron atom and the newly forming chiral center.

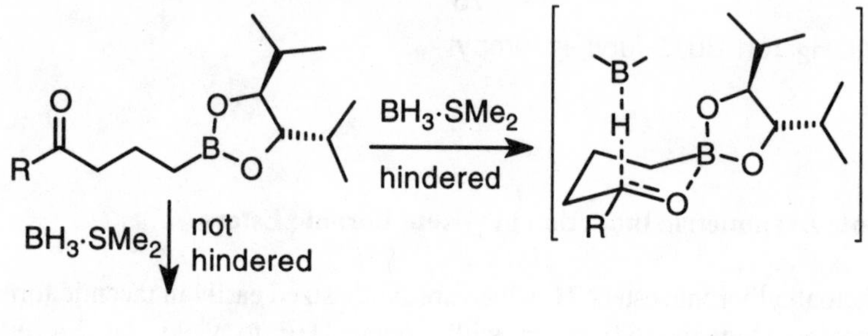

74   **75**

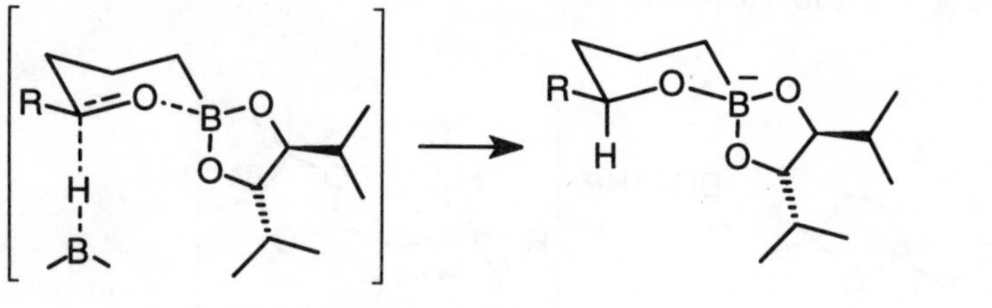

76

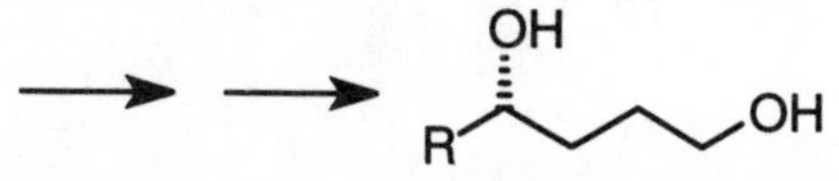

77 (R = Me, n-C5H11, cyclohexyl, Ph, Cl(CH2)3, etc.)

9.3.5 Borohydride Type Reducing Agents

The use of hindered trialkylborohydride reagents to achieve diastereoselective reduction is well known, and a general review of such borohydride reductions is beyond the scope of this book. Lithium tri-sec-butylborohydride, sold by Aldrich Chemical Company as "L-Selectride®," is a well known example. Reductions by this and related reagents transfer hydride to the less sterically hindered face of a carbonyl group. An example has been mentioned in Section 9.2.2, the use of "L-Selectride®" to make the diastereomer of steroidal propargylic alcohol **7** [11]. A recent example is the reduction of allenyl ketones substituted by a bulky *tert*-butyldimethylsilyl group (**78**) to diastereomer **79**, diastereomeric ratios 99:1 if R = Me or MeOCH₂OCH₂, 84:16 when R = Bu [70].

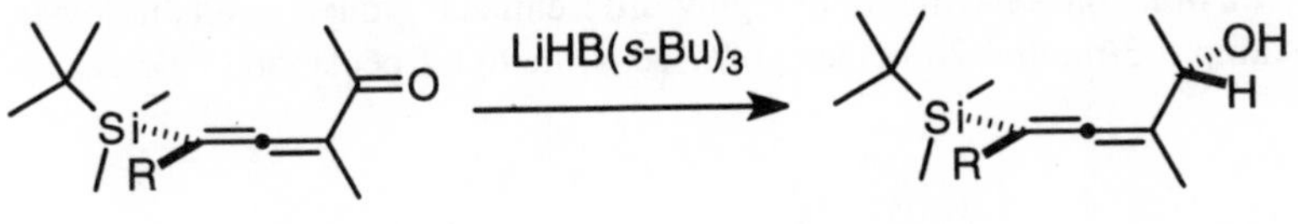

78 (racemic)   **79**

Commercial sodium or potassium hydride reacts sluggishly with trialkylboranes, but activation with lithium aluminum hydride in THF permits rapid trialkyl-borohydride formation [71, 72], to the point where activated KH will reduce primary alkyl halides to hydrocarbons in the presence of a catalytic amount of trialkylborane, and the very hindered KHB(CH₂SiMe₃)₃ can be made [71].

Another useful way to achieve diastereoselective reduction is to chelate boron to a hydroxyl group at an asymmetric center, as in the highly stereoselective reduction of β-hydroxy ketones such as **80** with tetramethylammonium triacetoxyborohydride to form *anti* 1,3-diols [73]. The preferred transition state is presumed to have a chair conformation with large groups equatorial, as in **81**. This reaction bears some resemblance to the reduction of 2-alkyl-4-ketoalkylboronic esters **71** with borane (see Section 9.3.4), but it appears that reduction of **71** involves attack of an external B—H bond on a cyclic boronic ester, while reduction of **80** involves intramolecular delivery of hydride from boron to the carbonyl carbon.

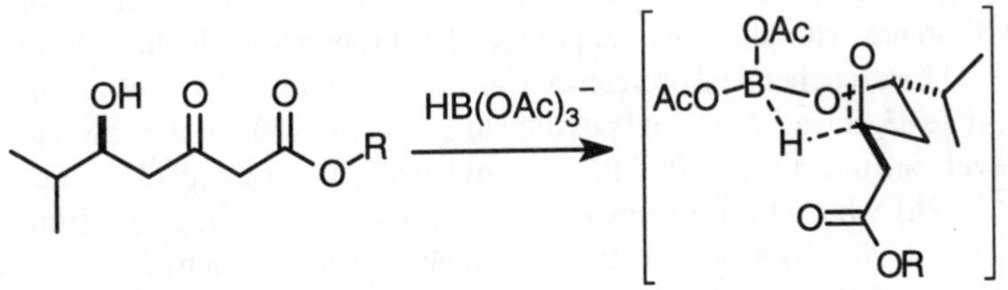

80 [R = (CH₂)₃Ph]   **81**

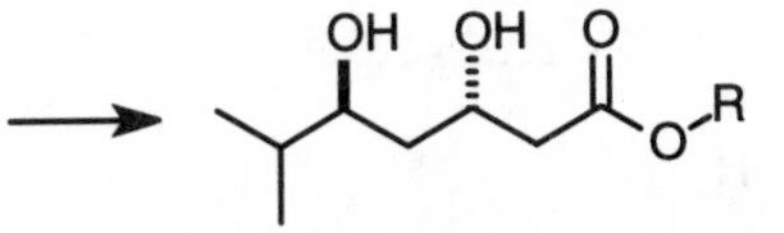

(*anti/syn* ratio 95:5)

Hindered asymmetric borohydrides can act as enantioselective reducing agents. A particularly useful reagent **82** is prepared via hydroboration of nopol benzyl ether with 9-BBN followed by reduction to the borohydride with *tert*-butyllithium. This borohydride is commercially available under the trade name "NB-Enantride" (Aldrich). Reduction of 2-octanone with **82** yields (*S*)-2-octanol (**83**) in 79% ee, and the reagent even reduces 2-butanone to (*S*)-2-butanol in 76% ee [74]. However, reduction by **82** of ketones having more highly differentiated groups results in lower ee's, for example, 3-methyl-2-butanone 68% ee, 4-methyl-2-pentanone 30% ee.

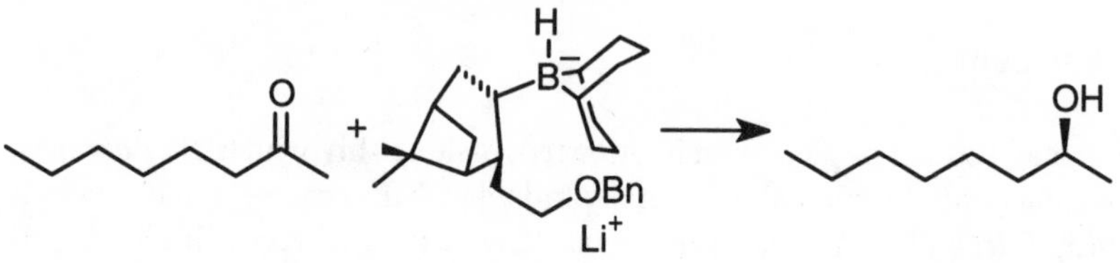

82   **83** (79% ee)

9.4 Other Borane Catalyzed Reactions

9.4.1 Alkynylation of Aldehydes

The asymmetric alkynylation of aldehydes recently reported by Corey and Cimprich is a new and different type of reaction that is in a class by it self. The oxazaborolidinechiral director **84** (R^3 = Bu) guides and assists the reaction of alkynyldimethylboranes with aldehydes to produce (*R*)-propargylic alcohols **86** in 90-97% ee [75]. The stereochemical preference was explained on the basis of postulated transition state **85**. It was also found possible to use **84** (R^3 = Ph) as a catalyst at the 25 mol % level, and under these conditions phenylethynylborane with aldehydes yielded **86** (R^2 = Ph) which from hexanal (R^1 = n-C_5H_{11}) was in 93% ee, from benzaldehyde (R^1 = Ph), 97% ee, or from cyclohexanecarboxaldehyde (R^1 = cyclohexyl), 85% ee [75]. It was also noted that the requisite amino alcohol for making **84** is commercially available.

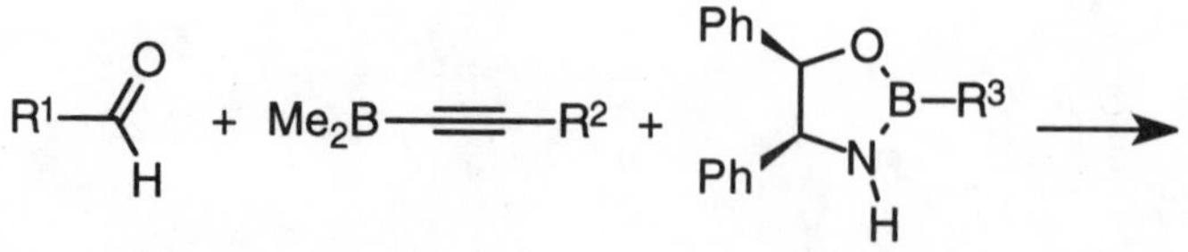

84

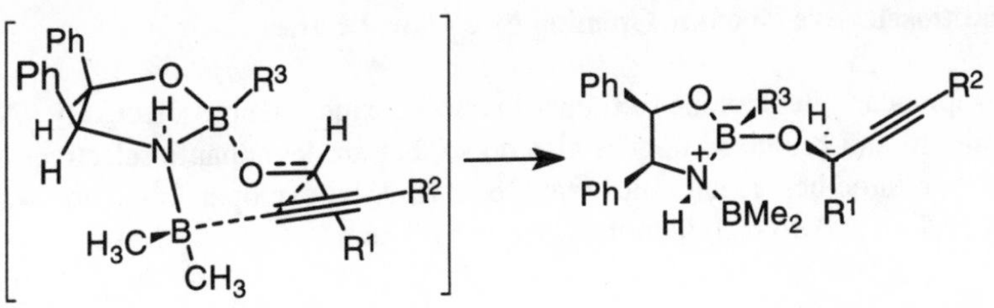

85

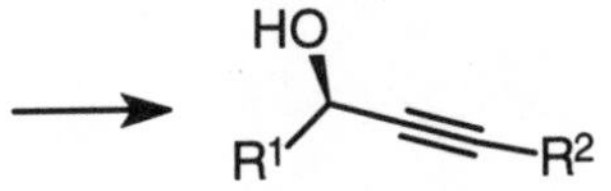

86

(R^1 = n-pentyl, cyclohexyl, t-Bu, Ph; R^2 = Ph, n-pentyl)

9.4.2 Addition of Diethylzinc to Aldehydes

The reaction of diethylzinc with aromatic aldehydes can be catalyzed by oxazaborolidine **87** with good asymmetric induction [76]. Both enantiomers of **87** are readily available, and the reagent is easier to make than other types that do not contain boron. Furfural only gave 66% ee, heptanal 52% ee, but it is typical that such reactions work best with aromatic aldehydes. It was noted that **87** acts as a catalyst in monomeric form, in contrast to the catalysts prepared from amino alcohols and the zinc reagent, which incorporate two molecules of the chiral director [77, 78].

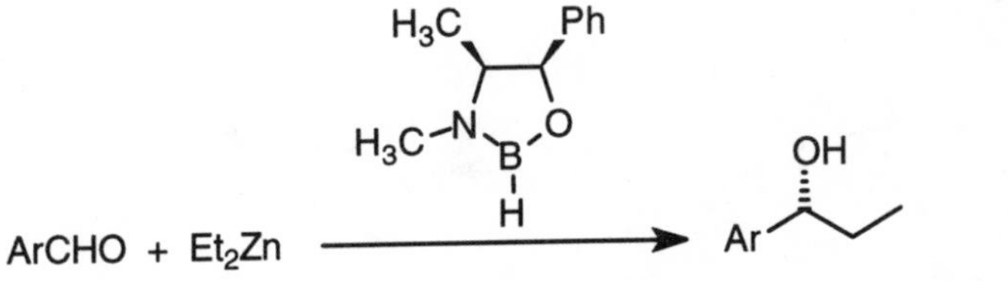

ArCHO + Et$_2$Zn $\longrightarrow$

87 (5 mol %)   ee's 91-96%

(Ar = Ph, p-ClC$_6$H$_4$, p-MeOC$_6$H$_4$, o-MeOC$_6$H$_4$)

9.4.3 Enantioselective Epoxide Opening by a Chiral Borane

B-Iododiisopinocampheylborane (**88**) opens *meso* epoxides enantioselectively [79, 80]. The chloro and bromo analogues also do so, but are less enantioselective. In addition to the three best results illustrated, ee's 78-95%, cyclopentene oxide was opened in 52% ee and *cis*-2,3-diethyloxirane in 69% ee.

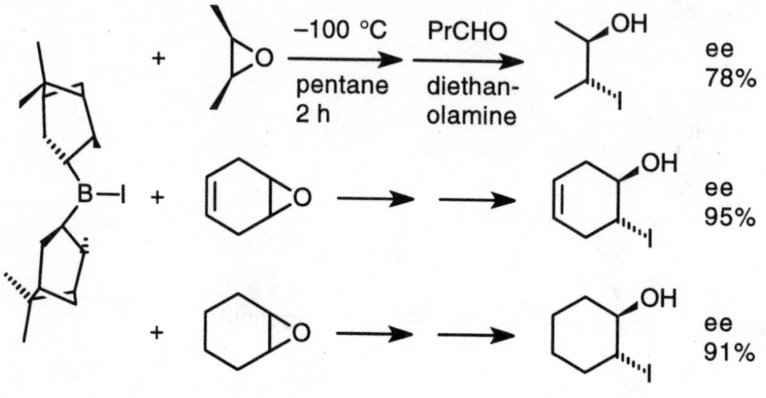

88

9.4.4 Asymmetric Epoxidation

The tartrate ester *tert*-butyl peroxyborate **89**, prepared from the tartrate ester methoxyborate and *tert*-butyl hydroperoxide in cyclohexane, has been tested as an asymmetric epoxidizing agent [81]. The ee's were mediocre with *trans*-stilbene and positively poor with less hindered substrates, but hope was expressed that further experimentation might lead to improvement, as there are several structural parameters that might be adjusted.

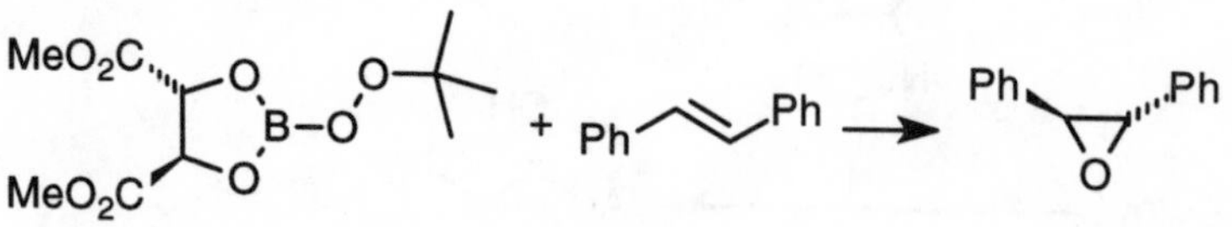

89   (51% ee)

9.5 Asymmetric Hydrozirconation

Another new reaction that does not fit into any of the previous classes is asymmetric hydrozirconation, in which an asymmetric 2-alkenyl-1,3,2-oxazaborolidine provides chiral direction for addition of the zirconium reagent. Hydrozirconation of (1R,2S)-N-neopentylnorephedrine derivatives of boronic acids (**90**) followed by deuteration with D_2O yields α-deutero alcohols **92** in 80-93% ee [82]. The postulated intermediate is the α-zirconylborane **91**.

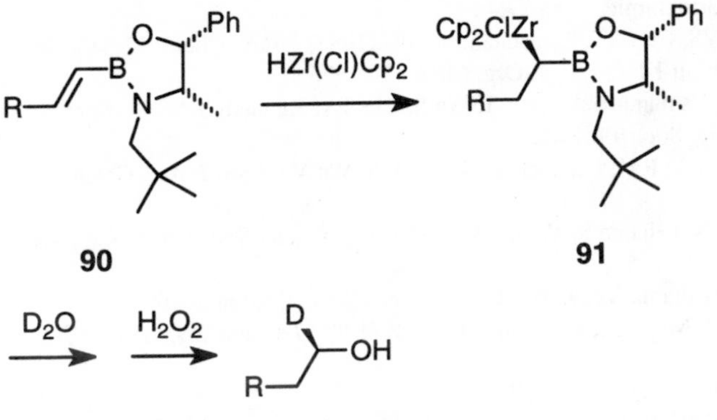

90   **91**

R = n-butyl, tert-butyl, cyclopentyl, 3-chloropropyl, 3-phenylpropyl

9.6 References

1. Midland MM (1989) Chem. Rev. 89:1553
2. Deloux L, Srebnik M (1993) Chem. Rev. 93:763
3. Mikhailov BM, Bubnov YuN, Kiselev VG (1966) J. Gen. Chem. USSR (Engl. Transl.) 36:65
4. Midland MM, Tramontano A, Zderic SA (1977) J. Am. Chem. Soc. 99:5211
5. Midland MM, Greer S, Tramontano A, Zderic SA (1979) J. Am. Chem. Soc. 101:2352
6. Midland MM, McLoughlin JI (1984) J. Org. Chem. 49:1316
7. Midland MM, McLoughlin JI, Gabriel J (1989) J. Org. Chem. 54:159
8. Midland MM, McDowell DC, Hatch RL, Tramontano A (1980) J. Am. Chem. Soc. 102:867
9. Midland MM, Tramontano A, Kazubski A, Graham R, Tsai DJS, Cardin DB (1984) Tetrahedron 40:1371
10. Midland MM, Nguyen NH (1981) J. Org. Chem. 46:4107
11. Midland MM, Kwon YC (1984) Tetrahedron Lett. 25:5981
12. Midland MM, Petre JE, Zderic SA, Kazubski A (1982) J. Am. Chem. Soc. 104:528
13. Brown HC, Pai GG (1982) J. Org. Chem. 47:1606
14. Brown HC, Pai GG (1983) J. Org. Chem. 48:1784

15. Midland MM, Lee PE (1985) J. Org. Chem. 50:3237
16. Roush WR, Sciotti RJ (1994) J. Am. Chem. Soc. 116:6457
17. Midland MM, Kazubski A (1982) J. Org. Chem. 47:2814
18. Chandrasekharan J, Ramachandran PV, Brown HC (1985) J. Org. Chem. 50:5446
19. Chandrasekharan J, Ramachandran PV, Brown HC (1988) J. Am. Chem. Soc. 110:1539
20. Brown HC, Chandrasekharan J, Ramachandran PV (1986) J. Org. Chem. 51:3394
21. Ramachandran PV, Teodorović AV, Rangaishenvi MV, Brown HC (1992) J. Org. Chem. 57:2379
22. Soderquist JA, Anderson CL, Miranda EL, Rivera I, Kabalka GW (1990) Tetrahedron Lett. 31:4677
23. Brown HC, Ramachandran PV (1992) Acc. Chem. Res. 25:16
24. Dhar RK (1994) Aldrichimica Acta 27:43
25. Brown HC, Park WS, Cho BT, Ramachandran PV (1987) J. Org. Chem. 52:5406
26. Molander GA, Bobbitt KL (1994) J. Org. Chem. 59:2676
27. Imai T, Tamura T, Yamamuro A, Sato T, Wollmann TA, Kennedy RM, Masamune S (1986) J. Am. Chem. Soc. 108:7402
28. Masamune S, Kennedy RM, Petersen JS, Houk KN, Wu Y (1986) J. Am. Chem. Soc. 108:7404
29. Hirao A, Itsuno S, Nakahama S, Yamazaki N (1981) J. Chem. Soc., Chem. Commun. 315
30. Itsuno S, Ito K, Nakahama S (1983) J. Chem. Soc., Chem. Commun. 469
31. Itsuno S, Nakano M, Miyazaki K, Masuda H, Ito K (1985) J. Chem. Soc., Perkin Trans. 1 2039
32. Itsuno S, Ito K (1984) J. Org. Chem. 49:555
33. Itsuno S, Sakurai Y, Ito K, Hirao A (1987) Bull. Chem. Soc. Jpn. 60:395
34. Itsuno S, Ito K, Hirao A, Nakahama S (1984) J. Chem. Soc., Perkin Trans. 1 2887
35. Itsuno S, Nakano M, Ito K, Hirao A, Owa M, Kanda N, Nakahama S (1985) J. Chem. Soc., Perkin Trans. 1 2615
36. Corey EJ, Bakshi RK, Shibata S (1987) J. Am. Chem. Soc. 109:5551
37. Corey EJ, Bakshi RK, Shibata S, Chen CP, Shing VK (1987) J. Am. Chem. Soc. 109:7925
38. Corey EJ, Link JO (1989) Tetrahedron Lett. 30:6275
39. Corey EJ, Azimioara M, Sarshar S (1992) Tetrahedron Lett. 33:3429
40. Corey EJ, Link JO, Bakshi RK (1992) Tetrahedron Lett. 33:7107
41. Corey EJ, Bakshi RK (1990) Tetrahedron Lett. 31:611
42. Corey EJ, Bakshi RK, Shibata S (1988) J. Org. Chem. 53:2861
43. Mathre DJ, Jones TK, Xavier LC, Blacklock TJ, Reamer RA, Mohan JJ, Turner Jones ET, Hoogsteen K, Baum MW, Grabowski EJ (1991) J. Org. Chem. 56:751
44. Jones TK, Mohan JJ, Xavier LC, Blacklock TJ, Mathre DJ, Sohar P, Turner Jones ET, Reamer RA, Roberts FE, Grabowski EJ (1991) J. Org. Chem. 56:763
45. Corey EJ, Link JO (1992) Tetrahedron Lett. 33:4141
46. Chavant P, Vaultier M (1993) J. Organomet. Chem. 455:37
47. Corey EJ, Chen CP, Reichard GA (1989) Tetrahedron Lett. 30:5547
48. Corey EJ, Rao KS (1991) Tetrahedron Lett. 32:4623
49. Corey EJ, Su WG (1988) Tetrahedron Lett. 29:3423
50. Corey EJ, Gavai AV (1988) Tetrahedron Lett. 29:3201
51. Corey EJ, Reichard GA (1989) Tetrahedron Lett. 30:5207
52. Corey EJ, Jardine PDS, Mohri T (1988) Tetrahedron Lett. 29:6409
53. Corey EJ, Jardine PDS (1989) Tetrahedron Lett. 30:7297
54. Corey EJ, Link JO (1990) Tetrahedron Lett. 31:601

55. Corey EJ, Kigoshi H (1991) Tetrahedron Lett. 32:5025
56. Corey EJ, Cimprich KA (1992) Tetrahedron Lett. 33:4099
57. Corey EJ, Cheng XM, Cimprich KA, Sarshar S (1991) Tetrahedron Lett. 32:6835
58. Corey EJ, Cheng XM, Cimprich KA (1991) Tetrahedron Lett. 32:6839
59. Corey EJ, Yi KY, Matsuda SPT (1992) Tetrahedron Lett. 33:2319
60. Corey EJ, Lee J (1993) J. Am. Chem. Soc. 115:8873
61. Corey EJ, Link JO (1992) J. Am. Chem. Soc. 114:1906
62. Corey EJ, Link JO (1992) Tetrahedron Lett. 33:3431
63. Corey EJ, Helal CJ (1993) Tetrahedron Lett. 34:5227
64. Corey EJ, Link JO, Shao Y (1992) Tetrahedron Lett. 33:3435
65. Bringmann G, Hartung T (1992) Angew. Chem., Internat. Ed Engl. 31:761
66. Brunel JM, Maffei M, Buono G (1993) Tetrahedron: Asymmetry 4:2255
67. Knochel P (1990) J. Am. Chem. Soc. 112:7431
68. Molander GA, Bobbitt KL, Murray CK (1992) J. Am. Chem. Soc. 114:2759
69. Molander GA, Bobbitt KL (1993) J. Am. Chem. Soc. 115:7517
70. Marshall JA, Ting Y (1993) J. Org. Chem. 58:3233
71. Soderquist JA, Rivera I (1988) Tetrahedron Lett. 29:3195
72. Hubbard JL (1988) Tetrahedron Lett. 29:3197
73. Evans DA, Chapman KT, Carreira EM (1988) J. Am. Chem. Soc. 110:3560
74. Midland MM, Kazubski A, Woodling RE (1991) J. Org. Chem. 56:1068
75. Corey EJ, Cimprich KA (1994) J. Am. Chem. Soc. 116:3151
76. Srebnik M, Joshi NN, Brown HC (1989) Tetrahedron Lett. 30:5551
77. Oguni N, Matsuda Y, Kaneko T (1988) J. Am. Chem. Soc. 110:7877
78. Kitamura M, Okada S, Suga S, Noyori R (1989) J. Am. Chem. Soc. 111:4028
79. Joshi NN, Srebnik M, Brown HC (1988) J. Am. Chem. Soc. 110:6246
80. Srebnik M, Joshi NN, Brown HC (1989) Israel J. Chem. 29:229
81. Manoury E, Mouloud HAH, Balavoine GGA (1993) Tetrahedron: Asymmetry 4:2339
82. Zheng B, Srebnik M (1994) Tetrahedron Lett. 35:6247

Index

Author Index

Druck: Mercedesdruck, Berlin
Verarbeitung: Buchbinderei Lüderitz & Bauer, Berlin